Software Engineering for Modern Web Applications:
Methodologies and Technologies

Daniel M. Brandon
Christian Brothers University, USA

INFORMATION SCIENCE REFERENCE
Hershey · New York

Acquisitions Editor:	Kristin Klinger
Development Editor:	Kristin Roth
Senior Managing Editor:	Jennifer Neidig
Managing Editor:	Jamie Snavely
Assistant Managing Editor:	Carole Coulson
Copy Editor:	Brenda Leach
Typesetter:	Carole Coulson
Cover Design:	Lisa Tosheff
Printed at:	Yurchak Printing Inc.

Published in the United States of America by
Information Science Reference (an imprint of IGI Global)
701 E. Chocolate Avenue, Suite 200
Hershey PA 17033
Tel: 717-533-8845
Fax: 717-533-8661
E-mail: cust@igi-global.com
Web site: http://www.igi-global.com

and in the United Kingdom by
Information Science Reference (an imprint of IGI Global)
3 Henrietta Street
Covent Garden
London WC2E 8LU
Tel: 44 20 7240 0856
Fax: 44 20 7379 0609
Web site: http://www.eurospanbookstore.com

Library of Congress Cataloging-in-Publication Data

Software engineering for modern Web applications : methodologies and technologies / Daniel M. Brandon, editor.

p. cm.

Summary: "This book presents current, effective software engineering methods for the design and development of modern Web-based applications"--Provided by publisher.

Includes bibliographical references and index.

ISBN 978-1-59904-492-7 (hardcover) -- ISBN 978-1-59904-494-1 (ebook)

1. Application software--Development. 2. Internet programming. 3. Web site development. 4. Software engineering. I. Brandon, Dan, 1946-

QA76.76.A65.S6588 2008

005.1--dc22

2008008470

British Cataloguing in Publication Data
A Cataloguing in Publication record for this book is available from the British Library.

Table of Contents

Detailed Table of Contents

Chapter I discusses our dependence and reliance on the Web which has increased dramatically over the years. As a result, the development of Web applications has become more complex and challenging than most of us think. In many ways, it is also different and more complex than traditional software development. But, currently, the development and maintenance of most Web applications is chaotic and far from satisfactory. To successfully build and maintain large, complex Web-based systems and applications, Web developers need to adopt a disciplined development process and a sound methodology. The emerging discipline of Web engineering advocates a holistic, disciplined approach to successful Web development. In this chapter, we articulate and raise awareness of the issues and considerations in large, complex Web application development, and introduce Web engineering as a way of managing complexity and diversity of large-scale Web development.

Chapter II discusses Process is an important element in the success of any information systems development project especially in academia where typically an undergraduate term project needs to go through the development phases within the space of a semester. The omission of customer feedback results in students completing "toy" projects without significant real-world experience; efforts to incorporate artificial customer interactions have not been very successful either. WebHelix has been recently introduced as a practical process for Web engineering that helps students gain valuable real-world experience without sacrificing project and product management phases. In this chapter, we propose the augmented WebHelix process that augments the WebHelix in three ways: provides an option at the end of each slice of the helix to both release the current version and continue to the next slice of development, and provides a qualitative evaluation framework, called the project evaluation framework (PEF); that provides a systematic approach for evaluating the status of the project; and the ability to evaluate the project at the end of each phase in a slice of the helix. In this chapter, we describe the augmented WebHelix process and demonstrate its applicability to both academia and industry with examples.

Chapter III defines a methodical approach, named model-centric architecting process (MAP), to effectively cope with the architecture design complexity and manage the architecting process and lifecycle of information systems development in a service-oriented paradigm. This comprehensive method comprises four dimensions of architecting activities: requirement analysis, specification, validation, and planning (RSVP). The process is broken down to 9 interrelated models: meta architecture, conceptual architecture, logical architecture, physical architecture, deployment architecture, management architecture, information architecture, aspect architecture, and component architecture. This systematic framework may be customized in different formats to design various information systems in different industries.

Service-oriented software systems (SOSS) are discussed in Chapter IV. It is preferable to understand service orientation not to be limited to Web services and Internet only. It is shown that there are several variants of SOSS having different application domains, different user properties, different development processes, and different software engineering properties. The conditions implying advantageous user properties of SOSS are presented.

The Web and its underlying platform technologies have often been used to integrate existing software and information systems. In Chapter V the Web context, where the Web platform is used to integrate different organisations or software systems, additionally the problem of heterogeneity arises. We introduce a specific data integration solution for Web applications such as Web-enabled information systems. Our contribution is an integration technology framework for Web-enabled information systems comprising, firstly, a data integration technique based on the declarative specification of transformation rules and the construction of connectors that handle the integration and, secondly, a mediator architecture based on information services and the constructed connectors to handle the integration process.

Chapter VI discusses the development and maintenance of Web applications from an engineering perspective. A methodology, termed as POWEM, for deploying patterns as means for improving the quality of Web applications is presented. The role of a process, the challenges in making optimal use of patterns, and feasibility issues involved in doing so, are analyzed. The activities of a systematic selection and application of patterns are explored. Following a top-down approach to design, examples illustrating the use of patterns during macro- and micro-architecture design of a Web application are given. Finally, the implications towards Semantic Web applications and Web 2.0 applications are briefly outlined.

The software engineering community touts component-based software systems as a potential silver bullet for many of its woes: reducing cycle time; reducing cost;, increasing productivity;, allowing easier integration - to name just a few. Indeed, many Web-based systems are now built with open-source and vendor provided component technologies. While these advances have led to improvements in the development process, they have also led to a great deal of pressure on downstream processes as these systems must be deployed, tuned, and supported. The complexities in deploying and supporting component-based software for distributed and Web-based applications are not understood in the academic or professional communities. Chapter VII stresses the need for addressing this problem by presenting component-based software for Web applications from a deployment perspective, characterizing the issues through real-world experiences with highly component-based applications, and presents strategies and directions for the community to pursue.

Chapter VIII discusses how Web application software architecture has evolved from the simple beginnings of static content, through dynamic content, to adaptive content and the integrated client-server technologies of the Web 2.0. It reviews how various technologies and standards have developed in a repeating cycle of innovation, which tends to fragment the Web environment, followed by standardization, which enables the wider reach of new technologies. It examines the impact of the Web 2.0, XML, Ajax and mobile Web clients on Web application architectures, and how server side processes can support increasingly rich, diverse and interactive clients. It provides an overview of a server-side Java-based architecture for contemporary Web applications that demonstrates some of the key concepts under discussion. By outlining the various forces that influence architectural decisions, this chapter should help developers to take advantage of the potential of innovative technologies without sacrificing the broad reach of standards based development.

Chapter IX discusses the introduction of agile practices into software organizations which may cause unhealthy tensions between the developers and data professionals. The underlying reason is that when agile methodologies are employed, the two communities use incompatible approaches, namely simple design and iterative development, which are practices associated with all agile methodologies, and big design up front (BDUF), a popular database technique. BDUF is inflexible, as once the database foundation is set, it is difficult to make changes throughout the software development life cycle. This chapter describes a database development method for a Web environment. The result is that the database development becomes more iterative and incremental. This has the added benefit of supporting rapid application development in a dynamic environment, a fundamental characteristic of most Web applications.

Service-oriented architecture (SOA) is an emerging architectural style for developing and structuring business applications, especially enterprise resource planning (ERP) systems. However, current composition standards like BPEL have no ability to interact with users. Therefore, we propose in Chapter X a mechanism for including user interaction descriptions into the composition and extending the composition platform for generating user interfaces. In our case study, a federated ERP (FERP) system, this mechanism has been implemented in a prototype based on yet another workflow language (YAWL) dynamically generating Web pages for accessing the ERP system. Because every aspect including the user interfaces can be manipulated through the service composition, such systems are highly flexible yet maintainable.

Chapter XI addresses issues, alternatives, and best practices for prototyping in Web development. The chapter's primary objective is to provide a clear and concise overview of key concepts and best practices for practitioners and students, as well as other audiences. The chapter focuses on graphical user interface (UI) prototyping for Web development, but many of the principles apply to non-UI prototyping and other sorts of software development. First, we introduce the chapter, and review the major objectives, benefits and risks, and classifications of prototypes. Second, we describe the major approaches to prototyping. Finally, we conclude with future trends and a summary of best practices.

Chapter XII introduces several modern methods and tools for creating tests and integrating testing in Web applications, as well as presenting some practices that expand the role of testing in software development. Teams adopting practices such as test-driven development (TDD) and acceptance (customer) testing have shown significant gains in the quality and correctness of the code produced. These practices encourage a more holistic view of testing by integrating a sound testing solution into current software development life cycle models. Furthermore, tests can play an important role in gathering requirements and providing documentation. In this chapter, in addition to offering an initiation to some of the modern testing methods and tools, the authors hope to motivate readers to consider testing as a multi-purpose tool to be used throughout all stages of development.

Chapter XIII addresses issues, alternatives, and best practices that apply when outsourcing Web development. The chapter's primary objective is to provide a concise overview of key concepts and best practices for practitioners and students, as well as other audiences. First, we introduce the chapter, provide background, and present three key ideas that are expanded and developed in the two subsequent sections. The first describes four steps to help executives and upper management address strategic issues and decisions in outsourcing. The second describes four more steps to help managers, team leaders, and development teams address more tactical issues. We conclude with future trends and implications, and a summary of the best practices.

Chapter XIV discusses conventional desktop software applications which are usually designed, built, and tested on a platform similar to the one on which they will be deployed and run. Wireless mobile application development, on the other hand, is more challenging because applications are developed on one platform (like UNIX or Windows) and deployed on a totally different platform like a cellular phone. While wireless applications can be much smaller than conventional desktop applications, developers should think in small terms of the devices on which the applications will run and the environment in which they will operate instead of the amount of code to be written. This chapter presents a systematic approach to engineering wireless application and offers practical guidelines for testing them. What is unique about this approach is that it takes into account the special features of the new medium (mobile devices and wireless networks), the operational environment, and the multiplicity of user backgrounds; all of which pose new challenges to wireless application development.

The process involved with the development of web applications is significantly different from the process of developing applications on older platforms. This is a difference not only in technologies but in the overall business process and associated methodology, in other words the project management. Web applications generally manage content and not just data, and many web applications are document centric versus data centric. In addition, there are many more people involved in the definition, design, development, testing, and approval for Web applications. The pace of business life is much quicker today than in the past, thus Web applications need to be deployed much quicker that earlier IT application and they need to be engineered for more flexibility and adaptability in terms of changing requirements, changing content, changing presentation, mass user customization. In addition, security concerns are more prevalent in Web applications since the end users are outside as well as inside the corporate virtual perimeter. Web applications can serve a global audience and thus there are more diverse stakeholders for these applications. Issues such as language, culture, time zones, weights and measures, currency, logistics, and so forth, need to be considered. This chapter re-examines the project management issues involved with Web applications.

Chapter XVI provides an extensive listing of resources for Web software engineering as well as an author sponsored link to more detailed and more current resources as they become available.

Preface

Not since the industrial revolution have people all over the world experienced such dramatic business and lifestyle changes as are now occurring due to information technology (IT) in general and to the Internet specifically. According to the RAND organization (Hundley, 2004):

Advances in information technology are affecting most segments of business, society, and governments today in many if not most regions of the world. The changes that IT is bringing about in various aspects of life are often collectively called the "information revolution.

THE INFORMATION AND INTERNET REVOLUTION

The current IT revolution is not the first of its kind. Historians and nations may debate the exact time and place of previous information revolutions, but they were:

- The invention of writing, first in Mesopotamia or China about 3000B.C.
- The invention of the written book in China or Greece about 1000B.C.
- Gutenberg's printing press and engraving about 1450A.D.

One is reminded of the opening sentence from *A Tale of Two Cities* by Charles Dickens: "It was the best of times, it was the worst of times." Dickens was referring to the French Revolution, but today in the 21st century we are well into the "IT Revolution."

All major revolutions help some people and organizations, and these may think it is the "best of times" but they also hurt some, and these organizations and people think it to be the "worst of times." With big revolutions there always will be big winners and big losers. As an example, when the printing press was invented, the largest occupation in Europe was the hand copying of books in thousands of monasteries each of which was home to hundreds of monks; fifty years later the monks had been completely displaced. The impact to society was enormous, not because of the displacement of monks by other craftsmen and machines, but because the price of books dropped so drastically that common people could now afford to educate themselves.

For many this new IT revolution is bringing great things with unprecedented improvements in the quality and efficiency of all we do as organizations and as individuals. However for others, IT is a double-edged sword bringing about many problems, disturbances, and unresolved issues such as the creation of a "Digital Divide" between the IT the "haves" and the "have-nots." In addition IT security and privacy problems are getting out of control as evidenced by computer viruses, worms, e-mail fraud and spam, the compromise of personal and private digital information, spyware, piracy of intellectual property, identity theft, IP hijacking, hacking, and other computer crimes. Today, there are major and numerous security "holes" in most software that corporations and individuals use on an everyday basis.

The single most important technology of this information revolution has to be the Internet, which is the combination of several underlying technologies. Consider the penetration rate (in the time to reach 50 million users) of milestone information technologies of the recent past compared to the Internet:

- It took the telephone 40 years to reach 50 million users.
- It took radio 38 years to reach 50 million users.
- It took cable TV 10 years to reach 50 million users.
- It only took the Internet only 5 years to reach 50 million users!

It seems that most of the really cool and innovative IT products and projects today involve Web applications, one way or another; such products and technologies as Amazon, Google, YouTube, MapQuest, eBay, Priceline, and iPhone immediately come to mind. However, as well as these glamorous innovations, many more everyday business applications are migrating to the Web. This is true for both intra-company applications and extra-company applications.

But the Internet and related technologies are beginning to cause significant industrial disruptions including:

- Internet shopping is disrupting traditional sales channels for hard goods;
- Internet sharing and distribution is disrupting traditional intellectual property rights and sales of soft goods (print, audio, video, multi-media);
- "Voice Over IP" combined with ultra high speed optical and wireless media will start to disrupt traditional telecommunications;
- Open source software with community on-line support will start to disrupt the traditional software marketplace;
- Separation of "work" from "workplace" will disrupt corporate and personal real estate and related business sectors;
- As national barriers (political, physical, economic, and temporal) are removed, massive "globalization" will allow the free flow of both work and product;
- The need for retraining and "lifetime learning" coupled with "distance education" is transforming the traditional higher education landscape.

The process and results of these disruptions has been called "creative destruction" by the RAND Corporation, and this results in the "economic eclipse" of organizations not embracing the new IT world. Traditional mechanisms of government (i.e., jurisdiction, taxation, regulation, permits and licenses, etc.) will also be subject to significant disruption in response to these other changes, as will the insurance and finance industries.

BETTER CHEAPER FASTER

Today's commerce landscape is defined by fierce global competition, thus the "battle cry" of the modern business world is:

Better !

Cheaper !

Faster !

That battle cry starts in the board room then down through the management chain as these themes are the crux of market positioning (quality, cost, and time to market) as illustrated in Figure I-1. As Tom Cruise said in the movie *Top Gun*: "I feel a need, a need for speed." That "need for speed" has been emphasized to IT project managers and software development teams by upper management.

To produce better and/or cheaper products or services and get them to market quicker requires better, cheaper, faster processes as illustrated in Figure I-2. In today's world, information systems play a key role and an ever increasing role in the overall process of producing/delivering products or providing services. Today almost every aspect in the design, creation, delivery, and support of products or services today is highly dependent on IT.

The IT platform of choice for implementing business applications has evolved over the last 50 years mainly by mutations rather than by the continuous change of natural selection. This is illustrated in Figure I-3. Today's Internet and Web applications specifically offer business a way to tap into a global market very quickly. Both large and small organizations can easily, cheaply, and quickly put forth a global presence and/or storefront via today's Internet.

The initial simplicity of Web-based HTTP/HTML applications provided a rapid proliferation of simple internet applications. In those early days of the Internet, Web-user interface designers may have felt a little envious of their colleagues who created desktop (PC) client-server software. PC and client-server desktop applications had a richness and responsiveness that initially seemed out of reach for early

Figure I-1. Marketing dimensions

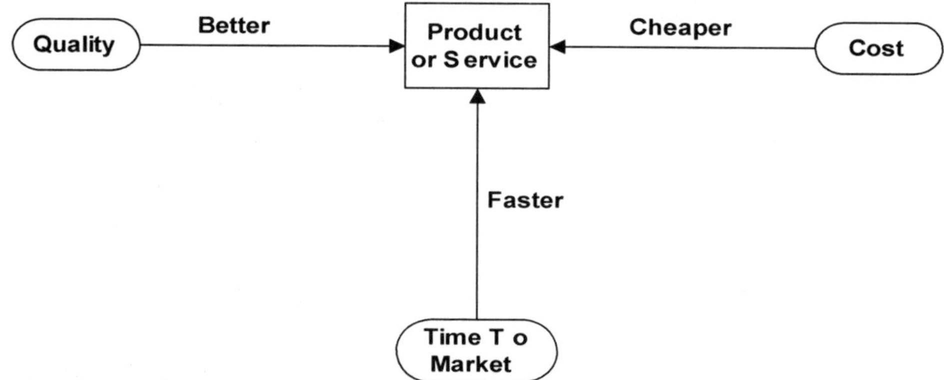

Figure I-2. Products and processes

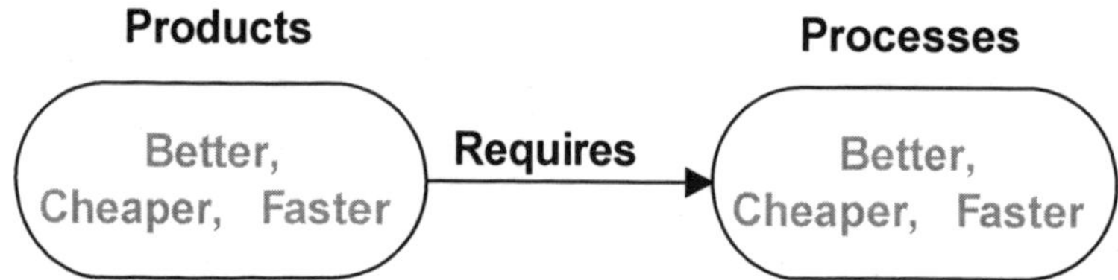

HTML Web software. However, today software technology stacks such as DHTML and Ajax now allow rich and responsive Web applications to be built which increase the depth and breadth of possible Web application penetration and proliferation.

Web-based applications have continued to evolve as more and newer technolgies become available. We are now into the "Web 2" era with these rich and dynamic Web pages, and we are beginning to experiment with "Web 3" technologies and methodologies to create a hyperspace not only of documents, but also of multi-dimensional database objects.

WEB SOFTWARE ENGINEERING

From a software engineering perspective, and also from a project management perspective, developing Web-based applications is different from developing traditional applications based on mainframe

Figure I-3. Application platform evolution

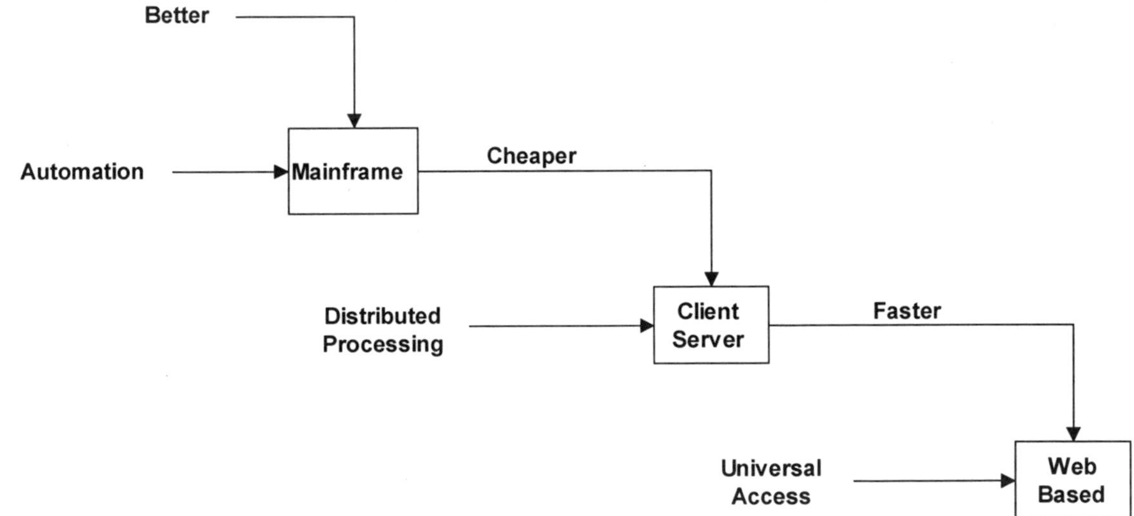

Figure I-4. Web evolution

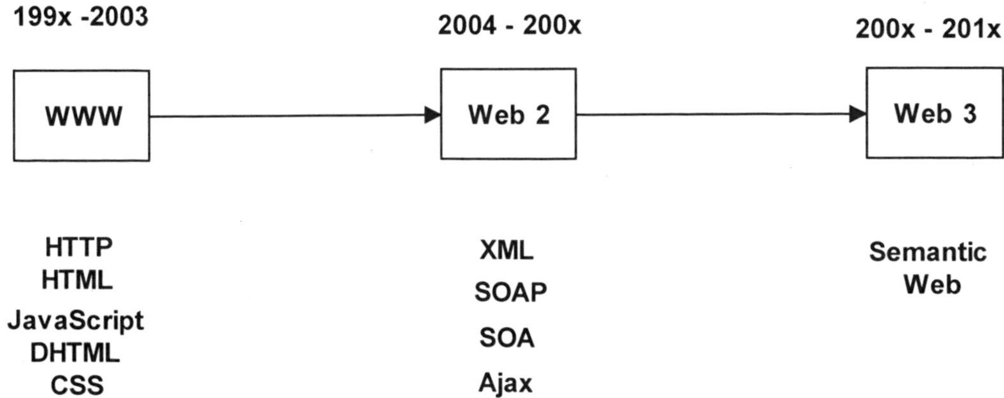

and/or client-server technologies. Some of these differences arise due to changes in the technologies involved, some differences are due to the business climate today, and some differences are due to the scope and nature of the Web applications themselves.

The pace of business life is much quicker today than in the past, stimulated in a large extent due to global competition and the rate of new technology advances. Thus Web applications need to be *deployed much quicker* than earlier IT applications. Also due to the rapid business pace today, Web applications need to be engineered for more *flexibility and adaptability* in terms of *changing requirements, changing content, and changing presentation*. Business growth can be rapid for some products and services, thus Web applications need to be more *scalable* to accommodate large and rapid growth in terms of number of users and number of "hits." Many Web applications are also for products that feature the concept of "mass customization" which often translates into *user customization* of the Web user interface as well as the product and/or service being delivered.

Web applications can serve a *global audience* and thus there are more *diverse stakeholders* for these applications. Issues such as *language, culture, time zones, weights and measures, currency, logistics,* and so forth, need to be considerd. *Norms and even laws* may be different where Web applications are defined as opposed to where Web applications are utilized. Also because of the larger and more diverse stakeholder sets, Web applications are subject to much *more external scrutiny*.

Web applications are physically or virtually delivered over a network rather than installed as was the case with mainframe or desktop applications. Thus network considerations such as *responsiveness, throughput, and security* are much greater.

Older mainframe, client-server, and/or desktop applications generally involved a single programming language (and toolset), at least for each distinct application/program. Mainframe applications were typically written in COBOL, RPG, or FORTRAN and client-server/desktop applications were written in C, C++, or Visual Basic. However, producing Web applications generally involves *multiple languages and tool sets* including HTML, JavaScript, and CSS on the browser (client) side, and Java, PHP, ASP, or Perl on the server side. New technologies may also be intermixed on both sides including XML, SOAP, and Ajax. For business applications, a common denominator between older applications and Web applications has been the use of SQL for database queries and manipulation.

Web applications generally manage content and not just data; further, many Web applications are *document centric versus data centric*. There are many *more people involved* in the definition, design, development, testing, and approval for Web applications as opposed to older IT applications, both technical types and business types. This is illustrated in Figure I-5, which shows that many Web applications have not only a software development lifecycle, but also a content management lifecycle.

This book addresses many of the above differences between Web software engineering and traditional software engineering platforms, and introduces and discusses methodologies and technologies for successful application development in the Web environment:

Chapter I: Web Engineering: Introduction and Perspectives. Web-based systems and applications now deliver a complex array of functionality to a large number of diverse groups of users. As our dependence and reliance on the Web has increased dramatically over the years, their performance, reliability and quality have become paramount importance. As a result, the development of Web applications has become more complex and challenging than most of us think. In many ways, it is also different and more complex than traditional software development. But, currently, the development and maintenance of most Web applications is chaotic and far from satisfactory. To successfully build and maintain large, complex Web-based systems and applications, Web developers need to adopt a disciplined development process and a sound methodology. The emerging discipline of Web engineering advocates a holistic,

Figure I-5. Dual cycles of Web engineering

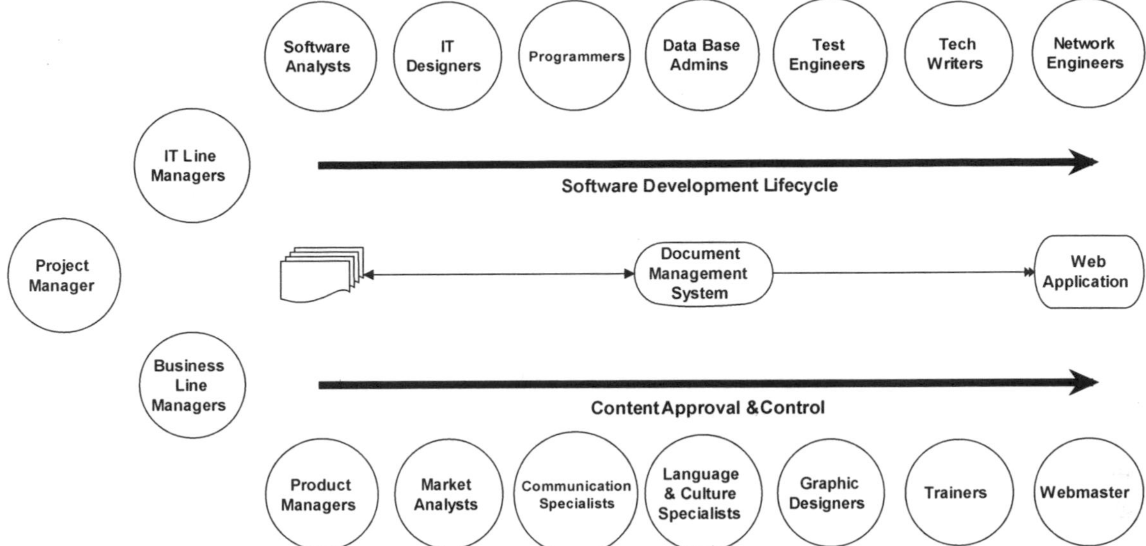

disciplined approach to successful Web development. In this chapter, we articulate and raise awareness of the issues and considerations in large, complex Web application development, and introduce Web engineering as a way of managing complexity and diversity of large-scale Web development.

Chapter II: Augmented WebHelix: A Practical Process for Web Engineering. Process is an important element in the success of any information systems development project especially in academia where typically an undergraduate term project needs to go through the development phases within the space of a semester. Traditionally academic processes have been adapted versions of well-known industrial processes with one major exception— lack of customer feedback in the process. This omission of customer feedback results in students completing "toy" projects without significant real-world experience; efforts to incorporate artificial customer interactions have not been very successful either. It is our opinion that the industry processes cannot be simply copied in academia; what is required is a process that will better equip the students to face real-world challenges. WebHelix has been recently introduced as a practical process for Web engineering that helps students gain valuable real-world experience without sacrificing project and product management phases. In this chapter, we propose the Augmented WebHelix process that augments the WebHelix in three ways: provides an option at the end of each slice of the helix to both release the current version and continue to the next slice of development, and provides a qualitative evaluation framework, called the project evaluation framework (PEF); that provides a systematic approach for evaluating the status of the project; and the ability to evaluate the project at the end of each phase in a slice of the helix. The first augmentation provides the ability to release and continue which is more practical than the go/no-go approach adopted by WebHelix; the second augmentation, the PEF, allows different factors besides the return-on-investment as in WebHelix to be considered for evaluating the current phase and status of the project, and the third augmentation provides the ability to ensure the project is on track. In this chapter, we describe the augmented WebHelix process and demonstrate its applicability to both academia and industry with examples.

Chapter III: Model-Centric Architecting Process. This chapter defines a methodical approach, named model-centric architecting process (MAP), to effectively cope with the architecture design complexity and manage the architecting process and lifecycle of information systems development in a service-oriented paradigm. This comprehensive method comprises four dimensions of architecting activities: requirement analysis, specification, validation, and planning (RSVP). The process is broken down to 9 interrelated models: meta architecture, conceptual architecture, logical architecture, physical architecture, deployment architecture, management architecture, information architecture, aspect architecture, and component architecture. A 2-D matrix serves as a blueprint to denote a step-by-step procedure to produce and manage the architectural artifacts and deliverables in the lifecycle of systems architecture design, development and governance. The holistic framework provides a multidisciplinary view of the design principles, tenets, idioms, and strategies in the IT architecting practices. The characteristics and features of the constituent elements in the MAP approach are articulated in great detail. Recommendations and future trends are also presented in the context. It helps build high-quality service-oriented solutions focused on different domains, and in the meantime keeps the agility, flexibility and adaptiveness of the overall method. This systematic framework may be customized in different formats to design various information systems in different industries.

Chapter IV: Architecture, Specification, and Design of Service-Oriented Systems. Service-oriented software systems (SOSS) are becoming the leading paradigm of software engineering. The crucial elements of the requirements specification of SOSSs are discussed as well as the relation between the requirements specification and the architecture of SOSS. It is preferable to understand service orientation not to be limited to Web services and Internet only. It is shown that there are several variants of SOSS having different application domains, different user properties, different development processes, and different software engineering properties. The conditions implying advantageous user properties of SOSS are presented. The conditions are user-oriented interfaces of services, the application of peer-to-peer philosophy, and the combination of different technologies of communication between services (seemingly the obsolete ones inclusive), and autonomy of the services. These conditions imply excellent software engineering properties of SOSSs as well. Service orientation promises to open the way to the software as a proper engineering product.

Chapter V: Data Integration Through Service-Based Mediation for Web-Enabled Information Systems. The Web and its underlying platform technologies have often been used to integrate existing software and information systems. Traditional techniques for data representation and transformations between documents are not sufficient to support a flexible and maintainable data integration solution that meets the requirements of modern complex Web-enabled software and information systems. The difficulty arises from the high degree of complexity of data structures, for example in business and technology applications, and from the constant change of data and its representation. In the Web context, where the Web platform is used to integrate different organisations or software systems, additionally the problem of heterogeneity arises. We introduce a specific data integration solution for Web applications such as Web-enabled information systems. Our contribution is an integration technology framework for Web-enabled information systems comprising, firstly, a data integration technique based on the declarative specification of transformation rules and the construction of connectors that handle the integration and, secondly, a mediator architecture based on information services and the constructed connectors to handle the integration process.

Chapter VI: Using Patterns for Engineering High-Quality Web Applications. In this chapter, we view the development and maintenance of Web applications from an engineering perspective. A methodology, termed as POWEM, for deploying patterns as means for improving the quality of Web applications is presented. To that end, relevant quality attributes and corresponding stakeholder types are identified. The role of a process, the challenges in making optimal use of patterns, and feasibility issues involved in doing so, are analyzed. The activities of a systematic selection and application of patterns are explored. Following a top-down approach to design, examples illustrating the use of patterns during macro- and micro-architecture design of a Web application are given. Finally, the implications towards Semantic Web applications and Web 2.0 applications are briefly outlined.

Chapter VII: Component-Based Deployment for Web Applications. The software engineering community touts component-based software systems as a potential silver bullet for many of its woes: reducing cycle time; reducing cost;, increasing productivity;, allowing easier integration - to name just a few. Indeed, many Web-based systems are now built with open-source and vendor provided component technologies. While these advances have led to improvements in the development process, they have also led to a great deal of pressure on downstream processes as these systems must be deployed, tuned, and supported. The complexities in deploying and supporting component-based software for distributed and Web-based applications are not understood in the academic or professional communities. This chapter stresses the need for addressing this problem by presenting component-based software for Web applications from a deployment perspective, characterizing the issues through real-world experiences with highly component-based applications, and presents strategies and directions for the community to pursue.

Chapter VIII: Evolving Web Application Architectures, from Model 2 to Web 2. This chapter explores how Web application software architecture has evolved from the simple beginnings of static content, through dynamic content, to adaptive content and the integrated client-server technologies of the Web 2.0. It reviews how various technologies and standards have developed in a repeating cycle of innovation, which tends to fragment the Web environment, followed by standardization, which enables the wider reach of new technologies. It examines the impact of the Web 2.0, XML, Ajax and mobile Web clients on Web application architectures, and how server side processes can support increasingly rich, diverse and interactive clients. It provides an overview of a server-side Java-based architecture for contemporary Web applications that demonstrates some of the key concepts under discussion. By outlining the various forces that influence architectural decisions, this chapter should help developers to take advantage of the potential of innovative technologies without sacrificing the broad reach of standards based development.

Chapter IX: Applying Agility to Database Design. Agile methods are flexible, allowing software developers to embrace changes during the software development life cycle. But the introduction of agile practices into software organizations may cause unhealthy tensions between the developers and data professionals. The underlying reason is that when agile methodologies are employed, the two communities use incompatible approaches, namely simple design and iterative development, which are practices associated with all agile methodologies, and big design up front (BDUF), a popular database technique. BDUF is inflexible, as once the database foundation is set, it is difficult to make changes throughout the software development life cycle. This chapter describes a database development method for a Web environment. Using this method, a data professional divides the database into loosely coupled partitions and resolves the above conflicts by applying certain agile practices. The result is that the database development becomes more iterative and incremental. This has the added benefit of supporting rapid application development in a dynamic environment, a fundamental characteristic of most Web applications.

Chapter X: Automatic Creation of GUI's for Web Based Systems. Service-oriented architecture (SOA) is an emerging architectural style for developing and structuring business applications, especially enterprise resource planning (ERP) systems. SOA applications are composed of small, independent and network-accessible software components, named services. The service composition is normally based on the enterprise's business processes. However, current composition standards like BPEL have no ability to interact with users. Therefore, we propose a mechanism for including user interaction descriptions into the composition and extending the composition platform for generating user interfaces. In our case study, a federated ERP (FERP) system, this mechanism has been implemented in a prototype based *on yet* another workflow language (YAWL) dynamically generating Web pages for accessing the ERP system. Because every aspect including the user interfaces can be manipulated through the service composition, such systems are highly flexible yet maintainable.

ChapterXI: Prototyping in Web Development. This chapter addresses issues, alternatives, and best practices for prototyping in Web development. The chapter's primary objective is to provide a clear and concise overview of key concepts and best practices for practitioners and students, as well as other audiences. The chapter focuses on graphical user interface (UI) prototyping for Web development, but many of the principles apply to non-UI prototyping and other sorts of software development. First, we introduce the chapter, and review the major objectives, benefits and risks, and classifications of prototypes. Second, we describe the major approaches to prototyping. Finally, we conclude with future trends and a summary of best practices.

Chapter XII: Testing Methods for Web Applications. This chapter introduces several modern methods and tools for creating tests and integrating testing in Web applications, as well as presenting some practices that expand the role of testing in software development. Teams adopting practices such as test-driven development (TDD) and acceptance (Customer) testing have shown significant gains in the quality and correctness of the code produced. These practices encourage a more holistic view of testing by integrating a sound testing solution into current software development life cycle models. Furthermore, tests can play an important role in gathering requirements and providing documentation. In this chapter, in addition to offering an initiation to some of the modern testing methods and tools, the authors hope to motivate readers to consider testing as a multi-purpose tool to be used throughout all stages of development.

Chapter XIII: Outsourcing Issues in Web Development. This chapter addresses issues, alternatives, and best practices that apply when outsourcing Web development. The chapter's primary objective is to provide a concise overview of key concepts and best practices for practitioners and students, as well as other audiences. First, we introduce the chapter, provide background, and present three key ideas that are expanded and developed in the two subsequent sections. The first describes four steps to help executives and upper management address strategic issues and decisions in outsourcing. The second describes four more steps to help managers, team leaders, and development teams address more tactical issues. We conclude with future trends and implications, and a summary of the best practices.

Chapter XIV: Engineering Wireless Mobile Applications. Conventional desktop software applications are usually designed, built, and tested on a platform similar to the one on which they will be deployed and run. Wireless mobile application development, on the other hand, is more challenging because applications are developed on one platform (like UNIX or Windows) and deployed on a totally different platform like a cellular phone. While wireless applications can be much smaller than conventional desktop applications, developers should think in small terms of the devices on which the applications will run and the environment in which they will operate instead of the amount of code to be written. This chapter presents a systematic approach to engineering wireless application and offers practical guidelines for testing them. What is unique about this approach is that it takes into account the special features of the new medium (mobile devices and wireless networks), the operational environment, and the multiplicity of user backgrounds; all of which pose new challenges to wireless application development.

Chapter XV: Project Management and Web Software Engineering. The process involved with the development of Web applications is significantly different from the process of developing applications on older platforms. This is a difference not only in technologies but in the overall business process and associated methodology, in other words the project management. Web applications generally manage content and not just data, and many Web applications are document centric versus data centric. In addition, there arc many more people involved in the definition, design, development, testing, and approval for Web applications. The pace of business life is much quicker today than in the past, thus Web applications need to be deployed much quicker that earlier IT application and they need to be engineered for more flexibility and adaptability in terms of changing requirements, changing content, changing presentation, mass user customization. In addition, security concerns are more prevalent in Web applications since the end users are outside as well as inside the corporate virtual perimeter. Web applications can serve a global audience and thus there are more diverse stakeholders for these applications. Issues such as language, culture, time zones, weights and measures, currency, logistics, and so forth, need to be considered. This chapter re-examines the project management issues involved with Web applications.

Chapter XVI: Resources on Web-Centric Computing. This chapter provides an extensive listing of resources for Web software engineering as well as an author sponsored link to more detailed and more current resources as they become available

REFERENCE

Hundley, R., et al. (2004), *The Global Course of the Information Revolution: Recurring Themes and Regional Variations*, RAND Corporation, www.rand.org/publications/MR/MR1680/

Chapter I
Web Engineering:
Introduction and Perspectives

San Murugesan
Southern Cross University, Australia

Athula Ginige
University of Western Sydney, Australia

ABSTRACT

Web-based systems and applications now deliver a complex array of functionality to a large number of diverse groups of users. As our dependence and reliance on the Web has increased dramatically over the years, their performance, reliability and quality have become paramount importance. As a result, the development of Web applications has become more complex and challenging than most of us think. In many ways, it is also different and more complex than traditional software development. But, currently, the development and maintenance of most Web applications is chaotic and far from satisfactory. To successfully build and maintain large, complex Web-based systems and applications, Web developers need to adopt a disciplined development process and a sound methodology. The emerging discipline of Web engineering advocates a holistic, disciplined approach to successful Web development. In this chapter, we articulate and raise awareness of the issues and considerations in large, complex Web application development, and introduce Web engineering as a way of managing complexity and diversity of large-scale Web development.

INTRODUCTION

Within a decade, the World Wide Web has become ubiquitous, and it continues to grow unabated at exponential rate. Web-based systems and applications now deliver a complex array of varied content and functionality to a large number of heterogeneous users. The interaction between a Web system and its backend information systems has also become more tight and complex.

As we now increasingly depend on Web-based systems and applications, their performance, reliability and quality have become paramount importance, and the expectations of and demands placed on Web applications have increased significantly over the years. As a result, the design, development, deployment and maintenance of Web-based systems have become more complex and difficult to manage.

Though massive amounts of Web development and maintenance continue to take place, most of them are carried out in ad hoc manner, resulting in poor quality Web systems and applications. Problems such as outdated or irrelevant information, difficulties in using the Web site and finding relevant information of interest, slow response, Web site crashes, and security breaches are common. We encounter these kinds of problems because Web developers failed to address users' needs and issues such as content management, maintenance, performance, security, and scalability of Web applications. They also often overlook important non-technical considerations such as copyright and privacy.

Many Web developers seem to think that Web application development is just simple Web page creation using HTML or Web development software such as Front Page or Dreamweaver and embodying few images and hyperlinking documents and Web pages. Though certain simple applications such as personal Web pages, seminar announcements, and simple online company brochures that call for simple content presentation and navigation fall into this category, many Web applications are complex and are required to meet an array of challenging requirements which change and evolve. There is more to Web application development than visual design and user interface. It involves planning, Web architecture and system design, testing, quality assurance and performance evaluation, and continual update and maintenance of the systems as the requirements and usage grow and develop.

Hence, ad hoc development is not appropriate for large, complex Web systems, and it could result in serious problems: the delivered systems are not what the user wants; they are not maintainable and scalable, and hence have short useful life; they often do not provide desired levels of performance and security; and/or most Web systems are often much behind schedule and overrun the budget estimates.

More importantly, many enterprises and organisations cannot afford to have faulty Web systems or tolerate downtime or inconsistent or stale content/information. The problems on the Web become quickly visible and frustrate the users, possibly costing the enterprises heavily in terms of financial loss, lost customer and loss of reputation. As is often said, "We cannot hide the problems on the Web."

Unfortunately, despite being faced with these problems and challenges, most Web application development still continues to be ad hoc, chaotic, failure-prone, and unsatisfactory. And this could get worse as more inherently complex Web systems and applications that involve interaction with many other systems or components pervade us and our dependence on them increases.

To successfully build large-scale, complex Web-based systems and applications, Web developers need to adopt a disciplined development process and a sound methodology, use better development tools, and follow a set of good guidelines.

The emerging discipline of Web engineering addresses these needs and focuses on successful development of Web-based systems and applications, while advocating a holistic, disciplined approach to Web development.

Web Engineering uses scientific, engineering, and management principles and systematic approaches to successfully develop, deploy, and maintain high-quality Web systems and applications (Murugesan et al., 1999). It aims to bring Web-based system development under control, minimise risks and improve quality, maintainability, and scalability of Web applications.

The essence of Web engineering is to successfully manage the diversity and complexity of Web application development, and hence, avoid potential failures that could have serious implications.

This chapter aims to articulate and raise awareness of the issues and considerations in large-scale Web development and introduce Web engineering as a way of managing complexity and diversity of large-scale Web development.

Following a brief outline of the evolution of the Web and the categorisation of Web applications based on their functionality, this chapter examines current Web development practices and their limitations, and emphasises the need for a holistic, disciplined approach to Web development. It then presents an overview of Web engineering, describes an evolutionary Web development process, discusses considerations in Web design and recommends ten key steps for successful development. In conclusion, it offers perspectives on Web Engineering and highlights some of the challenges facing Web developers and Web engineering researchers.

EVOLUTION OF THE WEB

The Web has become closely ingrained with our life and work in just a few years. From its initial objective of facilitating easy creation and sharing of information among a few scientists using simple Web sites that consisted primarily of hyperlinked text documents, the Web has grown very rapidly in its scope and extent of use, supported by constant advances in Internet and Web technologies and standards. In 10 years, the number of Web sites dramatically has grown from 100 to over 45 million (Figure 1).

Figure 1. Growth of Web sites

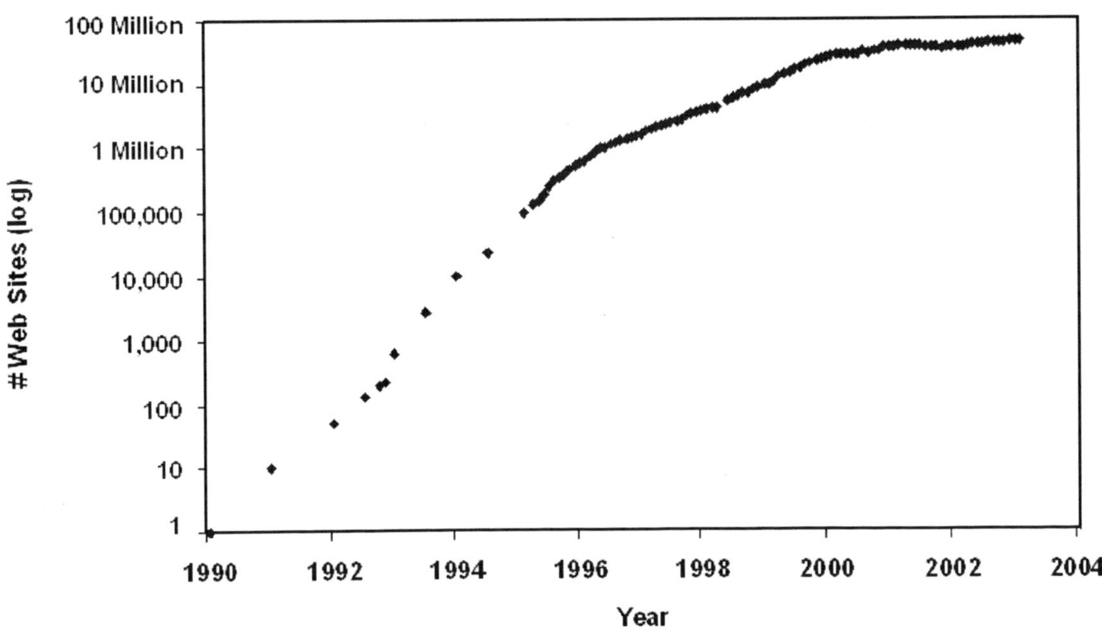

Note: Web Sites = Number of Web servers; one host may have multiple sites by using different domains or port numbers.
Source: Hobbes' Internet Timeline, 2004, www.zakon.org/robert/internet/timeline/

Enterprises, travel and hospitality industries, banks, educational and training institutions, entertainment businesses and governments use large-scale Web-based systems and applications to improve, enhance and/or extend their operations. E-commerce has become global and widespread. Traditional legacy information and database systems are being progressively migrated to the Web. Modern Web applications run on distributed hardware and heterogeneous computer systems. Furthermore, fuelled by recent advances in wireless technologies and portable computing and communication devices, a new wave of mobile Web applications are rapidly emerging. The Web has changed our lives and work at every level, and this trend will continue for the foreseeable future.

The evolution of the Web has brought together some disparate disciplines such as media, information science, and information and communication technology, facilitating easy creation, maintenance, sharing, and use of different types of information from anywhere, any time, and using a variety of devices such as desktop and notebook computers, pocket PCs, personal digital assistants (PDAs), and mobile phones. Contributions of each of these disciplines to the evolution and growth of the Web are:

- **Media:** Integration of different types of media such as data, text, graphics, images, audio and video, and their presentation (animation, 3D visualisation); different types of interaction and channels of communications (one-to-one, one-to-many, many-to-one, and many-to-many).
- **Information science:** Information organisation, presentation, indexing, retrieval, aggregation, and management; and collaborative and distributed content creation.
- **Information and communication technology and networking:** Efficient and cost-effective storage, retrieval, processing, and presentation of information; infrastructures that facilitate transfer and sharing of data and information; wired and wireless Internet communication; and personalised and context-aware Web applications.

Many new Web technologies and standards have emerged in the last couple of years to better support new, novel Web applications: XML, Web services, the Semantic Web, Web personalisation techniques, Web mining, Web intelligence, and mobile and context-aware services.

The advances in Internet and Web technologies and the benefits they offer have led to an avalanche of Web sites, a diverse range of applications, and phenomenal growth in the use of the Web.

Table 1. Categories of Web applications based on functionality

Functionality/Category	**Examples**
Informational	Online newspapers, product catalogues, newsletters, manuals, reports, online classifieds, online books
Interactive	Registration forms, customized information presentation, online games
Transactional	Online shopping - ordering goods and services, online banking, online airline reservation, online payment of bills
Workflow oriented	Online planning and scheduling, inventory management, status monitoring, supply chain management
Collaborative work environments	Distributed authoring systems, collaborative design tools
Online communities, marketplaces	Discussion groups, recommender systems, online marketplaces, e-malls (electronic shopping malls), online auctions, intermediaries

Categories of Web Applications

The scope and complexity of Web applications vary widely: from small scale, short-lived (a few weeks) applications to large-scale enterprise applications distributed across the Internet, as well as via corporate intranets and extranets. Web applications now offer vastly varied functionality and have different characteristics and requirements. Web applications can be categorised in many ways—there is no unique or widely accepted way. Categorisation of Web applications based on functionality (Table 1) is useful in understanding their requirements and for developing and deploying Web-based systems and applications.

WEB DEVELOPMENT PRACTICES

Web development has a very short history, compared to the development of software, information systems, or other computer applications. But within a period of few years, a large number of Web systems and applications have been developed and put into widespread use.

The complexity of Web-based applications has also grown significantly—from information dissemination (consisting of simple text and images to image maps, forms, common gateway interface [CGI], applets, scripts, and style sheets) to online transactions, enterprise-wide planning and scheduling systems, Web-based collaborative work environments, and now multilingual Web sites, Web services and mobile Web applications.

Nevertheless, many consider Web development primarily an authoring work (content/page creation and presentation) rather than application development. They often get carried away by the myth that "Web development is an art" that primarily deals with "media manipulation and presentation." Sure, like the process of designing and constructing buildings, Web development has an important artistic side. But Web development also needs to follow a discipline and systematic process, rather than simply hacking together a few Web pages.

Web applications are not just Web pages, as they may seem to a causal user. The complexity of many Web-based systems is often deceptive and is not often recognised by many stakeholders — clients who fund the development, Web development managers and Web developers—early in the development.

Several attributes of quality Web-based systems such as usability, navigation, accessibility, scalability, maintainability, compatibility and interoperability, and security and reliability often are not given the due consideration they deserve during development. Many Web applications also fail to address cultural or regional considerations, and privacy, moral and legal obligations and requirements. Most Web systems also lack proper testing, evaluation, and documentation.

While designing and developing a Web application, many developers fail to acknowledge that Web systems' requirements evolve, and they do not take this into consideration while developing Web systems. Web-based systems development is not a one-time event as perceived and practiced by many; it is a process with an iterative lifecycle.

Another problem is that most Web application development activities rely heavily on the knowledge and experience of individual (or a small group of) developers and their individual development practices rather than standard practices.

Anecdotal evidence and experience suggest that the problems of ad hoc development (outlined above and in the Introduction section) continue to be faced by developers, users, and other stakeholders. As a result, these are increasing concerns about the manner in which complex Web-based systems are created as well as the level of performance, quality, and integrity of these systems.

Many organisations are heading toward a Web crisis in which they are unable to keep the system updated and/or grow their system at the rate that is needed. This crisis involves the proliferation of quickly 'hacked together' Web systems that are kept running via continual stream of patches or upgrades developed without systematic approaches. (Dart, 2000)

Poorly developed Web-based applications have a high probability of low performance and/or failure. Recently, large Web-based systems have had an increasing number of failures (Williams, 2001). In certain classes of applications such as supply-chain management, financial services, and digital marketplaces, a system failure can propagate broad-based problems across many functions, causing a major Web disaster. The cost of bad design, shabby development, poor performance, and/or lack of content management for Web-based applications has many serious consequences.

The primary causes of these failures are a lack of vision, shortsighted goals, a flawed design and development process, and poor management of development efforts—not technology (Ginige & Murugesan, 2001a). The way we address these concerns is critical to successful deployment and maintenance of Web applications.

Therefore, one might wonder whether development methodologies and processes advocated over the years for software or information systems development and software engineering principles and practices could be directly used for developing Web applications. Though the valuable experiences gained and some of processes and methodologies used in software engineering (and other domains) could be suitably adapted for Web development as appropriate, they are not adequate, as Web development is rather different from software development in several aspects.

Web Development is Different

It is important to realise that Web application development has certain characteristics that make it different from traditional software, information system, or computer application development (Deshpande et al., 2002; Deshpande & Hansen, 2001; Ginige & Murugesan, 2001a, 2001b; Glass, 2001; Lowe 2003; Murugesan et al., 1999; Pressman, 2001 and 2004).

Web applications have the following characteristics:

- Web applications constantly evolve. In many cases, it is not possible to fully specify what a Web site should or will contain at the start of the development process, because its structure and functionality evolve over time, especially after the system is put into use. Further, the information contained within and presented by a Web site will also change. Unlike conventional software that goes through a planned and discrete revision at specific times in its lifecycle, Web applications continuously evolve in terms of their requirements and functionality (instability of requirements). Managing the change and evolution of a Web application is a major technical, organisational and management challenge—much more demanding than a traditional software development.
- Further, Web applications are inherently different from software. The content, which may include text, graphics, images, audio, and/or video, is integrated with procedural processing. Also, the way in which the content is presented and organised has implications on the performance and response time of the system.
- Web applications are meant to be used by a vast, variable user community—a large number of anonymous users (could be many millions like in the cases of eBay and the 2000 Sydney Olympics Web site) with varying requirements, expectations, and skill sets. Therefore, the user interface and usability features have to meet the needs of a diverse, anonymous user community to whom we cannot offer training sessions, thus complicating human-Web interaction

(HWI), user interface, and information presentation.

- Nowadays, most Web-based systems are content-driven (database-driven). Web-based systems development includes creation and management of the content, as well as appropriate provisions for subsequent content creation, maintenance, and management after the initial development and deployment on a continual basis (in some applications as frequently as every hour or more).
- In general, many Web-based systems demand a good "look and feel," favouring visual creativity and incorporation of multimedia in presentation and interface. In these systems, more emphasis is placed on visual creativity and presentation.
- Web applications have a compressed development schedule, and time pressure is heavy. Hence, a drawn-out development process that could span a few months to a year or more is not appropriate.
- Ramifications of failure or dissatisfaction of users of Web-based applications can be much worse than conventional IT systems.
- Web applications are developed by a small team of (often young) people with diverse backgrounds, skills, and knowledge compared to a team of software developers. Their perception of the Web and the quality of Web-based systems also differ considerably, often causing confusion and resulting in misguided priorities.
- There are rapid technological changes—constant advances in Web technologies and standards bring their own challenges—new languages, standards, and tools to cope with; and lots of errors and bugs in early versions of new mark-up languages, development tools, and environments (technology instability).
- Web development uses cutting-edge, diverse technologies and standards, and integrates numerous varied components, including traditional and non-traditional software, interpreted scripting languages, HTML files, databases, images, and other multimedia components such as video and audio, and complex user interfaces (Offurt, 2002).
- The delivery medium for Web applications is quite different from that of traditional software. Web applications need to cope with a variety of display devices and formats, and supporting hardware, software, and networks with vastly varying access speeds.
- Security and privacy needs of Web-based systems are more demanding than that of traditional software.
- The Web exemplifies a greater bond between art and science than generally encountered in software development.

These unique characteristics of the Web and Web applications make Web development different and more challenging than traditional software development.

WEB ENGINEERING

Web engineering is way of developing and organising knowledge about Web application development and applying that knowledge to develop Web applications, or to address new requirements or challenges. It is also a way of managing the complexity and diversity of Web applications.

A Web-based system is a living system. It is like a garden — it continues to evolve, change, and grow. A sound infrastructure must be in place to support the growth of a Web-based system in a controlled, but flexible and consistent manner. Web engineering helps to create an infrastructure that will allow evolution and maintenance of a Web system and that will also support creativity.

Web engineering is application of scientific, engineering, and management principles and disciplined and systematic approaches to the successful development, deployment and maintenance of high quality Web-based systems and applications (Murugesan et al., 1999).

It is a holistic and proactive approach to the development of large Web-based systems, and it aims to bring the current chaos in Web-based system development under control, minimise risks, and enhance the maintainability and quality of Web systems.

Since its origin and promotion as a new discipline in 1998 (Deshpande, Ginige, Murugesan & Hansen, 2002; Murugesan, 1998), Web engineering is receiving growing interest among the stakeholders of Web-based systems, including developers, clients, government agencies, users, academics, and researchers. In addition, this new field has attracted professionals from other related disciplines such as multimedia, software engineering, distributed systems, computer science, and information retrieval.

Web Engineering is Multidisciplinary

Building a large, complex Web-based system calls for knowledge and expertise from many different disciplines and requires a diverse team of people with expertise in different areas. Web engineering is multidisciplinary and encompasses contributions from diverse areas: systems analysis and design, software engineering, hypermedia/ hypertext engineering, requirements engineering, human-computer interaction, user interface, information engineering, information indexing and retrieval, testing, modelling and simulation, project management, and graphic design and presentation.

"Contrary to the perception of some professionals, Web Engineering is not a clone of software engineering, although both involve programming and software development" (Ginige & Murugesan, 2001a). While Web Engineering uses software engineering principles, it encompasses new approaches, methodologies, tools, techniques, and guidelines to meet the unique requirements of Web-based systems. As previously stated, development of Web-based systems is much more than traditional software development. There are subtle differences in the nature and lifecycle of Web-based and software systems, as well as the way in which they're developed and maintained. "Web development is a mixture between print publishing and software development, between marketing and computing, between internal communications and external relations, and between art and technology" (Powell, 2000).

Evolution of Web Engineering

Web Engineering is progressively emerging as a new discipline addressing the unique needs and challenges of Web-based systems development. Since 1998, when the First Workshop on Web Engineering was held in Brisbane, Australia, in conjunction with the World Wide Web Conference (WWW7), there has been series of workshops and special tracks at major international conferences (WWW conferences 1999-2005, HICS 1999-2001, SEKE 2002 and 2003 and others), and a dedicated annual International Conference on Web Engineering (ICWE) 2002-2005.

There also have been a few special issues of journals on topics related to Web Engineering. There are two new dedicated journals, Journal of Web Engineering (www.rintonpress.com/journals/jweonline.html) and Journal of Web Engineering and Technology (www.inderscience.com), as well as an edited book, Web Engineering: Managing Diversity and Complexity of Web Application Development (Murugesan & Deshpande, 2001).

The bibliography at the end of this chapter gives details of special issues, conferences, books, and journal articles on Web engineering and other related areas.

New subjects and courses on Web engineering are now being taught at universities, both at undergraduate and postgraduate levels, and more research is being carried out on various aspects of Web engineering. Also, not surprisingly, there is growing interest among Web developers in using Web engineering approaches and methodologies.

EVOLUTIONARY WEB DEVELOPMENT

Web-applications are evolutionary. For many Web applications, it is not possible to specify fully what their requirements are or what these systems will contain at the start of their development and later, because their structure and functionality will change constantly over time. Further, the information contained within and presented by a Web site often changes — in some applications as often as every few minutes to a couple of times a day. Thus, the ability to maintain information and to scale the Web site's structure (and the functions it provides) is a key consideration in developing a Web application.

Given this Web environment, it seems the only viable approach for developing sustainable Web applications is to follow an evolutionary development process where change is seen as a norm and is catered to. And, this also mandates adoption of a disciplined process for successful Web development.

Web Development Process

A Web development process outlines the various steps and activities of Web-based systems development. It should clearly define a set of steps that developers can follow and must be measurable and trackable (Ginige & Murugesan, 2001c).

Characteristics of Web applications that make their development difficult—and uniquely challenging—include their real-time interaction, complexity, changeability, and the desire to provide personalised information. In addition, the effort and time required to design and develop a Web application is difficult to estimate with a reasonable accuracy.

Based on our practical experience in building Web applications, we recommend an evolutionary process for Web development, shown in Figure 2. This process assists developers in understanding the context in which the application will be deployed and used; helps in capturing the requirements; enables integration of the know-

Figure 2. Web development process

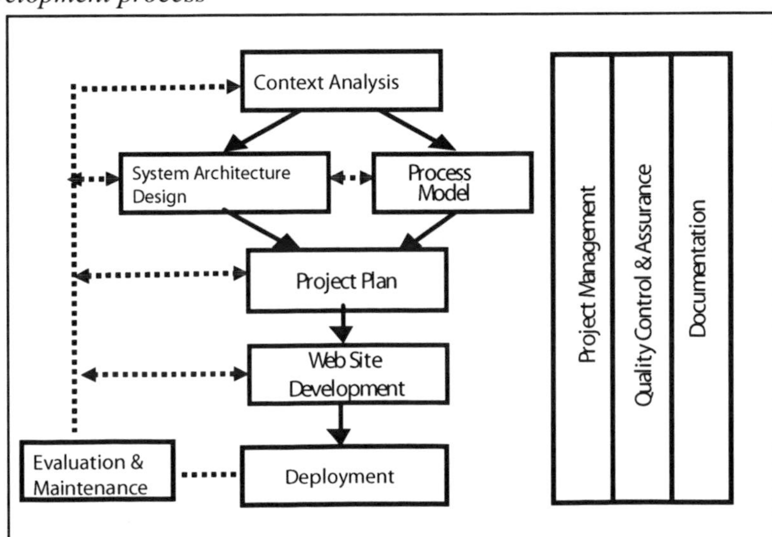

how from different disciplines; facilitates the communication among various members involved in the development process; supports continuous evolution and maintenance; facilitates easier management of the information content; and helps in successfully managing the complexity and diversity of the development process (Ginige & Murugesan 2001c).

Context Analysis

The first essential step in developing a Web-based system is "context analysis," where we elicit and understand the system's major objectives and requirements, as well as the needs of the system's typical users and the organisation that needs the system. It is important to realise at this stage that requirements will change and evolve — even during system development and after its deployment. It is also important to study briefly the operation for which a Web application is to be developed, and the potential implications of introduction of the new system on the organisation. This study should normally include: how information (to be made available on the Web) is created and managed; organisational policy on ownership and control (centralised or decentralised) of information; its current and future plans and business objectives; possible impact of the introduction of Web-based applications on the organisation; the resulting changes in its business and business processes; and emerging trends in the industry sector.

As the Web applications evolve and need to be modified to cater to new requirements — some of which arise from changes or improvements in the business process as a result of deployment of the new Web-based system — an understanding of a big picture about the organisation and its information management policies and practices is a prerequisite for successful design, development, and deployment of Web-based applications.

Before starting Web development, therefore, developers need to elicit and understand the system's major objectives and requirements, gather information about the operational and application environment, and identify the profile of typical system users.

In addition to the functional requirements, potential demands on the scalability, maintainability, availability, and performance of the system need to be specifically elicited and understood by the developers at the beginning of the development process. Based on this information, developers then arrive at the system's functional, technical, and non-technical requirements, which, in turn, influence the system's architectural design.

For instance, if the information content and the system's functions are going to evolve considerably, like in most e-business systems, the system needs to be designed for scalability. On the other hand, if the information changes frequently — like in weather reports, special sales offerings, job vacancies, product price list, brochures, and latest news or announcements — to keep the information current and consistent, the system needs to be designed for easy information maintainability (Merialdo et al., 2003). Moreover, where the application demands very high availability and needs to cater for high peak or uncertain demands, the system may be required to run on multiple Web servers with load balancing and other performance enhancement mechanisms (Almedia & Menasce, 2002; Menasce & Almedia, 2002; Oppenheimer & Patterson, 2002). Examples of this category of applications are online stock trading, online banking, and high volume near-real-time sports and entertainment Web sites such as the Olympics, Wimbledon, and Oscar Web sites.

Thus, it is very important to recognise that scalability, maintainability, and/or performance need to be built into the initial system architecture. It would be very hard, or impossible, to incorporate these features if the initial architecture is not designed to support them. To illustrate this, consider an e-business Web site that provides product information, such as price and availability, which appears on many different pages and changes frequently. If the Web site is designed as static Web pages, then every time a product's information changes, one has to incorporate the change in every page that contains this information. This is a cumbersome and laborious task, and often changes are only made to a few pages, instead of all relevant pages. As a consequence of this, the same information appearing on different pages will be inconsistent.

A better approach to ensure consistency of information across all Web pages is to automatically retrieve the information, when and where needed, from a single information source. If product information is stored in a single central database, then by extracting the relevant information from this database, we can dynamically create various Web pages that contain this information. In the database-driven approach, we need to change the information only in one place: the database. Further, the database-driven Web sites can have a back-end system to allow an authorised person, who may not be skilled in Web page development, to make information changes easily through a Web interface, from anywhere. A database-driven Web site requires a completely different architecture than a Web site that has only static Web pages. Hence, an appropriate architecture that would meet the system's requirements needs to be chosen early in the system development.

Thus, as highlighted in Table 2, the objective of context analysis is to capture and derive the key information required to develop the Web application. In addition, it can also identify non-technical issues that have to be addressed for successful implementation and application of the system. These may include reengineering of business processes where required, organisational and management policies, staff training, and legal, cultural and social aspects.

Context analysis can minimise or eliminate the major problems plaguing large Web-based system development. But, many developers and project managers overlook this essential first step in Web system development and face the problems later when it is hard to correct them.

Based on the context analysis, we then arrive at the system's technical and non-technical requirements (Lowe, 2003), which, in turn, influence the system architecture design.

Architecture Design

In system architecture design, we decide on various components of the system and how they are linked. At this stage, we design:

- An overall system architecture describing how the network and the various servers (Web servers, application servers and database servers) interact;

Table 2. Objectives of context analysis of Web applications

The objectives of context analysis, the first step in Web development, are to:
- Identify the stakeholders and their broader requirements and experiences.
- Identify the functions the Web site needs to provide (immediately, and in the short, medium, and long term).
- Establish what information needs to be on the Web site, how to get this information, and how often this information may change.
- Identify the corporate requirements in relation to look and feel, performance, security, and governance.
- Get a feel of the number of users (typical and peak) and anticipated demands on the system.
- Study similar (competitive) Web sites to gain an understanding of their functionalities, strengths, and limitations.

- An application architecture depicting various information modules and the functions they support; and
- A software architecture identifying various software and database modules required to implement the application architecture.

Table 3 summarises the means of fulfilling some of the requirements of Web-based applications (Ginige & Murugesan, 2001c).

We then decide on an appropriate development process model (Uden, 2002; Pressman, 2004) and develop a project plan. To successfully manage Web development, a sound project plan and a realistic schedule are necessary. Progress of development activities must be monitored and managed. Project planning and scheduling techniques that are commonly used in other disciplines can be used for Web development. Following this, the various components of the system and Web pages are designed, developed and tested.

Web Page Design

Web page design is an important activity; it determines what information is presented and how it is presented to the users. A prototype usually contains a set of sample pages to evaluate the page layout, presentation, and navigation (within and among different pages). Based on the feedback from the stakeholders, the page design is suitably modified. This process may go through a few iterations until the stakeholders and designers are satisfied with the page layout, presentation and the navigation structure.

Web page content development needs to take into consideration the stakeholders' requirements, users' cognitive abilities (Cloyd, 2001), technical issues and considerations, nontechnical issues, earlier experiences of developers and users, and lessons learned from similar Web applications (Figure 3).

If the Web system is intended for global use, by users from different countries, the Web content and presentation may have to be localised; there

Table 3. Means of fulfilling the requirements of Web application

Requirement	Means of Fulfilment
Uniform look and feel across all the Web pages that can easily be modified	Creation of Web pages using templates and style sheets
Consistency of information that may appear in different places or pages	Storing information in a single place (in a database or as an XML file) - without duplication of information in different places or databases – and retrieving the required information for presentation where and when needed
Ease of information update and maintenance	Provision of a back-end system to edit information in a data repository; could have Web interface for easy access from anywhere
Ability to add new Web pages easily	Dynamic generation of navigational links, rather than predetermined static navigational links
Decentralised system administration	Provision of a multi-user login system to access back-end systems and inclusion of a "user administration system" that can assign specific functions and data sets to content managers and other developers/administrators
Mechanisms for quality control and assessing the relevance of information	Inclusion of metadata for Web pages; use of a Web robot for gathering salient information, processing the information gathered and taking appropriate action(s) for ensuring quality or relevance of information presented.
Increased probability of being found through search engines	Using meta tags and registering with search engines

Figure 3. Web page design

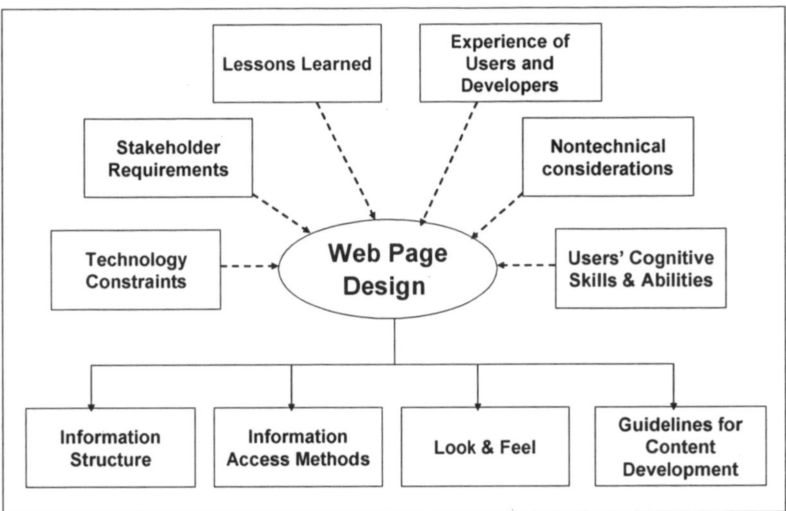

also may be a need for multilingual Web sites (for details, see Becker & Mottay, 2001; Collins, 2002). Also, the Web site's content and usability have to be designed from a global perspective and be responsive to cultural sensitivity in language along with appropriate use of colour, presentation, and animation (Becker & Mottay, 2001).

Web Maintenance

After a Web-based system is developed and deployed online for use, it needs to be maintained. As outlined earlier, content maintenance is a continual process. We need to formulate content maintenance policies and procedures, based on the decision taken at the system architecture design stage on how the information content would be maintained, and then we need to implement them. Further, as the requirements of Web systems grow and evolve, the system needs to be updated and also may be redesigned to cater to the new requirements.

It is important to periodically review Web-based systems and applications regarding the currency of information content, potential security risks, performance of the system, and usage patterns (by analysing Web logs), and take suitable measures to fix the shortcomings and weaknesses, if any.

Project Management

The purpose of project management is to ensure that all the key processes and activities work in harmony. Building successful Web-based applications requires close coordination among various efforts involved in the Web development cycle. Many studies, however, reveal that poor project management is the major cause of Web failures both during development and subsequently in the operational phase. Poor project management will defeat good engineering; good project management is a recipe for success. Successfully managing a large, complex Web development is a challenging task requiring multidisciplinary skills and is, in some ways, different from managing traditional IT projects.

Quality control, assurance and documentation are other important activities, but they are often neglected. Like project management, these activities need to spread throughout the Web development lifecycle.

Steps to Successful Development

Successful development of Web systems and applications involves multiple interactive steps which influence one another. We recommend the following key steps for successful development and deployment of Web applications (Ginige & Murugesan, 2001c):

1. Understand the system's overall function and operational environment, including the business objectives and requirements, organisation culture and information management policy.
2. Clearly identify the stakeholders—that is, the system's main users and their typical profiles, the organisation that needs the system, and who funds the development.
3. Elicit or specify the (initial) functional, technical, and nontechnical requirements of the stakeholders and the overall system. Further, recognise that these requirements may not remain the same; rather, they are bound to evolve over time during the system development.
4. Develop overall system architecture of the Web-based system that meets the technical and nontechnical requirements.
5. Identify subprojects or subprocesses to implement the system architecture. If the subprojects are too complex to manage, further divide them until they become a set of manageable tasks.
6. Develop and implement the subprojects.
7. Incorporate effective mechanisms to manage the Web system's evolution, change, and maintenance. As the system evolves, repeat the overall process or some parts of it, as required.
8. Address the nontechnical issues, such as revised business processes, organisational and management policies, human resources development, and legal, cultural, and social aspects.
9. Measure the system's performance, analyse the usage of the Web application from Web logs, and review and address users' feedback and suggestions.
10. Refine and update the system.

WEB SYSTEM DESIGN: CHALLENGES

The Internet is an open platform that provides unparalleled opportunities. But it has virtually no control over visitor volume, or when and how they access a Web system. This makes developing Web applications that exhibit satisfactory performance even under a sudden surge in number of users a nebulous and challenging task.

Satisfying the expectations and needs of different types of users with varying skills is not easy. When users find a site unfriendly, confusing, or presented with too much information, they will leave frustrated. Worse yet, these frustrated users may spread the bad news to many others. Web site usability factors include good use of colours, information content, easy navigation, and many more. They also include evaluation from an international perspective so that you can reach a global audience. Web usability factors that impact the Web user experience are (Becker & Berkemeyer, 2002): page layout, design consistency, accessibility, information content, navigation, personalisation, performance, security, reliability, and design standards (naming conventions, formatting, and page organisation).

A Web-based system also has to satisfy many different stakeholders besides the diverse range of users, including: persons who maintain the system, the organisation that needs the system, and those who fund the system development. These may pose some additional challenges to Web-based system design and development.

Today's Web-savvy consumers do not tolerate much margin of error or failure. Web system slow down, failure, or security breach may cause a loss of its customers — probably permanently. A whopping 58 percent of first time customers would not return to a site that crashed (Electronic Hit and Run, USA Today, 10 Feb 2000). According to a study (Inter@ctive Week, 6 Sep 1999), US$4.35 billion may be lost in e-business due to poor Web download speeds alone.

As Web applications are becoming mission-critical, there is greater demand for improved reliability, performance, and security of these applications.

Poor design and infrastructure have caused many Web applications to be unable to support the demands placed on them, so they have therefore failed. Many Web sites have suffered site crashes, performance failures, security breaches, and outages — resulting in irate customers, lost revenue, devalued stocks, a tarnished reputation (bad publicity, lack of customer confidence), permanent loss of customers, and law suits (Williams, 2001). Stock prices have become inextricably linked to the reliability of a company's e-commerce site.

The recent major failures and their impact on enterprises have served as a forceful reminder of the need for capacity planning, and improved performance, quality, and reliability. Successful Web application deployment demands consistent Web site availability, a better understanding of its performance, scalability, and load balancing. Proactive measures are needed to prevent grinding halts and failures from happening in the first place.

Large-scale Web system design is a complex and a challenging activity as it needs to consider many different aspects and requirements, some of which may have conflicting needs (Ivory & Hearst, 2002; Siegel, 2003; Cloyd, 2001).

We use terms like scalability, reliability, availability, maintainability, usability, and security to describe how well the system meets current and future needs and service-level expectations. These -ilities characterise (Williams, 2000) a Web system's architectural and other qualities. In the face of increasingly complex systems, these system qualities are often more daunting to understand and manage.

Scalability refers to how well a system's architecture can grow, as traffic, demand for services, or resource utilisation grows. As Web sites grow, small software weaknesses that had no initial noticeable effects can lead to failures, reliability problems, usability problems, and security breaches. Developing Web applications that scale well represents one of today's most important development challenges.

Flexibility is the extent to which the solution can adapt as business requirements change. A flexible architecture facilitates greater reusability and quicker deployment.

Thus, the challenge is to design and develop sustainable Web systems for better:

- Usability—interface design, navigation (Becker & Mottay 2001),
- Comprehension,
- Performance—responsiveness,
- Security and integrity,
- Evolution, growth, and maintainability, and
- Testability.

WEB TESTING AND EVALUATION

Testing plays a crucial role in the overall development process (Becker & Berkemeyer, 2002; Hieatt & Mee, 2002; Lam, 2001). However, more often than not, testing and evaluation are neglected aspects of Web development. Many developers test the system only after it had met with failures or limitations have become apparent, resorting to what is known as retroactive testing. What is desired in the first place is proactive testing at various stages of the Web development lifecycle. Benefits of proactive testing include assurance of proper functioning and guaranteed performance levels, avoidance of costly retroactive fixes, optimal performance, and lower risk.

Testing and validating a large complex Web system is a difficult and expensive task. Testing should not be seen as a one-off activity carried out near the end of development process. One needs to take a broad view and follow a more holistic approach to testing — from design all the way to deployment, maintenance, and continual refinement.

The test planning needs to be carried out early in the project lifecycle. A test plan provides a roadmap so that the Web site can be evaluated through requirements or design stage. It also helps to estimate the time and effort needed for testing — establishing a test environment, finding test personnel, writing test procedures before any testing can actually start, and testing and evaluating the system.

Lam (2001) groups Web testing into the following broad categories and provides excellent practical guidelines on how to test Web systems:

- Browser compatibility
- Page display
- Session management
- Usability
- Content analysis
- Availability
- Backup and recovery
- Transactions
- Shopping, order processing
- Internalisation
- Operational business procedures
- System integration
- Performance
- Login and security

Experience shows that there are many common pitfalls in Web testing and attempts should be made overcome them (Lam, 2001). Testing and evaluation of a Web application may be expensive, but the impact of failures resulting from lack of testing could be more costly or even disastrous.

KNOWLEDGE AND SKILLS FOR WEB DEVELOPMENT

The knowledge and skills needed for large, complex Web application development are quite diverse and span many different disciplines. They can be broadly classified as:

- Technologies supporting and facilitating Web applications
- Design methods
  - Design for usability—interface design, navigation
  - Design for comprehension
  - Design for performance—responsiveness
  - Design for security and integrity
  - Design for evolution, growth and maintainability
  - Design for testability
  - Graphics and multimedia design
  - Web page development
- System architecture
- Web development methods and processes
- Web project management
- Development tools
- Content management
- Web standards and regulatory requirements

Web Development Team

As previously mentioned, development of a Web application requires a team of people with diverse skills and backgrounds (Hansen, 2004). These individuals include programmers, graphic designers, Web page designers, usability experts, content developers, database designers and administrators, data communication and networking experts, and Web server administrators. A Web development team is multidisciplinary, like a film production team, and must be more versatile than a traditional software development team.

Hansen et al. (2001) presents a classification of the participants in a Web development team and a hierarchy for their skills and knowledge. This classification helps in forming a team and in devising a strategy for successful reskilling of the development team.

CONCLUSION

Web engineering is specifically targeted toward the successful development, deployment and maintenance of large, complex Web-based systems.

It advocates a holistic and proactive approach to developing successful Web applications. As more applications migrate to the Web environment and play increasingly significant roles in business, education, healthcare, government, and many day-to-day operations, the need for a Web engineering approach to Web application development will only increase. Further, as we now place greater emphasis on the performance, correctness, and availability of Web-based systems, the development and maintenance process will assume greater significance.

Web Engineering is an emerging discipline having both theoretical and practical significance. It is gaining the interest among researchers, developers, academics, and clients. This is evidenced by increased research activities and publications in this area, hosting of dedicated international conferences and workshops, publication of new journals devoted to Web Engineering, and universities offering special courses and programmes on the subject. It is destined for further advancement through research, education, and practice.

To advance Web engineering, it is essential to define its core body of knowledge, to identify the areas in need of greater research and to develop a strategy to tackle the new technologies, new applications and the various technical, methodological, and societal issues that arise in tandem with such developments. (Deshpande, Olsina & Murugesan, 2002)

Some of the areas that need further study, in no particular order, include:

- Web application delivery on multiple devices — desktop and pocket PCs, mobile phones, PDAs, TVs and refrigerators
- Context-aware Web applications and context-sensitive responses
- Device-independent Web access and content presentation
- Modelling and simulation of Web applications and systems
- Performance evolution and enhancement
- Testing and validation of systems
- Effort and cost estimation
- Web personalisation
- Quality control and assurance

No Silver Bullet!

Web Engineering will not make the problems and the risks go away. But, it can help you plan, monitor, control, and cope with the challenging task of developing large, complex Web applications. It will also facilitate making more informed decisions and developing better quality and better-engineered Web systems and applications.

It is important to understand the wider context in which a Web-based system or application will be used, and design an architecture that will support the development, operation, and maintenance as well as evolution of the Web application in that context, addressing the key issues and considerations. We strongly recommend that Web developers and project managers move away from an ad hoc, hacker-type approach to a well-planned, systematic, and documented approach for the development of large, high-performance, evolutionary, and/or mission-critical Web sites and applications.

Our key recommendations for successfully developing and implementing large, complex Web application are to:

- Adopt a sound strategy and follow a suitable methodology to successfully manage the development and maintenance of Web systems.
- Recognise that, in most cases, development of a Web application is not an event, but a process, since the applications' requirements evolve. It will have a start, but it will not have a predictable end as in traditional IT/software projects.
- Within the continuous process, identify, plan, and schedule various development activities so that they have a defined start and finish.
- Remember that the planning and scheduling of activities is very important to successfully manage the overall development, allocate resources, and monitor progress.
- Consider the big picture during context analysis, planning, and designing a Web application. If you do not, you may end up redesigning the entire system and repeating the process all over again. If you address the changing nature of requirements early on, you can build into the design cost-effective ways of managing change and new requirements.
- Recognise that development of a large Web application calls for teamwork and shared responsibility among the team members, so motivate a team culture.

Web engineering has been successfully applied in a number of Web applications. A well-engineered Web system is:

- Functionally complete and correct
- Usable
- Robust and reliable
- Maintainable
- Secure
- Perform satisfactorily even under flash and peak loads
- Scalable
- Portable, where required perform across different common platforms; compatible with multiple browsers
- Reusable
- Interoperable with other Web and information systems
- Universal accessibility (access by people with different kinds disabilities)
- Well-documented

Time to deploy an online Web system, though still important, is no longer a dominant process driver, as more emphasis is now placed on quality Web systems in terms of functionally, usability, content maintainability, performance, and reliability.

Web engineering can help enterprises and developers to convert their Web systems and applications from a potential costly mess into powerful resource for gaining sustainable competitive advantage.

ACKNOWLEDGMENT

The authors would like to thank Yogesh Deshpande and Steve Hansen, both from University of Western Sydney, Australia, for their contribution in origination and development of the Web engineering discipline and for their input on various aspects of Web development reported in this chapter which evolved through our collaborative efforts over the years. We would also like to thank our graduate students Anupama Ginige and Indra Seher who contributed to formulation and presentation some of the ideas presented in this chapter.

REFERENCES

Almeida, V.A.F., & Menasce, D.A. (2002). Capacity planning for Web services: An essential tool for managing Web services. IT Professional, (July-August), 33-38.

Becker S., & Berkemeyer, A. (2002). Rapid application design and testing for usability. IEEE Multimedia, (Oct-Dec), 38-46.

Becker, S., & Mottay, F. (2001). A global perspective of Web usability for online business applications. IEEE Software, 18(1), 54-61.

Cloyd, M.H. (2001). Designing user-centered Web applications in Web time. IEEE Software, 18(1), 62-69.

Collins, R.W. (2002). Software localization for Internet software: Issues and methods. IEEE Software, (March/April), 74-80.

Dart, S. (2001). Configuration management: A missing link in Web engineering. Norwood, MA: Arttech House.

Deshpande, Y. et al. (2002). Web engineering. Journal of Web Engineering, 1(1), 3-17.

Deshpande, Y., Ginige, A., Murugesan, S., & Hansen, S., (2002). Consolidating Web engineering as a discipline. SEA Software, (April), 32-34.

Deshpande, Y., & Hansen, S. (2001). Web engineering: creating a discipline among disciplines. IEEE Multimedia, (April - June), 82-87.

Deshpande Y., Olsina, L., & Murugesan, S. (2002). Web engineering. Report on the Third ICSE Workshop on Web Engineering, ICSE2002, Orlando, FL, USA.

Ginige, A., & Murugesan, S. (2001a). Web engineering: An introduction. IEEE Multimedia, 8(1), 14-18.

Ginige, A. & Murugesan, S. (2001b). The essence of Web engineering: Managing the diversity and complexity of Web application development. IEEE Multimedia, 8(2), 22-25.

Ginige, A., & Murugesan, S. (2001c). Web engineering: A methodology for developing scalable, maintainable Web applications. Cutter IT Journal, 14(7), 24-35.

Glass, R. (2001). Who's right in the Web development debate? Cutter IT Journal, 14(7), 6-10.

Hansen, S. (2002). Web information systems: The changing landscape of management models and Web applications. Proceedings of the 14th international conference on software engineering and knowledge engineering (pp. 747-753). ACM.

Hansen, S., Deshpande, Y. & Murugesan S. (2001). A skills hierarchy for Web-based systems development. In S. Murugesan & Y. Deshpande (Eds.), Web Engineering – Managing Diversity and Complexity of Web Application Development (LNCS Vol 2016, pp. 223-235). Berlin: Springer.

Hieatt, E., & Mee, R. (2002). Going faster: Testing the Web application. IEEE Software, (March - April), 60-65.

Ivory, M.Y, & Hearst, M.A. (2002). Improving Web site design. IEEE Internet Computing, (March - April), 56-63.

Lam, W. (2001). Testing e-commerce systems: A practical guide. IT Professional, 3(2), 19-27.

Lowe, D. (2003). Web system requirements: An overview. Requirements Engineering, 8, 102-113.

Menasce, D.A, & Almeida, V.A.F. (2002). Capacity planning for Web services: Metrics, models, and methods. Upper Saddle River, NJ: Prentice Hall.

Merialdo, P. et al. (2003). Design and development of data-intensive Web sites: The Araneus Atzeni. ACM Transactions on Internet Technology, 3(1), 49-92.

Murugesan, S. (1998). Web engineering. Presentation at the First Workshop on Web Engineering, World Wide Web Conference (WWW7), Brisbane, Australia.

Murugesan, S. et al. (1999). Web engineering: A New Discipline for Development of Web-based systems. In Proceedings of the First ICSE Workshop on Web Engineering, Los Angeles (pp. 1-9).

Murugesan, S., & Deshpande, Y. (Eds) (2001). Web engineering: Managing diversity and complexity of Web application development. Lecture Notes in Computer Science – Hot Topics, 2016. Berlin: Springer Verlag.

Offutt, J. (2002). Quality attributes of Web software applications. IEEE Software, Special Issue on Software Engineering of Internet Software, 19(2), 25-32.

Oppenheimer, D., & Patterson, D.A. (2002). Architecture and dependability of large-scale Internet services. IEEE Internet Computing, September-October, 41-49.

Pressman, R.S. (2001). What a tangled Web we weave. IEEE Software, 18(1), 18-21.

Pressman, R.S. (2004). Applying Web Engineering, Part 3. Software Engineering: A Practitioner's Perspective (6th ed.). New York: McGraw-Hill.

Reifer, D.J. (2000). Web development: Estimating quick-to-market software. IEEE Software, 17(6), 57-64.

Siegel, D.A. (2003). The business case for user-centered design: Increasing your power of persuasion. Interactions, 10(3).

Uden, L. (2002). Design process for Web applications. IEEE Multimedia, (Oct-Dec), 47-55.

Williams, J. (2000). Correctly assessing the "ilities" requires more than marketing hype. IT Professional, 2(6), 65-67.

Williams, J. (2001). Avoiding CNN moment. IT Professional, 3(2), 68-70.

BIBLIOGRAPHY ON WEB ENGINEERING

For further information on many different aspects of Web development and Web Engineering, we have listed below some useful resources such as books, special issues, journal articles, and Web sites.

Books

Burdman, J. (1999). Collaborative Web development: Strategies and best practices for Web teams. Addison-Wesley.

Dart, S. (2001), Configuration management: A missing link in Web engineering. Norwood, MA: Arttech House.

Dustin, E., Rashka, J., & McDiarmid, D. (2001). Quality Web systems: Performance, security, and usability. Reading, MA: Addison-Wesley.

Friedlein, A. (2000). Web project management: Delivering successful commercial Web sites. Morgan Kaufmann.

Friedlein, A. (2003). Maintaining and evolving successful commercial Web sites. Morgan Kaufmann.

Gerrad, P. & Thompson, N. (2002). Risk-based e-business testing. Artech Publishers.

Hackos, J.T. (2002). Content management for dynamic Web delivery. John Wiley & Sons.

Lowe, D. & Hall, W. (1999). Hypermedia and the Web: An engineering approach. New York: John Wiley & Sons.

Menasce, D.A. & Almeida, V.A.F. (2002). Capacity planning for Web services: Metrics, models, and methods. Upper Saddle River, NJ: Prentice Hall.

Nakano, R. (2002). Web content management: A collaborative approach. Boston: Addison Wesley.

Nguyen, H. Q. (2001). Testing applications on the Web: Test planning for Internet-based systems. John Wiley.

Nielsen, J. (1999). Designing Web usability: The practice of simplicity. Indianapolis, IN: New Riders Publishing.

Powell, T.A. (1998). Web site engineering: Beyond Web page design. Upper Saddle River, NJ: Prentice Hall.

Powell, T.A. (2000). Web design: The complete guide. New York: McGraw-Hill.

Pressman, R.S. (2004). Applying Web engineering. In Software engineering: A practitioner's perspective. New York: McGraw-Hill.

Rosenfeld, L. & Morville, P. (2002). Information architecture for the World Wide Web: Designing large-scale Web sites. O'Reilly & Associates.

Scharl, A. (2000). Evolutionary Web Development. Springer.

Shklar, L. & Rosen, R. (2003). Web application architecture: Principles, protocols and practices. John Wiley & Sons.

Stottlemyer, D. (2001). Automated Web testing toolkit: Expert methods for testing and managing Web applications. John Wiley.

Vidgen, R. et al (2002). Developing Web information systems: From strategy to implementation. Butterworth Heinemann.

Wodtke, C. (2002). Information architecture: Blueprints for the Web. New Riders.

Journals

IEEE Internet Computing. www.computer.org/internet

IEEE Software. www.computer.org/software

Journal of Web Engineering, Rinton Press. www.rintonpress.com/journals/jwe

Journal of Web Engineering and Technology. www.inderscience.com

Web Information Systems Engineering. http://www.i-wise.org

World Wide Web, Kluwer Academic Publishers. http://www.kluweronline.com/issn/1386-145X

Special Issues

Engineering Internet Software, IEEE Software, March-April 2002.

Testing E-business Applications, Cutter IT Journal, September 2001.

Usability and the Web, IEEE Internet Computing, March-April 2002.

Usability Engineering, IEEE Software, January-February 2001.

Web Engineering, Cutter IT Journal, 14(7), July 2001.

Web Engineering, IEEE MultiMedia, Jan.–Mar. 2001 (Part 1) and April–June 2001 (Part 2).

Journal Articles

Almedia, V.A.F., & Menasce, D.A. (2002). Capacity planning for Web services: An essential tool for managing Web services. ITPro, July-August 2002, 33-38.

Arlitt, M., et al. (2001). Characterizing the scalability of a large Web-based shopping system.

ACM Transactions on Internet Technology, 1(1), 44-69.

Barnes, S. & Vidgen, R. (2002). An integrative approach to the assessment of e-commerce quality. Journal of Electronic Commerce Research, 3(3). http://www.webqual.co.uk/papers/jecr_published.pdf

Baskerville. et al. (2003). Is Internet-speed software development different? IEEE Software, Nov-Dec, 70-77.

Becker, S. & Mottay, F. (2001). A global perspective of Web usability for online business applications. IEEE Software, 18(1), 54-61.

Brewer, E.A. (2002). Lessons from giant-scale services. IEEE Internet Computing, July, 46-55.

Cardellini, V. et al. (1999). Dynamic balancing on Web server systems. IEEE Internet Computing, May-June, 2839.

Ceri, S., Fraternali, P., & Bongio, A. (2000, May). Web modelling language (WebML): A modelling language for designing Web sites. Proceedings of the World Wide Web WWW9 Conference, Amsterdam.

Cloyd, M.H. (2001). Designing user-centered Web applications in Web time. IEEE Software, 18(1), 62-69.

Collins, R.W. (2002). Software localization for Internet software: Issues and methods. IEEE Software.

Davison, B.D. (2002). A Web catching primer. IEEE Internet Computing.

Deshpande et al. (2002). Web engineering. Journal of Web Engineering, 1(1), 3-17.

Deshpande, Y. et al. (2002). Consolidating Web engineering as a discipline. SEA Software.

Deshpande, Y. et al. (2002, July). Web site auditing – The first step towards reengineering. Proc 14th International Conference on Software Engineering and Knowledge Engineering, Italy, 2002, pp. 731 – 737.

Deshpande, Y. & Hansen, S. (2002). Web Engineering: Creating a discipline among disciplines. IEEE Multimedia, 82-87.

Fewster, R. & Mendes, E. (2001, April 4-6). Measurement, prediction and risk analysis for Web applications. IEEE Seventh International Software Metrics Symposium London, England, pp. 338-348.

Ginige, A. & Murugesan, S. (2001) Web engineering: An introduction. IEEE Multimedia, 8(1), 14-18.

Ginige, A. & Murugesan, S. (2001). Web engineering: A methodology for developing scalable, maintainable Web applications. Cutter IT Journal, 14(7) 24–35.

Ginige, A. & Murugesan, S. (2001). The essence of Web engineering: Managing the diversity and complexity of Web application development. IEEE Multimedia, 8(2), 22-25.

Glass, R. Who's right in the Web development debate? Cutter IT Journal, 14(7), 6-10.

Goeschka, K.M. & Schranz, M.W. (2001). Client and legacy integration in object-oriented Web engineering. IEEE Multimedia, Special issues on Web Engineering, 8(1), 32-41.

Hieatt, E. & Mee, R. (2002). Going faster: Testing the Web application. IEEE Software, 60-65.

Ingham, D.B., Shrivastava, S.K., & Panzieri, F. (2000). Constructing dependable Web services. IEEE Internet Computing, 25-33.

Isakowitz, T., Stohr, E. & Balasubmmnian, P. (1995). RMM: A methodology for structured hypermedia design. Comm A CM, 38(8), 35-44.

Ivory, M.Y & Hearst, M.A. (2002). Improving Web site design. IEEE Internet Computing, 56-63.

Kirda, E., Jazayeri, M., Kerer, C. & Schranz, M. (2001). Experiences in engineering flexible Web services. IEEE Multimedia, Special issues on Web Engineering, 8(1), 58-65.

Lam, W. (2001). Testing e-commerce systems: A practical guide. IT Professional, 3(2), 19-27.

Liu, S., et al. (2001). A practical approach to enterprise IT security. IT Professional, 3(5) 35-42.

Lowe, D. (2003). Web system requirements: An overview. Requirements Engineering, 8, 102-113.

Lowe, D. & Henderson-Sellers, B. (2001). OPEN to change. Cutter IT Journal, 14(7), 11-17.

Maurer, F. & Martel, S. (2002). Extreme programming: Rapid development for Web-based applications. IEEE Internet Computing, 86-90.

Menasce, D.A. (1993). Load testing of Web sites. IEEE Internet Computing, 89-92.

Merialdo. P. et al. (2003). Design and development of data-intensive Web sites: The Araneus Atzeni. ACM Transactions on Internet Technology, 3(1), 49-92.

Mich, L. et al. (2003). Evaluating and designing Web site quality. IEEE Multimedia, 34-43.

Offutt, J. (2002). Quality attributes of Web software applications. IEEE Software, Special Issue on Software Engineering of Internet Software, 19(2), 25-32.

Olsina, L., Lafuente, G. & Rossi, G. (2001). Specifying quality characteristics and attributes for Websites. In S. Murugesan & Y. Deshpande (Eds), Web engineering – managing diversity and complexity of Web application development (pp. 266-278). Berlin: Springer.

Oppenheimer, D., & Patterson, D.A. (2002). Architecture and dependability of large-scale Internet services. IEEE Internet Computing, 41-49.

Perlman, G. (2002). Achieving universal usability by designing for change. IEEE Internet Computing, 46-55.

Powel, T.A. (1998). Web site engineering: Beyond Web page design. Prentice Hall.

Pressman, R.S. (2001). What a tangled Web we weave. IEEE Software, 18(1), 18-21.

Pressman, R.S. (2001). Can Internet-based applications be engineered? IEEE Software, 15(5), 104-110.

Reifer, D.J. (2000). Web development: Estimating quick-to-market software. IEEE Software.

Roe, V. & Gonik, S. (2002). Server-side design principles for scalable Internet systems. IEEE Software, 34-41.

Scalable Internet Services (2001). Internet Computing.

Schwabe, D. & Rossi, G. (1998). An object oriented approach to Web-based application design. Theory and Practice of Object Systems (TAPOS), special issue on the Internet, 4(4), 207-225.

Schwabe, D., Esmemldo, L., Rossi, G. & Lyardet, F. (2001). Engineering Web application for reuse. IEEE Multimedia, 8(1), 20-31.

Scott, D., & Sharp, R. (2002). Developing secure Web applications, IEEE Internet Computing, 38-45.

Siegel, D.A. (2003). The business case for user-centred design: Increasing your power of persuasion. Interactions, 10(3).

Upchurch, L. et al. (2001). Using card sorts to elicit Web page quality attributes. IEEE Software.

Williams, J. (2000). Correctly assessing the "ilities" requires more than marketing hype. IT Professional, 2(6), 65-67.

Web Sites

ACM SIGWEB: www.acm.org/sigweb

Jakob Nielsen's Website: www.useit.com

NIST Web Usability: zing.ncsl.nist.gov/WebTools/index.html

Universal Usability Guide: www.universalusability.org

Usability Professional Association: www.upassoc.org

Usable Web: www.usableweb.com

Web Engineering Resources, R.S. Pressman and Associates: www.ispa.com/spi/index.html#webe

Web Engineering.org Community Homepage: www.webengineering.org

Web Information System Development Methodology: www.wisdm.net

Web Information Systems Engineering: http://www.i-wise.org

Web Quality: www.webqual.co.uk

World Wide Web Consortium: www.w3.org

Conferences

International Conference on Web Engineering (ICWE) 2004 and 2005. www.icwe2004.org; www.icwe2005.org

Web Information Systems Engineering Conference. http://www.i-wise.org/

World Wide Web Conference. www.www2004.org; www.www2005.org

This work was previously published in Web Engineering: Principles and Techniques, edited by W. Suh, pp. 1-30, copyright 2005 by IGI Publishing, formerly known as Idea Group Publishing (an imprint of IGI Global).

Chapter II
Augmented WebHelix: A Practical Process for Web Engineering

Nary Subramanian
University of Texas at Tyler, USA

George Whitson
University of Texas at Tyler, USA

ABSTRACT

Process is an important element in the success of any information systems development project, especially in academia where typically an undergraduate term project needs to go through the development phases within the space of a semester. Traditionally academic processes have been adapted versions of well-known industrial processes with one major exception—lack of customer feedback in the process. This omission of customer feedback results in students completing "toy" projects without significant real-world experience; efforts to incorporate artificial customer interactions have not been very successful either. It is our opinion that the industry processes cannot be simply copied in academia; what is required is a process that will better equip the students to face real-world challenges. WebHelix has been recently introduced as a practical process for Web engineering that helps students gain valuable real-world experience without sacrificing project and product management phases. In this chapter we propose the Augmented WebHelix process that augments the WebHelix in three ways: provides an option at the end of each slice of the helix to both release the current version and continue to the next slice of development; provides a qualitative evaluation framework, called the project evaluation framework (PEF), that provides a systematic approach for evaluating the status of the project; and the ability to evaluate the project at the end of each phase in a slice of the helix. The first augmentation provides the ability to release and continue which is more practical than the go/no-go approach adopted by WebHelix; the second augmentation, the PEF, allows different factors besides the return-on-investment as in WebHelix to be considered for evaluating the current phase and status of the project, and the third augmentation provides the ability to ensure the project is on track. In this chapter we describe the augmented WebHelix process and demonstrate its applicability to both academia and industry with examples.

1. INTRODUCTION

A Web application is an information system that can deliver complex content and functionality to a broad population of end users [17] and consists of a set of Web pages that are generated in response to user requests [5]. Examples of Web applications include search engines, online stores, auctions, news sites, instructional sites, and games. Features of a Web application include substantial published content, a complex navigational model, a complex data design, many computational modules, and security considerations.

Most companies have a Web presence today. The initial Web presence for a company is often an informational Web site with little or no database use. Because of the simplicity of informational Web sites, companies often underestimate the difficulty of upgrading such a site to a database-driven Web application. They see little need to use a formal software engineering process and this generally produces a Web application that is behind schedule, not fully functional and almost impossible to upgrade. Experienced project managers realize that developing a Web application could be a complex effort and that some form of systems analysis and design is needed. But, projects usually have small budgets and short deadlines, so if too much time is spent in the design phase, projects usually do not get finished on time (this has been referred to as the analysis-paralysis problem in [26]).

A similar situation exists in academia as well. The majority of the courses in computer science (CS) and computer information systems (CIS) programs require term projects to be completed during the semester. Very often the upper-level courses require the term projects to be team projects as well with the expectation that the students will familiarize themselves with the real-world system and software development processes. All the projects, almost without exception, go through the broad phases of analysis, design, and implementation. Many of the courses stress to a certain degree the concepts such as software process, divide-and-conquer, establishment of phases, and project management, with the belief that some understanding of these core ideas will help the students complete the semester project while at the same time, in many cases, doing course work besides taking the full load of classes for the semester. Our experience in the past has been that very frequently students do not complete or in many cases skip altogether the important phases of analysis and design for the term projects defeating the very purpose of preparing the students for the industry.

Most of the courses focus on using one of the software development processes such as the waterfall, incremental, object-oriented, extreme programming (XP), rational unified process (RUP), or system development life cycle (SDLC) [17, 26, 27]. Our analysis for the reasons for many of the problems faced by the students in their project courses has led us to the conclusion that *simply adapting an industry-strength process does not help in academia.* One of the most important reasons seems to be that the current processes used in the industry and recommended for use in academia, suffer from one big requirement - *the feedback loops require actual customer participation.* Most academic projects do not have realistic customer feedback resulting in the following problems:

1. **Students imagine the feedback:** Students go through the motions of updating their artifacts based on imagined customer feedback; while role-playing and reviews have their uses in software development, their primary purpose is not to replace the customer and this imagination approach tends to make the feedback mechanism almost unimportant;
2. **Students ignore feedback:** The assumption is that "let us assume that the customer has approved the artifacts of the previous phase"—this assumption is very closely related to imagined feedback; the realistic

mechanism that fosters healthy communication and engenders team cohesiveness is completely lost;

3. **Traceability studies are often misleading:** In this method, the students establish the traceability of the artifacts of one phase with those of the preceding phase; in industry traceability is used as a technique to verify the artifacts of a phase but the correctness is usually ascertained with the help of customer feedback; therefore, traceability studies alone tend to mislead the students and remain an academic exercise;
4. **Lack of faith in the process:** Students begin to believe that analysis and design were done "just because the instructor wanted" and that in real life they needed to only code and their job was done; such an impression ill-prepares the students for the industry;
5. **Incomplete projects:** Students end up with incomplete projects due to the apparent lack of functioning of the process; students face one of the major problems faced by the industry due to inadequate process or inadequate application of an established process; in some cases this results in demoralized students and in the worst case results in students who change their majors.

In order to overcome these limitations the WebHelix process (WH) model was proposed [28]—WH is a helical process model for developing Web applications where each turn of the helix is referred to as the slice and each slice consists of five phases including analysis, design, coding, testing, and evaluation. The evaluation phase determines whether the current version of the Web application is ready for release to the customer or whether another iteration that takes the project through the next slice of the helix is mandated. Since evaluation phase is critical in making the release or continue decision, it is important that the evaluation takes into account all the factors important for the business and the project; however, WH does the evaluation only by computing the return-on-investment (ROI) and does not take other factors into account.

In this chapter we propose the augmented WebHelix process (AWH) that augments the WH by permitting a release and continue feature (as opposed to the release or continue feature) at the end of each slice, and by providing a qualitative evaluation framework, called the project evaluation framework (PEF), that provides a systematic approach for evaluating the status of the project at the end of each phase on a slice on the helix (as opposed to just once at the end of a slice in WH) by considering different factors including ROI. In this chapter we describe the AWH and demonstrate its applicability to both academia and industry with examples.

This chapter is organized as follows: *Current Web Application Development Processes* discusses some of the interesting processes for Web application development and evaluation, and provides a basis for the introduction of the AWH; *The Augmented WebHelix Process* section discusses the AWH; the *Project Evaluation Framework* section discusses the PEF used by AWH; the *Application of the Augmented WebHelix Process* section discusses the application of AWH to several Web application projects; the *Observations* section presents our observations on the use of the AWH process; and the *Conclusion and Future Work* section concludes the chapter and provides directions for future work.

2. CURRENT WEB APPLICATION DEVELOPMENT PROCESSES

Since a Web application includes a complete set of modules to perform business functions its construction needs to follow a systematic and practical methodology. Once this methodology has been established, some approach to project management also needs to be used in building the application. Many Web application processes have been proposed. Some are modifications of traditional software engineering processes, some are adjusted business re-engineering plans, and some are extensions of traditional Web page development tools. While all of the current Web application design processes have merit, none is universally accepted as "the" Web application process. In this section we briefly survey the current Web application process methodologies and evaluation approaches.

Most of today's Web application development processes are extensions of standard software engineering processes. The waterfall model process was perfect for developing a file maintenance program for mainframes, but far too restrictive a process for building a Web application. Similar to software development, Web application development is also an iterative process [27] and one such iterative process is the spiral [4] approach; but the exact steps at each cycle of the spiral are debated, as is the metric to be used to determine the completion of a cycle, while the iterated waterfall model, another iterative process, is considered too rigid an approach for developing Web applications [17].

Object-oriented development of Web applications has been proposed [8] using the .NET framework. Another object-oriented approach is to add stereotypes for Web navigation to the UML [13]. A more extensive use of UML is seen in the Web application extensions (WAE) [9]. In fact, much of the work of those who believe in model driven design (MDD) seems to be motivated by the UML [14]. The rational unified process (RUP) [27] is another process that makes extensive use of UML while at the same time being iterative and object-oriented. Domain-driven Web application development generalizes MDD in that it attempts to discover more general predevelopment patterns [11] and then automate the design process. All of the object-oriented Web application development processes, especially those based on the UML, have a large number of design diagrams and documents. Since Web applications generally need to be completed quickly, the substantial up-front time involved in UML-based design techniques may be difficult to achieve in practice. Agile UP (AUP) [29] has been proposed as a simplified version of RUP; however, the requirements for employing AUP include staff knowledge and tool independence, both of which, we believe, are intrinsically unsuited to academic projects.

Extreme programming (XP) has been popularized by numerous developers [2]. It is a natural fit for building Web applications because of its emphasis on minimum design, quick prototype development and acceptance testing. The more general agile programming paradigm stresses the importance of reducing the design overhead in project development and being able to modify the project development plan [1]. Web application processes derived from these technologies keep the design phase of Web application development small and produce limited documentation. While streamlining the process of developing the original application, extreme programming teams must be careful to produce adequate documentation for project maintenance and upgrading. It has been suggested that Scrum [19] together with XP will help develop academic projects quickly; however, basis for Scrum seems to be alien to students at most U.S. universities since rugby, which was the original basis for Scrum, is unknown to many students.

A few of today's Web application development processes have been derived from a business-oriented approach to applications development [21]. Most of these processes develop a business plan for the e-business associated with the Web application, sometimes re-engineering the business along the way, and use factors like return-on-investment (ROI) as a metric for the Web application development process. While the business success of the Web application is important, and we believe that business decisions should play a larger role in all stages of software development than seen in the traditional software engineering processes; we also believe that it cannot be the only metric used in the software process.

Some of today's Web application development processes have been derived from well-known Web content tools. Hypermedia application development [25] has been around for more than fifty years, and because of their database use, many of the hypermedia software development processes and tools adapt well to Web application development. Web page design tools, like Microsoft's FrontPage and Macromedia Studio, have migrated into good Web application development tools and some Web application development processes are based on these tools [12], and some are extensions [6].

A number of Web developers have coined the term, Web Engineering, to describe the theory of Web application development. In Web engineering, a balance is struck among the programming, publishing and business aspects of developing Web Applications [10]. There are a large number of different Web application processes that fall under the umbrella of Web engineering. Some of the features common to all Web engineering processes are:

- The Web site of the application has much informational content and the development of the site involves considerable attention to publishing this content;
- The Web application development process has an information management design, not just a database management design;
- While the system architecture is important, as in all software engineering processes, the navigational system is often closely related to the systems architecture;
- The Web application process tends to be spiral, agile (if not extreme) and of short duration.

In spite of the large number of Web engineering processes that have been developed over the past few years, none has been fully accepted as "the" Web engineering process neither in the industry nor in the academia. In order to complete Web-based system term projects academia needs a process that supports rapid prototyping while at the same time prepares the students for industry. It is our belief that a useful Web engineering process for academia should have the following attributes:

1. Supports rapid prototyping
2. Supports incremental development
3. Helps the students to evaluate the artifacts of a phase
4. Helps the students be on track to complete the project
5. Minimizes documentation

One such process is the WebHelix (WH) process that was introduced in [28]. The advantages of WH are:

1. Does a careful business analysis of the probable success of the Web application
2. Has a formal Web engineering process:
    a. That is helical except for some pre and post stages and each stage is exited when a "testing" and "profit" threshold are met
    b. Develops the systems architecture, navigational system and information model as quickly as possible
    c. Is agile with a small design step and produces no unnecessary model diagrams
3. Has an associated Project Management plan that:

    a. Creates a task list from a systems architecture diagram
    b. Uses a Gantt chart to synchronize the team workload

However, one of the issues with WebHelix, it has been pointed out, has been the lack of a evaluation framework and its over-dependence on ROI calculation as a technique to move from one slice to another which does not permit straightforward application in an academic setting. In order to overcome these drawbacks the WebHelix process was augmented with an evaluation framework, called the project evaluation framework, and the resulting Augmented WebHelix is described in the next section.

For evaluation of projects several measures and frameworks have been proposed. For example, for the Boehm's spiral process model [4], risk assessment is used in each turn of the spiral, and as suggested in [17], not everybody is comfortable with risk assessment techniques. WebQEM [15] provides an evaluation process with four phases: quality requirements definition and specification; elementary evaluation; global evaluation; and conclusion. For example, for an e-commerce Web site, usability has been defined to include global site understandability (which has been further broken down into global organization scheme, quality of labeling system, audience-oriented guided tour, and image map), feedback and help features (which has been broken down into quality of help features, addresses directory, link-based feedback, and form-based feedback), interface and aesthetic features (further decomposed into cohesiveness by grouping main control objects, presentation permanence and stability of main controls, and style issues), and miscellaneous features (broken down into foreign language support, Web site last update indicator, and screen resolution indicator). Some of the sub-factors have been further decomposed. Similar approach has been adopted for other non-functional requirements of a Web application such as functionality, reliability, and efficiency. However, these definitions seem rather rigid since the definition of usability, for example, of a Web application (including an e-commerce application) could vary significantly from project-to-project and from organization-to-organization (see [16] for a discussion of this point); for example, in [17] usability has been defined as "usability is a measure of how well a computer system … facilitates learning..." Therefore, an evaluation process that uses strict definitions for non-functional requirements seems to have limited practical benefit. In [18] a framework, called framework for e-metrics, has been suggested for evaluating business metrics of Web applications that uses the concept of functionality interaction which refers to the interaction between applications up-stream and down-stream of the value chain for a business. The e-metrics framework defines three levels of metrics for efficiency, effectiveness, and strategic importance of managerial decisions and each level is evaluated along the dimensions of planning, development, inbound, production, and outbound. As may be expected, the e-metrics framework may be useful only to evaluate the process management and product management aspects of Web application development, but does not seem to assist in the evaluation of technical and other business factors that need to be considered for releasing a version of the product or continuing with further development. In [20] the COBRA (cost estimation, benchmarking, and risk assessment) method is used for cost estimation of Web applications. This method uses data from previous projects for arriving at an estimate; however, as mentioned accurate cost data is hard to come by and cost usually is only one aspect that needs to be considered when making the decision to release the product or continue with further development. In [16] it has been pointed out that metrics vary from company-to-company and suggests several metrics that could be used by companies for evaluating the effectiveness of their Web applications including site performance, user efficiency, customer satisfaction, customer loyalty, and average time spent on their system. While these metrics could be used to evaluate the effectiveness of Web sites they should be used to complement other factors that affect the release/continue decisions. The project evaluation framework (PEF) introduced later in this chapter hopes to overcome the drawbacks of some of the current approaches by providing a systematic approach to making release/continue decisions.

3. THE AUGMENTED WEBHELIX PROCESS

Figure 1 shows the original WebHelix (WH) process [28]. WH uses a modified spiral approach to systems development. Some parts of the design process, like the business analysis and planning, are done once at the beginning of the process. Some parts of the process, like deployment and maintenance, are done once at the end of the process. The major creation of the Web application (coding) is done by using an identical set of steps repeatedly, producing a set of more and more complete prototypes, with the final prototype being the completed Web application.

It has been pointed out that WH suffers from a lack of metrics to determine when to release or to continue to the next slice of the helix and the fact that WH depends on ROI as the only means for moving from one slice of the helix to the next which makes WH somewhat unsuitable for straightforward application in academia. We therefore augmented the WH process, called the Augmented WebHelix process (AWH), as follows:

1. Provided a systematic framework to qualitatively evaluate whether to release the project in the current state and/or to continue to the next slice; this evaluation framework is called the project evaluation framework (PEF)
2. The evaluation is done at the end of each phase of a slice rather than once at the end of a slice; this helps to keep the project on track
3. In order to more accurately reflect the needs of the real world, the AWH gives the following choices for the project at the end of each slice:

    a. Release the project and do not continue to the next slice (go the deployment and maintenance phase of the project)
    b. Release the project and continue to the next slice (the released portion goes

Figure 1. The original WebHelix process

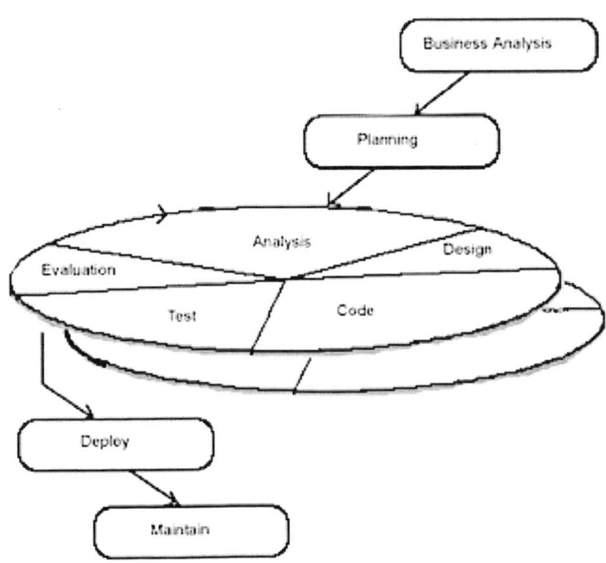

into the deployment and maintenance phase)
c. Do not release the project but go to the next slice to add additional features.

Figure 2 shows the Augmented WebHelix process. The process starts with the Business Analysis and Planning phases, respectively, the PreHelix1 and PreHelix2 phases. After the Planning phase, the project enters the first slice of the helix at the Analysis phase and then proceeds clockwise round the slice through the Design, Coding, and Test phases. After each phase in a slice there is an Evaluation phase (represented by ε) wherein the PEF is applied to decide whether to proceed to the next phase in the slice or not. During the evaluation following the Test phase the major decision of whether to release the product (Release) or to continue to the next slice of the helix (Continue) or both is determined. After the final slice the post helix phases of deployment (Deploy or PostHelix1) and maintenance (Maintain or PostHelix2) are performed. The details of activities performed in each phase are described below. If a Continue decision is made at the end of a slice then the project can proceed along two paths simultaneously—one taking it to the Analysis phase of the subsequent slice and the other taking the project to the deployment phase.

- **PreHelix 1** Business Analysis

1. Catalogue the business processes.
2. Capture real and virtual supply chains.
3. Develop a high-level business plan
   i. Add usual modules plus
   ii. Estimate potential internal/external profit as a return-on-investment (ROI) to determine whether to proceed with the project or not (typically used in industry projects).

- **PreHelix 2:** Planning

1. Select hardware and software
   i. Select Web application platform.
   ii. Select information storage technologies.
2. Determine project management plan and select a project management tool if needed.
3. Develop a detailed business plan with a better ROI estimate and again decide whether to continue with the project or not (typically used in industry projects)

- **Step 1:** Analysis

1. Develop or update the software requirements document.
2. Develop or update the complete menu/navigation system and some sample screens for application.
3. Develop or update the information design.
   i. Entities
   ii. Web, database and other storage systems.
4. Create detailed test criteria for project acceptance.

- **Step 1A**: Analysis Phase Evaluation

1. Develop factors for analysis phase evaluation
2. Apply the PEF to decide whether to go to Step 2 or not—if analysis phase is determined to be incomplete repeat the analysis phase (may be viewed as proceeding directly to the analysis phase of the next slice); continue on to the design phase.

- **Step 2:** Design

1. Develop or update a detailed system architecture diagram and an updated menu/navigation system. Also add annotations to capture any background processing needed.
2. Develop or update an object diagram for the main objects of the Web application.
3. Develop or update a complete information design system, including a detailed database, XML and Web site design.

Figure 2. The augmented WebHelix process

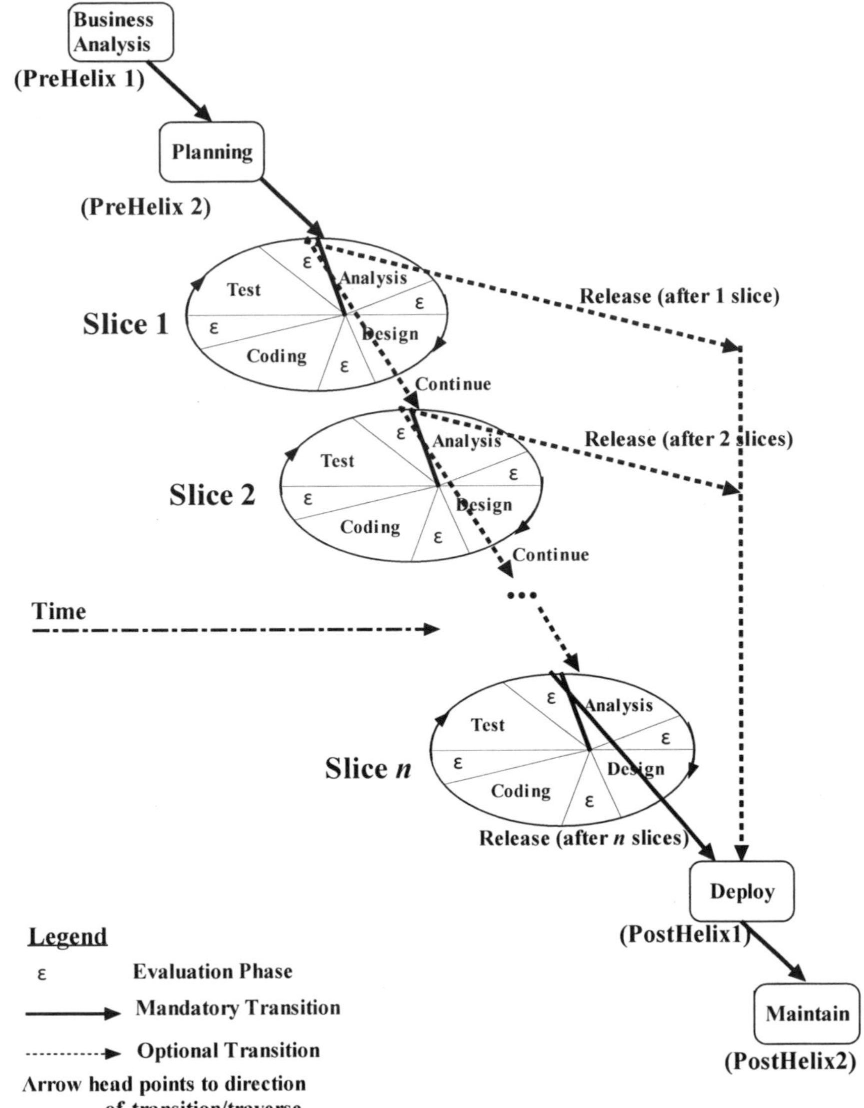

4. Develop or update a complete project management plan.
   i. From the systems architecture diagram make a complete list of tasks.
   ii. Create the programming team and make a list of personal skills.
   iii. Create a Gantt chart for the project.
5. Update the test criteria for project acceptance.

- **Step 2A**: Design Phase Evaluation

1. Develop factors for design phase evaluation
2. Apply the PEF to decide whether to go to Step 3 or not—if design phase is determined to be

incomplete repeat the design phase (may be viewed as proceeding directly to the design phase of the next slice); if the analysis phase is now determined to be incomplete repeat the analysis phase (may be viewed as proceeding directly to the analysis phase of the next slice); continue on to the coding phase.

- **Step 3:** Coding and Integration

1. Select components to be added to the system.
2. Develop Web site, write code, integrate components and unit test all code. We use teams of two during the code development and refactoring [2].
3. Update the test criteria for project acceptance.

- **Step 3A**: Coding Phase Evaluation

1. Develop factors for coding phase evaluation.
2. Apply the PEF to decide whether to go to Step 4 or not—if coding phase is determined to be incomplete repeat the coding phase (may be viewed as proceeding directly to the coding phase of the next slice); if the design phase is now determined to be incomplete repeat the design phase (may be viewed as proceeding directly to the design phase of the next slice); if the analysis phase is now determined to be incomplete repeat the analysis phase (may be viewed as proceeding directly to the analysis phase of the next slice); continue on to the testing phase.

- **Step 4:** Testing

1. Use structured walkthroughs to test Web design.
2. Use multimedia testing, including automation tools, to test code and integration.
3. Work with customer and complete the acceptance testing

- **Step 4A:** Testing Phase Evaluation

1. Develop factors for testing phase evaluation
2. Apply the PEF to decide whether testing phase is completed or not; if testing phase is determined to be incomplete repeat the testing phase (may be viewed as proceeding directly to the testing phase of the next slice) and do not go to the next step; if coding phase is determined to be incomplete repeat the coding phase (may be viewed as proceeding directly to the coding phase of the next slice) and do not go to the next step; if the design phase is now determined to be incomplete repeat the design phase (may be viewed as proceeding directly to the design phase of the next slice) and do not go to the next step; if the analysis phase is now determined to be incomplete repeat the analysis phase (may be viewed as proceeding directly to the analysis phase of the next slice) and do not go to next step; if testing phase is determined to be complete go to the next step.
3. Use PEF to determine whether to Release and/or Continue.
4. If PEF application results in a 'Yes' for Release then release the current prototype and go to PostHelix1 phase.
5. If PEF application results in a 'Yes' for Continue then begin Step 1 of the next slice.

- **PostHelix 1:** Deployment and Training

1. Deploy the Web application.
2. Set up a training program for using the Web application.

- **PostHelix 2:** Maintenance and Future Updates

The number of slices through the helix can vary depending on the project due to the following reasons:

1. During the planning phase it may be determined that the overall project could be

developed in a series of increments with each increment delivering greater functionality. In that case each increment may be developed as a separate slice.

2. Each increment may itself be developed over several slices; incomplete analysis phase signals the start of the analysis phase of the next slice, incomplete design phase could signal the start of the analysis or the design phase of the next slice, and so on.
3. Since AWH provides Release and Continue option, subsequent slices may optimize existing functionality or improve non-functional requirements such as reliability, performance, and modifiability.

Of course, the more slices traversed the longer the time taken to complete the project and more effort is expended on the project. Therefore, the number of slices should be taken from a project management and quality standpoint.

4. PROJECT EVALUATION FRAMEWORK

Project evaluation framework (PEF) helps to determine if the project in its current state is ready to proceed to the next phase of a slice, ready for release, and/or needs to continue to the next slice of AWH. PEF borrows concepts from the NFR Framework [7], where NFR stands for non-functional requirements, which is a qualitative, process-oriented and goal-oriented framework. The NFR Framework and its derivatives have been used for evaluating processes [24], software artifacts [23], and for establishing traceability [22].

At the end of each phase on a helix, PEF is applied to evaluate whether that phase is completed or not, called the phase-stage evaluation. If determined to be complete, the project may proceed to the next phase on the helix; the project may repeat the phase. At the end of the Testing phase on a helix two evaluations are made: first evaluation determines if the testing phase is complete (phase-stage) and if so the second evaluation as to whether the product may be released and/or the project should proceed to the next slice is determined (slice-stage); if the phase-stage evaluation determines that the testing phase has not completed, then the slice-stage evaluation does not take place.

At the end of application of the PEF at the slice-stage we will be able to determine if the Release (whether the product can be released) is a yes (Y) or a no (N) and if Continue (project needs to continue to the next slice) is a Y or a N. Table 1 gives the actions possible with the four combinations of Y's and N's. There is one more possibility with the PEF which is the U state—the Unknown state; in this state we do not have enough information currently to determine if the state is a Y or a N. Therefore Table 1 has nine possible determinations. The steps to help resolve some of the determinations in Table 1 are discussed later in this section.

The application of the steps of the PEF results in an AND-OR graph called the Factor Interdependency Graph or a FIG. FIG establishes the relationships between the factors affecting Release and Continue stages (also referred to as variables of the project). There can be three types of relationships: AND; OR; and EQUAL. In an AND relationship all the child factors need to be satisfied in order for the parent factor to be satisfied and even if one child factor is not satisfied then the parent factor is not satisfied. In an OR relationship satisfaction of even one child factor satisfies the parent factor. In an EQUAL relationship the parent factor takes the same level of satisfaction as the child. The creation of a FIG is also called decomposition (or clarification) of the Release (or Continue) stage into its constituent factors. EQUAL relationship is also called a refinement. PEF creates three types of FIGs: one type of FIG is used for evaluation at the phase-stage (the FIGs for evaluating the ends of Analysis, Design,

Coding, and Test phases could all be different); another FIG for evaluating Release; and the third FIG for evaluating Continue.

The steps of the PEF are iterative and include:

1. Determine the factors that affect the stage
2. Create a FIG for the stage by establishing the relationships (AND, OR, or EQUAL) among the factors determined in step 1
3. Apply the propagation rules to the FIGs to determine if the values of parent factors of a stage (completion of a phase, Release or Continue) is one of Y, N or U; the propagation rules are:
    a. If a factor has AND relationship with its child factors, then
        R1. If all the child factors are Y then the parent factor is Y
        R2. If all the child factors are U then the parent factor is U
        R3. If even one child factor is N then the parent factor is N
        R4. A combination of only Y and U among the child factors will have an effect of Y on the parent
    b. If a factor has OR relationship with its child factors, then
        R5. If all the child factors are N then the parent factor is N
        R6. If all the child factors are U then the parent is U
        R7. If even one child factor is a Y then the parent is Y
        R8. A combination of only N and U among the child factors will have an effect of N on the parent
    c. If a factor has only one child (EQUAL relationship), then

Table 1. Determination of course of action based on the values of release and continue variables

	Release	Continue	Action
Determination 1	Y	N	Release the Web application in its current state and do not go to the next slice of the helix.
Determination 2	N	Y	Do not release the Web application in its current state and continue on to the next slice of the helix.
Determination 3	N	N	We are unable to determine whether to continue or to release; this may be due to incomplete knowledge at present (discussed later in this section).
Determination 4	Y	Y	Here we have an option of either releasing the current version and/or continuing with further development in the next slice of the helix.
Determination 5	Y	U	We can make the release but we are not sure based on the current information whether to continue to the next slice of the helix (discussed later in this section).
Determination 6	N	U	We cannot make the release but we are not sure based on the current information whether to continue to the next slice of the helix (discussed later in this section).
Determination 7	U	Y	We can continue to the next slice of the helix but we are not sure based on the current information whether to release the Web application now (discussed later in this section).
Determination 8	U	N	We cannot continue to the next slice of the helix but we are not sure based on the current information whether to release the Web application now (discussed later in this section).
Determination 9	U	U	We are unsure of whether to release or to continue to the next slice of the helix (discussed later in this section).

R9. If child is Y then the parent is Y
R10. If child is N then the parent is N
R11. If child is U then the parent is U

4. For phase-stage evaluation, if the root factor (the top-most factor in a FIG) is a Y, then the phase is complete and the project can proceed to the next phase; however, if the root factor is an N or a U, then the project may have to repeat the current phase or may have to repeat from an earlier phase in the next slice
5. For slice-stage evaluation use Table 1 to determine the next course of action.

FIGs may be drawn using the notation of a circle for a factor, a line for a relationship, a single arc for AND-decomposition, and a double arc for OR-decomposition, and an example FIG for slice-stage is shown in Figure 4. A phase-stage FIG looks similar with the major difference being that during phase-stage we are concerned with factors that affect the phase while during slice-stage evaluation we are concerned with factors that affect Release and Continue variables. The ontology for the FIG is shown in Figure 3.

The FIG of Figure 4 is read as follows: the top-most factor (root factor) Release (on the left side of Figure 4) is AND-decomposed (indicated by the single arc) into three child factors—Testing Successful, Requirements Met, and Business Goals Achieved; this means that the factor Release is a Y only if all its child factors are each Y (that is the current version may be released if the testing is successful, the requirements are met, and the business goals are achieved). However, what do we mean by testing is successful, that is, the factor Testing Successful is met? This is clarified by further AND-decomposing (indicated by the single arc) the factor Testing Successful into its child factors Errors Acceptable and System Works, that mean, respectively, that the errors in the current version are acceptable as per organization's standards and that the system is accessible over the internet in an acceptable manner. The factor Requirements Met is AND-decomposed into factors NFRs Met and FRs Met, where NFR stands for non-functional requirements and FR stands for functional requirements—the requirements are met when both the functional and non-functional requirements are met. The factor NFRs Met is OR-decomposed (indicated by the double arc) into factors Reliability and Efficiency – this means that the factor NFRs Met is satisfied if either reliability or efficiency or both are satisfied for the Web application in its current version. Similarly, the factor FRs Met is OR-decomposed into factors Storefront and ProductInfo, which means that the functional requirements of the Web application are satisfied if the application has either a Storefront or provides ability to obtain product information (captured by the factor ProductInfo). Similarly the factor Business Goals Achieved is AND-decomposed into factors Profitable and Improves Customer Relationship which means that profitability and improved customer relationship are both required for achievement of business goals. The factor Profitable has been refined (EQUAL relationship) into the factor ROI Acceptable which means that the current version is profitable if the ROI is acceptable. In a similar

Figure 3. Ontology for FIGs

○ Factor	—— Relates Factors	‿ AND Decomposition
‿ OR Decomposition	! Priority	◌ Claim Factor

Figure 4. Example factor interdependency graph (FIG) for slice-stage

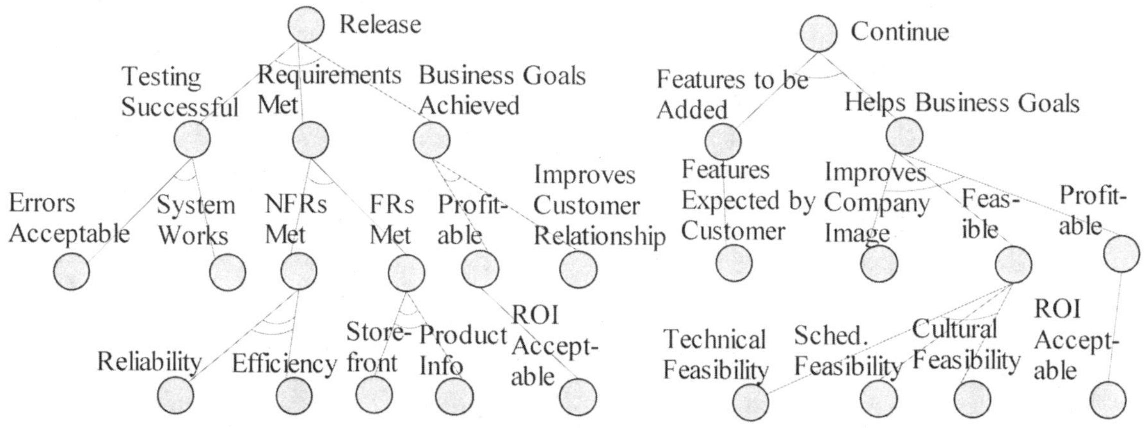

Release
Testing Successful
Requirements Met
Business Goals Achieved
Errors Acceptable
System Works
NFRs Met
FRs Met
Profit-able
Improves Customer Relationship
Reliability
Efficiency
Store-front
Product Info
ROI Accept-able

Continue
Features to be Added
Helps Business Goals
Features Expected by Customer
Improves Company Image
Feas-ible
Profit-able
Technical Feasibility
Sched. Feasibility
Cultural Feasibility
ROI Accept-able

Figure 5. A more practical excel representation of the FIG on the left in Figure 4

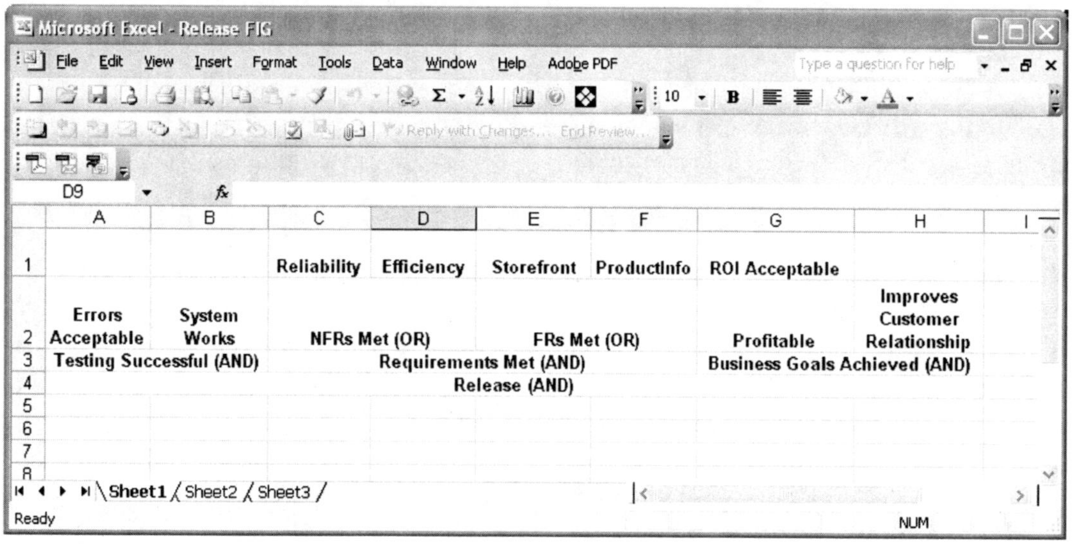

Microsoft Excel - Release FIG

	A	B	C	D	E	F	G	H
1			Reliability	Efficiency	Storefront	ProductInfo	ROI Acceptable	
2	Errors Acceptable	System Works	NFRs Met (OR)		FRs Met (OR)		Profitable	Improves Customer Relationship
3	Testing Successful (AND)		Requirements Met (AND)				Business Goals Achieved (AND)	
4				Release (AND)				
5								
6								
7								

manner the FIG for Continue (the FIG on the right side of Figure 4) may be read.

However, perhaps a more useful way will be to use an Excel spreadsheet as shown in Figure 5, where the FIG of on the left of Figure 4 is shown inverted with the leaf factors in the top-most row and proceeding to the root factor in a downward manner. The relationships are captured by annotating a factor with AND, OR, EQUAL within braces; if no annotation exists, then it means that the factor is a root or has an EQUAL relationship with it child.

One of the advantages of using Excel is that simple macros may be written with VB (Visual Basic) in Excel so that the state of the root factor Release or Continue is computed semi-automatically. For example, the code in Figure 6 is a simple example of a VB macro written using the

Microsoft VB editor that captures the state of the factor NFRs Met (in cell C2 of Figure 5) given the values (or states) of its child factors Reliability (cell C1) and Efficiency (cell D1).

This way of semi-automatically calculating the states of Release and Continue will considerably ease the burden on the project stakeholders during the evaluation phases, both phase-stage and slice-stage, of Augmented WebHelix.

What happens if during the evaluation using PEF we encounter states in rows corresponding to Determination 3 or Determination 5 through Determination 9 of Table 1? Whenever we encounter situations corresponding to these rows, it means that our current knowledge is incomplete and the reasons could be:

1. The decompositions for Release and Continue have not considered all the pertinent factors;
2. Further decompositions of some of the factors may be warranted—we do not know the state of a factor (i.e., it is U) but if we could find that factor's sub factors we may know a more definite answer for the state of the sub factors (i.e., a Y or a N);
3. Some factors may have higher priorities than others and we may need to apply the updated propagation rules given in the later part of Section 5;
4. Earlier stages may need to be repeated in the next slices: for example, if the design phase of the current slice uncovers an analysis error then instead of proceeding to the coding phase of the current slice, we proceed to the analysis phase of the next slice.

5. APPLICATION OF THE AUGMENTED WEBHELIX PROCESS

The AWH process has been applied to more than a dozen projects within the Department of Computer Science at the University of Texas at Tyler. In the senior level course, COSC 4360, Net-Centric Computing, we build a full Web application with an equal emphasis on the software engineering aspects of the application and how

Figure 6. A simple VB macro for partially evaluating FIG of Figure 5

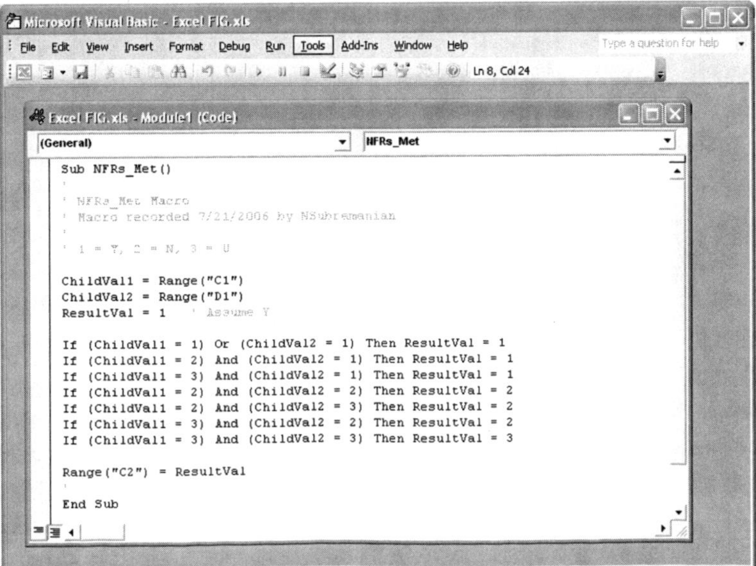

the underlying network protocols (HTTP) operate as the applications executes. The Web application is started early in the semester and AWH is introduced as a required design methodology to be used to complete the project. By using AWH, the students were able to finish fairly complex Web applications that implemented an informational computer security Web site including a message board, secure login and considerable database-driven content.

In COSC 4309, Design of Information Systems course, we build Web applications with more attention being paid to systems analysis and design than to the coding aspects of a Web application. In the course, students are given a booklet describing AWH and then the students develop complete Web applications using the AWH process. For example, in the Fall of 2005, we developed a CS Alumni site that contained an updatable newsletter, message board, e-portfolio for all current and former students and specialized search engine. By using AWH, students were able to savor many of the aspects of Web application design and still finish a complex project. In the Fall of 2006, three information systems were developed by student teams using AWH; one was a e-store for games, another a Web log, and the third a library checkout system.

In the Spring of 2006, for the course COSC3331, e-commerce programming, we again used AWH for five Web application projects developed by teams of two students each – these included an e-book store, an e-realty store, an e-jersey shop, an e-ball shop, and an e-T-shirt store.

Here we explain in detail how the AWH was used in the development of two projects, both of which developed e-commerce store, one for games and the other for books, one as part of COSC4309 and the other as part of COSC3331. In the games e-store project students focused on moving from one phase to the other, while in the e-store for books project students focused on moving from one slice to another.

5.1 Application of AWH for Developing Games E-Store

The Games E-Store project was developed by a team of three seniors for the COSC4309 Design of Information Systems course in the Fall of 2006. The application of different steps of AWH for this project is described below.

PreHelix 1: It was determined that the team will develop an e-store for video games called GameBuy. This activity took 1 day.

PreHelix 2: The resources needed for this project were determined—for example, the development environment (Microsoft Visual Studio), any special security features, the categories of the video games and the number of games in each category were determined. It was determined that there will be following phases: business requirements analysis, software requirements analysis, design, coding, and testing. This activity took 1 day.

Step 1 (Slice1)—Business requirements analysis: The initial business requirements for the project were developed as shown in Figure 7. This activity took 3 days.

Step 1A (Slice 1)—Business requirements evaluation: The Excel spread sheet used to evaluate the business requirements is shown in Figure 8. The corresponding FIG is a simple one as shown in Figure 9—the parent factor (the root factor) Business Requirements Completed is AND-decomposed into the five child factors: Services Provided by GameBuy, Benefits, Financial Advantages, Competitive Advantages, and Technical Advantages. As can be seen from the Figure 8, two of the factors are not satisfied (that is get a value of N) one (Competitive Advantages) due to the fact that one requirement may be considered a software requirement while the other (Technical Advantages) due to the fact that one requirement repeats. Using the propagation rules of PEF (specifically, rule R3) it may be determined that the business requirements are not completed

Figure 7. The initial business requirements for the games e-store project

Video Game Online Store
B1 E-Commerce (Online Store) B2 Up-to-date listing of products B3 Reduce overhead cost by replacing storefront B4 Reduce Staff B5 Customer Friendly B6 Gather customer Information cheaply B7 Less startup cost B8 Reduce customer serv ice support B9 Suggest products to users B10 Real-time inventory B11 Secure customer information B12 Process payments online

and, therefore, the project had to redo the business requirements as part of Slice 2.

Step 1 (Slice 2)—Business requirements analysis: The updated business requirements is shown in Figure 10. This activity took 1 day.

Step 1A (Slice 2)—Business requirements evaluation: Upon evaluating the business requirements with the FIG of Figure 9 we can determine that the phase is complete since all child factors are now satisfied.

The next phase is software requirements analysis. Since this is analysis phase as well we proceed to the analysis phase of the next slice, Slice 3.

Step 1 (Slice 3)—Software requirements analysis: During this phase the software requirements for the project were developed. This activity took 3 days.

Step 1A (Slice 3)—Software requirements evaluation: A FIG similar to Figure 9 was developed with one of the child factors being traceability of the software requirements to the business requirements. The phase-stage evaluation indicated completion of software requirements phase.

Step 2 (Slice 3)—Design: The user interfaces, database, and the classes to be used for this project were all designed. This activity took 3 weeks.

Step 2A (Slice 3)—Design evaluation: The phase-stage evaluation indicated completion of design phase.

Step 3 (Slice 3)—coding: All the components of the system were coded and integrated. This activity took 2 weeks.

Step 3A (Slice 3)—Coding evaluation: The phase stage evaluation indicated completion of the coding phase.

Step 4 (Slice 3)—Testing: The information system was tested. This activity took 1 week.

Step 4A (Slice 3)—Testing evaluation (Phase-Stage): The phase stage evaluation indicated completion of the testing phase.

Step 4A (Slice 3)—Testing evaluation (Slice-Stage): The slice stage evaluation indicated that the product was complete and there was not need to go to the next slice. The Continue was N and Release was Y—therefore from Table 1 (Determination 1), it was determined that the product was ready for release. A more complete application of the FIG for slice-stage determination is given in Section 5.2, for the second example of AWH application.

Figure 8. Evaluation of initial business requirements for the games e-store project

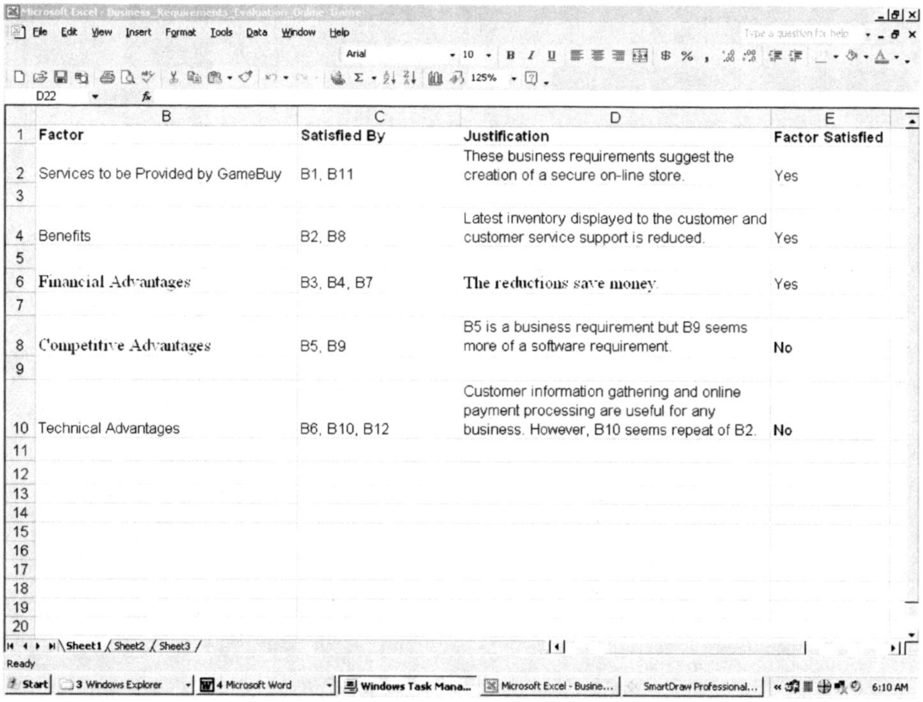

	B	C	D	E
1	Factor	Satisfied By	Justification	Factor Satisfied
2	Services to be Provided by GameBuy	B1, B11	These business requirements suggest the creation of a secure on-line store.	Yes
3				
4	Benefits	B2, B8	Latest inventory displayed to the customer and customer service support is reduced.	Yes
5				
6	Financial Advantages	B3, B4, B7	The reductions save money.	Yes
7				
8	Competitive Advantages	B5, B9	B5 is a business requirement but B9 seems more of a software requirement.	No
9				
10	Technical Advantages	B6, B10, B12	Customer information gathering and online payment processing are useful for any business. However, B10 seems repeat of B2.	No

Figure 9. FIG corresponding to the spreadsheet evaluation of Figure 8

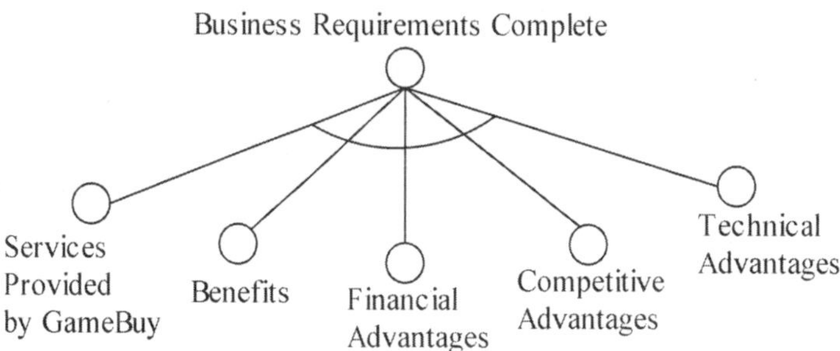

PostHelix1: The product was deployed and demonstrated. This activity took 2 days.

PostHelix2: The product entered the maintenance phase and is still being maintained.

Figure 11 shows the progress of this project through the AWH process; three slices were used some fully and some partially. The entire project was completed within a period of two months with evaluations taking less than a day to complete.

5.2 Application of AWH for Developing Book E-Store

This project was developed by a team of three seniors for the COSC3331 E-Commerce Programming in the Spring of 2006. The application of different steps of AWH for this project is described below.

PreHelix 1: Here the team developed a detailed business plan that included the following factors: mission, potential customers, market entry, competition, and profit avenues. The business plan also discussed the advantages of using e-commerce for this business. This phase lasted 2 weeks.

PreHelix 2: Here the team developed the software requirements for the Web application for the e-commerce storefront for Zebra Books. The constraints on the requirements included the

Figure 10. Updated business requirements for the Games E-Store Project

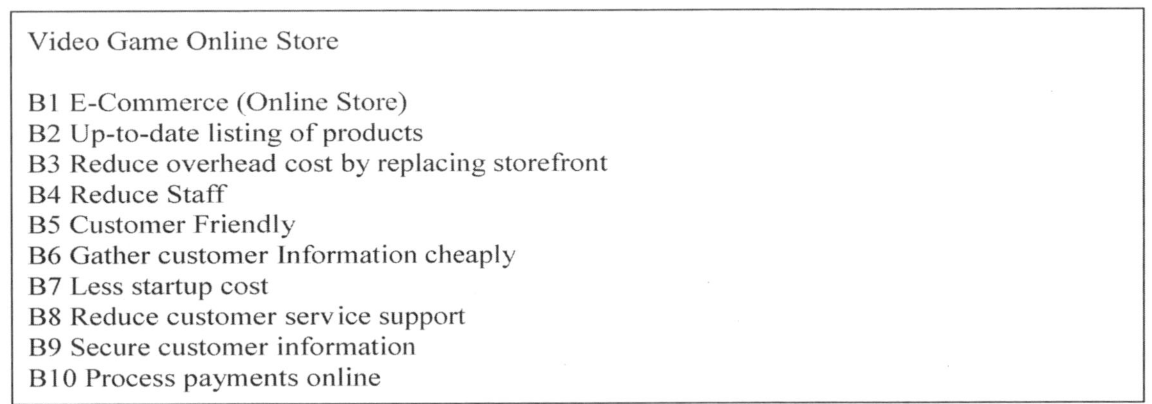

Video Game Online Store

B1 E-Commerce (Online Store)
B2 Up-to-date listing of products
B3 Reduce overhead cost by replacing storefront
B4 Reduce Staff
B5 Customer Friendly
B6 Gather customer Information cheaply
B7 Less startup cost
B8 Reduce customer serv ice support
B9 Secure customer information
B10 Process payments online

Figure 11. The flow of the Games E-Store Project through AWH process

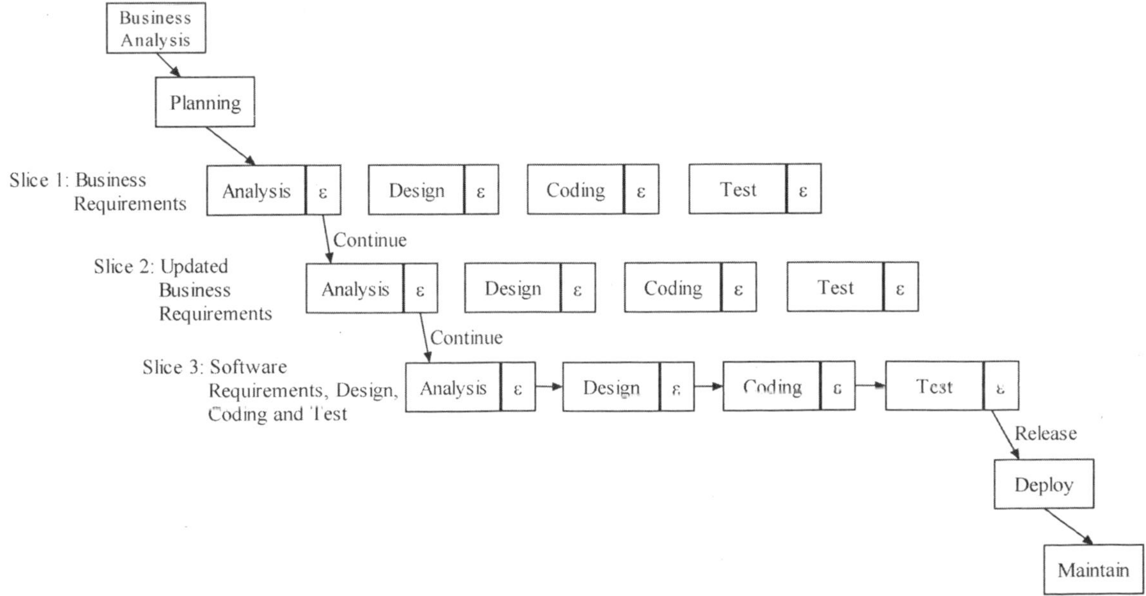

usage of Visual Studio 2003 as the IDE (integrated development environment) for application development, ASP.NET as the technology for the Web application, Visual Basic.NET as the language for the Web application, and Microsoft SQL for the database. During this stage, the number of iterations of the helix as well as the features needed for each iteration were also decided as follows:

1. **Iteration 1:** Development of product catalog—includes the development of the catalog database, the basic storefront and facilities (front-end) for viewing/updating the catalog;
2. **Iteration 2:** Catalog Search Facility, Paypal Transaction Ability, and Shopping Basket feature;
3. **Iteration 3:** Ability to Place Orders and accept Customer Details.

This phase took 1 week.

Each iteration was developed in its own slice, however, the decision whether to Release and/or Continue at the end of each slice was made using PEF (discussed later). For each slice Step 1 (Analysis) through Step 5 (Evaluation) were performed. We discuss these steps for Iteration 1 (Slice 1) below.

Step 1 (Slice 1)—Analysis: During this step all the items that need to be in the product catalog were determined including the item name, description, cost, and a picture (if available). The storefront look and feel was also determined. The features of the screens for viewing the catalog and/or update the catalog were also determined. This step took 3 days.

Step 1A (Slice 1)—Analysis evaluation: This was done similar to that for the project of Section 5.1 and the phase-stage evaluation determined the completion of this phase.

Step 2 (Slice 1)—Design: During this step the database schema was developed for the catalog and the screens for the storefront and database access/update were quickly prototyped in Visual Studio. Also the business classes that the code will use were designed at this step. This step took 3 days.

Step 2A (Slice 1)—Design evaluation: This phase-stage evaluation determined the completion of this phase.

Step 3 (Slice 1)—Coding: During this step the prototyped screens were provided with Visual Basic back end code and the database was populated with the catalog items. This step took 4 days.

Step 3A (Slice 1)—Coding evaluation: This phase-stage evaluation determined the completion of this phase.

Step 4 (Slice 1)—Testing: During this step the database and the front-end were tested to ensure all the database calls worked and that the contents of the database were correct such as for example, the prices were correctly input and there were no spelling mistakes in the item descriptions. The storefront look and feel was also evaluated to ensure that the requirements were met. This step took 3 days.

Step 4A (Slice 1)—Testing evaluation: This phase stage evaluation determined the completion of this phase.

Step 4A (Slice 1)—Testing evaluation (Slice-Stage): During this step PEF was applied to determine the states of Release and Continue variables currently for the project. The Excel spreadsheet shown in Figure 5 was used for the evaluation of the Release variable and macros developed as shown in Figure 6 were used to partially automate the evaluation process. For slice 1, Reliability and Efficiency (cells C1 and D1, respectively, of Figure 5) are both considered met since the partial product does have these properties[1]; likewise, the Storefront factor (cell E1) and ProductInfo factor (cell F1) are both met since at the end of this slice both exist. By propagation rule R7, since at least one child factor of NFRs Met factor (cell C2) is a Y, the NFRs Met is also Y. Likewise, by rule R7, the factor FRs Met (cell E2) is also a Y. By propagation rule R1, the parent factor Requirements Met is a Y as well. However, the factor ROI Acceptable (cell Gl) is not met since as is the system does not have the functionality to improve ROI – therefore, by rule R10, the factor Profitable (cell G2) is not met (that is, it gets an N by the propagation rule). The factor Improves Customer Relationship (cell H2) is not met (N) since the Web application currently does not provide much functionality for the customer. By propagation rule R3, the parent factor Business Goals Achieved (cell G3) is not met, that is, it gets an N). The factor Errors Acceptable (cell A2) was met (Y) since the errors were small; likewise the factor System Works is acceptable (Y) since the system is accessible over a browser. By propagation rule R1, the parent factor Testing Successful (cell A3) is a Y. By propagation rule R3, the parent factor Release (cell A4) is an N since one of its child factors (Business Goals Achieved) is an N—that is the product cannot be released in its current stage.

For deciding the value of Continue variable we used the Excel spreadsheet representation of the FIG on the right of Figure 4 (shown in Figure 12). The factor Technical Feasibility (cell Cl of Figure 12) is a Y since we have used the available technology successfully for this slice and we believe we can do so for the next slice as well; the factor Schedule Feasibility (cell Dl) is a Y as well since we have no reason to expect schedule slippage; and factor Cultural Feasibility (cell El) is a Y since the company (the development group of two students) does not have any issues in continuing further work on this project. By propagation rule R1, the parent factor Feasible (cell C2) is a Y. The factor ROI Acceptable (cell Fl) is a Y since as per our calculations the ROI will improve with more features that customers can use—by propagation rule R9, the parent factor Profitable (cell F2) is a Y. The factor Improves Company Image (cell B2) is a Y since a more usable Web site will help improve the company image; therefore, by propagation rule R1, since all the child factors of the parent factor Helps Business Goals (cell B3) is a Y, the parent factor is itself a Y. The factor Features Expected by Customer (cell A2) is a Y since the features planned for iteration 2 (slice 2) are usually provided by e-commerce sites - therefore, by propagation rule R9, the parent factor Features to be Added (cell A3) is a Y. Since both the child factors of the factor Continue (cell A4) is a Y, by

Figure 12. A more practical excel representation of the FIG on the right in Figure 4

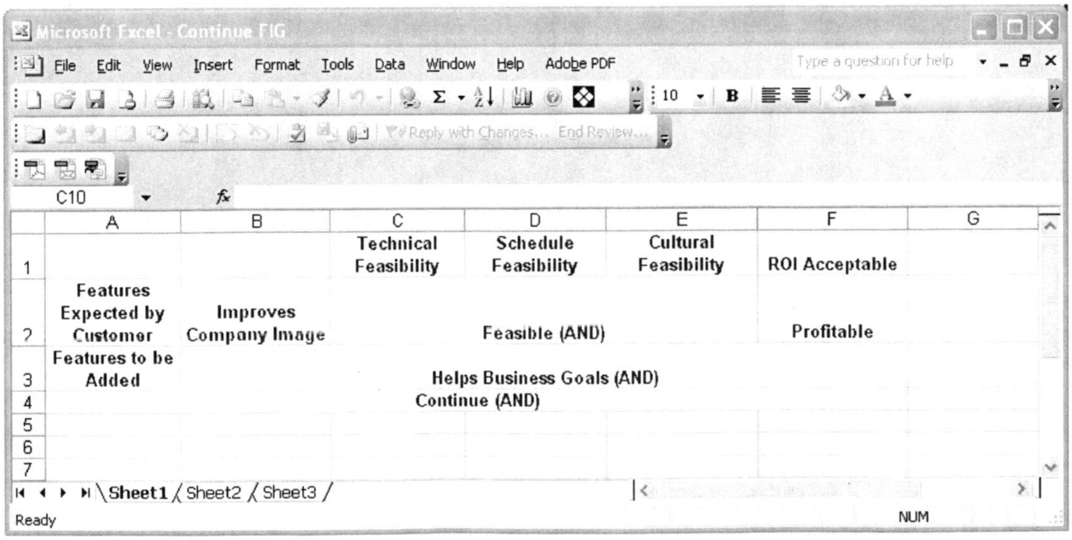

Microsoft Excel - Continue FIG

File Edit View Insert Format Tools Data Window Help Adobe PDF

C10

	A	B	C	D	E	F	G
1			Technical Feasibility	Schedule Feasibility	Cultural Feasibility	ROI Acceptable	
2	Features Expected by Customer	Improves Company Image		Feasible (AND)		Profitable	
3	Features to be Added			Helps Business Goals (AND)			
4				Continue (AND)			
5							
6							
7							

Sheet1 / Sheet2 / Sheet3

Ready NUM

propagation rule R1, the parent factor Continue is also a Y; that is, the project is good to go on to the next slice of the helix.

In a similar manner, Step1 through Step 4A of the AWH were applied for the next two slices; they together took another two weeks. However, for the final slice, during the evaluation phase, Release is a Y and Continue is an N which means that the project needs to be released in its current state and not continued to the next slice. The reason for Release to be a Y is that ROI Acceptable and Improves Customer Relationship factors (cells G1 and H2, respectively, of Figure 5) are now both Y, which makes its parent factor Business Goals Achieved (cell G3 of Figure 5) also Y, and this in turn makes Release a Y. However, for Continue, the factor Features Expected By Customer (cell A2 of Figure 12) is an N since most of the features commonly available in e-commerce sites are already there and, therefore, Continue is an N as well.

PostHelix1: The project was deployed and presented (took 3 days)

PostHelix2: The project was maintained (will be continued into the next semester by another group of students)

Here again the project was completed within about 2 months including the evaluation process. The progress of this project through the AWH process is shown in Figure 13.

6. OBSERVATIONS

In Section 5 we described in detail the application of Augmented WebHelix. In this section we discuss the lessons learned from the application and the merits/demerits of AWH.

6.1 Salient Features of AWH

AWH is an iterative process for rapidly developing Web applications with several repetitive slices each of consist of eight phases: analysis, analysis evaluation, design, design evaluation, coding, coding evaluation, testing, and testing

Figure 13. The flow of the book E-Store Project through AWH process

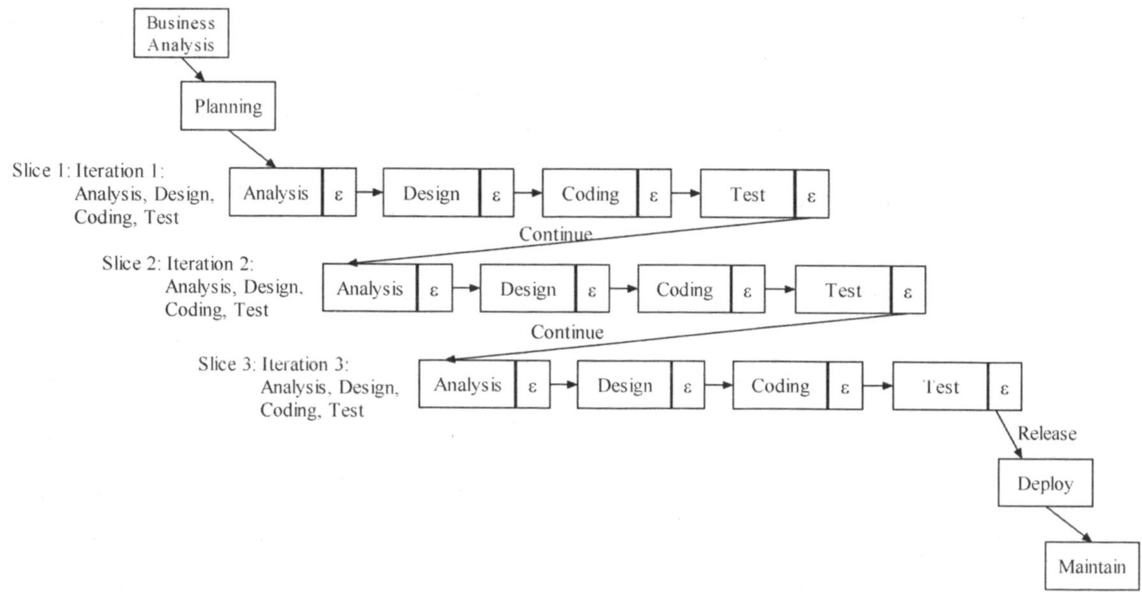

evaluation, where the evaluation phases use the PEF for accomplishing the task. Thus AWH is significantly different from the other processes described in Section 2. It is not a simple linear waterfall process since AWH is iterative and there are intervening evaluation phases; it is different from spiral since evaluation phase uses the systematic PEF unlike the somewhat vague risk assessment done by the spiral process; it is different from RUP since the evaluation phase of AWH serves as the milestone to proceed to the next phase unlike in RUP where there is no formal framework to decide the completion or otherwise of the current phase; and AWH is different from XP in that analysis and design are stressed but the evaluation phases ensure that documentation is kept to the bare minimum in the spirit of XP. We also believe that AWH is better than these other processes simply because of the success of AWH in rapidly developing Web applications in the academic environment. The PEF uses cardinal measurement of the achievement of each factor; this follows from the economic concept of satisficing [7] used by the NFR Framework [7] (from which the PEF is derived) which means that factors may not be achievable in the absolute sense but only in the relative sense; moreover, we found that students were able to easily assign cardinal measures more repetitively since they found some parallel with the grading system widely used in the universities. It has asked what happens if, for example, the design phase uncovers a requirements defect? This defect would be found during the design evaluation phase of the current slice and as per the AWH procedure we would go to the analysis phase of the next slice; therefore AWH is equipped to handle out-of-phase problems.

6.2 AWH Meets the Requirements for a Better Process

In Section 1 we stated some reasons why existing processes were unsuitable for our use in academia. In this subsection we highlight how AWH meets the requirements for a better process. Firstly, the students do not need to imagine the feedback—the PEF lists the factors that the project needs to fulfill and the students need to determine how satisfactorily the factors have been met (or otherwise) and document reasons; therefore, the feedback is provided in a systematic manner. Secondly, because of the systematic effort required by AWH, students cannot ignore feedback—the evaluation phase of AWH depends on using PEF to make go/no-go decisions. Traceability is established by making it one of the important factors in the PEF and ensuring the artifacts of the current phase are traceable (or not) with those of the previous phase and capturing justifications. Because of the practicality of AWH, students realize the invaluable help provided by a process. In fact they tend to become process-oriented at the end of the project which we feel is a good preparation for the industry. Finally, due to the continual guidance provided by AWH we have never had a failed project so far; students quickly discover deviations, if any, and their reasons and find time, if necessary, to keep their projects on track.

6.3 Number of Slices Through the Helix

In the two examples discussed in Section 5 both went through three slices of the helix: this is coincidental—different projects can through different numbers of slices through the helix. The number of slices is dependent on factors discussed at the end of Section 3.

6.4 Variability of Evaluation Factors Between Slices

In Section 5 we assumed the factors for evaluation to be static between slices. This need not be so in the real world; if factors change during development the FIG can be updated to reflect reality and during the next evaluation phase the updated FIG will be used for evaluation. Thus in the Zebra

Books project (Section 5.2) the factor FRs Met (Figure 4 or Figure 5) could be decomposed to reflect the functional requirements for each slice rather than just Storefront and ProductInfo which are both available in the first iteration itself.

6.5 Assigning Priority to Factors

In Section 5 we assumed all factors to be of the same priority. This need not be true in practice (for example, profitability may be of high priority in many cases); in order to indicate priority on a FIG the '!' symbols (see Figure 3) are used in the graphical representation and colored cells in the Excel representation, and the propagation rules are extended as below (where the evaluator can make the appropriate decisions):

In an AND-relationship involving high-priority factors:

R12. If a high-priority child factor is a Y and the remaining child factors are all N then the parent factor may be a Y.

R13. If a high-priority child factor is a U then the parent factor may be U.

In an OR-relationship involving high-priority factors:

R14. If a high-priority child factor is an N and the remaining child factors are all Y then the parent factor may be a N.

R15. If a high-priority child factor is a U then the parent factor may be U.

6.6 Capturing Justifications in the FIG

In Figure 8 we have written the justifications for the decisions, but these do not appear in the FIG of Figure 9. The justifications for Y and N (and U) may also be captured in the FIG by means of claim factors; graphical notation is a dotted circle (as shown in Figure 3) and in Excel it is a cell with the justification in italics. Thus for example, we claimed that the factor Cultural Feasibility (cell E1 in Figure 12) is a Y since "the company (the development group of two students) does not have any issues in continuing further work on this project." The statement within quotes is the justification and this can be captured by a claim factor.

6.7 Promotes Team Work

The students will have to work together for the evaluation phase since it is quite possible that different artifacts may satisfy the factors and no one student may have developed or be even aware of all the artifacts. This is a positive aspect of AWH for academic projects since, at least initially, students are wary of working with others in a team.

6.8 Captures Rationale for Decisions

This is perhaps one of the biggest strengths of AWH. Since PEF captures justifications and since the justifications encapsulate the reasons for the decisions taken at various phases, AWH implicitly promotes capture of rationale for the decisions. This helps the project team to quickly re-evaluate past decisions based on the rationale used for the decisions.

6.9 Simulates Customer Feedback

Another positive aspect of AWH is that the evaluation framework necessitates team work and rationale capture. The discussion within the team to complete the evaluation table very closely simulates that of the interaction with live customer but does not require an actual customer. For example, in Figure 8, when students are trying to determine the satisfaction (or otherwise) of the factor "Benefits" they will have to discuss among themselves for determining reasons and rationale for satisfaction of the factor; the rationale (or justification) will help capture customer feedback in practice and lead to better understanding of the problem domain. The team is fully aware of taking decisions and the reasons for taking them—exactly the feedback the students may expect had they interacted with a real customer.

6.10 Supports Rapid Prototyping

All projects that have used AWH have finished well within schedule. On those rare occasions where all the features could not be implemented, the project team was aware of the reasons for not implementing them. Besides, rapid prototyping supported by AWH is backed by upfront requirements and design so that students realize that implementation does not supplant these important activities. In fact, one of our undergraduate seniors happily informed us that he got a job only because he was able to stress the importance of documentation in the system development process.

6.11 Phases May Be Decomposed

As shown by the example of Section 5.1, the analysis phase of a slice was decomposed into business requirements analysis and software requirements analysis. Thus a generic phase may be decomposed as required by the project. Decomposition is viewed as another pass through the same generic phase in the next slice of the helix as shown in Figure 11.

6.12 Develop Understanding of System Evolution

An important point to note in Figure 13 is that each slice of the helix extends the existing system. This helps to teach the students about system evolution and how this is a natural property of systems.

6.13 Scalable to Industry Projects

AWH has been used in Capstone Projects where students complete information systems in the course of a semester for the local industry. The experience of the students has been successful with the major difference being that the evaluation phases are completed with input from their industry customers whenever necessary (for example, in Figure 8, the satisfaction of the factor "Services to be Provided by GameBuy" will need to be determined by customer feedback and captured as such in the justification). The authors have also used AWH for private projects with industry and likewise our experience has been positive as well. We therefore believe that AWH is scalable for industry projects; however, we need to apply this process on large teams before we can be confident of AWH's scalability.

6.14 Some Issues with AWH

While AWH has several advantages, based on the performance and the feedback from the students we found that there are still some issues that need to be resolved. Chief among them is the fact that AWH is not very easy to learn initially. The students had to learn from instructor's notes that are themselves evolving as the course is taught each time and too many of the concepts of AWH were somewhat involved. They had several questions on the process which when clarified helped them become comfortable with the process. Their answers to embedded questions in the exams also confirm this observation. AWH forces team work, especially the evaluation phases. This sometimes creates stress on the students who are working on teams for the first time. Students do not always appreciate the need for any documentation at all, and the capturing of justifications as part of evaluation is seen as unnecessary. However, we believe that development of better pedagogical aids including text books on AWH will help us overcome some of these problems; however, the benefits obtained from using AWH makes it a promising Web engineering process.

7. CONCLUSION AND FUTURE WORK

Process is important for developing Web-based systems. While several processes such as incremental, Rational unified process, object-oriented, extreme programming, and system development lifecycle are available for use by industry, their adaptation to the academic world has been troublesome, chiefly because of the *lack of feedback from actual customers* for several academic projects. The feedback from customers is taken for granted in most industry-oriented processes. WebHelix [27] was introduced as a process that is useful for academia and industry by providing a good balance between process formalities and practical constraints. WebHelix is a helical process for developing Web applications where each turn of the helix, called a slice, has five phases: analysis, design, coding, testing, and evaluation. The evaluation phase determines whether the current version of the Web application is ready for release to the customer or whether further development through another slice of the helix is required. The evaluation phase mostly uses the return-on-investment (ROI) factor to determine the next course of action which restricts its usefulness for academic projects. In this chapter we proposed the augmented WebHelix (AWH) process that augments the WebHelix in three ways:

1. Provides an option to both release the current version and continue with the next iteration in the next slice; this reflects reality better since in many projects iterative/incremental development process is used;
2. Provides a qualitative evaluation framework, called the project evaluation framework (PEF) for evaluating the project status; this framework allows factors affecting a particular project to be considered for evaluation by constructing AND-OR graphs called Factor Interdependency Graphs (or FIGs) and applying propagation rules on the FIGs;
3. PEF is applied at the end of each phase (phase-stage evaluation) of a slice in order to determine whether to proceed to the next phase of the same slice and at the end of slice (slice-stage evaluation) to determine whether or not to proceed to the next slice.

The FIGs may be drawn graphically or represented in Excel spreadsheets. The latter approach allows for the use of Visual Basic macros for automatic evaluation. The practicality of the AWH was confirmed by applying it for several Web application projects in the academia and industry. The lessons learned from applying AWH reinforced our belief that AWH is a practical process for Web engineering.

There are several directions for further research. In our application of AWH for the project of Section 5.1, we assumed that when a phase is decomposed into say, Business Requirements Analysis and Software Requirements Analysis, we simply move to the analysis phase of the subsequent slice; however, we did not apply PEF to do slice jump. We deliberately did this in order to keep the idea simple; however, we believe we need to modify AWH in order to make this slice jump legal and we need to study this concept further. Our belief that AWH is scalable to industry requirements is based on performance of the process on Capstone Projects that our students cooperatively develop with the local industry and the personal experience of the authors on their private projects with industry. However, we need to use AWH with large teams to confirm its general industrial applicability. PEF currently gives qualitative metrics; more specifically it helps make Yes/No decisions for both the phase-stage and slice-stage evaluations. It will be perhaps helpful if some form of quantitative metrics could be provided by the PEF. We have developed quantitative metrics for software measurement [22], and we hope to extend that work to PEF. Yet another line of research is to identify how AWH works in combination with other processes such as extreme programming, component-based development, and rational unified process, among others. This will help create hybrid strategies that organizations could use perhaps more effectively to migrate from their current process to the AWH process.

However, we believe that the augmented WebHelix is a useful and practical process for engineering Web applications by practitioners both in the industry and academia.

ACKNOWLEDGMENT

We express sincere gratitude to the anonymous referees of the earlier version of this chapter for their insightful comments and suggestions which helped us significantly improve the quality of the chapter.

REFERENCES

1. Aoyama, M. (1998).Web-based agile software development, *IEEE Software*, 57-65.

2. Beck, K., & Fowler, M. (2001). *Planning Extreme Programming,* Addison-Wesley..

3. Becker, S. A., & Bostelman, M. L. (1999). Aligning strategic and project measurement systems. *IEEE Software*, *16*(3), 46-51.

4. Boehm, B., & Hansen, W. J. (2001). *The spiral model as a tool for evolutionary acquisition.* Retrieved on September 27, 2007, http://www.stsc.hill.af.mil/crosstalk/2001/05/boehm.html

5. Boehm, A., & Lowe, D. (2006). *Murach's ASP.NET 2.0 Web programming with VB 2005.* Fresno, CA: Mike Murach & Associates.

6. Britton, C., & Bye, P. (2004).*IT architecture and middleware.* Addison-Wesley.

7. Chung, L., Nixon, B. A., Yu, E., & Mylopoulos, J. (2000) *Non-functional requirements in software engineering.* Boston:.Kluwer Academic Publishers.

8. Clark, B. (2004). *Enterprise application integration using .NET.* Addison-Wesley.

9. Conallen, J. (2003). *Building Web applications with UML, 2/E.* Addison-Wesley.

10. Deshpande, Y., & Hansen, S. (2001) Web engineering: Creating a discipline among disciplines. *IEEE Multimedia, April-June,* 82-87.

11. Evans, E. (2003). *Domain-driven design: Tackling complexity in the heart of software.* Addison-Wesley.

12. Hawthorne Software Engineering Company, *Software Development.* Retrieved on September 27, 2007, http://www.hawthornesoftware.com/Software.htm

13. Koch, N., & Kraus, A. (2002). *The expressive power of UML-based Web engineering.* Paper presented at the Second International Workshop on Web-Oriented Software Technology (IWWOST2), Malaga, Spain.

14. Mellor, S., & Balcer, M. (2002). *Executable UML: A foundation for model driven architecture.* Addison-Wesley.

15. Olsina, L., & Rossi, G. (2002) Measuring Web application quality with WebQEM. *IEEE Multimedia*, *9*(4), 20-29.

16. Patton, S. (2002). Web metrics that matter. *Computer World.* [Electronic version]. Retrieved on September 27, 2007, from http://www.computerworld.com/databasetopics/data/story/0,10801,76002,00.html

17. Pressman, R. S. (2005). *Software engineering: A practitioner's perspective.* New York: McGraw-Hill.

18. Riggins, F. J., & Mitra, S. (2003). *A framework for developing net-enabled business metrics*

through functionality interaction. Retrieved on September 27, 2007, from http://ids.csom.umn.edu/faculty/friggins/e-metrics.pdf

19. Rising, L & Janoff, N. S. (2000). The scrum software development process for small teams. *IEEE Software, July/August,* 2-8.

20. Ruhe, M., Jeffrey, R., & Wieczorek, I. (2003, May). Cost Estimation for Web Applications. In *Proceedings of the 25th International Conference on Software Engineering, IEEE Computer Society,* (pp. 285-294).

21. Standing, C. (2002). Methodologies for developing Web applications. *Information and Software Technology, 44*(3), 151-159.

22. Subramanian, N., Chung, L., & Song, Y-t. (2006, June). An NFR-Based framework to establish traceability between application architectures and system architectures. In *Proceedings of the 7th ACIS International Conference on Software Engineering, Artificial Intelligence, Networking, and Parallel/Distributed Computing (SNPD 2006),* IEEE Computer Society, Las Vegas.

23. Subramanian, N., & Chung, L. (2003, September). Process-oriented metrics for software architecture evolvability. In *Proceedings of the International Workshop on Principles of Software Evolution*, IEEE Computer Society, (pp. 65-70). Helsinki, Finland.

24. Subramanian, N., Puerzer, R., & Chung, L. (2005, September). A comparative evaluation of maintainability: A study of engineering department's Website maintainability. In *Proceedings of the International Conference on Software Maintenance*, (pp. 669-672). IEEE Computer Society. Budapest, Hungary.

25. White, B. (1996, May 6-11). *Web document engineering.* Talk given at 5th International World Wide Web Conference, Paris, France. [Electronic version]. Retrieved on September 27, 2007, http://www.slac.stanford.edu/pubs/slacpubs/7000/slac-pub-7150.html

26. Whitten, J. L., & Bentley, L. D. (2006). *Systems analysis and design methods.* McGraw Hill..

27. Schach, S., (2005). *Object-oriented and classical software engineering.* McGraw Hill.

28. Whitson, G. (2006) WebHelix: Another Web engineering process. *The Journal of Computing Sciences in Colleges, 21*(5), 21-27.

29. http://www.ambysoft.com/unifiedprocess/agileUP.html

ENDNOTE

1. One observation that could be made is how do we know that Reliability and Efficiency are met for these are themselves in some senses vague qualities—this can be better estimated by decomposing Reliability and Efficiency into their sub factors: for example, Reliability could mean "Non Failure of Web site" and "24x7 accessibility;" Efficiency could mean "All pertinent information available on the screen" and "ease of use" and evaluating these sub factors first.

Chapter III
Model-Centric Architecting Process

Tony C. Shan
IBM, USA

Winnie W. Hua
CTS, Inc., USA

ABSTRACT

This chapter defines a methodical approach, named model-centric architecting process (MAP), to effectively cope with the architecture design complexity and manage the architecting process and lifecycle of information systems development in a service-oriented paradigm. This comprehensive method comprises four dimensions of architecting activities: requirement analysis, specification, validation, and planning (RSVP). The process is broken down into nine interrelated models: meta architecture, conceptual architecture, logical architecture, physical architecture, deployment architecture, management architecture, information architecture, aspect architecture, and component architecture. A 2-D matrix serves as a blueprint to denote a step-by-step procedure to produce and manage the architectural artifacts and deliverables in the lifecycle of systems architecture design, development and governance. The holistic framework provides a multidisciplinary view of the design principles, tenets, idioms, and strategies in the IT architecting practices. The characteristics and features of the constituent elements in the MAP approach are articulated in great detail. Recommendations and future trends are also presented in the context. It helps build high-quality service-oriented solutions focused on different domains, and in the meantime keeps the agility, flexibility and adaptiveness of the overall method. This systematic framework may be customized in different formats to design various information systems in different industries.

INTRODUCTION

In today's on-demand business world, the electronic business models demand increasingly higher performance of information technology (IT) systems. We must do more with less, so as to provide a higher level of services at a lower cost for the business to compete and succeed. This means that IT has to build more complex, flexible, scalable, extensible, and forward-thinking technical solutions, to meet the ever-growing business needs.

In large organizations like worldwide financial institutions, virtually hundreds, if not thousands, of IT applications and systems have been built or purchased to provide electronic services for external customers and internal employees in the past years, utilizing heterogeneous technologies and architectures to satisfy diverse functional requirements from different lines of business. The banking industry is no exception. The business process operations generally contain different business sectors in retail, commercial, small business, wealth management, and capital management. In particular, services are delivered to different channels such as automated teller machines (ATMs), Web browsers, interactive voice response, agent assistance, emails, mobile devices, and so forth. To effectively manage the architecture complexity and optimize the design practices in such a disparate environment, a multidisciplinary design approach is crucial to abstract concerns, divide responsibilities, mitigate risks, encapsulate the complexity, and rationalize operating processes.

BACKGROUND

Previous studies in the last few decades have strived to address the issue of architecture design complexity, which has grown exponentially as the computing paradigm has evolved from a monolithic to a service-oriented architecture. The Zachman framework (Zachman, 1987) is a logical structure for classifying and organizing the descriptive representations of an enterprise IT environment that are significant to the management of the organization as well as to the development of the enterprise's information systems. It takes the form of the two-dimensional matrix, and has achieved a level of penetration in the domain of business and information systems architecture and modeling. It is mainly used as a planning or problem-solving tool. However, it tends to implicitly align with the data-driven approach and process-decomposition methods, and it operates above and across the individual project level. Extended enterprise architecture framework (E2AF) (IEAD, 2004) takes a very similar approach in the Zachman framework. Its scope contains business, information, system, and infrastructure in a 2-dimensional matrix. E2AF is more technology-oriented. Both of these approaches are heavyweight methodologies, which require a fairly steep learning curve to get started.

Rational unified process (RUP) (Kruchten, 2003) attempted to overcome these shortcomings by applying the unified modeling language (UML) in a use-case driven, object-oriented and component-based approach. The concept of 4+1 views interprets the overall system structure from multiple perspectives. RUP is more process-oriented, and rooted in a waterfall-like approach. RUP barely addresses software maintenance and operations, and lacks an in-depth coverage on physical topology and development/testing tools. It mainly operates at the individual project level. RUP has recently been expanded to enterprise unified process (EUP), part of which has become open source – OpenUP in Eclipse Process Framework (EPF) project (Eclipse, 2007).

Another heavyweight approach, The Open Group Architectural Framework (TOGAF) (The Open Group, 2007), is a detailed framework with a set of supporting tools for developing an enterprise architecture to meet the business and information technology needs of an organization. The three core parts of TOGAF are architecture development method (ADM), enterprise architecture continuum, and TOGAF resource base. The scope of TOGAF covers business process architecture, applications architecture, information architecture, and technology architecture. The focal point of TOGAF is at the enterprise architecture level, rather than the individual application architecture level. On the other hand, model-driven architecture (MDA) (OMG, 2007) takes an agile approach. MDA aims to separate business logic or application logic from the underlying platform technology. The core of MDA is the platform-independent model (PIM) and platform-specific model (PSM), which provide greater portability and interoperability as well as enhanced productivity and maintenance. MDA is primarily for the software modeling part in the development lifecycle process.

Other related works on IT architecture frameworks are largely tailored to particular domains. They are useful references when a team creates their own models for their organization. The C4ISR architecture framework (DoD, 1997) gives comprehensive architectural guidance for various Commands, Services, and Agencies within the U.S. Department of Defense, in order to ensure interoperable and cost effective military systems. The Federal Enterprise Architecture (FEA) framework (Federal Office, 2007) provides direction and guidance to U.S. federal agencies for structuring enterprise architecture. The Treasury Enterprise Architecture Framework (TEAF) (Treasury Department, 2000) is to guide the planning and development of enterprise architectures in all bureaus and offices of the Treasury Department. The Purdue Enterprise Reference Architecture (PERA) (Purdue University, 1989) is aligned to computer integrated manufacturing.

ISO/IEC 14252 (a.k.a. IEEE Standard 1003.0) is an architectural framework built on POSIX open systems standards. The ISO reference model for open distributed processing (RM-ODP) (Putman, 2001) is a co-coordinating framework for the standardization of open distributed processing in heterogeneous environments. It creates an architecture that integrates the support of distribution, interworking and portability, using five "viewpoints" and eight "transparencies". Scenario-based architecture analysis method (SAAM) (SEI, 2007) was developed to analyze a system for modifiability but is useful for testing any nonfunctional aspect. Architectures are examined via scenarios in SAAM with regard to achieving quality attributes. The Solution Architecture of N-Tier Applications (Shan, 2006) presents a multi-layer and multi-pillar model for Web-based applications.

Despite sporadic use of the methods previously discussed, most today's real-world practices of information system development are still ad hoc, manual, and error-prone. This type of immaturity inevitably leads to chaotic outcomes and failures in the project execution. A recent survey (Standish Group, 2007) reveals that a vast majority of information systems projects are behind schedule, over budget, or canceled. Inadequate systematic frameworks describing the key design practices and artifacts in the service-oriented paradigm are directly attributed to this situation.

A new framework is proposed in the next section, with more detailed descriptions of the key characteristics and features of the components in the section that follows. The best practice usage and future trends are presented in the subsequent sections, followed by the conclusions section.

COMPREHENSIVE METHOD

As discussed in the preceding section, most of the previous methods reveal the architectural aspects of a software application to some extent at a fairly high level or from a restricted perspective. The necessity of a comprehensive approach to architecting the end-to-end distributed solutions becomes more and more evident, demanding a systematic disciplined way. A highly structured mechanism is thus designed in this chapter to meet this ever-growing need, and present a comprehensive and holistic view of all architectural elements, components, knowledge, platforms, planning, and their interrelationships. Design procedures are established accordingly in this approach to facilitate the creation, organization, and management of the architecture assets and solutions at different levels in a large organization.

Design Philosophy

The development of this disciplined mechanism followed some key design principles, part of which were adapted from TOGAF (The Open Group, 2007) but significantly modified/expanded to be tailored to the services-oriented distributed application development process.

Business Principles

- **Primacy of principles:** All groups and stakeholders within an organization must follow these principles of solution architecting.
- **Maximize benefits:** Maximum benefits will be achieved to the entire organization.
- **Business Continuity:** Business operations are not interrupted in spite of system changes.
- **Active engagement:** All stakeholders participated in the process to accomplish business objectives.
- **Compliance with regulations:** The architecting processes comply with all relevant laws, policies, and regulations.
- **IT accountability:** Thc IT group accounts for owning and implementing IT processes and infrastructure that enable solutions to satisfy business requirements for functionality, service levels, cost, and delivery timelines.
- **Innovations:** The stimulation and protection of the corporate innovations is enforced in the IT architecture, management, and governance processes.

Technical Principles

- **Flexibility:** The technical model is agile and nimble to be adaptive in response to future business needs.
- **Responsive change management:** Changes to the corporate architecture/infrastructure environment are planned and implemented in a phased approach.
- **Requirement scope control:** Avoid scope creeping and waterfall approach.
- **Technology standardization:** Technological diversity is controlled to minimize immature and proprietary solutions and products.
- **Interoperability:** Software, hardware, network and infrastructure should conform to defined standards that promote compatibility for data, applications, services, communications, integration, security and technology.

Solution Principles

- **Ease of use:** Solutions are user friendly, with the underlying technology transparent to users, so they can concentrate on tasks at hand.
- **Technology independence:** Technical solutions are independent of specific technology choices and therefore can operate on a variety of technology platforms.
- **Common services and components:** Minimize the redundant development of similar functionalities to promote common service and components across the organization.

Data Principles

- **Data asset:** Data is an asset that has value to the enterprise and is managed accordingly.
- **Data ownership:** Each data element owned by an entity accountable for the data quality.
- **Common vocabulary and meta-data:** Data is defined consistently throughout the organization, and the meta-data are standardized and available to all users.
- **Shared data:** Data is shared across lines of business for individual applications and systems to perform their duties.
- **Data access:** Data is accessible for users to perform their functions
- **Data security:** Data is protected from unauthorized use and disclosure. In addition to the traditional aspects of national security classification, this includes, but is not limited to, protection of pre-decisional, sensitive, and proprietary information.

Service Principles

- **Encapsulation:** Service elements and details are enclosed inside larger, more abstract entities.
- **Loose coupling:** Services maintain a relationship that minimizes dependencies and only requires that they maintain an awareness of each other.
- **Contract:** Beyond what is described in the service contract, services hide logic from the outside world.
- **Reusability:** Logic is divided into services with the intention of promoting reuse.
- **Composability:** Collections of services can be coordinated and assembled to form composite services.
- **Autonomy:** Services have whole control over the logic they encapsulate.
- **Statelessness:** Services minimize retaining information specific to an activity.
- **Discoverability:** Services are designed to be outwardly descriptive so that they can be found and assessed via available discovery mechanisms.

Figure 1. RSVP dimensions in MAP

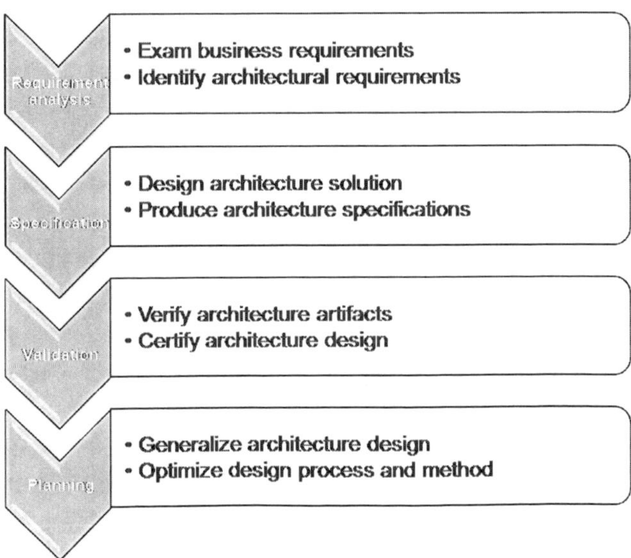

General Approach

The Model-centric Architecting Process (MAP) is designed in this work as a multi-disciplinary approach. It defines a comprehensive method to control the application design and development practices for the quality delivery of information systems. The *MAP* framework is a holistic discipline to help analyze and strategize the thought process, methods, tradeoffs, and patterns in the architecture design.

As illustrated in Figure 1, there are four dimensions defined in *MAP*: requirement analysis, specification, validation, and planning (RSVP), which aligns with the best practices in systems and software engineering. The *requirement analysis* is to thoroughly examine the business requirements and identify the architectural requirements. The *specification* dimension deals with designing the architecture based on the outcome of the analysis results and producing the architecture specifications. The *validation* dimension is for objective/subjective verification and certification of the architecture design completed. Lastly, the focus of the *planning* dimension is for both individual application architecture and portfolio architecture. The granularity varies at the enterprise, division, channel, business domain, and project levels. Benchmarks are used to measure and track the maturity of architecture models. The planning emphasizes the architecture reusability and generalization, which lead to architectural styles and patterns for an enterprise computing environment. The RSVP dimensions represent the major stages in the architecture lifecycle of solution designs.

MAP forms a common baseline to implement service-oriented architecture, integration, process and management in information systems. It primarily focuses on a broad range of architecture models, as illustrated in Figure 2: meta architecture, conceptual architecture, logical architecture, physical architecture, deployment architecture, management architecture, information architecture, aspect architecture, and component architecture.

- **Meta architecture:** The fundamental organization and formal description of an IT system portfolio, interrelationships, decisions and their associated rationales.
- **Conceptual architecture:** The grouping of functional capabilities and practices that make use of conceptualism in architecture.
- **Logical architecture:** The static and dynamic structure of layers, modules, and building blocks that are mapped to the server and technology platforms, which are the baseline containers for the applications.
- **Physical architecture:** The runtime environment that is mapped to software and hardware products, on which applications are installed and run in operations.
- **Deployment architecture:** The topological layout the solution system that will be deployed at different geographical locations connected via networks.
- **Management architecture:** The architecture and infrastructure to manage, govern, and administrate IT assets and enterprise-wise distributed computer systems.
- **Information architecture:** The practice of processing, storing, and utilizing information (data or knowledge), which is often structured in relation to its context in user interactions and large databases.
- **Aspect architecture:** Organization of architectural, control operational and development aspects, which are cross-cutting concerns and advice.
- **Component architecture:** An evolution from object-oriented design where a component of a program refers to a generic module composed of a set of elements or objects. The granularity of components varies, being primitive, aggregated, or composite. The structure of a component can become

Figure 2. Architecture levels in MAP

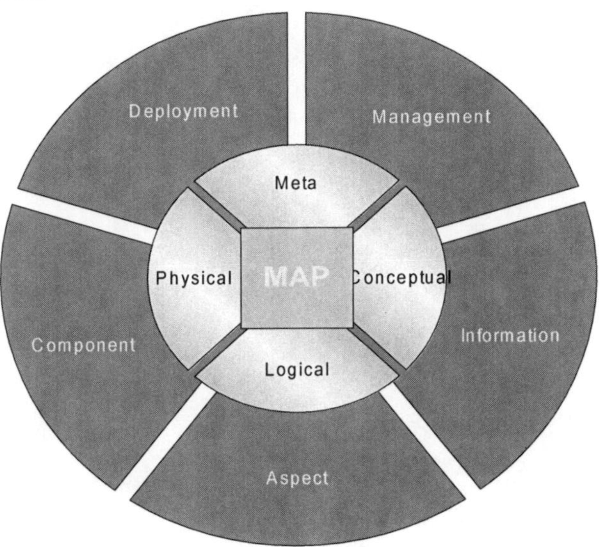

complex enough so that the component itself needs a micro-framework to organize the design artifacts.

MODEL-CENTRIC ARCHITECTING PROCESS

Some of the nine architecture models in *MAP* have been routinely used in the project architecting practices, such as conceptual, logical, physical, and information architectures. The *MAP* framework introduces the concept and application of new modeling disciplines, and intensifies unprecedented attention to the key design attributes for the service-oriented paradigm. The deployment architecture is distinguished from physical architecture, in order to further separate concerns and enhance loose coupling of the models. The management architecture becomes an independent module to take account of the increasing demand for quality of services and non-functional requirements. Lifting aspects and components to an architecture level brings forth a holistic view of the key architecture design artifacts from end to end, eliminating the disconnect between the architecture design and implementation design in the traditional approach. In addition, the meta model sheds light on the portfolio blueprint, cross-application concerns, roadmapping, and optimization to pave the way towards a service-oriented enterprise. The nine architecture types are articulated in detail in the following subsections.

Meta Architecture

As shown in Figure 3, the requirements of meta architecture are architecture vision, principles, tenets, idioms, guidance, and trends. It also includes the alignment of business and technology models, as well as the architecture portfolio rationalization and IT simplification. From the architecture specification perspective, the meta architecture specifies a number of key architecture artifacts across the application portfolio: architecture

Figure 3. Meta architecture in MAP

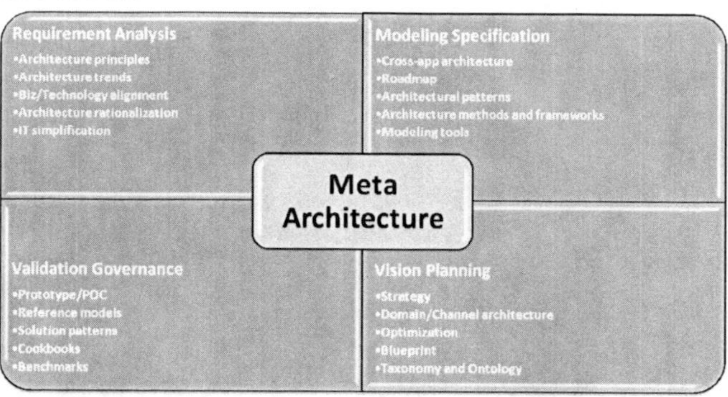

Figure 4. Conceptual architecture in MAP

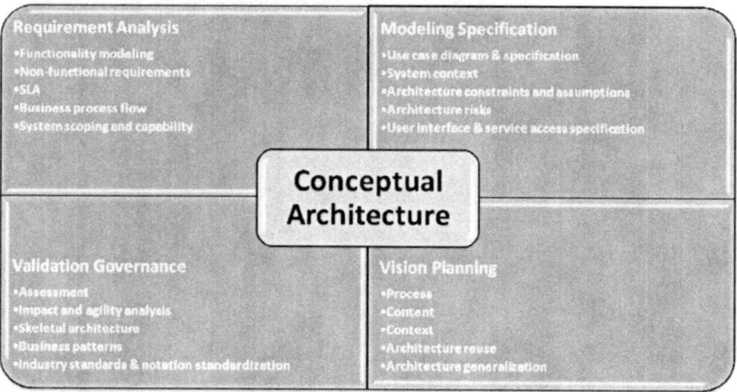

mission and vision, strategy, roadmap, methods, tools, frameworks, cross application architecture models, architectural patterns, business analysis patterns, and standards.

In the meta architecture validation stage, enterprise standards and polices are established, coupled with reference models and solution patterns. Prototype and proof-of-concept evaluations are used to justify the feasibility of new technologies and products. Best practice guidelines and development cookbooks are created as guiding documentation for the project teams. The meta architecture planning emphasizes on strategy, domain/channel architecture, architecting optimization, architecture blueprint, architecture taxonomy for classification and ontology for semantics.

Conceptual Architecture

The conceptual architecture, as depicted in Figure 4, deals with modeling the system functionalities, identifying non-functional requirements, determining the service-level agreements, analyzing the business process, and scoping the system capabilities. The deliverables of the conceptual architecture design are the use case diagram, use case specification, system scope, architecture constraints and assumptions, architecture risks, user interface specification, and service access specification (API and service interface).

In validating the conceptual architecture, the business patterns and skeletal architecture are assessed to conduct the impact and agility analysis. The industry standards are verified, and the notations are standardized. With regard to the

Figure 5. Logical architecture in MAP

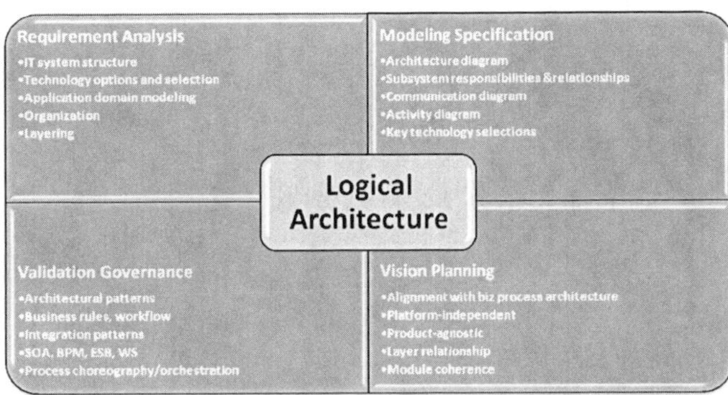

Figure 6. Physical architecture in MAP

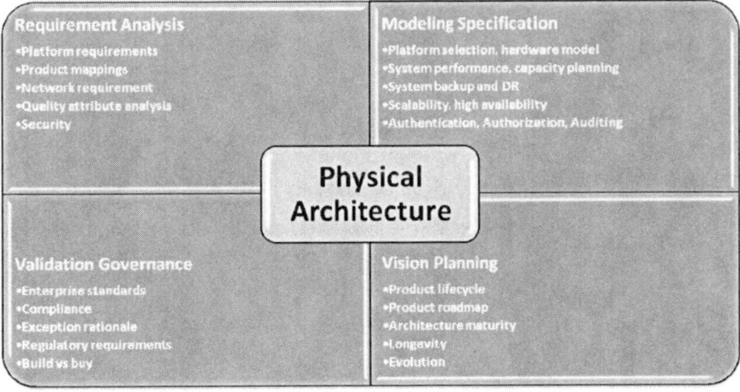

planning, the conceptual architecture deals with the architecture process, content, context, reuse and generalization.

Logical Architecture

As sketched in Figure 5, in terms of requirements, the logical architecture is with regard to the overall system structure, technology options and selection criteria, module organization and application domain modeling. In the logical architecture, the design is specified via the architecture diagram, subsystem responsibilities and relationships, communication diagram, activity diagram, key technology selection, architecture layering, and interoperability.

The focus of the logical architecture validation is on the architectural patterns, business rules, workflow, integration patterns, and reference models. From the architecture planning standpoint, the logical architecture must align closely with the business process architecture. The model should be platform-independent and product-agnostic, allowing for openness and flexibility for long-term stability in the design.

Physical Architecture

The physical architecture, as drawn in Figure 6, handles the platform selection criteria, product mapping approach, capacity planning, and quality attribute analysis. The physical architecture specification captures the platform selection, hardware model, system management, system backup and disaster recovery, scalability, high availability, authentication & authorization, security, and network requirement.

A key decision justification in the physical architecture is build versus buy. Enterprise standards are checked with regard to the products used in the

Figure 7. Deployment architecture in MAP

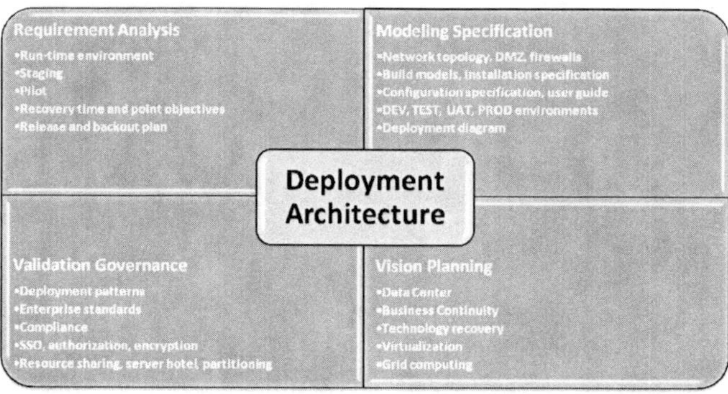

Figure 8. Management architecture in MAP

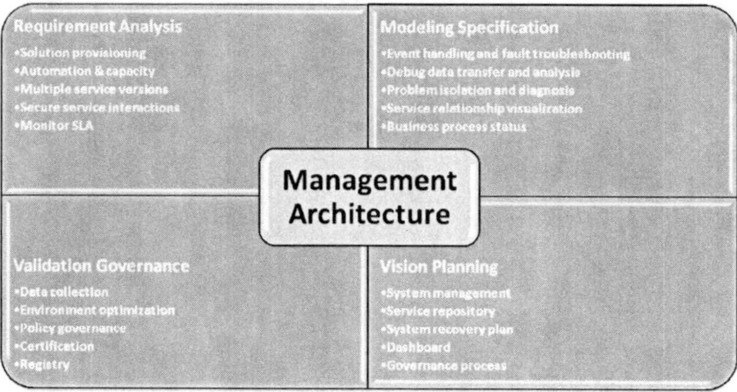

projects. Regulatory requirements and compliance are evaluated. Exception rationales and tradeoffs are carefully deliberated. The product lifecycle and roadmap have a significant impact on the physical architecture planning. Additionally, the architecture maturity and longevity quantify the evolution of physical architecture design.

Deployment Architecture

As represented in Figure 7, the requirement scope of the deployment architecture contains the run-time environment to host the individual applications with the topological organization in the geographical data centers. It also takes account of staging, pilot, recovery time objective, recovery point objectives, release and backout plans. The key elements in the deployment architecture specification are the network topology, hosting farm or hotel, deployable build models, configuration specification, development/test/production environments, deployment diagram, installation specification, user guide, and data center operations.

In the deployment architecture verification, the deployment patterns should have been considered as the engineering implementation approaches. Other artifacts are also examined, such as compliance, SSO, authorization, encryption, and resource sharing via server hotels and partitioning. Regarding the planning, the data center design and environment buildout will meet the needs of the business continuity and timely technology recovery. Virtualization and grid computing are leveraged to maximize the reuse of infrastructure.

Figure 9. Information architecture in MAP

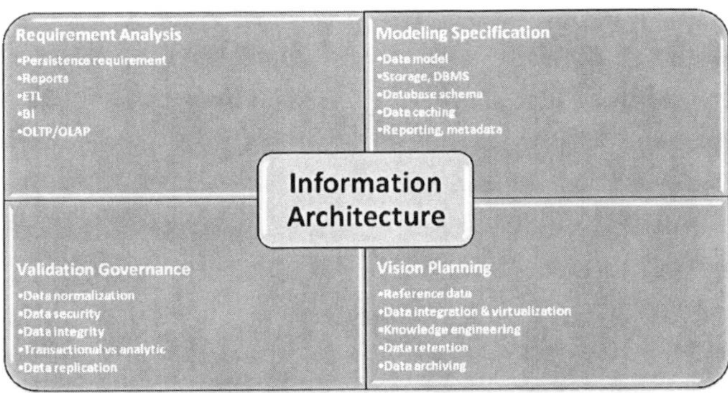

Figure 10. Aspect architecture in MAP

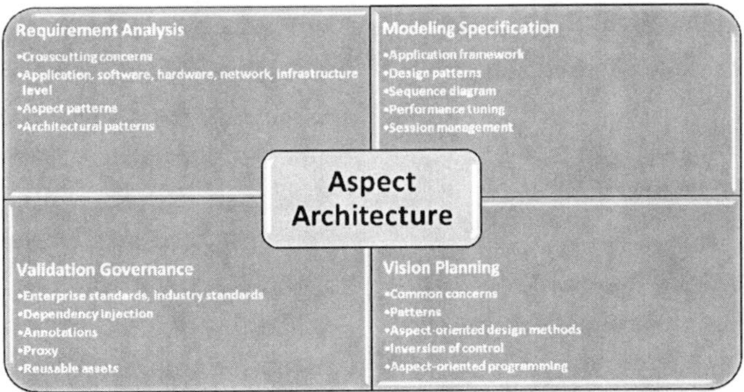

Management Architecture

The elements in management architecture requirements, as pictured in Figure 8, are solution provisioning, automation, capacity, service versioning, service interaction security, and SLA monitoring. The management architecture specification covers event handling, fault troubleshooting, product trace, debug data transfer and analysis, problem isolation and diagnosis, service relationship visualization, and business process status.

From the management architecture validation viewpoint, the key artifacts are inspected such as data collection, analysis, persistence, visualization, environment optimization, policy governance, certification, and registry. The management architecture planning focuses on the overall system management, service repository, system recovery plan, dashboard, and governance process.

Information Architecture

As illustrated in Figure 9, the information architecture handles the data and content requirements, such as persistence, reporting, data integration, data quality, analytics, business intelligence, data store, and data extract-transform-load. The information architecture specification defines the data model, storage, DBMS, database schema, replication, data transfer, data backup, data caching, reporting, metadata, query, workload priority, and access control.

The information architecture validation consists of evaluations of data normalization, security, integrity, transactional versus analytic, reuse, replication, quality, analysis, stream flow, and audit. The information architecture planning deals with reference data, data integration & virtualization, knowledge engineering, data retention and archiving.

Aspect Architecture

The aspect architecture, as described in Figure 10, treats crosscutting concerns as aspects in systems using the architectural patterns. The aspect architecture pertains to the application framework, design patterns, sequence diagram, collaboration diagram, performance tuning, session management, and open source products.

The primary artifacts in the aspect architecture validation are enterprise standards, compliance, dependency injection, annotations, proxy, industry standards, and reusable assets. The aspect architecture planning handles common concerns, patterns, aspect-oriented design methods, and inversion of control.

Component Architecture

As portrayed in Figure 11, the component architecture is at the micro level of design and construction in application software. The analysis of the component needs and progressive decomposition from the process down to the method level help identify reusable components and toolkits to be leveraged, either commercial off-the-shelf (COTS) or open source products. Regarding the specification, the component architecture covers the component specification, CRC cards, class diagram, design patterns, technology, package, and open source utilities.

The component architecture validation is related to standards, compliance, open source asset repository, interface, component composition and orchestration. In the component architecture planning, the focus is on the component hierarchy, maintainability, conventions, component catalog and repository, and reuse efficiency.

APPLICATION AND FUTURE TRENDS

Table 1 is a 2-D matrix, which summarizes the key artifacts in the MAP. The matrix serves as a reference card of major activities and deliverables defined in the architecting lifecycle. The process can be leveraged as a baseline to construct information systems in a round-trip engineering process. The nine models provide a progressive granularity for architecting and design that was not addressed in such a comprehensive manner in previous studies. The framework also clarifies the synergies and linkage between the models, facilitating the forward-engineering and reverse-engineering activities and responsibilities.

From a design perspective, the framework provides an architectural skeleton of a distributed application. Various elements in the MAP are candidates to be selected to build a solutions architecture model. For example, a Web application framework can be chosen as the micro-structure

Figure 11. Component architecture in MAP

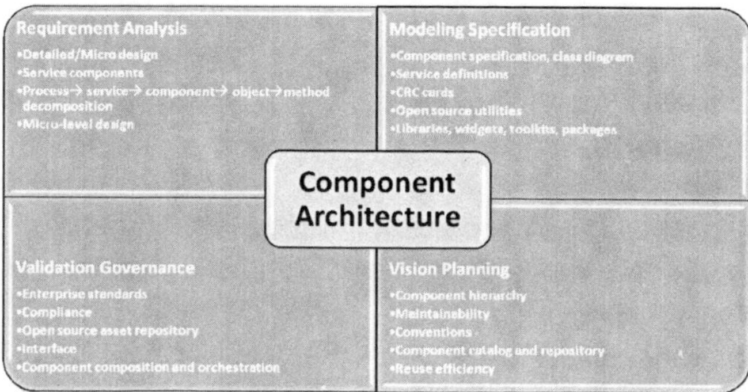

Table 1. Reference card of model-centric architecting process

	Meta Architecture	Conceptual Architecture	Logical Architecture	Physical Architecture	Deployment Architecture	Management Architecture	Information Architecture	Aspect Architecture	Component Architecture
Architecture Requirements (analysis)	-Architecture principles -Architecture trends -Biz and Technology alignment -Architecture rationalization and simplification	-Functionality modeling -Non-functional requirements -SLA -Business process flow -System scoping -Capability	-IT system structure -Technology options and selection -Application domain modeling -Organization	-Platform requirements -Product mappings -Network requirement -Quality attribute analysis	-Run-time environment -Staging -Pilot -Recovery time and point objectives -Release and backout plan	-Solution provisioning -Automation & capacity -Multiple service versions -Secure service interactions -Monitor SLA	-Persistence requirement -Reports -ETL -BI -Data warehouse -Data mart -Risk data -OLTP/OLAP	-Crosscutting concerns -Application, software, hardware, network, infrastructure level -Aspect patterns	-Detailed/ Micro design -Service components -Operations → Process → service → component → object → method decomposition
Architecture Specification (modeling)	-Cross-app architecture -Roadmap -Architectural patterns -Business analysis patterns -Architecture methods -Architecture standards -Architecture frameworks -Modeling tools	-Use case diagram & specification -System context -Architecture constraints and assumptions -Architecture risks -User interface specification -Service access specification -Computation-independent	-Architecture diagram -Subsystem responsibilities and relationships -Communication diagram -Activity diagram -Key technology selections -Architecture layering -Inter-operability	-Platform selection -Hardware model -System performance -System backup and DR -Scalability -High availability -Authentication -Security -Capacity planning	-Network topology -Farm/Hotel -Build models -Configuration specification -DEV, TEST, UAT, PROD environments -Deployment diagram -Installation specification -User guide -DMZ and Firewalls	-Event handling and fault troubleshooting -Product trace and debug data transfer and analysis -Problem isolation and diagnosis -Service relationship visualization -Business process status	-Data model -Storage -DBMS -Database schema -Replication -Data transfer -Data backup -Data caching -Reporting -Metadata -Query -Workload priority -Access control	-Application framework -Design patterns -Sequence diagram -Collaboration diagram -Performance tuning -Session management -Open source products -System & UI navigation	-Component specification -Service definitions -CRC cards -Class diagram -Design patterns -Technology -Package -Open source utilities -Libraries -Widgets -Toolkits -API
Architecture Validation (governance)	-Appraisal -Prototype and POC -Reference models -Solution patterns -Best-practice guidelines -Cookbooks -Standards -Benchmarks	-Assessment -Impact analysis -Skeletal architecture -Business patterns -Industry standards -Notation standards -Agility analysis	-Review -Architectural patterns -Business rules -Workflow -Integration patterns -SOA, BPM, ESB, WS -Biz process choreography	-Evaluation -Enterprise standards -Compliance -Exception rationale -Reference models -Regulatory requirements -Build vs. buy	-Examination -Deployment patterns -Enterprise standards -Compliance -SSO and authorization -Encryption -Resource sharing	-Inspection -Data collection -Analysis -Persistence -Visualization -Environment optimization -Policy governance -Certification -Registry	-Normalization -Data security -Data integrity -Transactional vs. analytic -Data reuse -Data replication -Data quality -Data analysis -Data stream flow & audit	-Justification -Enterprise standards -Compliance -Dependency injection -Annotations -Proxy -Industry standards -Reusable assets	-Review -Enterprise standards -Compliance -Open source asset repository -Interface -Component composition -Component orchestration
Architecture Planning (vision)	-Strategy -Domain /Channel architecture -Optimization -Blueprint -Taxonomy -Ontology	-Process -Content -Context -Architecture reuse -Architecture generalization	-Model alignment with biz process architecture -Platform-independent & tech-agnostic	-Product lifecycle -Product roadmap -Architecture maturity -Longevity	-Data Center planning -Business Continuity -Technology recovery -Virtualization	-System management -Service repository -System recovery plan -Dashboard	-Reference data -Information integration & virtualization -Knowledge engineering -Data retention and archiving	-Common concerns -Patterns -Aspect-oriented design methods -Inversion of control	-Component hierarchy -Maintain-ability -Conventions -Component catalog and repository
RSVP	*Portfolio level*			*System level*				*Software level*	

of the application software organization. And the aspect solutions can be used to address application-wise concerns in the design, such as the session management, logging, and MVC pattern.

From a development standpoint, the solution modules cataloged in the MAP facilitates the development activities. Various tools listed can be employed to semi-automate the development lifecycle. The UML tool is used to document use cases and activity diagrams, as well as conduct object-oriented analysis and design. The IDE tools are used to code and implement software services and components. The testing tools are used to validate and profile application units and components regressively. The build tools automate the compilation and packaging of the deployable as well as the installer. The version control tools assure the effective configuration management and quality control. The community tools help coordinate the collaborations among the resources that may be geographically located beyond the continent boundaries in today's globalized development environment.

From an implementation point of view, a lightweight Linux-Apach-MySQL-PHP (LAMP) combination can be directly employed in small- and medium-size applications. Other engines and servers presented in the MAP are used for large-scale complex systems, such as portal server, process engine, rule engine, middleware, embedded database, and service registry/repository. Various open source solutions can be directly deployed to monitor and manage the system operations.

From a project management viewpoint, various products specified in the method help manage the development tasks and issues efficiently. dotProject is a project management tool designed to assist in the management of defined project work. The defect tracking system such as Bugzilla provides a collaborative environment to resolve development issues and bugs. POM-based tools like Maven can manage a project's build, reporting and documentation from a central piece of information. Products like Wrike facilitate the online collaborative planning, which decentralizes control and responsibility for overall plans and permit online access to plans that is equal for all related participants.

As new enabling technologies are emerging at an unprecedented pace and the standards/specifications in SOA are maturing, the attributes and contents of the MAP framework tend to expand drastically and some of them become obsolete much faster. Timely refreshment and frequent reassessment are mandatory to keep pace with the latest advances in the field. The strategic planning and execution process also need to be revalidated accordingly. Future work includes building domain-oriented reference models, implementation patterns, best practice styles, closer alignment with business process strategy/architecture, capability maturity, and governance policies/program. Since the difficulties in SOA architecting practices are beyond technical challenges, large organizations must make sociological changes in mindset to adopt the engineering practices and overcome the cultural barriers during the transition. Other disciplines must be leveraged and incorporated as well, to establish an overarching methodical approach to sustainable architecture designs.

CONCLUSION

To effectively manage the architecture complexity and organize diverse technical assets in information systems, a comprehensive method is a necessity to abstract concerns, define roles, separate responsibilities, and present a holistic view of the architectural aspects in a highly structured way. The Model-centric Architecting Process (MAP) approach introduced in this chapter is a holistic framework to facilitate architecting distributed applications in a systematic fashion. It provides comprehensive taxonomy of the architectural artifacts from both development and runtime perspectives. It builds concrete architecture solutions focused on different domains, and in the meantime keeps the agility, flexibility and adaptiveness of the overall method.

The design principles of the framework are presented in the context. The general approach comprises an array of multiple dimensions: *requirement, specification, validation, andplanning* (RSVP). *MAP* is composed of a range of architecture models: *meta architecture, conceptual architecture, logical architecture, physical architecture, deployment architecture, management architecture, information architecture, aspect architecture,* and *component architecture.* The elements of each module are discussed and the features are articulated in detail.

Thanks to its comprehensiveness, this overarching framework has been extensively used in the service-oriented application development in one form or another. It can be streamlined as an agile variant tailored to specific domains or individual projects. The approach has proven to be a great success in the real-world project work in terms of cost saving, productivity, standardization, and communications. Moreover, this method is scalable and flexible for dynamic extensions and expansions, which can serve as a meta-framework to incorporate other general or specialized frameworks.

REFERENCES

DoD C4ISR Architecture Working Group (1997). *C4ISR architecture framework.* Version 2. Retrieved May 18, 2007, from http://www.fas.org/irp/program/core/fw.pdf

Eclipse (2007). *Eclipse process framework.* Retrieved May 18, 2007, from http://www.eclipse.org/epf

Federal Office of Management and Budget (2007). *Federal enterprise architecture framework.* Retrieved May 18, 2007, from http://www.whitehouse.gov/omb/egov/a-2-EAModelsNEW2.html

IEAD (Institute for Enterprise Architecture Developments) (2004). *Extended enterprise architecture framework.* Retrieved May 18, 2007, from http://www.enterprise-architecture.info

Kruchten, P. (2003). *The rational unified process: An introduction.* 3rd Edition. Massachusetts: Addison Wesley.

OMG (Object Management Group) (2007). *Model driven architecture.* Retrieved May 18, 2007, from http://www.omg.org/mda

Purdue University (1989). *The Purdue enterprise reference architecture.* Retrieved May 18, 2007, from http://pera.net

Putman, J.R. (2001). *Architecting with RM-ODP.* New Jersey: Prentice Hall PTR.

Shan, T.C., & Hua, W.W. (2006). Solution architecture of N-Tier applications. In *Proceedings of 3rd IEEE International Conference on Services Computing* (pp. 349-356). California: IEEE Computer Society.

Software Engineering Institute (SEI) at CMU (2007). *Scenario-based architecture analysis method.* Retrieved on May 18, 2007, from http://www.sei.cmu.edu/architecture/scenario_paper

Standish Group (2007). *The Standish Group Chaos Report 2006.* Retrieved May 18, 2007, from http://www.standishgroup.com

The Open Group (2007). *The Open Group architecture framework.* Retrieved May 18, 2007, from http://www.opengroup.org/togaf

Treasury Department CIO Council (2000). *Treasury enterprise architecture framework.* Version 1. Retrieved May 18, 2007, from http://www.eaframeworks.com/TEAF/teaf.doc

Zachman, J.A. (1987). A framework for information systems architecture. *IBM Systems Journal, 26*(3), 276-295.

Chapter IV
Architecture, Specification, and Design of Service-Oriented Systems

Jaroslac Král
Charles University, Czech Republic

Michal Žemlička
Charles University, Czech Republic

ABSTRACT

Service-oriented software systems (SOSS) are becoming the leading paradigm of software engineering. The crucial elements of the requirements specification of SOSSs are discussed as well as the relation between the requirements specification and the architecture of SOSS. It is preferable to understand service orientation not to be limited to Web services and Internet only. It is shown that there are several variants of SOSS having different application domains, different user properties, different development processes, and different software engineering properties. The conditions implying advantageous user properties of SOSS are presented. The conditions are user-oriented interfaces of services, the application of peer-to-peer philosophy, and the combination of different technologies of communication between services (seemingly the obsolete ones inclusive), and autonomy of the services. These conditions imply excellent software engineering properties of SOSSs as well. Service orientation promises to open the way to the software as a proper engineering product.

INTRODUCTION

Service orientation (SO) is becoming the central topic of software engineering. There is an explosive growth in the number of conferences, products, and articles discussing and using the principles of SO and service-oriented architectures (SOA). Service-oriented software systems

(SOSS) are of different types depending on the character of the functions the system provides, the system environment (for example, e-commerce or a decentralized international enterprise), and the way the system is developed. The common property of SOSS is that their components behave like the services in real life mass service systems. The SOSS must then be virtual peer-to-peer (p2p) networks of autonomous components (services). The services can have various properties; they need not be Web services in the sense of W3C (2002) and need not therefore use standard communication protocols, compare Barry and Associates (2003) and Datz (2004).

We shall show that the software engineering properties as well as the user-oriented properties of any SOSS strongly depend on the properties of the service interfaces and that user interfaces of the system should be implemented as specific services (peers of the network) as well. All these issues are related to the architecture of the system. We will discuss how the properties of the architecture influence the set of feasible functions, development (especially the requirements specifications), feasible development techniques (for example, agile ones), standards, politics of IT management, and marketing strategies of software vendors and/or system integrators (Figure 1). The feasible functions of SOSSs include the functions important for user top-management.

Feasible functions of any large system depend on its architecture. The decision as to what architecture is to be used must therefore be formulated in early stages of the system life cycle. On the other hand, the structure, techniques, and content of requirements specifications are influenced by the properties of the system architecture and the details of its implementation. We shall show that SOSS should use a combination of various techniques developed during the software history (for example, message passing, object orientation, common databases, and, sometimes, batch-oriented systems). All these issues should be addressed in the specifications of SOSSs. SO is a paradigm new for many software people. It implies some problems with the application of SO.

PEER-TO-PEER INFORMATION SYSTEMS (P2PIS)

Large information systems must often be developed as a network of loosely coupled autonomous components—services (possibly information systems) integrated using peer-to-peer principle (further P2PIS). The change of the architecture

Figure 1. Central role of system architecture

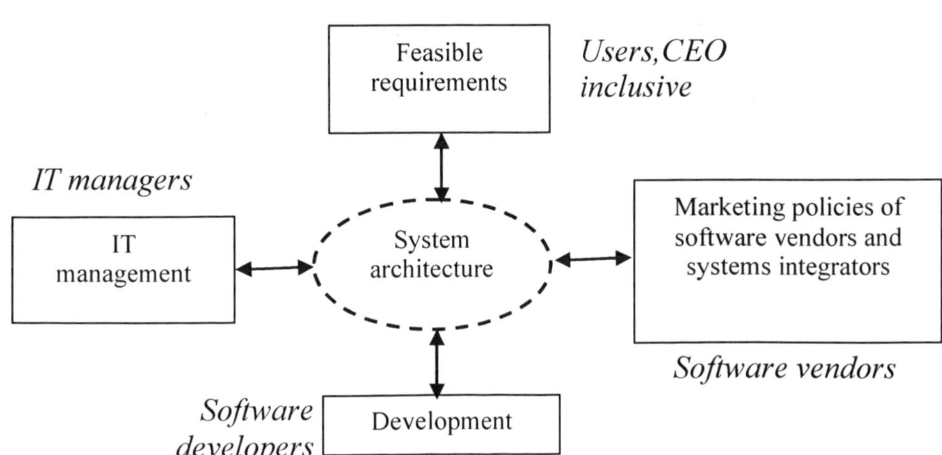

should be accompanied by changes in requirements specification that should reflect the service-oriented structure of the system.

The specification of P2PIS starts from the specification of system user interface (portal) and from the specifications of the services. The specification of services starts from the definition of their interfaces. It can be accompanied by specification of the services of the infrastructure (message formats, communication protocols, middleware services, in general). Services in P2PIS can be newly developed applications, encapsulated legacy systems, or third party products. P2PIS enables new types of requirements (for example, the requirement that a P2PIS should support decentralized and flexible organization of a global enterprise, see Král & emlika, 2003) and makes achievable software engineering properties like reusability, flexibility, openness, maintainability, the use of legacy systems and third party products, or the reduction of development costs and duration. Experience shows that such systems can be extremely stable (Král, 1995).

There are two main variants of P2PIS. The first one is used in e-commerce where the service starting a communication must first look for communication partners. The partners must offer their interfaces (typically specified by WSDL). This schema implies the use of Internet and international standards like SOAP. We shall call such systems *(software) alliances*.

The systems formed by stable sets of services knowing their permanent communication partners will be called *(software) confederations*. Confederations occur often. Examples are:

- Information systems of international enterprises having the structure of a network of autonomous organizational units (divisions). The information systems are formed by a peer-to-peer network of the information systems of the divisions and by some additional components serving the whole enterprise (for example, portals). Such an architecture simplifies the integration of new divisions and/or of newly purchased enterprises as well as the selling out or outsourcing of some divisions or splitting the enterprise into smaller ones.
- Information systems of e-government built as a network of the information systems of particular offices[1] (Král & emlika, 2001).
- A long-term collaboration between the information system of an enterprise and the information systems of its business partners needed for supply chain management (SCM) (Lowson, King & Hunter, 1999) and customer relationship management (CRM) (Dyché, 2002).
- An open association of health organizations (physicians, hospitals, laboratories, health database services, and so forth) forming an information system intended to simplify, enhance, and speed up health care.
- Process control systems (soft real-time systems) supporting, e.g. computer integrated manufacturing. Such systems were the first systems having main properties of service-oriented systems. They have proved for the first time the advantages of service orientation.

If a system *S* has p2p architecture, it must have structure allowing its peers/services to collaborate. The services must be equipped by gates connecting them to a middleware. The system *S* must usually be equipped by a user interface (portal). There can be several portals. Alliances need not have any portals (Figure 2).

The properties of P2PIS depend substantially on the properties of the interfaces provided by the gates and by the functions of the middleware. The most important property of the interfaces is how much they vary. Stable interfaces increase the stability of P2PIS, reduce the development and maintenance costs, and hide the implementation details and philosophy of the component. It is shown below that the gates need not trans-

Figure 2. Architecture of a service oriented system (G is a gate, UC is an user interface service (portal))

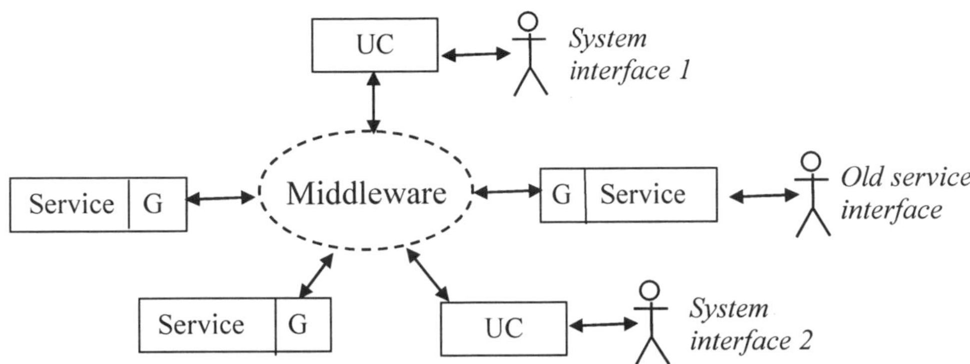

form components into Web services and that the middleware in confederations need not be Internet based. On the other hand, Web services and Internet-oriented middleware are necessary in alliances as the use of worldwide standards and tools is the precondition of e-commerce and of the communication between partners unknown to each other before the communication starts.

Note that SOSSs are usually built starting from the specification of the interfaces and that the components providing the services are mainly integrated as black boxes.

Alliances

A crucial issue of the development of alliances is the standardization of communication protocols. From the technical point of view, the dialog between partners (peers) in a P2PIS can easily be fully automated, but the business agreements should be (and, due to business and legislative reasons, usually are) controlled (supervised) personally by representatives of the communicating parties.

The dialog is driven by a businessperson (initiator) searching for business partners. The businessperson applies knowledge about the history of collaboration with the partners and about the current situation (that is, credibility) on the market. The partner should evaluate the enterprise of the initiator similarly. The partners have to check whether to conclude a contract. Supervision is necessary as someone must be responsible for any business operations. The business documents produced in this way are often to be formally confirmed by people. It holds for P2PIS in general, but it is especially needed in alliances.

Current practice is to establish the cooperation in alliances via Web services in the sense of W3C. The dialog of business partners then starts using UDDI, WSDL standards, and continues in SOAP.

The coordination of the business processes of several partners can be a complicated task. Optimal solution of the supervision of the stages of business processes is a cornerstone of alliance requirements specification. It is to some degree true for confederations as well. The business processes are often developed using development tools like .NET or J2EE. There are discussions about what choice is the best (see the discussion on e-services in *Communications of the ACM*, July 2003, pp. 24-69).

The main advantage of alliances is their flexibility, generality, and standardized solutions. The disadvantage is the problems with efficiency, stability, the size of the used standards, and problems with integration of legacy systems and third party products.

A deeper problem is that SOAP is not too user friendly as it is based on remote procedure calls (RPC) close to programmers' knowledge domain (for example, object orientation) and not to the user problem domains. It enlarges the problems with requirements specification via WSDL and UDDI. UDDI is a centralized service. Centralized services are not good in p2p frameworks. It is confirmed by the experiences with UDDI systems. SOAP, like object-oriented languages, requires many method calls and, therefore, also many messages during even very simple dialogs. It causes efficiency problems, problems with prototyping and understanding the communication by human beings.

The systems using RPC (for example, SOAP) are better suited to the business operative than to business process analysis, usually based on a common data tier.

Alliances are suited to operative tasks in the global market. Important decisions should often be, however, preceded by analysis of the market and the history of cooperation with a given partner. It implies user involvement, massive data analysis, and the tools specific for it. The tools often do not fit into RPC/SOAP frameworks.

A very important advantage of alliances is that communicating software parties can be in the framework of the SOAP/Web services developed individually like programs serving, for example, terminals.

It can happen that a business case (process) fails for some reason. In this case, the reasons of the failure and people responsible for it can be detected via the analysis of the messages logged during the corresponding business process. The analysis can be used as evidence at court. The messages stored in a log memory must therefore be understandable for users and experts in economy and even for lawyers, who should be user-oriented. It is not clear whether messages in SOAP format can fulfill this requirement. It indicates that a proper message presentation tier enhancing communication legibility should be available.

Software Confederations

E-Government: Confederation via Integration

We shall demonstrate some typical software confederation (SWC) related issues on the example of e-government. The engine of e-government —state information system, SIS—is one of the largest software systems to be built in any country. Let us now give the list of the most important requirements on SIS:

- SIS should service citizens, enterprises, and state and municipal offices. SIS should be able to communicate with information systems of private companies and/or (potentially) of citizens. To fulfill it, SIS must have a complex subsystem (portal) providing the interface for citizens. Such an interface should be flexible in its functionality depending on the rights/profiles of specific groups of citizens and/or state officers/clerks. There should be one or more user interface gates (portals) providing an integrated interface making the internal structure of the system invisible (transparent).
- SIS should support the collaboration of all state offices and majorities. Examples are the collaboration during the investigation of car robberies and/or document verification.
- SIS should reflect the frequent changes in laws and in the structure of state administration.
- SIS should use autonomous tools, often third-party products for data filtering, mining, and

analyzing. It is likely that many new tools will be added in the future.

As there are many systems used by individual offices, it is very difficult to rewrite them in time. The existing system should therefore be integrated without any *substantial* change of its functions. The systems must be easily integrated into SIS without any substantial reconstruction. There is yet another, maybe substantially important, reason for these requirements. No office will take any responsibility for a (sub)system if there is any doubt it works correctly—there must be the *feeling of ownership* of the (sub)system. It can reduce the resistance of users and/or politicians caused by their apprehensions about their positions and power.

As it is highly desirable that the IS of various offices should be at the local level (in a particular office), used without substantial changes (for example, *business as before*). It usually implies that the interfaces of constituent autonomous information systems (autonomous components/services) tend to be user knowledge domain oriented and coarse grained. We shall see that it offers substantial software engineering advantages as well as many benefits for users. A properly specified and designed software confederation increases the dynamics of the system structure and openness of the system.

The conclusions can be illustrated on the following example. People responsible for SIS of the Czech Republic wanted to redevelop the SIS as a monolithic system from scratch. Practical experiences induced them to accept that the SIS must be a P2PIS.

The number of peers in confederations is not too large (compared with e-commerce) and the peers are known. The collection of peers (services) does not vary too quickly. The communication protocols between the peers can then be suited to particular needs; they can be based on nonstandardized tools/solutions without any substantial penalty (compare to Demetriades, 2003). It allows use of various turns known from the history of computing for specific tasks like data reconstruction, data- or object-oriented design for the development of peers.

Manufacturing System: Decomposition and Integration

Systems supporting e-government are the systems developed mainly via integration of existing systems, possibly equipped by appropriate gates and transformed so that they can work as services (peers in a p2p system). Some SOSSs are, however, developed from scratch via decomposition of the system into services. Then the services, user interfaces, and middleware are developed and integrated. It is typical for (soft) real-time systems, for example, in manufacturing. Such systems have shown many advantages of service orientation.

Figure 3 shows the interface of the manager of a flexible manufacturing system producing parts of machine tools (Král, 1995). The manufacturing of the parts is defined by linear sequences of manufacturing operations. A generalization to more complex workflows (for example, assembling) is possible but the interface becomes more complex. The workshop manager chooses the central (actual) operation D.i, and the system shows the previous and next operations in the technological sequence D and in the workplace queues. Note that the manager felt the interface as a support for the standard management activities that were familiar. The manager could add/modify the technological sequence and rearrange the queues. The required data could also be filled by a scheduler from the enterprise level. If the scheduler produced right data, no actions from the manager were needed. We call such types of business processes reconstruction (BPR) the soft one. It should be used as often as possible.

The interface of the manager was data-oriented—generated from a database. Other parts of the system communicated via commands (for

Figure 3. Interface of the manager of the manufacturing system

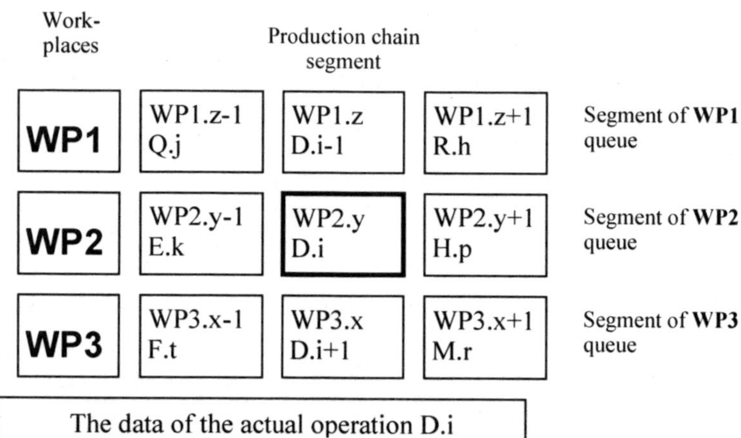

example, store product P in warehouse place WP), so different attitudes (not only RPC) had to be used. It was important that the structure of software services reflected the structure (and interfaces) of real-life services.

These properties of the system were the reasons why the flexible manufacturing system (FMS), an island of automation, was very successfully used for more than twenty years in an enterprise having several successive enterprise information systems. FMS was used without any substantial changes and almost without any maintenance. It was thanks to the fact that the interface has corresponded to the manager intuition, used his knowledge, and supported his skills. To generalize, EAI (service interface) should support the intuition and knowledge of users and should be user and problem oriented. The user should have a chance to influence the design of the interface. User orientation of interfaces offers the possibility to simulate (even substitute) the services (components) not yet existing by communication via portals. We say then that such interfaces are *user performable*. It substantially enhances prototyping tools.

Middleware Enhancement in Confederations

The decision whether to use standard or proprietary message formats should be based on a proper (service-oriented) analysis of the partnerships of autonomous services. The standards in their current form are difficult to use. It can be reasonable for the dynamic enterprises to choose a proprietary solution of message formats. It need not be too difficult to adapt the proprietary solutions to future stabilized standards using the tools like XSLT and PHP. Using the tools like XSLT and PHP, we can build new types of services called *front-end gate* (FEG). FEG is a service used as a front-end part of the gate G of a service S (Figures 4 and 5) or as a router. FEG transforms the formats of input and output messages of S into forms acceptable by the partners of S and hides undesirable properties of the gate G, like disclosing the implementation details of S. The problem is that, according to our experiments, XSLT is awfully ineffective and unstable today for more complex tasks.

Figure 4. Three-tier service (encapsulated information system) with gate G and its communication links

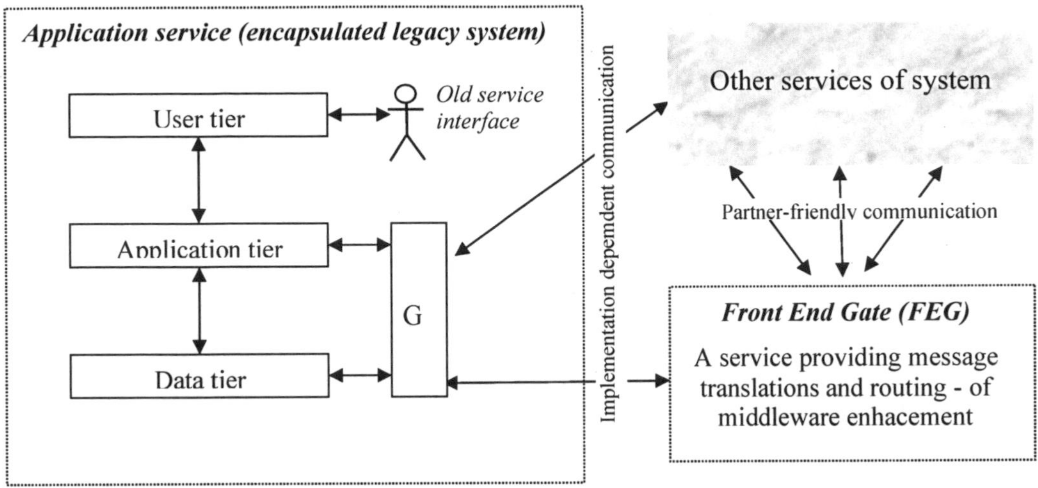

Figure 5. System with front-end gates

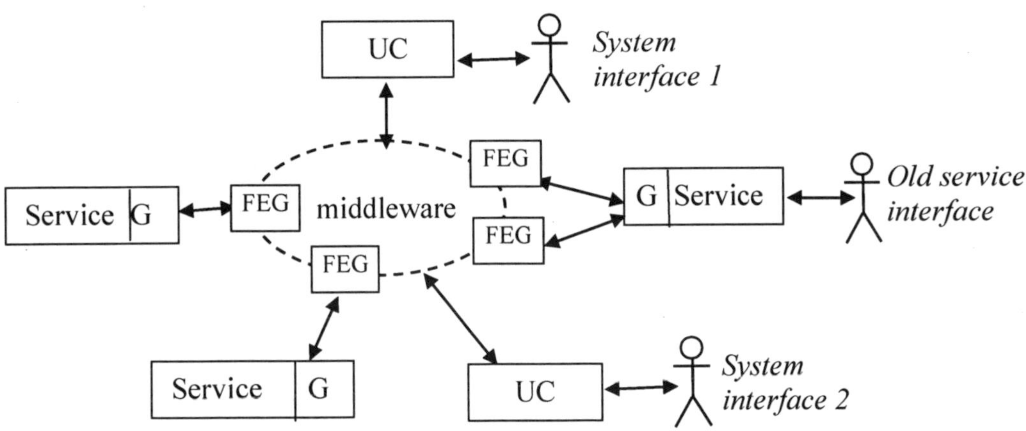

Our two examples show the main extreme situations in the confederation development. In e-government, the system is mainly built via integration of existing systems. In this case, we can define the interfaces only. In the case of manufacturing, the integration stage is preceded by decomposition. In this case, we must specify the *boundaries* of services. They should be intuitively "similar" to the boundaries of real world services.

A specific issue in confederations is that we must often also apply attitudes/philosophies known from the software development history. Middleware can be implemented via Internet or via a common database with triggers. CORBA can be used in some situations as well. All the attitudes must be in practice combined, depending on the type of the functionality (for example, commands on operational level and data analysis on management level), so different philosophies can and must be applied and combined in confederations. It increases the complexity of development.

Business Processes in Confederations

Business processes in confederations (BPC) must be composed from steps—functions/actions of peers offered by their interfaces. We have seen that business services must usually be supervised by users—the owners of the processes—or there should be such a possibility. It is also necessary to offer users a tool to define business processes. Both requirements imply that the languages understood by the gates should be based on user knowledge domain languages and notions. They should be *user-oriented.*

The data defining business processes are best to store in system user interface (portals) and the business steps are supervised via a portal or by a temporary service S controlling the particular process. An extreme solution is that there are no business processes defining data. Business data analysis (via, for example, OLAP) must be based on a data-oriented view. In this case, the users (usually managers) must have a transparent access to data tiers of the components. It is a very complicated problem (Lenzerini, 2001).

The notions and languages specifying the actions of components (for example, bookkeeping) in a user-oriented way are remarkably stable. The same can therefore hold for the languages of the component interfaces. It is then possible that the interface of a component need not vary if the component implementation varies. The stability of interfaces is important for the communication partners of the given component and for the stability of SOSSs. It has further substantial software engineering advantages (modifiability, outsourcing opportunities, reduction of number of messages, integration of products of various vendors, openness, and so forth).

HISTORICAL TECHNOLOGIES USED IN SERVICE-ORIENTED SYSTEMS

As mentioned above, it is not feasible to limit the philosophy and the software architecture to one choice, so the crucial point of the requirements specification is what philosophy or combination of philosophies is to be used. This issue is not addressed in existing CASE tools and only a little by research. We shall attempt to find some criteria for the choice of philosophy using historical knowledge.

Due to an improper, too concise coding of date, almost all programs written in COBOL had to be rewritten during the 1990s. It is known as the Y2K problem. It appeared that many enterprises used COBOL programs for years without having any COBOL programmer. Such programs had to be very stable as they were used without any maintenance. Systems written in COBOL are like SOSSs designed as virtual networks of COBOL applications. The communication between the applications is implemented via data stores/files. If input data stores of an application *A* are ready, *A* can be started and it works autonomously until input data stores are processed. Thanks to this property, COBOL systems can be developed incrementally provided that input data stores can be easily generated. Each application can be specified as a data transformer and programmed and tested autonomously. It led to specific techniques. COBOL systems usually contained various input filters, data format transformers, report generators, and so forth.

The ability of components to be developed autonomously is the crucial qualitative property of the components of large systems. Batch COBOL systems must often be combined with services of confederations. An example is massive data reconstruction.

Data-oriented systems (DOS) appeared after the success of modern database systems. It was common at that time that the main aim was to computerize operative levels of enterprises (warehouses, bookkeeping, business data). Data types were known and various known operations could be defined over the same data. Moreover, properly designed data structures can be used for many other potential operations — that is, the data enabled many further operations not specified yet. It was and still is supported by the power of SQL language. The data orientation must be applied in confederations when the services are intelligent, especially if management activities must be supported. Technically, it implies data analysis and presentation tools (Figure 3). There arc, however, problems with common database schema and data replication (Lenzerini, 2001). Elements of DOS must be applied in subsystems providing online/interactive data analysis (for example, OLAP/ROLAP).

IMPLEMENTATION ISSUES

Integration of Legacy Systems

Technically, a legacy system *LS* is integrated into the system via adding a gate *G* to *LS*. *LS* then becomes an autonomous service *AC* (Figure 4). *G* connects *AC* to a middleware. *AC* is also reconstructed to be a permanent process if needed. The software confederation then has the structure from Figure 2. We will often say that *AC* is an autonomous component if we want to express its implementation character.

Although the choice of the p2p architecture seems to be a technical matter, it should be included into requirements specification as it substantially influences user properties of the system and the structure of the specifications. If an autonomous component (encapsulated legacy system) *AC belongs* to some department/division (for example, its local information system), the department is satisfied with *AC*, and *AC* can provide all required information/services to other collaborating autonomous components, then *AC* can and often should be integrated without any substantial reconstruction. It reduces coding effort as substantial parts of the system are reused. The people using *AC* feel that they still *own* it. It is well known that a feeling of ownership is advantageous and often necessary.

Confederative architecture can support new requirement types raised by CEO, like the transformation of the enterprise organization structure into a decentralized form, selective outsourcing of some parts of the information system, selling out some divisions/departments, integration of newly purchased firms/divisions, support for supply chain management, and forming various business coordination groups (Král & emlika, 2002).

Front-End Gates

Requirements specifications of software confederations must be based on the properties of the interfaces of autonomous services/components (AC). The AC is used as *black box*, that is, its interface is known only to its partners. This is very important, as interfaces, especially the user-oriented ones, are usually more stable than the implementations (Král & emlika, 2003).

Interfaces should hide the implementation details and philosophy of autonomous services. If we, however, need the gate *G* from Figure 5 to offer an access to all functionality of the corresponding component, the gate must usually disclose the main features of the implementation. So, under such a condition, the message formats of *G* must be, to some degree, implementation oriented. As such, it is not stable enough or optimal for some (usually many) partners of AC. We have seen that we can use *front-end gates* (FEG) to solve the problem. FEG is again an autonomous service (peer) with the properties similar to user components. FEG must usually be developed — it is a white box.

Front-end gates play several roles. They stabilize and generalize the interfaces of components and make them partner-friendly. They can provide several different interfaces of a given AC when different groups of its partners need different interfaces. Different FEGs can provide different security levels or different middleware services for different partners. In some situations, one FEG can provide a parallel access to more than one application component. The condition is that some application components offer similar or complementary functions (for example, in a lecture/test reservation system of a university, different application components may handle different faculties or subject groups).

FEG can direct messages to the components that are less loaded at a given moment. Some services can be replicated. It enables load balancing and distribution of services. As FEG is used as a *white box,* the properties of its input language L and its output language L_I must be included in the requirements specifications. It is similar to the specification and design of user interface components (portals), compare Král and emlika (2003), and to the generation of temporary services controlling individual business processes. The proper specification of gates and front-end gates substantially simplifies the system specification, documentation, development, and maintenance.

FEG can be viewed as an enhancement of the middleware services as the developers develop, in this case, rather the middleware than the applications (Figure 5, see Král, 1999).

The interfaces of software services should mirror the interfaces of real-life services if possible. P2p systems enable the incremental development strategy starting from the most useful services. The relation of the services to their real-life counterparts enables a reliable estimation of what services are the most useful ones. According to Pareto 80-20 rule, it enables the achievement of 80% usefulness of the system consuming 20% of effort only.

Petri Nets

FEG are based on specific tools and methods having common features with compiler construction and formal language translation. Methodologically, they are similar to user interface services (portals). They can also play the role of message routers. The time of execution of FEG is negligible as there is no waiting on answers from users and/or technologies. FEG can compose several messages into one message and decompose one message into several messages. FEG is a solution of the interoperability problem mentioned by Schoder and Fischbach (2003). FEG can therefore be viewed as a generalization of places in Petri nets with colored tokens (Petersen, 1997). Tokens are messages, whereas application services behave like processes in temporal colored Petri nets. It is open as to how to generalize Petri nets so that they enable a proper modeling (diagramming inclusive) and simulation of confederations. Petri nets describe, in some sense, atoms of communications in networks of static structure. Workflows must therefore be defined using some other tools. Petri nets are used in manufacturing control systems (Vondrák et al., 2001). It again indicates links between software confederations and manufacturing systems (and, generally, soft real-time systems). Confederations are necessary but there is not enough experience with them among computer professionals at enterprise level.

REQUIREMENT SPECIFICATION ISSUES

The requirement specifications for confederations must often be oriented toward the use of existing systems and their interfaces. The specifications should take into account the properties of the interfaces and their dynamics. It appears that the people with the experience with low-level process control systems could today help a lot at the top-most tier of enterprise systems now having confederative architecture, as the confederative orientation is common for the soft real-time system developers. They should therefore take part in requirements specification of large software systems.

The requirements can now include the new system function types formulated by CEOs like enterprise decentralization, boundaries of divisions, various forms of in- and outsourcing, CRM, purchase coalitions, and so forth. Many such requirements are feasible only if the enterprise information system is confederated and has an appropriate structure. The top management should be aware about it, and the developers should know something about management and its needs. So, there should be no high wall between developers and users (Král & emlika, 2003). It contradicts the recommendations from "Recipe for Success" by Standish Group, available at *www.standishgroup.com*. Note that agile programming proposes permanent contacts between developers and users. Software confederations offer the opportunity to use agile programming in the development of large systems provided they are service-oriented.

Confederations are challenge and issues for CIO like the balance between centralized and decentralized agreement of message formats. The concept of software confederations and e-commerce is a large challenge for software vendors and system integrators. They have to change their business strategies.

OTHER ISSUES

Software confederations can solve the problem of Reorg Cycle (Armour, 2003) saying that the enterprises are permanently reorganized. As during a lengthy reorganization, the conditions on the market change and a new reorganization is necessary. SO is a solution, as it simplifies and shortens the reorganization.

The first SOSS have long ago manifested their advantages like easy prototyping, incremental development, stability, and flexibility. The mainstream of software engineering has then not been faced with the necessity to build confederations at enterprise level. There was no appropriate technology and no strong need to leave the design viewing the system as one possibly distributed logical unit containing no large black boxes.

The situation has changed due to the progress in hardware and software (Internet). SOSS became technically feasible. At the same time, globalization has generated the need for confederated global enterprises and, therefore, for SOSS.

The services can be cloned and made movable like software agents. It can simplify the design of mobile systems. There are issues common with grid systems.

Paradigm Shift

The construction of any system as a collection of services mirroring the structure of real-world services has substantial advantages:

- As the interfaces are user-oriented, the system specification is simpler due to easier involvement of users and a better structure of the specification. It simplifies the use of the system by users, as they understand what the system offers. The users can then easily modify business processes.
- It supports preferable software engineering properties like openness, maintainability, modifiability, and so forth.
- The services can be specified by their interfaces. It opens the opportunity to apply agile programming (Beck et al., 2001) or extreme programming (Beck, 1999) in large projects.

The decomposition of the activities into autonomous services is a very important invention. It is likely that the use of human-like behaving services in service-oriented systems brings the flexibility and power known from human society.

The main barrier of a wider application of service orientation in our sense is that SO is a new paradigm for the majority of software developers. The acceptance and governance of SO will therefore be a long-term process (remember the case of object orientation). There is no general agreement, even in the definition of the content of service orientation (Barry and Associates, 2003). Many antipatterns from Brown et al. (1998), like "Islands of Automation," "Function Decomposition," and so forth, need not be antipatterns in SOSS anymore. Other antipatterns like "Lava Flow" are not dangerous.

The main attributes of good SOSSs are: (i) system is designed as a peer-to-peer network of services; (ii) the peers mirror real-life services and have user-oriented interfaces; and (iii) it is preferable to have user performable service interfaces.

Information Systems of Manufacturing and Information Systems of Global Enterprises

The automated manufacturing systems have the main features of software confederations; they are (partly) service-oriented. SO has therefore been applied in manufacturing systems for many years. Object orientation (OO) is common now at the enterprise level. OO CASE systems based on UML (OMG, 2001b) and model-driven architecture (MDA) (OMG, 2001a) are used there. SO is common for the lowest enterprise management (manufacturing) systems and should be common on the top management level (management of international enterprises), as well. The middle management (local factories) usually rely on OO methodology (compare Table 1).

The format mismatch of enterprise application interface (EAI) can be resolved by front-end gates. Note that EAI is not primarily intended to support B2B (compare Pinkston, 2002). The fact that manufacturing systems are service-oriented has important consequences. The middle management, usually managing the local units of international enterprises, should not insist on the use of OO methods as a *golden hammer* applicable everywhere as it can lead to the application of OO philosophy outside its applicability. The shift to SOSS on the upper enterprise level is, however, not easy as there is lack of SO experts. It is not optimal to involve here only the people from middle management level. They are not service-oriented. It is difficult for them to *convert* their object-oriented thinking into the service-oriented one. The obstacle is their mental barrier. The nice OO design patterns (Gamma et al., 1993) are, to a high degree, useless in SOSS; OO people often have the feeling that the confederative philosophy is a step back. Such a barrier is exceptional for people having experience with systems including technological process control. The difficulty of the acceptance of SO thinking could be for OO people even more difficult than the conversion

Table 1. Application domains of object and confederative orientations

Software Technique	Area of Application
Service orientation (possibly partly)	*Manufacturing control level*: CIM components, real-time systems
Object orientation (for example, UML, MDA) still suffices	*Monolithic enterprise level*: middle management, divisions of an international enterprise, highly centralized organizations
Service orientation desirable, necessary, EAI	*Global (world-wide) enterprise level*: international enterprises, state administration ...
SO philosophy necessary, EAI, B2B	*World-wide business*: some health network services, coalition of car vendors, e-business

from structural thinking to OO thinking (compare Nelson, Armstrong & Ghods, 2002) several decades ago.

CONCLUSION

Important functions, especially the functions supporting CEO activities, depend on application of service orientation. It implies that the system must have a proper architecture as a p2p network of autonomous services having user-oriented interface, system user interface services, and eventually, some newly developed infrastructure services (FEG). The SO philosophy therefore influences requirements specification more substantially than other philosophies. This issue is not understood enough. Architectures are very difficult to change. They must therefore be chosen early during the requirements specifications. Vice versa, the requirements must reflect the properties of the architecture.

User-oriented interfaces of services simplify the collaboration between users and developers, enable the development, exploiting the possibility that services can be simulated via user interface services (portals). It simplifies the design of business processes and enhances the software engineering properties of the system. Good interfaces can easily be designed if the services mirror the real-life services.

The user-oriented interfaces can serve as a specification of the services. It is achievable only if the collaboration between users and developers should be tight, and the developers should be able to understand user knowledge domains. It must be trained. The user-oriented interfaces enable the use of different implementation technologies in different services and the agile forms of development in large projects.

Properly chosen system architecture influences and often determines the tools and processes of the system development. An issue is that there are no good modeling tools for SOSS. The main obstacle here, however, is the inability or unwillingness of many developers to apply SO. It is the consequence of the fact that it is as any paradigm shift: a long-term process. Solution can be in the engagement of people already having the service-oriented feeling, for example, of the developers of soft real-time systems.

SO changes the tasks of IT management that should facilitate the agreements of the details of system architecture. IT management can be less dependent on software vendors as it now has a greater freedom of what to buy from whom and what to develop to achieve a competitive advantage. Good software engineering properties of SOSS simplify many tasks of IT management (selective outsourcing, development process control, modifications and maintenance, and so forth). SO requires changes of marketing strategies and methods of software vendors and system integrators.

SOSS developers must often apply data- and object-oriented techniques and even integrate batch applications. The developers must be able to understand and use the knowledge of users from all the levels of organization hierarchy. It is not easy, and it should be taken into account in the education of software experts, usually too proud of their narrow and detailed computer-oriented knowledge.

SO is a philosophy influencing the whole software industry and practice. It promises to open the way to the software of the quality known from the other branches of industry (no "Warranty Disclaimer").

ACKNOWLEDGMENT

This work has been supported by Czech Science Foundation by grants No. 201/02/1456 and 201/04/1102.

REFERENCES

Armour, P. (2003). The Reorg Cycle. *Communications of the ACM, 46,* 19-22.

Barry and Associates. (2003). Retrieved August 15, 2004, from *http://www.service-architecture.com*

Beck, K. (1999). *Extreme programming explained: Embrace change.* Boston: Addison-Wesley.

Beck, K., Beedle, M., van Bennekum, A., Cockburn, A., Cunningham, W., Fowler, M., et al. (2001). *Agile programming manifesto.* Retrieved August 15, 2004, from *http://www.agilemanifesto.org/*

Bray, I. K. (2002). *An introduction to requirements engineering.* Harlow, UK: Addison-Wesley.

Brown, W. J., Malveau, R. C., McCormick, I. H. W., & Mowbray, T. J. (1998). *AntiPatterns: Refactoring software, architectures, and projects in crisis.* New York: John Wiley & Sons.

Datz, T. (2004, January 15). What you need to know about service-oriented architecture. *CIO Magazine.* Retrieved August 15, 2004, from *http://64.28.79.79/archive/011504/soa.html*

Demetriades, J. T. (2003). Does IT still matter. *Business Integration Journal,* 20-23.

Donnay Software Designs (1999). *Mature, portable, data-driven systems.* Retrieved August 15, 2004, from *http://www.dclip.com/datadr.htm*

Dyché, J. (2002). *The CRM handbook: A business guide to customer relationship management.* Boston: Addison-Wesley Professional.

Gamma, E., Helm, R., Johnson, R., & Vlissides, J. (1993). *Design patterns. Elements of reusable object-oriented software.* Boston: Addison-Wesley.

Král, J. (1995). *Experience with the development and use of flexible manufacturing systems.* Unpublished manuscript.

Král, J. (1999). Middleware orientation: Inverse software development strategy. In W. Wojtkovski, W. G. Wojtkowski, S. Wrycza, & J. upani (Eds.), *Systems development methods for databases, enterprise modeling, and workflow management* (pp. 385-396). New York: Kluwer Academic/Plenum.

Král, J., & emlika, M. (2000). Autonomous components. In V. Hlavá, K. G. Jeffery, & J. Wiedermann (Eds.), *SOFSEM 2000: Theory and practice of informatics, Vol. 1963 LNCS* (pp. 375-383). Berlin: Springer-Verlag.

Král, J., & emlika, M. (2001). Electronic government and software confederations. In A. M. Tjoa, & R. R. Wagner (Eds.), *Twelfth International Workshop on Database and Experts System Application* (pp. 125-130). Los Alamitos, CA: IEEE Computer Society.

Král, J., & emlika, M. (2002). Component types in software confederations. In M. H. Hamza, (Ed.), *Applied informatics* (pp. 125-130). Anaheim: ACTA Press.

Král, J., & emlika, M. (2003). Software confederations - An architecture for global systems and global management. In S. Kamel, (Ed.), *Managing globally with information technology* (pp. 57-81). Hershey, PA: Idea Group.

Lenzerini, M. (2001). Data integration is harder than you thought. Retrieved August 15, 2004, from *www.science.unitn.it/coopis, choice videos/slides*

Lowson, B., King, R., & Hunter, A. (1999). *Quick response: Managing the supply chain to meet consumer demand.* New York: John Wiley & Sons.

Nelson, J., Armstrong, D. A., & Ghods, M. (2002). Old dogs and new tricks. *Communications of the ACM, 45*(10), 132-136.

OMG. (2001a). Model driven architecture. Retrieved August 15, 2004, from *http://www.omg.org/mda*

OMG. (2001b). Unified Modeling Language. Retrieved August 15, 2004, from *www.omg.org/ technology /documents/formal/uml.htm*

Peterson, J. L. (1997). Petri nets. *ACM Computing Surveys*, *9*(3), 223-251.

Pinkston, J. (2002). The ins and outs of integration, how EAI differs from B2B integration. *e-I Journal*, 48-52.

Rowe, D. (2002). E-government motives and organizational framework. In J. Pour, & J. Voíšek (Eds.), *Systems integration 2002, Conference presentations* (pp. 93-99). Prague University of Economics, Prague, Czech Republic.

Rumbaugh, J., Blaha, M., Premerlani, W., Eddy, F., & Lorensen, W. (1991). *Object-oriented modeling and design*. Englewood Cliffs, NJ: Prentice Hall.

Schoder, D., & Fischbach, K. (2003). Peer-to-peer prospects. *Communications of the ACM*, *46*, 27-29.

Vondrák, I., Kruel, M., Matoušek, P., Szturc, R., & Beneš, M. (2001). From business process modeling to workflow management. In M. Bieliková (Ed.), *DATAKON 2001* (pp. 241-248). Brno.

W3C. (2001). Web service definition language. A proposal of W3 Consortium. Retrieved August 15, 2004, from *http://www.w3.org/TR/wsdl*

W3C. (2002). Web services activity. Retrieved August 15, 2004, from *http://www.w3.org/2002/ ws/*

Yourdon, E. (1988). *Modern structured analysis* (2nd ed.). Prentice Hall.

ENDNOTES

1. Other solutions are not feasible for technical as well as for practical reasons (Král & emlika, 2001, 2003; Rowe, 2002).

This work was previously published in Service-Oriented Software System Engineering: Challenges and Practices, edited by Z. Stojanovic and A. Dahanayake, pp. 182-200, copyright 2005 by IGI Publishing, formerly known as Idea Group Publishing (an imprint of IGI Global).

Chapter V
Data Integration Through Service-Based Mediation for Web-Enabled Information Systems

Yaoling Zhu
Dublin City University, Ireland

Claus Pahl
Dublin City University, Ireland

ABSTRACT

The Web and its underlying platform technologies have often been used to integrate existing software and information systems. Traditional techniques for data representation and transformations between documents are not sufficient to support a flexible and maintainable data integration solution that meets the requirements of modern complex Web-enabled software and information systems. The difficulty arises from the high degree of complexity of data structures, for example in business and technology applications, and from the constant change of data and its representation. In the Web context, where the Web platform is used to integrate different organisations or software systems, additionally the problem of heterogeneity arises. We introduce a specific data integration solution for Web applications such as Web-enabled information systems. Our contribution is an integration technology framework for Web-enabled information systems comprising, firstly, a data integration technique based on the declarative specification of transformation rules and the construction of connectors that handle the integration and, secondly, a mediator architecture based on information services and the constructed connectors to handle the integration process.

INTRODUCTION

The Web and its underlying platform technologies have often been used to integrate existing software and information systems. Information and data integration is a central issue in this context. Basic techniques based on XML for data representation and XSLT for transformations between XML documents are not sufficient to support a flexible and maintainable data integration solution that meets the requirements of modern complex Web-enabled software and information systems. The difficulty arises from the high degree of complexity of data structures, for example, in business and technology applications, and from the constant change of data and its representation. In the Web context, where the Web platform is used to integrate different organisations or software systems, the problem of heterogeneity arises also. This calls for a specific data integration solution for Web applications such as Web-enabled information systems.

The advent of Web services and service-oriented architecture (SOA) has provided a unified way to expose the data and functionality of an information system. Web services are provided as-is at certain location and can be discovered and invoked using Web languages and protocols. SOA is a service-based approach to software application integration. The use of standard technologies reduces heterogeneity and is, therefore, central to facilitating application integration. The Web services platform is considered an ideal infrastructure to solve the problems in the data integration domain such as heterogeneity and interoperability (Orriens et al., 2003; Haller et al., 2005; Zhu et al., 2004). We propose a two-pronged approach to address this aim: firstly, data integration and adaptivity through declarative, rule-based service adaptor definition and construction; and, secondly, a mediator architecture that enables adaptive information service integration based on the adaptive service connectors. Abstraction has been used successfully to address flexibility problems in data processing; database query languages are a good example here.

XML as a markup language for document and data structuring has been the basis of many Web technologies. XML-based transformation languages like XSLT, the XML Stylesheet Transformation Language, XML-based data can be translated between formats. With recent advances in abstract, declarative XML-based data query and transformation languages beyond the procedural XSLT, this technology is ready to be utilised in the Web application context. The combination of declarative abstract specification and automated support of the architecture implementation achieves the necessary flexibility to deal with complexity and the maintainability of constantly changing data and system specifications.

Our objective is to explore and illustrate solutions to compose a set of data integration services. The data integration services deliver a unified data model built on top of individual data models in dynamic, heterogeneous, and open environments. The presentation of this technology framework aims to investigate the practical implications of current research findings in Web information systems technology.

A lightweight mediated architecture for Web services composition shall be at the centre of our solution. Data integration is a central architectural composition aspect. The flexibility of the architecture to enable information integration is essential in order to separate the business process rules from the rest of the application logic. Therefore, the data transformation rules are best expressed at the abstract model level. We apply our solution to the Web Services platform in the context of information technology services management in the application service providers ASP (on demand) business area. We focus on this context to illustrate problems and solutions. Portals, provided by ASPs, are classical examples where data might come from different sources that motivate our research. In order to consume the information, the data models and representation needs to be understood by all participants. The ASP maintains the application, the associated infrastructure, and the customer's data. The ASP also ensures that systems and data are available when needed.

The chosen area demonstrates the need to support deployment of Web service technology beyond toy examples (Stern & Davies, 2004). It is a specific, but important area due to the need to find solutions to accommodate constant structural changes in data representations. Two central themes shall be investigated:

- To identify data model transformation rules and how to express these rules in a formal, but also accessible and maintainable ways are central to the data integration problem and its automation;
- Service composition to enable interoperability through connector and relationship modelling based on workflow and business processes is central.

Our contribution based on these themes is an integration technology framework for Web-enabled information systems comprising:

- A data integration technique based on the declarative specification of transformation rules and the construction of connectors that handle the integration in a software system;
- A mediator architecture based on information services and the constructed connectors to handle the integration process.

We start our investigation by providing some data integration background. We then present the principles of our declarative data integration technique. The mediator architecture that realises the data integration technique for Web services is subsequently presented. A larger application scenario will then be discussed. We end with some conclusions.

BACKGROUND

Data Integration Context

The application service provider or ASP business model, which has been embraced by many companies, promotes the use of software as a service. Information systems (IS) outsourcing is defined as the handing over to third party the management of IT and IS infrastructure, resources and/or activities (Willcocks & Lacity, 1998). The ASP takes primary responsibility for managing the software application on its infrastructure, using the Internet as the delivery channel between each customer and the primary software application. The ASP maintains the application and ensures that systems and data are available when needed. Handing over the management of corporate information systems to third party application service providers in order to improve the availability of the systems and reduce costs is changing the ways that we manage information and information systems.

Information integration aims at bringing together various types of data from multiple sources such that it can be accessed, queried, processed, and analysed in an integrated and uniform manner. In a large modern enterprise, it is inevitable that different parts of the organization will use different systems to produce, store, and search their critical data.

Recently, service-based platforms have been used to provide integration solutions for ASP applications. Data integration in these types of collaborating systems is necessary. This problem has been widely addressed in component-based software development through adaptor and connector approaches (Crnkovic & Larsson, 2000; Szyperski, 2002). In the service-based Web applications context, the data in XML representation retrieved from the individual Web services needs to be merged and transformed to meet the integration requirements. The XML query and transformation rules that govern the integration may

Figure 1. Example of data integration in adaptive service architectures—two data schemas that need to be transformed into one another

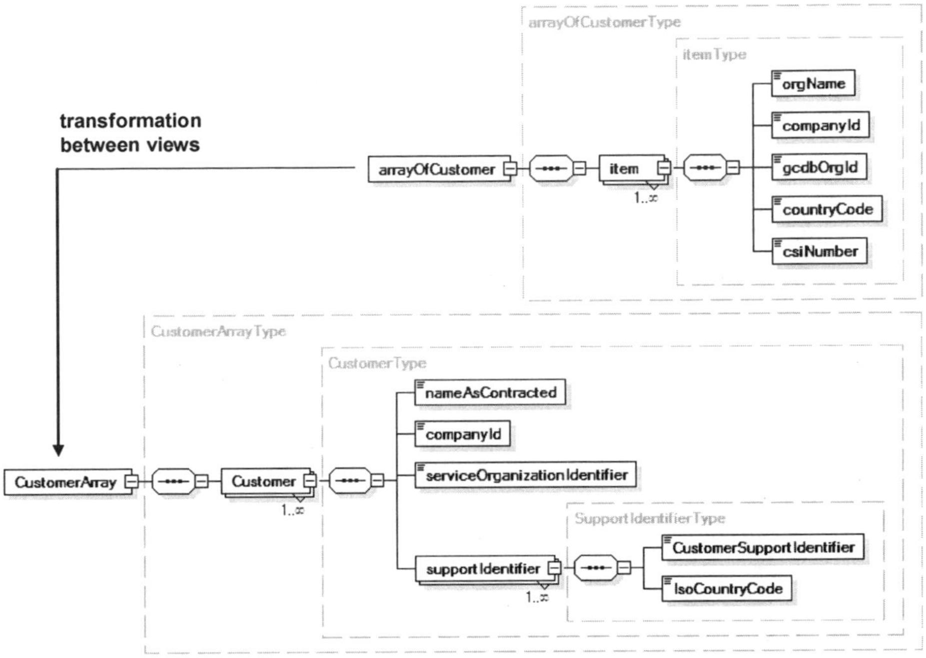

change; therefore, the programs for building up the connectors that facilitate the connection between integrated Web services and data service providers need to be adjusted or rewritten. As with schema integration, the schema-mapping task cannot be fully automated since the syntactic representation of schemas and data do not completely convey the semantics of different data sources. As a result, for both schema mapping and schema integration, we must rely on an outside source to provide some information about how different schemas (and data) correspond. For instance, a customer can be identified in the configuration management repository by a unique customer identifier; or, the same customer may be identified in the problem management repository by a combination of a service support identifier and its geographical location. In this case, a transformation might be necessary; see Fig. 1 for a visualisation of the customer identifier example.

Data Integration Principles

Information integration is the problem of combining heterogeneous data residing at different sources, and providing the user with a unified view (Lenzerini, 2002). This view is central in any attempt to adapt services and their underlying data sources to specific client and provider needs. One of the main tasks in information integration is to define the mappings between the individual data sources and the unified view of these sources and vice versa to enable this required adaptation, as the example in Figure 1 illustrates. The data integration itself is defined using transformation languages.

There are two major architectural approaches to the data integration problem that provide the infrastructure for the execution of transformations (Widom, 1995).

- Data warehousing is an eager or in-advance approach that gathers data from the appropriate data sources to populate the entities in the global view. A data warehousing approach to integration is suitable for data consumers wanting to access to local copies of data so that it can be modified and calculated to suite the business needs by nature.

- In contrast, the mediated approach extracts only data from export schemas in advance. A mediated approach to integration is suitable for information that changes rapidly, for service environments that change, for clients in need tailored data, for queries that operate over large amounts of data from numerous information sources, and most importantly, for clients with the need of the most recent state of data.

XSLT Shortcomings

XSLT is the most widely used language for XML data integration, but these XSLT transformations are difficult to write and maintain for large-scale information integration. It is difficult to separate the source and target parts of the rules as well as the filtering constraints. The verbosity of XML makes manual specifications of data and transformations difficult in any case. With this difficulty in mind, we propose a declarative query and transformation approach yielding more expressive power and the ability to automatically generate query programs as connectors to improve the development of services-based data integration in Web-based information systems.

XSLT does work well in terms of transforming data output from one Web service to another in an ad hoc manner. XSLT code is, however, difficult to write and almost impossible to reuse in a large enterprise integration solution. The syntactical integration of the query part and construction part of a XSLT transformation program is hard to read and often new programs are needed even when a small portion of the data representation changes. XSLT does not support the join of XML documents. We would in our context need to merge several source XML documents into one document before it can be transformed into another document according to an over-arching general schema.

A DECLARATIVE DATA INTEGRATION AND TRANSFORMATION TECHNIQUE

A declarative, rules-based approach can be applied into the data transformation problem (Orriens et al., 2003). A study by Peltier et al. (2001) introduces the MTRANS language that is placed on top of XSLT to describe data model transformations. XSLT is generated from an MTrans specification. The transformation rules are expressed in the form of MTrans and then parsed using a generator. Peltier et al. argue that the data transformation rules are best expressed declaratively at the abstract model level rather than at the concrete operational level in order to reduce the complexity of the transformation rules.

A data integration engine for the Web services context can be built in the Web service business process execution language WS-BPEL, which is another example of the benefits of abstraction in transformation and integration. A common over-arching information model governs what types of services are involved in the composition. In (Rosenberg & Dustdar, 2005), a business rule engine-based approach has been introduced to separate the business logic from the executable WS-BPEL process.

These two examples illustrate current work in this context. Now, a detailed discussion shall elicit the specific requirements for service-based information integration.

Requirements for Mediated Integration

The flexibility of the architecture in which information integration is to be realised is essential in order to separate the business logic from the rest of the application logic. Therefore, the data transformation rules are best expressed at an abstract business model level. These rules, stored in a repository, can be used to dynamically create XSLT-based transformations using a connector or integration service as the mediator. These integration services are the cornerstones of a mediator architecture that processes composite client queries that possibly involve different data sources provided by different Web services. We start our investigation by discussing the properties of suitable integration and transformation languages.

XML data might be provided without accompanying schema and sometimes is not well-formed; XML data often contains nested structures. Therefore, transformation techniques need more expressive power than traditional database languages such as relational algebra or SQL. The characteristics of an XML query language have been studied extensively (Jhingran et al., 2002; Lenzerini, 2002; Peltier et al., 2002). However, these investigations often focus on the features to query an XML or semi-structured data repository in the spirit of database query languages rather than constructing a new XML document in the context of the data integration. The following principles, which are inspired by the data integration literature such as (Lenzerini, 2002), aim to provide a comprehensive requirements list.

- The language should support both querying and restructuring XML Data.
- The language must enable the generation of query programs by other programs.
- The language should be capable of expressing the following operations in addition to the ones existing in database languages (such as projection, selection, and joins): restructuring (constructing a new set of element instances based on variable bindings and the global schema); combination (merging two or more element instances into one); and reduction (to express transformation rules that exclude parts of the data from the result).
- Compositionality is an essential feature for an XML query and transformation language to support query composition.

A rule-based, declarative language enables developers to concentrate on the integration logic rather than on implementation details and enables the required compositionality and expressiveness.

Most XML and semi-structured data query languages have been proposed to extract XML data from the XML databases or the Web. A comparative analysis of existing languages has been done by Reynaud et al. (2001). A language is generally designed to suit the needs for a limited application domain such as database querying or data integration; some languages are designated only for semi-structured data that predated the XML-format. A query language should be able to query data sources using complex predicates, joins, and even document restructuring. We add the following criteria specifically for the context of Web-based data integration:

- **Join:** The language must support joins of multiple XML data sources. A join condition is necessary to compare attributes or elements in any number of XML documents. In data integration systems, data is most likely to come from more than one source.
- **Data model:** The queries and their answers are the instances of a data model. Sometimes, a rich data model is needed to support the functionality of some query languages. The underlying framework plays a major role in determining a data model for a query language.
- **Incomplete query specification:** XML and semi-structured data is not as rigid as relational data in term of schema definitions and data structure. Therefore, it is

important that a query language is capable of expressing queries in incomplete form, such as by using wildcard and regular expressions—also called partially-specified path expressions.

- **Halt on cyclic query terms:** If a language supports querying with incomplete query specification by wildcard and regular expression, it might cause termination problems. Therefore, features to detect cyclic conditions are required.
- **Building new elements:** The ability to construct a new node added to the answering tree is an important feature for data integration systems.
- **Grouping:** Grouping XML nodes together by some conditions by querying the distinct values is another important feature in data integration. Some languages use nested queries to perform grouping operations; in contrast, some more powerful languages have built-in constructors.
- **Nested queries:** Nested queries are common in relational database languages for joining different data elements by their values. In logic-based languages, the construction part and the selection part are separated.
- **Query reduction:** Query reduction allows users to specify what part of the elements or what nodes in the query conditions will be removed from the resulting XML tree.

A number of potential candidates shall briefly be discussed in the context of these requirements:

- **XQuery** is a W3C-suppported query language that aims at XML-based database systems. XQuery is an extension of XPath 2.0 adding functionalities needed by a full query language. The most notable of these functionalities are support of sequences, the construction of nodes and variables, and user-defined functions.
- **UnQL** (the unstructured query language) is a query language originally developed for querying semi-structured data and nested relational databases with cyclic structures. It has later been adapted to query XML documents and data. Its syntax uses query patterns and construction patterns and a query consists of a single select or traverse rule that separates construction from querying. Queries may be nested, in which case the separation of querying and construction is abandoned. UnQL was one of the first languages to propose a pattern-based querying (albeit with subqueries instead of rule chaining).
- **XML-QL** uses query patterns and path expressions to select data from XML sources. These patterns can be augmented by variables for selecting data. XML-QL uses query patterns containing multiple variables that may select several data items at a time instead of path selections that may only select one data item at a time. Furthermore, variables are similar to the variables of logic programming, that is, joins can be evaluated over variable name equality. Since XML-QL does not allow one to use more than one separate rule, it is often necessary to employ subqueries to perform complex queries.

The shortcomings of these widely known and used languages in the context of the given requirements and the language comparisons have led us to choose a fully declarative language called Xcerpt (Bry & Schaffert, 2002) that satisfies all criteria that we have listed earlier on. However, other recently developed and well-supported transformation languages such as ATL and QVT are similarly suitable candidates. While QVT satisfies the criteria, it is currently not as well supported through tools and accessible tutorial material.

Xcerpt is a query language designed for querying and transforming both data on the standard

Web (e.g., XML and HTML data) and data on the Semantic Web (e.g., RDF data). Xcerpt not only allows one to construct answers in the same data formats as the data queries like XQuery, but also allows further processing of the data generated by this same query program. One of the design principles is to strictly separate the matching part and the construction part in a query. Xcerpt follows a pattern-based approach to querying the XML data. A similar approach has been proposed in the languages UnQL and XML-QL. However, Xcerpt has extended the pattern-based approach in the following aspects. Firstly, the query patterns can be specified by incomplete query specifications in three dimensions. Incomplete query specifications can be represented in depth, which allows XML data to be selected at any arbitrary depth in breadth, which allows querying neighbouring nodes by using wildcards, and in order. Incomplete query specifications allow the pattern specifications to be specified in a more flexible manner but without losing accuracy. Secondly, the simulation unification computes answer substitutions for the variables in the query pattern against underlying XML terms—similar to UnQL, but strict unification is used in UnQL.

Declarative Transformation Rules

We have adapted Xcerpt to support the construction of the service connectors, which is our central objective:

- From the technical point of view, in order to promote code reuse, the individual integration rules should not be designed to perform the transformation tasks alone. The composition of rules and rule chaining demand the query part of service connector to be built ahead of the construction part of the service connector.
- From the business point of view, the data presentation of the global data model changes as element names change or elements are being removed. These should not affect the query and integration part of the

Figure. 2. Declarative query and transformation specification of customer array element in Xcerpt

```
CONSTRUCT
   CustomerArray [
      all Customer[
         nameAsContracted[var Name],
         companyId[var CompanyId],
         serviceOrganizationIdentifier[var OrgId],
         all supportidentifier[
            CustomerSupportIdentifier [var Code],
            ISOCountryCode [var CSI]
         ]
      ]
   ]
FROM
   arrayOfCustomer[[
      item [[
         orgName[var Name],
         companyId[var CompanyId],
         gcdbOrgId [var OrgId],
         countryCode[var Code],
         csiNumber[var CSI]
      ]]
   ]]
```

logic. Only an additional construction part is needed to enable versioning of the global data model.

Grouping and incomplete query specifications turn out to be essential features.

Xcerpt is a document-centric language which is designed to query and transform XML and semi-structured documents. Therefore, the ground rules, which read data from the document resources, are tied with at least one resource identifier. This is a bottom up approach in terms of data population because the data are assigned from the bottom level of the rules upward until the rule application reaches the ultimate goal of a complex, hierarchically structured rule. These rules are defined through an integration goal at the top level and structured into sub-rules down to ground rules, which address individual data elements.

Figure 2 shows a transformation example for a customer array based on Figure 1. Figure 1 is a graphical illustration of XML-based data structures. The upper structure provides the data schema of the input document; the lower structure is the target data schema that a transformation needs to map onto. The graphical representation allows us to avoid the verbosity of XML-based data representations for this investigation. An output customer in CustomerArray is constructed based on the elements of an item in an arrayOfCustomer by using a pattern matching approach, identifying relevant attributes in the source and referring to them in the constructed output through variables. For instance, the Name variable is used to declare nameAsContracted and OrgName as semantically equal elements in both representations that are syntactically different.

This original Xcerpt approach is unfortunately not feasible in an information integration solution because the resource identifiers can not be hard coded in the ground rules in our setting. A wrapper mechanism has been developed to pass the resource identifiers from the goal level all the way down to the ground rules. In addition to the original Xcerpt approach, we propose a mediator-based data integration architecture where the Xcerpt-based connectors are integrated with the client and provider Web services. WS-BPEL code is generated by a transformation generator within the mediator service (see Figure 4 below, which is explained in a separate section).

Implementation of Connector Construction

The construction of Xcerpt-based connectors, which specify integration through declarative rules, can be automated using rule chaining. Ground rules are responsible for querying data from individual Web services. Intermediate composite rules are responsible for integrating the ground rules to render data types that are described in global schemas. The composite rules are responsible for rendering the data objects described in the interfaces of the mediator Web services based on demand. Therefore, exported data from a mediator service is the goal of the corresponding connector (i.e., a query program); see Figure 3. Figure 1 defines again the respective input and output data schemas. The `CONSTRUCT .. FROM` clauses in Figure 3 define the individual rules. Here, information from `ArrayOfCustomers` and `Customers` is selected to construct the `SupportIdentifier`.

We apply backward goal-based rule chaining in this adapted implementation to execute complex queries based on composite rules. Figure 3 shows an example of this pattern matching-based approach that separates a possibly partial query based on resource and construction parts. This transformation rule maps the supportIdentifier element of the customer example from Figure 1. Figure 3 is a composite rule based on the SupportIdentifier construction rule at a lower level.

These rules are saved in a repository. When needed, a rule will be picked and the backward rule chaining enables data objects to be populated to

Figure 3. Transformation specification in Xcerpt based on goal chaining

```
GOAL
   Out { Resource {"file:SupportIdentifier_Customer.xml"},
         SupportIdentifier [ All var SupportIdentifier ] }
FROM
   Var SupportIdentifier -> SupportIdentifier {{}}
END

CONSTRUCT
   SupportIdentifier [var Code, optional Var Cname, Var Code]
FROM
in { Resource {"file:customer1.xml"},
     ArrayOfCustomer [[
         customer [[ optional countryName [var CName],
                     countryCode [var Code]
                     csiNumber [var CSI]  ]]  }
END

CONSTRUCT
   SupportIdentifier [var Code, Var Cname, optional Var Code]
FROM
in { Resource {"file:customer2.xml"},
     Customers [[ customer [[
                     countryName [var CName],
                     optional countryCode [var Code]
                     csiNumber [var CSI]  ]] }
END
```

answer transformation requests. This architecture will be detailed in the subsequent section.

MEDIATOR ARCHITECTURE

Motivation

Zhu et al. (2004) argue that traditional data integration approaches such as federated schema systems and data warehouses fail to meet the requirements of constantly changing and adaptive environments. We propose, based on (Haller et al., 2005; Sheth & Larson, 1990; Wiederhold, 1992; Zhu et al., 2004), a service-oriented data integration architecture to provide a unified view of data on demand from various data sources. A service-oriented data integration architecture is different from business process integration as the latter is concerned with integrating the business process rather than data. The proposed integration architecture uses Web services to enable the provision of data on demand whilst keeping the underlying data sources autonomous.

There is consequently a need for mediators in an architecture that harmonise and present the information available in heterogeneous data sources (Stern & Davies, 2003). This harmonisation comes in the form of identification of semantic similarities in data while masking their syntactic differences; see Figure 1. Relevant and related data is then integrated and presented to a higher layer of applications. The sourcing, integration, and presentation of information can be seen as logically separated mediator rules for integration, implemented by mediator services, which shall form the basis for the presented mediator architecture.

Garcia-Molina et al. (1997) identify that the following requirements are essential in order to build a mediator architecture. Firstly, it must be based on a common data model that is more flexible than the models commonly used for the database management systems. Secondly, it must be supported by a common query language. Finally, there must be a tool to make the creation of new mediators and mediator systems more cost-effective than building them from scratch.

Architecture Definition

The mediator architecture transforms local XML documents into documents based on a global schema. Figure 4 illustrates this architecture with a few sample information services—Customer Data, E-business System, Request Logging and Analysis Service—that a client might access. The data integration engine is built based on a composition of individual services using WS-BPEL, where component invocation orders are predefined in the integration schemas. These service orchestrations are defined by specifying the order in which operations should be invoked.

The proposed Web services-based mediator architecture, Figure 4, contains the following components:

- **Schema repository:** Each object within the model is a logical representation of the entity and will often be populated with data sourced from more than one repository. The advantage of having a unified view of data is to make sure that the customers will have a consistent view of data and to avoid duplication.
- **Information services:** These provide source data retrieved from the underling data repositories to clients and other services. The signature of the Web service interfaces such as input parameters and data output is agreed in advance by business domain experts from both client and provider sides. The benefit of asking the data sources to provide a Web service interface is to delegate the responsibility and cut down the effort spent on developing data access code and understanding the business logic.
- **Data integration and mediation services:** A common data model can be implemented as an XML schema. Two basic approaches have been proposed for the mappings between the export schemas and the feder-

Figure 4. Mediator architecture for adaptive service-based information systems with sample information services

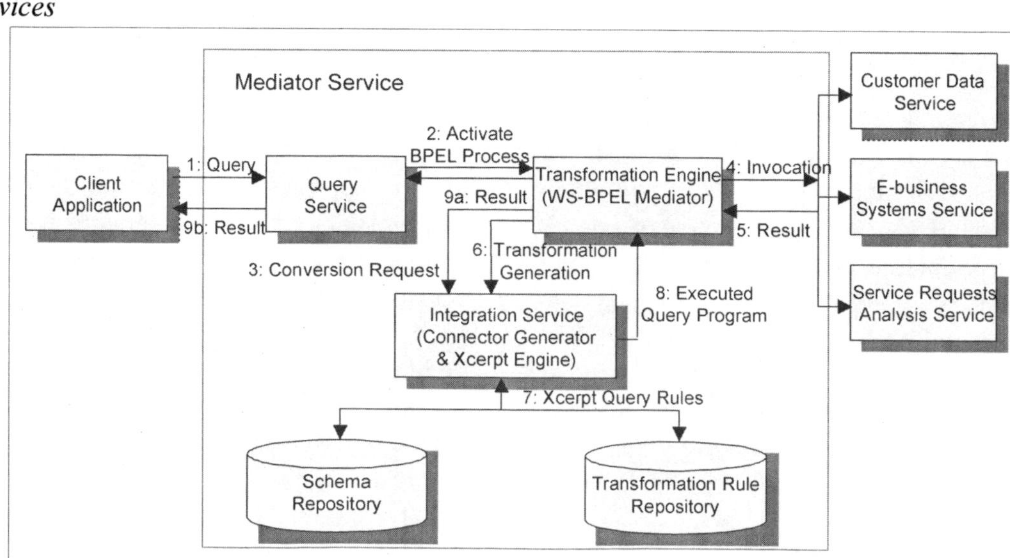

ated schema— called global-as-view and local-as-view in (Lenzerini, 2002). The former approach defines the entities in the global data model as views over the export schemas whereas the latter approach defines the export schemas as views over the global data model. In this work, a data integration service will be treated as a mediator in the mediator architecture. We introduce a novel approach to ease and improve the development of the mediators. There are two quite different styles of transformation: procedural, with explicit source model traversal and target object creation and update; and declarative, with implicit source model traversal and implicit target object creation. Therefore, an approach based on a declarative rule mark up language to express the data transformation rules and a rule engine have been chosen. The mapping should be conducted at the abstract syntax mappings level, leaving the rendering of the result to a separate step at runtime to the BPEL engine.

- **Query component:** The query service is designed to handle inbound requests from the application consumer side. The application developers build their applications and processes around common objects and make successive calls to the mediated Web services. Therefore, the interfaces of individual Web service providers are transparent to the application customers; they may send any combinations of the input parameters to the query service. In order to facilitate these unpredicted needs, the query service has to decompose the input messages into a set of pre-defined WS-BPEL flows. Normally a BPEL flow belongs to a mediator that delivers a single common object. Occasionally, two or more mediators need to be bundled together to deliver a single object.

Each of the components can in principle be offered as a service by a (potentially different) provider. Within the composite mediator service, both transformation and connector generation services are separated, and only loosely coupled.

Developer Activities

The architecture in Figure 4 explains the runtime view from the client and user perspective. In order to complete the picture, the development perspective shall also be addressed. Figure 5 illustrates development activities, looking at the developers of architecture, rules, and services—and their respective activities. A number of actors including service provider engineers, application software engineers, integration business analysts, integration software architects, and integration software engineers are distinguished. These are associated with the activities they are involved in. In particular, the integration team is involved with Xcerpt-based rule definition and application. Activities are also related among themselves. The participation of different roles from possibly different organisations (application customer, service provider, integration team) demonstrates the need for common understanding and maintainability of the integration problem, which can be achieved through abstract and declarative rule specifications (here in Xcerpt format), shared by service provider developers, integration business analysts, and integration software developers.

APPLICATION SCENARIO AND DISCUSSION

The presented data integration technique and the mediated architecture are complemented by an incremental, evolutionary process model. Some pragmatic aspects of this process shall now be addressed. In the proposed architecture, the unified data model (over-arching schema) is maintained manually. The schema for large enterprise integration solutions might consist of a large number of data aspects. From the development point of view, it is only reasonable to deliver the data integration services on a phased basis such as one data aspect for one release cycle. A mediator consists of the following components: the individual provided Web services; a WS-BPEL workflow; and one or more service connectors, as illustrated in Figure 4. Mediators in our solution are used to deliver these data aspects according to the unified schema. This schema is available to the customers so that these can decide which mediator to call based on the definition of the unified schema.

The focus of this investigation is not on the automatic composition of Web services, rather on how the data output from multiple Web services can be automatically integrated according to a global data model and sent back to users. Therefore, in terms of the WS-BPEL process flow, a static approach with respect to the orchestration of the involved Web services can be taken. These can be orchestrated together in form of a WS-BPEL flow built in advance.

During the development phase, the mappings between the global model and the local models will be expressed at the abstract model level, for instance in the widely used MOF (meta object facility) framework for modelling language definition. Model transformation between different metamodels can then be automatically carried out. The inputs are the source XML schema definitions and the transformation rules. The output is an XSLT transformation file.

In the proposed process model illustrated in Figure 5, the unified data model and the creation of rules are the responsibility of the business solution analysts, not necessarily the software architect. The rules are merely mappings from the elements exposed by Web service providers to the elements in the unified data model. We assume here that the semantic similarity is determined manually. In the literature on data model transformation, the automation of the mapping is often limited to transforming the source model and the destination model rather than integrating more than one data model into a unified data model. Even in the case of source to destination model mapping, the user's intervention is needed to select one from more than one set of mappings that are generated. In our proposed architecture, the service connectors can be generated on the fly by rule composition. The sacrifice is that semantic similarity is not taken into consideration.

The data integration rules are created at the higher level than the Xcerpt ground query programs themselves, as the following schematic example demonstrates (Figure 3 shows an example of a composite rule like A below).

Rule A: $A(a, b)$:= $B(a, b), C(b)$
Rule B: $B(a ,b)$:= $D(a), E(b)$
Rule C: $C(b)$:= $E(b), F(b)$

Each of the above rules would be implemented in the Xcerpt language. In this example, rule *A* is a composite rule, based on *B* and *C*. This could be used to answer a user's query directly, but internally referring to subordinated rules dealing with the extraction and transformation of specific data aspects. The resource identifiers in form of variables and the interfaces for the data representation such as version number of the unified data model will be supplied to the transformation generator. The rule mappings in the transformation generator serve as an index to find the correct Xcerpt queries for execution. As a result, a query program including both query part and construction part is being executed to generate the XML output, which is sent back to the transformation generator.

In terms of examples, we have so far only addressed complex transformations based on compositional rules within data provided by one Web service—the customer information service. Queries could of course demand to integrate data from different services. For instance, to retrieve all services requests by a particular customer would target two services, based on several composite integration and transformation rules.

FUTURE TRENDS

Adaptivity in service-based software systems is emerging as a crucial aspect beyond the discussed area of service-based ASP infrastructures and on-demand information systems. Adaptability of services and their infrastructure is necessary to reconcile integration problems that arise in particular in dynamic and changing environments.

We have excluded the problem of semantic interoperability from our investigation. Mappings between schemas might still represent the same semantic information. The recently widely investigated semantic Web services field, with

Figure 5. Overview of different developer roles and their activities involved

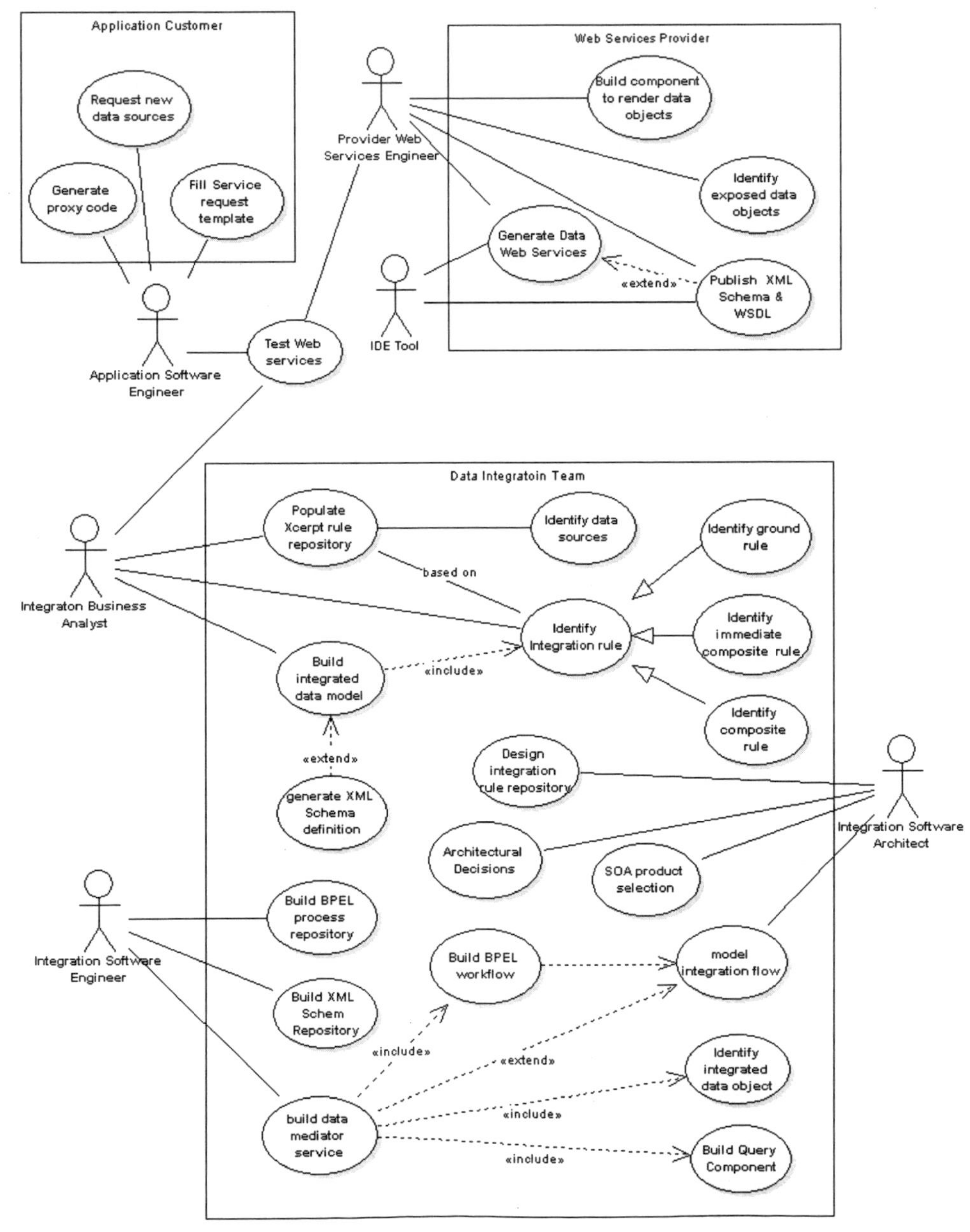

ontology-based domain and service models, can provide input for some planned extensions in this direction (Haller et al., 2005).

Re-engineering and the integration of legacy systems is another aspect that we have not addressed. The introduction of data transformation techniques for reengineering activities can improve the process of re-engineering legacy systems and adopting service-oriented architecture to manage the information technology services (Zhang & Yang, 2004). Business rules often change rapidly, requiring the integration of legacy systems to deliver a new service. How to handle the information integration in the context of service management has not yet been exploited in sufficient detail in the context of transformation and re-engineering.

CONCLUSION

The benefit of information systems on demand must be supported by corresponding information service management systems. Many application service providers are currently modifying their technical infrastructures to manage information using a Web services-based approach. However, how to handle information integration in the context of service-based information systems has not yet been fully exploited.

The presented framework utilises information integration technologies for service-oriented software architectures. The crucial solutions for the information integration problem are drawn from mediated architectures and data model transformation, allowing the data from local schemas to be transformed, merged, and adapted according to declarative, rule-based integration schemas for dynamic and heterogeneous environments. We have proposed a declarative style of transformation, with implicit source model traversal and implicit target object creation. The development of a flexible mediator service is crucial for the success of the service-based information systems architecture from the deployment point of view.

REFERENCES

Alonso, G., Casati, F., Kuno, H. & Machiraju, V. (2004). *Web services—Concepts, architectures and applications*. Springer Verlag.

BPEL Coalition (2006). *Business process execution language for Web services version 1.1*. Retrieved on April 14, 2008 from http://www.ibm.com/developerworks/library/ws-bpel/

Bry, F. & Schaffert, S. (2002). Towards a declarative query and transformation language for XML and semistructured data: Simulation unification. In *Proceedings Intl. Conference on Logic Programming*. LNCS 2401, Springer-Verlag.

Crnkovic, I. & Larsson, M. (2000). A case study: Demands on component-based development. In *Proceedings of the 2nd International Conference on Software Engineering*, (pp. 23-31). ACM Press.

Garcia-Molina, H., Papakonstantinou, Y., Quass, D., Rajaraman, A., Sagiv, Y., Ullman, Y. D., Vassalos, V. & Widom, J. (1997). The TSIMMIS approach to mediation: Data models and languages. *Journal of Intelligent Information Systems*, *8*(2), 117-132.

Haller, A., Cimpian, E., Mocan, A., Oren, E. & Bussler, C. (2005). WSMX - A semantic service-oriented architecture. In *Proceedings of the Inernational Conference on Web Services ICWS'05*.

Jhingran, A.D., Mattos, D. & Pirahesh, N.H. (2002). Information integration: A research agenda. *IBM System Journal 41*(4), special issue on information integration. [Electronic version] Retrieved on April 14, 2008 from www.research.ibm.com/journal/sj/414/jhingran.pdf

Lenzerini, M. (2002). Data integration: A theoretical perspective. In *Proceedings of the Principles of Database Systems Conference PODS'02*, (pp. 233-246). ACM.

Orriens, B., Yang, J. & Papazoglou, M. (2003). A framework for business rule driven Web service composition. In M. A. Jeusfeld & O. Pastor, (Eds). *Proceedings of the ER'2003 Workshops,* LNCS 2814, (pp. 52-64). Springer-Verlag.

Peltier, M., Bezivin, J & Guillaume, G. (2001). MTRANS: A general framework, based on XSLT, for model transformations. In *Proceedings of the Workshop on Transformations in UML WTUML'01.*

Peltier, M., Ziserman, F. & Bezivin. (2002). On levels of model transformation. In *Proceedings of the XML Europe Conference* (pp. 1-17). Paris, France: Graphic Communications Association.

Reynaud, C., Sirot, J. P. & Vodislav, D. (2001). Semantic integration of XML heterogeneous data sources. In *Proceedings of the IDEAS Conference* (pp. 199-208).

Rosenberg, F. & Dustdar, S. (2005). Business rules integration in BPEL - A service-oriented approach. In *Proceedings of the 7th International IEEE Conference on E-Commerce Technology.*

Sheth A. P. & Larson J. A. (1990). Federated database systems for managing distributed, heterogeneous, and autonomous databases. *ACM Computing Surveys, 22*(3), 183.

Seltsikas, P. & Currie, W.L. (2002). Evaluating the application service provider (ASP) business model: The challenge of integration. In *Proceedings of the 35th Annual Hawaii International Conference,* (pp. 2801-2809).

Stern, A & Davis, J. (2003). A taxonomy of information technology services: Web services as IT services. In *Proceedings of the First International Conference on Service Oriented Computing.*

Stern, A. & Davis, J. (2004). Extending the Web services model to IT services. In *Proceedings of the IEEE International Conference on Web Services,* (pp. 824-825).

Szyperski, C. (2002). *Component software: Beyond object-oriented programming,* 2nd Ed. Addison-Wesley.

Widom, J. (1995). Research problems in data warehousing. In *Proceedings of the 4th International Conference on Information and Knowledge Management.*

Wiederhold, G. (1992). Mediators in the architecture of future information systems. *IEEE Computer, 25.* March, 38-49.

Willcocks, L. P. & Lacity, M. C. (1998). The sourcing and outsourcing of IS: Shock of the new? In L. P. Willcocks & M. C. Lacity (Eds.) *Strategic sourcing of information technology: Perspectives and practices.* Wiley.

Zhang, Z. & Yang, H. (2004). Incubating services in legacy systems for architectural migration. In *Proceedings of the 11th Asia-Pacific Software Engineering Conference (APSEC'04)* (pp. 196-203).

Zhu, F., Turner, M., Kotsiopoulos, I., Bennett, K., Russell, M., Budgen, D., Brereton, P., Keane, J., Layzell, P., Rigby, M. & Xu, J. (2004). Dynamic Data Integration Using Web Services. In *Proceedings 2nd International Conference on Web Services (ICWS'04).*

Chapter VI
Using Patterns for Engineering High-Quality Web Applications

Pankaj Kamthan
Concordia University, Canada

ABSTRACT

In this chapter, we view the development and maintenance of Web applications from an engineering perspective. A methodology, termed as POWEM, for deploying patterns as means for improving the quality of Web applications is presented. To that end, relevant quality attributes and corresponding stakeholder types are identified. The role of a process, the challenges in making optimal use of patterns, and feasibility issues involved in doing so, are analyzed. The activities of a systematic selection and application of patterns are explored. Following a top-down approach to design, examples illustrating the use of patterns during macro- and micro-architecture design of a Web application are given. Finally, the implications towards Semantic Web applications and Web 2.0 applications are briefly outlined.

INTRODUCTION

In the past decade, the Internet and the Web have opened new vistas for many sectors of society including education, businesses, and government. Indeed, Web applications have played an increasingly integral role in our daily activities of communication, information, and entertainment, and continue to do so. In retrospect, the sustainability of the successes of Web applications is brought into question due to their failures (Nguyen, Johnson, & Hackett, 2003), many of which are related to issues of quality (Pertet & Narasimhan, 2005; Vijayaraghavan, 2003).

In order to continue providing the desirable services to the consumers, it is the responsibility of the providers to ensure the quality of Web applications. There have been several approaches from different viewpoints for understanding and managing the issue of the quality of Web applications. However, as discussed later, most focus on (1) quality as an afterthought rather than as an integral consideration to be embraced early and carried throughout the development process, and (2) preventative rather than curative means for addressing quality. The purpose of this chapter is to motivate the use of patterns (Appleton, 1997) within a systematic approach to the development of "high-quality" Web applications and to point out the benefits and challenges in doing so (Kamthan, 2008).

The rest of the chapter is organized as follows. We first outline the background and state-of-the-art necessary for the discussion that follows and state our position in that regard. This is followed by a discussion of the suitability of a process model, the presentation of the quality model that includes quality attributes at a granular level for representations in Web applications, selection and application of patterns as means for addressing the quality attributes in the quality model, and supporting examples. Next, challenges and directions for future research are outlined and, finally, concluding remarks are given.

BACKGROUND

In this section, we present a synopsis of Web Engineering, quality in Web applications, and patterns.

Characteristics of Web Applications

There are certain uniquely defining social and technical characteristics of Web applications that bring enormous benefits to the users. They also pose a variety of new challenges that the providers must deal with.

Specifically, Web applications differ from traditional software in many ways including that they are largely document-centric (rather than data-centric), they are delivered to a user over the network (rather than locally installed), their users often have little control over the behavior or rendering of information, they need to increasingly compete for visibility, the laws of jurisdictions from where they are being served and where they being consumed can be quite different, they are based on user interfaces that currently lack standards for presentation and interaction, they are expected to be an exemplar of universality (delivered to anybody, anywhere, at any time, on virtually any device), and so on. These characteristics naturally manifest in their development, operation, and maintenance.

Engineering of Web Applications

The need for managing increasing size and complexity of Web applications and the necessity of a planned development was realized in the late 1990s (Coda et al., 1998; Powell, Jones, & Cutts, 1998). This led to the discipline of Web Engineering (Ginige & Murugesan, 2001), which has been treated comprehensively in recent years (Kappel et al., 2006; Mendes & Mosley, 2006; Rossi et al., 2008).

Web Engineering is defined as a discipline concerned with the establishment and use of sound scientific, engineering and management principles and disciplined and systematic approaches to the successful development, deployment, and maintenance of high-quality Web applications. It relies and draws upon other engineering disciplines including computer engineering, document engineering, hypermedia engineering, information systems engineering, media engineering, network engineering, software engineering, and systems engineering for its existence.

For the rest of the chapter, a Web application will mean a Web site that behaves like an interactive software system specific to a domain in a distributed client-server environment. A Web application will in general require programmatic ability and may deploy additional software (such as application servers, media servers, or database servers) for some purpose (such as dynamic delivery of resources).

Quality of Web Applications

The unique nature of Web applications as compared to traditional software makes the need for "high-quality" all the more critical. That Web applications exhibit "high-quality" is critical to all stakeholders involved. If unaddressed, there is a potential for a resource in a Web application to be rendered unreadable on a user agent of a customer, be inaccessible to someone who is visually impaired, or be prohibitive to adaptive maintenance by an engineer (say, for transformation from one environment to another for a certain purpose). This can impact the *sine qua non* of any organization, whether it is commercial, educational, or governmental.

There have been various initiatives for addressing the quality of Web applications: listing, organizing, and discussing relevant quality attributes (Brajnik, 2001; Dustin, Rashka, & McDiarmid, 2001; Hasan & Abuelrub, 2006; Offutt, 2002; Ziemer & Stålhane, 2004), including in some cases from a user's perspective (Ross, 2002); and providing a means for evaluation (Mich, Franch, & Gaio, 2003; Olsina & Rossi, 2002). However, these efforts are limited by one or more of the following issues: although quality attributes relevant to Web applications are given, the means of addressing them are either suggested casually or not at all, and the focus is less on assurance (prevention) and more on evaluation (cure). For example, the emphasis ostensibly is on testing rather than on inspections.

Patterns for Web Applications

A pattern is defined as a proven solution to a recurring problem in a given context (Appleton, 1997). The existence of *rationalized* and *proven solutions* based on established principles, that are specific to *problems* in a given *context* in a structured form, often makes patterns more practical in their applicability compared to other means (Wesson & Cowley, 2003), such as guidelines (Vijayaraghavan, 2003), for an improvement in quality.

A pattern is typically described (Meszaros & Doble, 1998) using an ordered list of elements labeled as (pattern) *name*, *author*, *context*, *problem*, *forces*, *solution*, *example*, and *related patterns*. At times, these assigned labels may vary across a domain or a community of users. Furthermore, optional elements, such as those related to metadata, may be included to enrich the description. In the rest of the chapter, the elements of a pattern are highlighted in italics.

In general, patterns do not exist in isolation and are usually part of an overall vocabulary (namely, a communication and organization structure such as a "pattern system" or a "pattern language") that attempts to solve a larger problem than an individual pattern.

Patterns were formally introduced in the urban planning and architecture domain. Since then, patterns have permeated into computer science and software engineering in general and the development of Web applications in particular. It has been shown that patterns can be successfully applied to both academic (Kendall, 1998) and industrial (Beck et al., 1996) contexts.

In particular, patterns have been discovered in the following domains of interest to us: navigation design (Gillenson, Sherrell, & Chen, 2000; Rossi, Schwabe, & Lyardet, 1999); hypermedia design (German & Cowan, 2000; Rossi, Lyardet, & Schwabe, 1999); and Web applications in general (Montero, Lozano, & González, 2002; Rossi & Koch, 2002; Van Duyne, Landay, & Hong, 2003; Weiss, 2003).

Indeed, patterns have been used for the development of certain Web applications (Garzotto et al., 1999; Montero, Lozano, & González, 2002; Montero, López-Jaquero, & Molina, 2003). However, in these cases the relation of patterns to the underlying development process or to the improvement of quality is not discussed explicitly.

TOWARDS A SYSTEMATIC INTEGRATION OF PATTERNS IN THE ENGINEERING OF LARGE-SCALE WEB APPLICATIONS

In this section, we propose a methodology for pattern-oriented Web Engineering, namely POWEM, for addressing the quality of Web applications. The construction of POWEM is driven by characteristics unique to Web applications and inspired by situational method engineering (Kumar & Welke, 1992) in such a way that acknowledges the significance of quality assurance in Web applications and the role of patterns in addressing it.

POWEM consists of the following interrelated and non-linear sequence of broadly labeled non-atomic activities:

1. Selecting the development process model.
2. Identifying and organizing quality concerns from a semiotics viewpoint.
3. Selecting and applying suitable patterns.

The aforementioned activities need to be satisfied as a *collective*. Furthermore, each of the activities needs to be feasible. While pursuing one activity, the non-linearity allows us to revisit other activities and make appropriate modifications if deemed necessary.

In the following sections, we expound on the work involved in each of these activities.

Selecting the Development Process Model

The inclusion of patterns in the development of Web applications cannot be ad hoc or an afterthought. In fact, the deployment of patterns in the development of Web applications needs to take place within the auspices of a suitable process model.

A pattern-based software development process based on case-based reasoning (CBR) has been discussed and applied to an Internet chat application (Wentzlaff & Specker, 2006). However, it does not explicitly suggest any implications towards quality. Also, besides *problem* and *solution*, it does not take other mandatory elements of a pattern into consideration. For example, it does not discuss the significance of the *context* in which a *problem* occurs and the *forces* that its *solution* resolves.

The selection and adoption of a process model depends on several factors including the following non-mutually exclusive factors that we deem indispensable:

- **Flexibility:** It is well-known that the development of Web applications is uniquely sensitive to changes in the market and other uncertainties (Ginige & Murugesan, 2001; Ziemer & Stålhane, 2004). This requires that the development process of Web applications be flexible. The flexibility can be achieved by allowing revisitations of previous phases and by facilitating parallel development.
- **Support for quality improvement:** There must be an explicit provision in the process specification for improvement of the quality of the underlying product, namely Web applications, using patterns.
- **User-centricity:** It is well-known that Web applications are inherently interactive in nature, and the users (and their computing environments) vary broadly in their capabilities. Therefore, any selection and adoption of a process model must especially be sensitive to the users.

The other factors could for example be low learning curve of the process model, the familiarity of engineers with the process model, maturity demonstrated by successful/proven use, and cost-effective, broad, and readily available tool support.

A flexible user-centric process aiming for the development of "high-quality" Web applications will typically be non-linear (iterative and incremental), and address aspects of both the analysis and the synthesis. Furthermore, both analysis and synthesis will typically have their own set of workflows. During analysis, an understanding and specification of the problem domain will take place, leading to artifacts such as the domain model and use model, and based on them, the requirements specification. During synthesis, an understanding and specification of the solution domain will take place, leading to artifacts for the macro-architecture design and micro-architecture design, followed by the implementation of source code and data.

The Agility-Discipline Spectrum

We recommend three different types of process environments for the development of Web applications. They are all in agreement with the aforementioned criteria, are customizable, and their suitability depends on the (team) size of the project.

- **XP:** Extreme programming (XP) (Beck & Andres, 2005) is a broadly-used and well-documented agile methodology (Highsmith, 2002) for test-driven software development. XP is suited for small-to-medium size projects.
- **UP:** The unified process (UP) (Jacobson, Booch, & Rumbaugh, 1999) is an archetype of model-based and use case-driven configurable process framework, of which the Rational Unified Process (RUP) (Kruchten, 2004) is an instance. RUP is suited for large size projects.
- **OPEN:** The object-oriented process, environment, and notation (OPEN) process framework (Firesmith & Henderson-Sellers, 2002) is a configurable process framework for the development of object-oriented and component-based systems. OPEN defines a process meta-model that can be instantiated to suit large size projects.

Each of XP (Wallace, Raggett, & Aufgang, 2002), RUP (Kappel et al., 2006), and OPEN via its dialect Web OPEN (Henderson-Sellers, Lowe, & Haire, 2002) have been "adapted" to Web applications. There is explicit support for the use of patterns during design in XP, UP, and OPEN and, by reference, their extensions/instantiations for Web applications.

The benefits and drawbacks of agile and disciplined approaches to development have been pointed out and a heterogeneous combination that balances agility and discipline is suggested (Boehm & Turner, 2004). Further discussion of this issue is beyond the scope of this chapter.

Feasibility of Development Process Model

The adoption of the process model for the development of a Web application will evidently depend on the organizational process maturity (Paulk et al., 1995). This in turn involves several factors, including budget, availability of qualified personnel, submission time line, nature of the application domain (for example, established or new), and available tool support.

Identifying and Organizing Quality Concerns from a Semiotics Viewpoint

A Web application should be able to communicate with humans and with machines. We adopt the theory of semiotics (Stamper, 1992) for this purpose. For the sake of this chapter, we focus on the semiotic quality of the representations in Web applications. Among the proposed approaches for quality, we adopt and extend the treatment in (Lindland, Sindre, & Sølvberg, 1994).

The steps of the construction are as follows:

1. **View:** From a semiotics perspective, we can view a Web application on six interrelated levels: physical, empirical, syntactic, semantic, pragmatic, and social. In this chapter, we shall restrict ourselves to the pragmatic level, which is responsible for the relation of signs to their interpreters. The interpreters in our case are the *stakeholders*, namely the producers and the consumers of the Web application.
2. **Decompose:** We contend that quality is a multi-dimensional concept, and decompose it into granular levels that consist of known attributes that can be addressed directly or indirectly (Fenton & Pfleeger, 1997). These quality attributes could for example manifest themselves as non-functional requirements of a Web application. For the definitions of these quality attributes, we adopt the IEEE Standard 1061-1998 and the ISO/IEC 9126-1 Standard.
3. **Assign:** Among the possible means, we choose patterns for improving the quality attributes of a Web application.

The aforementioned construction summarized in Table 1.

We contend that the quality attributes in Table 1 are necessary, however, make no claim of their sufficiency. The relevance of these quality attributes that justifies their inclusion is discussed later.

The quality attributes within the same tier in Table 1 not necessarily mutually exclusive. For example, the steps taken towards improving reliability (say, fault tolerance) may lead to redundant source code or data (that can be unfavorable to maintainability) and but enable ease-of-use (that can be favorable to usability).

The quality attributes in Tier 2 depend on that in Tier 1. For example, if an engineer cannot comprehend the information in a Web application, he/she may not be able to maintain it to its desired expectations.

The Pragmatic Quality-Stakeholder Contract

For the sake of this chapter, we view pragmatic quality as a *contract* between a Web application and a stakeholder. For the sake of simplicity, we will limit ourselves to the discussion of (not necessarily mutually exclusive) stakeholders of the type end-user and engineer.

The relevance of quality attributes in Table 1 varies with respect to stakeholder types. The qual-

Table 1. A model for addressing the semiotic quality of Web applications by means of patterns

Semiotic Level	Quality Attributes	Means for Quality Assurance
Social Quality Concerns		Patterns
Pragmatic	[Tier 2] Maintainability, Usability	
	[Tier 1] Comprehensibility, Performance, Readability, Reliability	
Physical, Empirical, Syntactic, and Semantic Quality Concerns		

ity attributes of direct concern to an end-user at the level Pragmatic-Tier 1 are comprehensibility, performance, readability, and reliability. The quality attributes of direct concern to an engineer at the level Pragmatic-Tier 1 is comprehensibility.

The quality attributes of direct concern to an end-user at the level Pragmatic-Tier 2 is usability. We will view accessibility as a special case of usability (Mendes & Mosley, 2006). The quality attributes of direct concern to an engineer at the level Pragmatic-Tier 2 is maintainability. We will consider modifiability, portability, and reusability as special cases of maintainability (Buschmann et al., 1996).

Finally, we note that the significance and priority of quality attributes will likely vary across different types of Web applications (Selmi, Kraïem, & Ghézala, 2005). For example, the quality needs of an education portal will vary from that of a Web application providing weather information.

Feasibility of Quality Attributes

It is assumed that user modeling at an early stage in the process has led to user profiles, which can then be used to derive user requirements pertaining to quality. However, in a distributed environment of a Web application that aims to serve remote users, a precise realization of such profiles can be challenging.

The expectations of improving the quality attributes of a Web application must be feasible in order to be practical. We contend that the pragmatic quality attributes in Table 1 cannot, with respect to the stakeholders, be *completely* satisfied. For example, an a priori guarantee that a Web application will be usable to *all* users at *all* times in *all* task-specific or environment-specific situations that the users can find themselves in, is simply unrealistic. Therefore, the quality requirements of a Web application must reflect the fact that certain attributes can only be *satisficed* (Simon, 1996).

Selecting and Applying Suitable Patterns

In general, the relationship between a quality attribute and a pattern is many-to-many (as, for example, is evident from Table 2). This leads to the need for selection of patterns.

There are some patterns available specifically for addressing maintainability concerns of Web applications (Weiss, 2003). However, in some cases the *solutions* are highly technology-specific, not all the mandatory elements of a pattern are appropriately documented, and the integration of patterns into any development process is not mentioned. There are also some patterns available for addressing usability concerns of Web applications (Graham, 2003; Perzel, & Kane, 1999). However, usability is viewed as an atomic (non-decomposable) concept, the patterns are strongly oriented towards user interface design, and their integration into any user-centered development process is not shown explicitly.

The underlying *problem* at hand along with the *context* in which it occurs will play a crucial role in selecting desirable patterns. Although there are preliminary results on automation such as an expert system-based decision analysis (McPhail & Deugo, 2001), the selection of patterns appropriate for a task largely remains a manual process.

There are two main non-mutually exclusive concerns in the application of patterns: (1) the understanding of the pattern description, and (2) the order in which patterns are applied. The understanding of the underlying *problem* the *context* in which it occurs, and the trade-offs and consequences of the proposed *solution,* is imperative. The appropriate use of patterns depending on the context of use is particularly critical.

A well-documented pattern description will have *context* and *related patterns* elements that may give an indication of the order of application of patterns. The patterns that precede the pattern under consideration will usually be mentioned in the *context* element and the patterns that succeed the pattern under consideration will usually be mentioned in the *related patterns* element. The order in which patterns should be selected and applied is straightforward when a set of patterns is organized as pattern system or a pattern language, as these cases the relationships among patterns are explicit.

Selecting and Applying Patterns to the Design of Web Applications

In the following, we will limit ourselves to addressing the role of patterns in the design phase. In the design phase, the patterns for high-level design are applied first, followed by the patterns for low-level design.

The selection of patterns in our case is based on generality, neutrality with respect to any specific application domain, broad availability, parity to the quality attribute at hand, suitability of the *context* and the *forces* (where available), and the reputation of the authors.

Often, a pattern is referred to by its *name*. In the following, the *name* of a pattern is listed in uppercase in order to distinguish it from the main text.

Macro-Architecture Design of Web Applications

The macro-architecture design is the place where high-level design decisions, independent of any implementation paradigm or technology, are made.

The macro-architecture patterns that we suggest are based on the notion that Web applications are a class of distributed request-response-type interactive systems. Specifically, the applicable patterns are the CLIENT-SERVER pattern (Schmidt et al., 2000) followed by the APPLICATION SERVER pattern (Manolescu & Kunzle, 2001), which in turn is followed by the MODEL-VIEW-CONTROLLER (MVC) pattern (Buschmann et al., 1996). Figure 1 presents an abstract view of these macro-architecture design patterns.

The CLIENT-SERVER pattern supports maintainability. For example, a server or resources on the server-side could be modified without impacting the client. Also, a single server can support multiple clients simultaneously, or a client could make simultaneous requests for resources residing on multiple servers. For instance, an extensible markup language (XML) document could be

Figure 1. A view of the macro-architecture design patterns in the development of Web applications

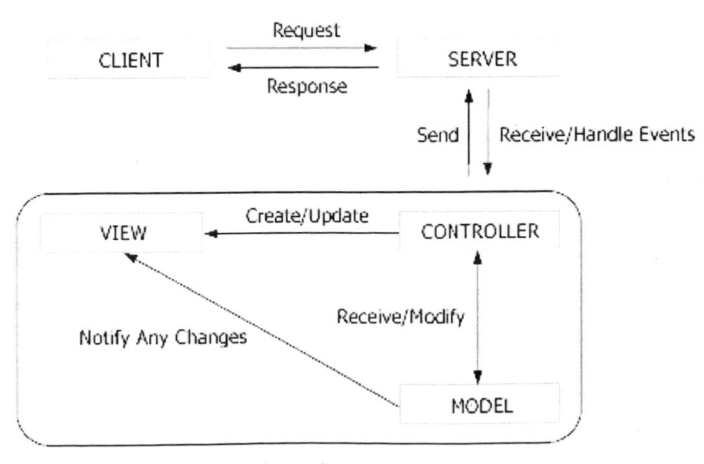

located on one server while an image graphic on another and a cascading style sheet (CSS) document on yet another.

The APPLICATION SERVER pattern also supports maintainability; it isolates the Web application from other aspects on the server-side such that the communication between the application itself and the Web server takes place via the SINGLE POINT OF ACCESS (Yoder & Barcalow, 1997) pattern. This separation allows the Web application to evolve independently.

The separation of structure of content in a markup document from its presentation is one of the principles of Web Architecture (Jacobs & Walsh, 2004). By adopting this principle and an appropriate use of MVC, leads to a separation of semantically-different aspects into three components, namely model, view, and controller, and a theoretical minimization of coupling between those components. Thus, modifications to one component are localized and the propagation of changes to other components is minimal. Furthermore, the same model in a MVC could also be used with multiple views and multiple controllers. For example, the same information could be transformed and delivered to different browser environments or user needs. This invariably improves the maintainability of a Web application.

We note that a true separation of model, view, and controller at the macro-architecture level alone is hard to realize in practice. In an object-oriented environment, it is with the help of micro-architecture design patterns (Gamma et al., 1995) such as OBSERVER, COMPOSITE, and STRATEGY, that a separation is achieved.

There are several implementations of MVC available in a variety of programming languages such as Java and PHP Hypertext Preprocessor (PHP), and application frameworks like Asynchronous JavaScript and XML (AJAX) (Mahemoff, 2006) and Rails (Tate & Hibbs, 2006).

Reliability Design

For addressing reliability (specifically, availability) concerns, the macro-architecture design of server-side components of a Web application could use a number of available patterns (Ahluwalia & Jain, 2006). For example, extra measures (unrelated to the functionality of the Web application) to support the availability of a Web application could be included by using the INTRODUCE REDUNDANCY pattern, and if and when the need arises, a failure message could be relayed using the FAILURE NOTIFICATION pattern. In retrospect, redundancy also increases maintenance responsibilities.

Micro-Architecture Design of Web Applications

The micro-architecture design is the place where low-level design decisions that must be easily implemented are cast. In the following, we will focus only on the design aspects that impact pragmatic quality. As such, our attention is geared more towards client-side rather than server-side concerns.

Interaction design (Preece, Rogers, & Sharp, 2002) is an approach to the design of interactive systems that focuses on the human as well as the computer aspects. Its goal is to make both content and user interfaces useful, easy-to-use, and enjoyable. Many of the patterns available for interaction design in general (Tidwell, 2005) are also applicable to Web applications.

We now consider three major interaction design aspects of Web applications, namely information design, navigation design, and search design. The discussion is essentially independent of any specific domain.

Information Design

Information design is concerned with the structure and behavior of the information served to the user.

The heterogeneity of information is pervasive in Web applications. It is often the case that the information presented on a single "Web Page" is aggregated from several sources. For example, the "Home Page" of a news organization served from the main source may include a latest news ticker from one server and weather information from another server, the stock market information from a financial Web Service, and periodically changing advertisements from yet another source. This can be systematically realized by the use of the WHOLE-PART pattern (Buschmann et al., 1996), which enables a hierarchical organization of objects. Since each of these objects can be modified or replaced independently, the WHOLE-PART pattern supports maintainability. Also, since a "part" can correspond to more than one "whole," the WHOLE-PART pattern also supports reusability. However, multiple indirections stemming from client requests and responses for fulfilling them can lead to a loss of performance, particularly when each "part" itself is structured as WHOLE-PART.

The classification of information is a conventional approach by humans to understanding information. The information organization patterns (Van Duyne, Landay, & Hong, 2003), when use appropriately, aid comprehensibility and usability. For example, the GRID LAYOUT pattern that suggests the organization of information in a single document ("Web Page") into a grid of rows and columns where every atomic information element is made to fit within this grid. The WHAT'S NEW PAGE pattern that provides newly added information to a Web application could include the CHRONOLOGICAL ORGANIZATION pattern. A document in a Web application that provides event proceedings could contains a list of publications and/or their authors based on the ALPHABETICAL ORGANIZATION pattern.

The users of a Web application can vary in their capabilities and preferences, and may find one view of information to be more usable than another. The MIRRORWORLD pattern (German & Cowan, 2000) provides two or more views of the same information. Specifically, information in these views could be presented (Tidwell, 2005) in TWO-PANEL SELECTOR pattern when we have two different views that are to be presented simultaneously, or CLOSABLE PANELS or CARD STACK patterns when we have several different views to be presented in such as way that only one view is visible at a time in each panel or stack, respectively.

Tables are often used to structure information in two dimensions. However, the presence of many columns (or multiple lines to a row) can adversely affect readability, as it becomes increasingly hard to separate the entries visually. The ROW STRIPING pattern (Tidwell, 2005) suggests the use of two similar shades to alternately color the backgrounds of the table rows.

Now, documents in a Web application may contain images for presenting some information such as the corporate logo or product pictures. The FAST-DOWNLOADING IMAGES pattern suggests creation of images optimized for color and size in an appropriate format, and thus aids accessibility and performance. The REUSABLE IMAGES pattern suggests caching images that appear at multiple places in a Web application, and thereby aids performance.

To improve usability, there should be a provision in the information design to support internal locus of control (thereby provide options to a user) and for users to recover, say, from inadvertent errors. The MULTI-LEVEL UNDO pattern (Tidwell, 2005) provides a way to easily reverse a series of actions performed by the user.

Navigation Design

Navigation design concerned with the linear or nonlinear traversal in information space for some purpose by a user. Usually, sophisticated intra- and inter-document navigation within the context of Web application is realized by the use of hypermedia (Nelson, 1984).

Over the years, various patterns for navigating through a Web Application have been proposed

(Lyardet & Rossi, 1998; Van Duyne, Landay, & Hong, 2003). These navigation patterns, when use appropriately, aid usability. For example, the BREADCRUMBS pattern could be used to inform the user of his/her location and the FLY-OUT MENU pattern could be used to present content organized in a "compound" menu where each menu item itself has a sub-menu. The FLY-OUT MENU pattern could itself be arranged horizontally or vertically as suggested by the HORIZONTAL NAVIGATION or VERTICAL NAVIGATION patterns (Marks & Hong, 2006), respectively. The CLEAR ENTRY POINTS pattern presents only a few entry points into the interfaces, which can restrict the navigation to a specific category and make it task-oriented.

Any navigation design must take exceptional behavior into consideration to support usability. The SESSION pattern (Weiss, 2003) can help maintain the state of the Web application in the event of an interruption of navigation flow. The MISSING LINK pattern (German & Cowan, 2000) informs the user that certain hyperlink does not exist and suggests alternatives.

There are navigation design patterns that aid comprehensibility (Tidwell, 2005). For example, the WIZARD pattern leads the user through the interface step by step for carrying out tasks in a prescribed order. The RESPONSIVE DISCLOSURE pattern starts with a very minimal interface, and guides a user through a series of steps by showing more of the interface as the user completes each step. These two patterns could, for example, be used for carrying out a registration process.

Search Design

Search design is concerned with the ease, relevancy, and speed of finding information requested by a user. Searching is not native to Web applications, but has become ever more challenging as the amount of information to be searched through increases.

Over the years, various patterns for searching Web applications have been proposed (Lyardet, Rossi, & Schwabe, 1999; Van Duyne, Landay, & Hong, 2003). The searching patterns, when use appropriately, can aid performance. The use of STRAIGHTFORWARD SEARCH FORMS with a SIMPLE SEARCH INTERFACE that require minimal technical background on part of the user, SELECTABLE SEARCH SPACE that can restrict the search to a specific category, SELECTABLE KEYWORDS that based on the past experience can suggest keywords for improving subsequent search results, and ORGANIZED SEARCH RESULTS that present a summary of the most relevant search results, improve the effectiveness of the searching activity.

Examples

Figure 2 gives an abstract illustration of some of the interaction design patterns mentioned previously. The numbers indicate the order of application and the FLY-OUT MENU items are shown in dashed lines to exemplify non-permanence (as the items "hide-and-show" and are on the canvas only temporarily). Since, in general, many of the patterns (such as those that are behavioral) in the development of Web applications are not "visual" in nature, Figure 2 is *not* representative but rather presents only one view.

Figure 3 gives a concrete example of the use of some interaction design patterns in the "Products Page" at Adobe Systems. A close examination can lead to the discovery of other patterns not highlighted in the snapshot (to prevent crowding) or even to other patterns not mentioned in this chapter.

Feasibility of Selecting and Applying Patterns

The adoption and subsequent deployment of patterns needs to be viable. The challenges in the patterns' selection process stem from a variety of factors that we now discuss.

Figure 2. A collection of interaction design patterns in the development of Web applications

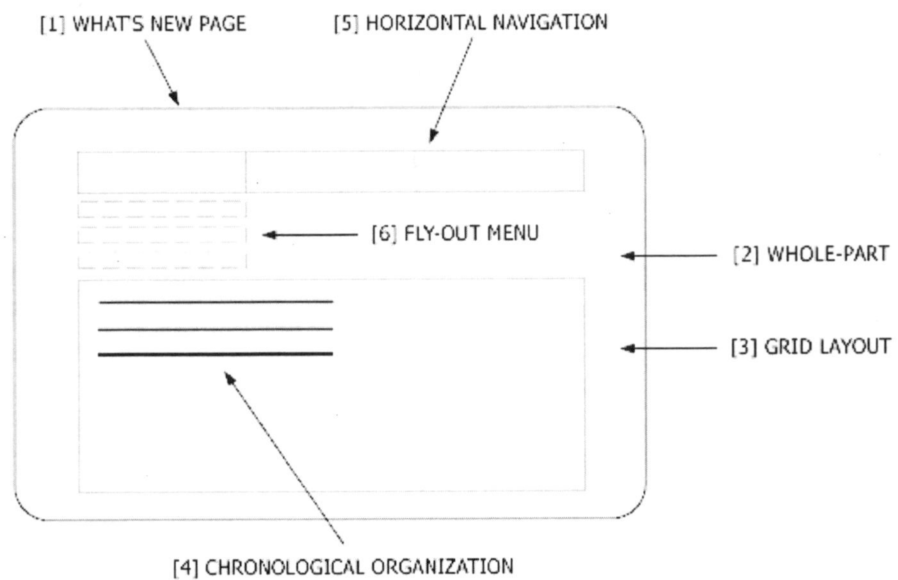

Figure 3. An assembly of some interaction design patterns in the development of Web applications

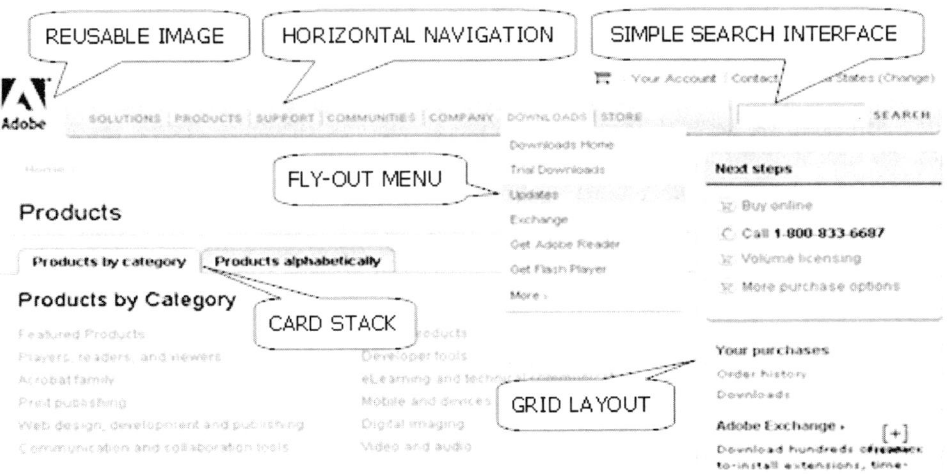

Availability of Patterns

For an adoption of a pattern-based approach to the development of Web applications, it is important that there be design and implementation patterns that can sufficiently "map" the solution space. There is no *apriori* guarantee that for every quality related problem there will be suitable pattern(s) available to solve it.

Locatability of Patterns

Even when it is ascertained that for a given problem, a pattern does exist, that pattern needs to be located. There are a few obstacles in locating desirable patterns, of which representation and classification of patterns are critical.

Representation. There is at present no unique way of representing patterns. There are a variety of notations with varying degree of formality in use for representing patterns. Furthermore, the "quality" of representing patterns varies significantly. Not all patterns make proper use of mandatory elements of a pattern, while others make seldom use of optional elements. It has been pointed out (Meszaros & Doble, 1998) that the *name* assigned to a pattern by its author is crucial. However, there are some patterns that have obtuse rather than evocative names (Meszaros & Doble, 1998) that may not be familiar to a typical reader. Also, there are patterns similar or same *names* but semantically different functionality, and patterns in different collections with similar intent or functionality but with different *names*. For example, the EASY UNDO pattern from one collection (Rossi, Lyardet, & Schwabe, 2000) is similar to the MULTI-LEVEL UNDO pattern from another collection (Tidwell, 2005). These patterns may have been (re)discovered independently.

Classification. There is currently no unique classification scheme for organizing patterns. There are some patterns that may be classified into familiar categories (like "structural" or "behavioral") while others may be presented as a linear and loosely related collection. These collections may be only print-based (say, in books), be only electronically-based (say, on the Web), but may only rarely be a combination thereof. In spite of some isolated efforts, patterns are currently also not organized by quality attributes. This could adversely impact the task of locating desirable patterns (Segerståhl & Jokela, 2006).

Cost of Selecting Patterns

One of the benefits of pattern *solutions* is conceptual reuse. However, reuse of any knowledge, including the use of patterns in the development of Web applications, is neither automatic, nor free. There is a cost in terms of time, effort, and resources of learning and adaptation involved in any reuse (Boehm et al., 2001), and patterns are no exception.

For example, there is a learning curve involved in aspects such as understanding the pattern description at large, checking if and how accurately the context of the pattern matches with that of the problem of the Web application under development at hand, and the constraints under which the solution suggested by the pattern exists. The trade-offs and consequences associated with the

Figure 4. Patterns help the engineering of large-scale Web applications, and conversely, Web applications help the distribution of patterns to it's global community of users

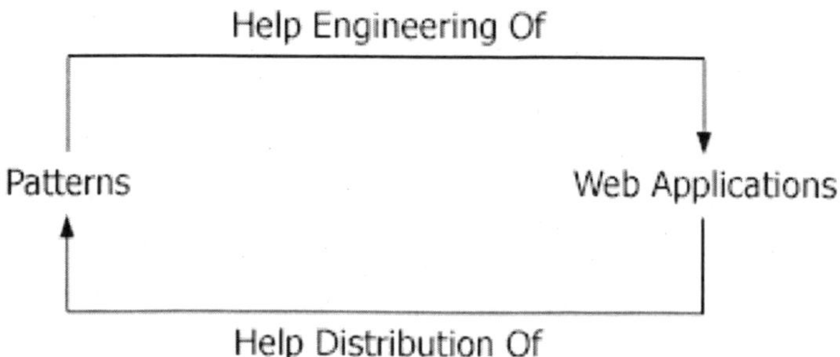

solution proposed by the pattern can help determine the suitability of the *solution* with respect to the required time, effort, and resources. It is crucial that the team responsible for the Web application must have the knowledge and skills to be able to understand the constraints of a pattern, including the *forces* that must be overcome to implement a pattern in the available technologies-of-choice.

In general, patterns can be related in many ways. We discuss two cases here. The application of a pattern may require that certain pattern(s) have already been applied (similar to a pre-condition(s)). Also, the application of a pattern places the system under development in a new *context* (similar to a post-condition(s)). Indeed, due to such *context-driven* relationships among them, the use of one pattern can necessitate the use of other pattern(s). That means, the decision of selecting patterns cannot be *singular*; it must take into account the application of patterns as a collective that is usually a directed acyclic graph (DAG).

POWEM in Software Engineering Education

POWEM at present has been put into practice in academia only. In this section, we briefly describe one experience of integrating patterns in software engineering education.

Under author's supervision, a "lightweight" variation of POWEM was followed in a software engineering course project titled *Patterns for Web, Web for Patterns (P4W4P): A Pattern Management System*. P4W4P was given in multiple undergraduate- and graduate-level courses, each spanning about four months, and was carried out by multiple teams of size 10 or so.

The basic idea of P4W4P was the following. The evolution of the Web coincides with the ascent of patterns, and the crosspollination of the two has proven fruitful. P4W4P would include a cohesive and carefully selected collection of patterns for designing Web applications and make them publicly available. This collection, namely Patterns for Web Applications (PWA), would explicitly acknowledge original sources, be navigable, searchable, and evolvable. Users of P4W4P would have context-dependent help and opportunities for feedback available to them at all times. P4W4P would also provide maintenance facilities such as modifying the information of an existing pattern and adding/deleting a pattern. The access to these administrative services would be restricted. Moreover, P4W4P would apply PWA to itself. In other words, P4W4P would help PWA, and in turn, PWA would help P4W4P. Figure 4 illustrates this symbiosis.

An initial, albeit informal and elementary, feasibility study for P4W4P was performed. The development of P4W4P followed an instance of UP with minor variations such as more emphasis on NFRs and on interaction design, and less on lengthy documentation. Based on instructor's formative assessment, some of these deliverables were revisited and revised for compliance with future submissions. Many of the patterns mentioned previously were included in PWA and used during the development of P4W4P.

Learning Implications

The reactions and learning outcomes of students at the end of P4W4P were mixed. In general, the students found the P4W4P experience to be worthwhile. However, they also faced and overcame various challenges. The students initially found the notion of a pattern rather abstract and hard to comprehend, particularly when it was introduced in a domain-independent fashion, but stimulating once exposed to real-world examples. They found the selection from a large number of patterns and application of seemingly similar patterns available on certain topics to be, at times, overwhelming. On the other hand, this made them appreciate the necessity of a well-designed navigation and precise searching in P4W4P. The students found that, in some cases, the descriptions of certain patterns were inadequate (such as absence of some mandatory elements) in order for them to make an informed decision about their selection. In retrospect, this also helped them "rediscover" and include certain key aspects such as the *context* of pattern when creating the pattern base PWA.

For patterns to continue being useful as entities of knowledge (May & Taylor, 2003), they must be adequately described and documented, readily available and findable, and evolve with the needs of the Web application. In other words, the efforts towards improving the quality of Web applications will inevitably depend on the quality of patterns themselves. In retrospect, this makes early initiatives like P4W4P only natural.

POWEM in Perspective

It is obvious that POWEM is not absolute in its realization. Indeed, the discussion on the feasibility of each of the steps in POWEM places constraints and invariably determines the scope of POWEM.

POWEM also does not come with an explicit guarantee of an improvement in quality. It is expected that engineers must be trained to make appropriate use of patterns, which often comes with experience. For example, not being able to understand the *context* in which a *problem* occurs can lead to a misapplication of the *solution* of the pattern and hence a non-optimal or even undesirable result.

It is known that to improve is to be able to measure. However, there is currently a lack of established metrics to quantify an improvement in quality via the use of patterns for any class of applications.

FUTURE TRENDS

The work presented in this chapter can evolve in a few different directions, which we now briefly discuss.

Potential Directions of POWEM Evolution

The evolution of POWEM is motivated (but not necessarily mandated) by the directions of evolution of the Web. POWEM can be extended in at least two orthogonal directions.

One possible extension of the model presented in Table 1 is increasing the granularity of the quality attributes at the level Pragmatic-Tier 1, and thereby adding another level (say, Tier 0) underneath it. In that case, for example, fault tolerance and availability could be two quality attributes that belong to the level Pragmatic-Tier 0.

Other possible extensions of the model are the use of patterns for improving higher- or lower-level semiotic quality concerns.

Patterns in Social Web Engineering

In recent years, the pseudonym Web 2.0 (O'Reilly, 2005) has been used to describe the apparent "humanization" and "socialization" of the Web as it moves towards becoming a means of participation and collaboration. An elaboration of the social level of Table 1 could help towards accommodating applications for Web 2.0. For example, one extension of interest would be addressing the social quality concerns, namely credibility, legality, privacy, and security.

A mature collection of patterns for addressing credibility concerns pertaining to Web applications (Kamthan, 2007) are yet be discovered. However, patterns for legality (specifically, licensing) (Kaminski & Perry, 2007; Perry & Kaminski, 2005), patterns for privacy (Hafiz, 2006; Romanosky et al., 2006; Van Duyne, Landay, & Hong, 2003; Weiss, 2003) and security (Schumacher et al., 2006; Van Duyne, Landay, & Hong, 2003; Weiss, 2003) have been proposed recently.

Since steps taken towards the aforementioned social quality concerns are not always favorable to usability, care would need to be taken in selecting and applying these patterns in the development of Web applications. Furthermore, such extensions would also require that the aspects of micro-architecture design to which the patterns are applied, are essentially different than those dealt with in this chapter. For example, in case of applying patterns for privacy and security, the attention would be more on the server-side rather than on client-side components.

Patterns in Semantic Web Engineering

The original goal of the Web was the invention of a medium for both human- and machine-consumption of information. This vision, particularly the latter, however, has not been completely realized yet.

The Semantic Web has recently emerged as an extension of the current Web that adds technological infrastructure for better knowledge representation, interpretation, and reasoning (Hendler, Lassila, & Berners-Lee, 2001). It thus makes information far more amenable to machine-consumption than that is possible within the current Web.

An ontology is an explicit formal specification of a conceptualization that consists of a set of concepts in a domain and relations among them (Gruber, 1993). Ontologies provide precise means of representing knowledge and, arguably, form one of the most important layers in the Semantic Web infrastructure. By enabling better (perhaps even "intelligent") opportunities for organization and inferencing from given information (via navigation or searching), ontologies can play a crucial role in Web Engineering of the future (Kamthan & Pai, 2006).

In other words, pattern-oriented Semantic Web Engineering, when practiced well, could spawn a new generation of Web applications that are developed by taking past, present, and future into consideration. An elaboration of the semantic level of Table 1 could help towards accommodating applications for the Semantic Web. For that to come to a realization, however, the efforts towards a systematic approach towards addressing the quality of ontologies for the Semantic Web (Burton-Jones et al., 2005) and patterns for developing ontologies (Gangemi, 2005) will be crucial.

For the sustainability of the architecture of the Web and for the sake of mutual benefit, it is essential that the extensions of the Web evolve harmonically. Therefore, the Semantic Web initiative and the 'Social Web' efforts such as Web 2.0 need to co-exist (Lassila & Hendler, 2007; Shadbolt, Hall, & Berners-Lee, 2006). This will inevitably be crucial for the evolution of POWEM and for the newer generation of Web applications.

Anti-Patterns for Quality of Web Applications

An *anti-pattern* (Brown et al., 1998) is defined as a combination of two aspects. First, for a recurring problem in a given context, there is a solution such that if this solution is applied, the system is placed in a state worse than it was previously. Second, to ameliorate this, the "negative" solution is refactored to a "positive" solution. If a pattern reflects a "best practice," then an anti-pattern reflects a "lesson learned" (Appleton, 1997).

A pattern applied to an inappropriate *context* can compromise the benefit it offers and give the appearance of an anti-pattern. For example, patterns that suggest solutions involving the use of color will not be applicable in situations where the underlying monitor does not support it (such as when the screen is monochrome) or the user is color blind.

There is currently an apparent lack of stable and mature collection of anti-patterns for Web applications. However, there is much negative guidance available such as for establishing trust (Sillence et al., 2006) that could be captured and structured as anti-patterns for Web applications. This will inevitably strengthen POWEM.

Patterns for Mobile Web Engineering

The use of mobile devices continues to increase and the future of mobile applications appears promising. A pattern-oriented systematic approach for a development of mobile applications would be of interest. There have been initiatives for addressing the quality (Spriestersbach & Springer, 2004) and patterns proposed for the development of mobile applications (Ahlgren & Markkula, 2005; Ballard, 2007; Roth, 2002).

However, 'recasting' POWEM for mobile applications would neither be trivial, nor automatic. The constrained environment of mobile applications is uniquely different from that of the stationary (desktop) applications (Ocampo et al., 2003), and must to be taken into consideration.

CONCLUSION

The society has increasingly become dependent upon the services provided by Web applications. If the surveys are any indication, this trend is likely to continue. However, the services provided by a Web application may not be unique: for a producer, this means competition; for a consumer, this means choice. Therefore, for a Web application, meeting stakeholder expectations of "high" quality is not only a reflection of the ethical and moral obligation of its products but an imperative for its survivability.

The demands and expectations from users of Web applications over the last decade have evidently had an impact on how these systems have been developed. Indeed, the development environment of Web applications is constantly facing technological and social challenges posed by new implementation languages, variations in user agents, demands for new services, and user classes from different cultural backgrounds, age groups, and capabilities. This motivates the need for a methodical approach towards the development life cycle and maintenance of "high-quality" Web applications.

In conclusion, patterns provide the rigor necessary for Web Engineering to mature as an "engineering" discipline. They provide a practical and preventative (rather than curative) means to addressing the quality of Web applications, if they are located, adopted and applied with care, taking feasibility issues into consideration. By adopting and investing in a user- and quality-centric and pattern-oriented approach as manifested in POWEM, all stakeholders can benefit in the long-term.

ACKNOWLEDGMENT

The author would like to thank his students for their dedicated participation in P4W4P, and the reviewers for their feedback and suggestions for improvement.

REFERENCES

Ahlgren, R., & Markkula, J. (2005, June 13-15). *Design patterns and organisational memory in mobile application development.* Paper presented at the Sixth International Conference on Product Focused Software Process Improvement (PROFES '05), Oulu, Finland.

Ahluwalia, K. S., & Jain, A. (2006, October 21-23). *High availability design patterns.* Paper presented at the 13th Conference on Pattern Languages of Programs (PLoP '06), Portland, USA.

Ballard, B. (2007). *Designing the mobile user experience.* John Wiley and Sons.

Beck, K., & Andres, C. (2005). *Extreme programming explained: Embrace change* (2nd Edition). Addison-Wesley.

Beck, K., Crocker, R., Meszaros, G., Vlissides, J., Coplien, J. O., Dominick, L., & Paulisch, F.

(1996, March 25-29)). *Industrial experience with design patterns.* Paper presented at the 18th International Conference on Software Engineering (ICSE 1996), Berlin, Germany.

Boehm, B. W., Abts, C., Brown, A. W., Chulani, S., Clark, B. K., Horowitz, E., Madachy, R., Reifer, D., & Steece, B. (2001). *Software cost estimation with COCOMO II.* Prentice Hall.

Boehm, B., & Turner, R. (2004). *Balancing agility and discipline: A guide for the perplexed.* Addison Wesley.

Brajnik, G. (2001, June 4-6). *Towards valid quality models for Web sites.* Paper presented at the Seventh Conference on Human Factors and the Web (HFWeb '01), Madison, USA.

Brown, W. J., Malveau, R. C., McCormick, H. W., & Mowbray, T. J. (1998). *AntiPatterns: Refactoring software, architectures, and projects in crisis.* John Wiley and Sons.

Burton-Jones, A., Storey, V. C., Sugumaran, V., & Ahluwalia, P. (2005). A semiotic metrics suite for assessing the quality of ontologies. *Data and Knowledge Engineering*, *55*(1), 84-102.

Buschmann, F., Meunier, R., Rohnert, H., Sommerlad, P., & Stal, M. (1996). *Pattern oriented software architecture, volume 1: A system of patterns.* John Wiley and Sons.

Coda, F., Ghezzi, C., Vigna, G., & Garzotto, F. (1998, April 16-18). *Towards a software engineering approach to Web site development.* Paper presented at the Ninth International Workshop on Software Specification and Design (IWSSD-9), Ise-shima, Japan.

Dustin, E., Rashka, J., & McDiarmid, D. (2001). *Quality Web systems: Performance, security, and usability.* Addison-Wesley.

Fenton, N. E., & Pfleeger, S. L. (1997). *Software metrics: A rigorous & practical approach.* International Thomson Computer Press.

Firesmith, D., & Henderson-Sellers, B. (2002). *The OPEN process framework: An Introduction.* Addison-Wesley.

Gamma, E., Helm, R., Johnson, R., & Vlissides, J. (1995). *Design patterns: Elements of reusable object-oriented software.* Addison-Wesley.

Garzotto, F., Paolini, P., Bolchini, D., & Valenti, S. (1999, November 15-18). *Modeling-by-patterns of Web applications.* Paper presented at the International Workshop on the World Wide Web and Conceptual Modeling (WWWCM '99), Paris, France.

German, D. M., & Cowan, D. D. (2000, January 4-7). *Towards a unified catalog of hypermedia design patterns.* Paper presented at the 33rd Hawaii International Conference on System Sciences (HICSS '00), Maui, USA.

Gilb, T. (1988). *Principles of software engineering management.* Addison-Wesley.

Gillenson, M., Sherrell, D. L., & Chen, L. (2000). A Taxonomy of Web site traversal patterns and structures. *Communications of the AIS*, *3*(4), 2000.

Ginige, A., & Murugesan, S. (2001). Web engineering: An introduction. *IEEE Multimedia*, *8*(1), 14-18.

Graham, I. (2003). *A pattern language for Web usability.* Addison-Wesley.

Gruber, T. R. (1993). Toward principles for the design of ontologies used for knowledge sharing. In N. Guarino & R. Poli (Eds.) *Formal ontology in conceptual analysis and knowledge representation.* Kluwer Academic Publishers.

Hafiz, M. (2006, October 21-23). *A Collection of privacy design patterns.* Paper presented at the 13th Conference on Pattern Languages of Programs (PLoP '06), Portland, USA.

Hasan, L. R., & Abuelrub, E. (2006, June 19-21). *Criteria for evaluating quality of Websites.*

Paper presented at the Sixth IBIMA Conference on Managing Information in Digital Economy, Bonn, Germany.

Henderson-Sellers, B., Lowe, D., & Haire, B. (2002). OPEN process support for Web development. *Annals of Software Engineering, 13*(1-4), 163-201.

Hendler, J., Lassila, O., & Berners-Lee, T. (2001). The Semantic Web. *Scientific American, 284*(5), 34-43.

Highsmith, J. (2002). *Agile software development ecosystems.* Addison-Wesley.

Jacobs, I., & Walsh, N. (2004). *Architecture of the World Wide Web, volume one.* W3C Recommendation. World Wide Web Consortium (W3C).

Jacobson, I., Booch, G., & Rumbaugh, J. (1999). *The unified software development process.* Addison-Wesley.

Kaminski, H., & Perry, M. (2007, May 27-30). *Open source software licensing patterns.* Paper presented at the Sixth Latin American Conference on Pattern Languages of Programming (SugarLoafPLoP '07), Porto de Galinhas, Brazil.

Kamthan, P. (2007). Towards a systematic approach for the credibility of human-centric Web applications. *Journal of Web Engineering, 6*(2), 99-120.

Kamthan, P. (2008). Patterns for improving the pragmatic quality of Web information systems. In C. Calero, M. Á. Moraga, & M. Piattini (Eds.), *Handbook of research on Web information systems quality.* Hershey, PA: Idea Group Publishing.

Kamthan, P., & Pai, H.-I. (2006, May 21-24). *Semantic Web-enabled Web engineering: The case of patterns.* Paper presented at the Seventeenth Annual Information Resources Management Association International Conference (IRMA ;06), Washington, D.C.

Kappel, G., Pröll, B., Reich, S., & Retschitzegger, W. (2006). *Web engineering.* John Wiley and Sons.

Kendall, E. A. (1998). Utilizing patterns and pattern languages in education. *Annals of* Software *Engineering, 6*(1-4), 281-294.

Kruchten, P. (2004). *The rational unified process: An introduction* (Third Edition). Addison-Wesley.

Kumar, K., & Welke, R. J. (1992). Methodology engineering: A proposal for situation-specific methodology construction. In W. W. Cotterman & J. A. Senn (Eds.), *Challenges and strategies for research in systems development,* (pp, 257-269). John Wiley and Sons.

Lassila, O., & Hendler, J. (2007). Embracing "Web 3.0". *IEEE Internet Computing, 11*(3), 90-93.

Lindland, O. I., Sindre, G., & Sølvberg, A. (1994). Understanding quality in conceptual modeling. *IEEE Software, 11*(2), 42-49.

Lyardet, F., & Rossi, G. (1998, August 11-14). *Patterns for designing navigable information spaces.* Paper presented at the Fifth Conference on Pattern Languages of Programs (PLoP '98), Monticello, USA.

Lyardet, F., Rossi, G., & Schwabe, D. (1999, July 8-10)). *Patterns for adding search capabilities to Web information systems.* Paper presented in the Fourth European Conference on Pattern Languages of Programming and Computing (EuroPLoP 1999), Irsee, Germany.

Mahemoff, M. (2006). *Ajax design patterns.* O'Reilly Media.

Manolescu, D., & Kunzle, A. (2001, September 11-15). *Several patterns for eBusiness applications.* Paper presented at the Eighth Conference on Pattern Languages of Programs (PLoP '01), Monticello, USA.

May, D., & Taylor, P. (2003). Knowledge Management with Patterns. *Communications of the ACM, 46*(7), 94-99.

McPhail, J. C., & Deugo, D. (2001, June 4-7). *Deciding on a pattern.* Paper presented at the 14th International Conference on Industrial and Engineering Applications of Artificial Intelligence and Expert Systems (IEA/AIE 2001). Budapest, Hungary.

Mendes, E., & Mosley, N. (2006). *Web engineering.* Springer-Verlag.

Meszaros, G., & Doble, J. (1998). A pattern language for pattern writing. In R. C. Martin, D. Riehle, & F. Buschmann (Eds.). *Pattern languages of program design 3.* Addison-Wesley, (pp 529-574).

Mich, L., Franch, M., & Gaio, L. (2003). Evaluating and designing Web Site quality. *IEEE Multimedia. 10*(1), 34-43.

Montero, F., López-Jaquero, V., & Molina, J. P. (2003, September 1-2). *Improving e-Shops environments by using usability patterns.* Paper presented at the Second Workshop on Software and Usability Cross-Pollination, Zürich, Switzerland.

Montero, F., Lozano, M., & González, P. (2002, August 5-7). *Designing Web sites by using design patterns.* Paper presented at the Second Latin American Conference on Pattern Languages of Programming (SugarLoafPLoP '02), Rio de Janeiro, Brazil.

Nelson, T. H. (1984). *Literary machines.* Mindful Press.

Nguyen, H. Q., Johnson, R., & Hackett, M. (2003). *Testing applications on the Web: Test planning for mobile and Internet-based systems* (2nd Edition). John Wiley and Sons.

Ocampo, A., Boggio, D., Münch, J., & Palladino, G. (2003). Towards a reference process for developing wireless Internet services. *IEEE Transactions on Software Engineering, 29*(12), 1122-1134.

Offutt, J. (2002). Quality attributes of Web software applications. *IEEE Software, 19*(2), 25-32.

Olsina, L., & Rossi, G. (2002). Measuring Web application quality with WebQEM. *IEEE Multimedia, 9*(4), 20-29.

O'Reilly, T. (2005). *What is Web 2.0: Design patterns and business models for the next generation of software.* O'Reilly Network, September 30, 2005.

Paulk, M. C., Weber, C. V., Curtis, B., & Chrissis, M. B. (1995). *The capability maturity model: Guidelines for improving the software process.* Addison-Wesley.

Perry, M., & Kaminski, H. (2005, July 6-19). *A pattern language of software licensing.* Paper presented at the Tenth European Conference on Pattern Languages of Programs (EuroPloP '05), Irsee, Germany.

Pertet, S. M., & Narasimhan, P. (2005). *Causes of failure in Web applications.* PDL Technical Report PDL-CMU-05-109. Carnegie Mellon University, Pittsburgh, USA.

Powell, T. A., Jones, D. L., & Cutts, D. C. (1998). *Web site engineering.* Prentice-Hall.

Perzel, K., & Kane, D. (1999, August 15-18). *Usability patterns for applications on the World Wide Web.* Paper presented at the Sixth Conference on Pattern Languages of Programs (PLoP '99), Monticello, USA.

Preece, J., Rogers, Y., & Sharp, H. (2002). *Interaction design: Beyond human-computer interaction.* John Wiley and Sons.

Romanosky, S., Acquisti, A., Hong, J., Cranor, L. F., & Friedman, B. (2006, October 21-23). *Privacy patterns for online interactions.* Paper presented at the 13th Conference on Pattern Languages of Programs (PLoP '06), Portland, USA.

Ross, M. (2002). Quality in Web design for visually impaired users. *Software Quality Journal, 10*(4), 285-298.

Rossi, G., & Koch, N. (2002, July 3-7). *Patterns for adaptive Web applications.* Paper presented at the Seventh European Conference on Pattern Languages of Programs (EuroPLoP '02). Irsee, Germany.

Rossi, G., Lyardet, F. D., & Schwabe, D. (1999). Developing hypermedia applications with methods and patterns. *ACM Computing Surveys. 31*(4es).

Rossi, G., Lyardet, F., & Schwabe, D. (2000, July 5-9). *Patterns for E-commerce applications.* Paper presented at the Fifth European Conference on Pattern Languages of Programs (EuroPLoP 2000), Irsee, Germany.

Rossi, G., Pastor, O., Schwabe, D., & Olsina, L. (2008). *Web engineering: Modelling and implementing Web applications.* Springer-Verlag.

Rossi, G., Schwabe, D., & Lyardet, F. (1999, May 11-14). *Improving Web information systems with navigational patterns.* Paper presented at the Eighth International World Wide Web Conference (WWW8), Toronto, Canada.

Roth, J. (2002). Patterns of mobile interaction. *Personal and Ubiquitous Computing. 6*(4), 282-289.

Schmidt, D. C., Stal, M., Rohnert, H., & Buschmann, F. (2000). *Pattern-oriented software architecture, Volume 2: Patterns for concurrent and networked objects.* John Wiley and Sons.

Schumacher, M., Fernandez-Buglioni, E., Hybertson, D., Buschmann, F., & Sommerlad, P. (2006). *Security patterns: Integrating security and systems engineering.* John Wiley and Sons.

Segerståhl, K., & Jokela, T. (2006, April 22-27). *Usability of interaction patterns.* Paper presented at the CHI 2006 Conference on Human Factors in Computing Systems, Montréal, Canada.

Selmi, S. S., Kraïem, N., & Ghézala, H. H. B. (2005, July 27-29). *Toward a comprehension view of Web engineering.* Paper presented at the Fifth International Conference on Web Engineering (ICWE '05), Sydney, Australia.

Shadbolt, N., Hall, W., & Berners-Lee, T. (2006). The Semantic Web revisited. *IEEE Intelligent Systems, 21*(3), 96-101.

Sillence, E., Briggs, P., Harris, P, & Fishwick, L. (2006). A framework for understanding trust factors in Web-based health advice. *International Journal of Human-Computer Studies, 64,* 697-713.

Simon, H. (1996). *The Sciences of the Artificial* (3rd Edition). The MIT Press.

Spriestersbach, A., & Springer, T. (2004). Quality attributes in Mobile Web Application Development. In F. Bomarius & H. Iida (Eds.). *Product Focused Software Process Improvement.* (pp. 120-130).Springer-Verlag.

Stamper, R. (1992, October 5-8). *Signs, Organizations, Norms and Information Systems.* Paper presented at the Third Australian Conference on Information Systems, Wollongong, Australia.

Tate, B. A., & Hibbs, C. (2006). *Ruby on Rails: Up and Running.* O'Reilly Media.

Tidwell, J. (2005). *Designing interfaces: Patterns for effective interaction design.* O'Reilly Media.

Van Duyne, D. K., Landay, J., & Hong, J. I. (2003). *The design of sites: Patterns, principles, and processes for crafting a customer-centered Web experience.* Addison-Wesley.

Vijayaraghavan, G. V. (2003). *A taxonomy of E-commerce risks and failures.* Master's Thesis. Florida Institute of Technology. Melbourne, USA.

Wallace, D., Raggett, I., & Aufgang, J. (2002). *Extreme programming for Web projects.* Addison-Wesley.

Weiss, M. (2003, September 8-12). *Patterns for Web applications.* Paper presented at the Tenth Conference on Pattern Languages of Programs (PLoP '03), Urbana, USA.

Wentzlaff, I., & Specker, M. (2006, July 10). *Pattern based development of user friendly Web applications.* Paper presented at workshop on Model-Driven Web Engineering (MDWE '06), Palo Alto, CA, USA..

Wesson, J., & Cowley, L. (2003, September 1-2). *Designing with patterns: Possibilities and pitfalls.* Paper presented at the Second Workshop on Software and Usability Cross-Pollination, Zürich, Switzerland.

Yoder, J., & Barcalow, J. (1997, September 3-5). *Architectural patterns for enabling application security.* Paper presented at the Fourth Conference on Pattern Languages of Programs (PLoP 1997), Monticello, USA.

Ziemer, S., & Stålhane, T. (2004, July 27). *The use of trade-offs in the development of Web applications.* Paper presented at the International Workshop on Web Quality (WQ 2004). Munich, Germany.

APPENDIX

Table 2 summarizes the patterns mentioned in this chapter. It clearly reflects that the mapping between patterns and quality attributes is many-to-many. The following rating scheme is used: a (+) label adjacent a pattern name reflects a positive impact on the corresponding quality attribute, whereas a (−) label reflects a negative impact. The list of patterns is subject to evolution. (The purpose of the chapter was not to provide a definitive list of patterns or point out every single pattern for a given quality attribute.) The rating scheme can also evolve to become more granular.

Table 2. Pragmatic quality attributes of a Web application and corresponding patterns with ratings

Pragmatic Quality Attribute	Pattern(s)
Comprehensibility	ALPHABETICAL ORGANIZATION (+) CHRONOLOGICAL ORGANIZATION (+) RESPONSIVE DISCLOSURE (+) WIZARD (+)
Maintainability	APPLICATION SERVER (+) CLIENT-SERVER (+) INTRODUCE REDUNDANCY (−) MODEL-VIEW-CONTROLLER (+) WHOLE-PART (+)
Performance	FAST-DOWNLOADING IMAGES (+) ORGANIZED SEARCH RESULTS (+) SELECTABLE KEYWORDS (+) SELECTABLE SEARCH SPACE (+) SIMPLE SEARCH INTERFACE (+) STRAIGHTFORWARD SEARCH FORMS (+) WHOLE-PART (−)
Readability	FLY-OUT MENU (+) GRID LAYOUT (+) HORIZONTAL NAVIGATION (+) VERTICAL NAVIGATION (+) ROW STRIPING (+)
Reliability	FAILURE NOTIFICATION (+) INTRODUCE REDUNDANCY (+)
Usability	BREADCRUMBS (+) CARD STACK (+) CLEAR ENTRY POINTS (+) CLOSABLE PANELS (+) FAST-DOWNLOADING IMAGES (+) MISSING LINK (+) MIRRORWORLD (+) MULTI-LEVEL UNDO (+) SESSION (+) TWO-PANEL SELECTOR (+) WHAT'S NEW PAGE (+)

Chapter VII
Component-Based Deployment for Web Applications:
Experiences with Duct Tape and Glue

Kevin Gary
Arizona State University, USA

Harry Koehnemann
Arizona State University, USA

ABSTRACT

The software engineering community touts component-based software engineering as a solution for many of its woes including reducing cycle time, reducing costs, increasing productivity, allowing easier integration to name just a few. Indeed, many Web-based systems are now built with open-source and vendor provided component technologies. While these advances have led to improvements in the development process, they have also led to a great deal of pressure on downstream processes as these systems must be deployed, tuned, and supported. The complexities in deploying and supporting component-based software for distributed and Web-based applications are not understood in the academic or professional communities. This chapter stresses the need for addressing this problem by presenting component-based software for Web applications from a deployment perspective, characterizing the issues through real-world experiences with highly component-based applications, and presents strategies and directions for the community to pursue.

INTRODUCTION AND MOTIVATION

Discussion of the engineering of modern Web-based systems inevitably focuses on many of today's popular buzzwords such as services-oriented architecture (SOA), n-tier architecture patterns, lightweight and Agile methodologies, AJAX, and so on. These discussions are deserved, and show the rapid pace at which the understanding of engineering such applications has evolved. These technologies are realized through a proliferation of methods and frameworks giving architects literally dozens of tools to construct applications to deploy on the Web.

However, this chapter considers a topic that has not received fair attention or debate. That topic is the downstream issues and costs involved in deploying (and redeploying) component-based software for distributed and Web-based applications. The main body of attention in Web-based software engineering is from point of view of the architects and developers, with ample debate on the merits of methods, patterns, and frameworks. While development teams tend to fall in love with these technologies, adequate consideration is rarely given to the ramifications of supporting them once the application is "thrown over the wall" to the application support team.

Component-based software engineering (CBSE) is a relatively new driver in Web-based software engineering and has enjoyed rapid acceptance in both industry and the research community. Spurred by the advent of distributed computing via the Internet, academic institutions and professional communities alike now teach, train, and mentor in the well-founded principles of object-oriented design and programming, and extend this model to coarser-grained component-based computing. Many (if not most) Web developers today espouse the benefits of component-based computing including reusability, flexibility, loose coupling, modularity, separation of concerns, and so forth, without necessarily understanding how component-based computing delivers on these promises. We by no means imply component-based Web applications are "wrong;" but we do feel there are significant gaps and assumptions made about the "goodness" of these principles that fails to address significant issues that may arise, specifically in the areas of deployment and post-deployment support. CBSE, while demonstrating value-add for software development processes, is not yet proven across the full set of business processes supporting enterprise class software. CBSE may in fact imply greater complexity, and therefore higher costs for supporting deployment and release processes for Web-based systems.

More to the point, do we hear our application support personnel espousing the benefits of flexible and collaborating components as frequently as we hear from developers? Rhetorically the answer is no as the application support team's members are too busy writing scripts, trying to understand deployment descriptors, figuring out what component is in what package, discovering undocumented hidden dependencies in components by tracing through stack traces, and finding a way to get the entire system talking in a particular Web application server production environment.

The application support team, which includes assemblers, deployers, and operators, cannot remain the forgotten partner for long. Configuration management, release management, testing, deployment, and other business functions will continue to raise issues related to CBSE:

- **Configuration management (CM):** The componentization of the software system means more possible permutations of components (configurations) to manage and support. For systems of any size there can be an explosion in permutations as development decompose systems into components.
- **Release management (RM):** Closely related to CM, RM is also concerned with coordinating all business processes related to getting software out the door, as well as how to provide updates, patches and migration to new releases. Given the potentially dynamic and "plug-n-play" nature of CBSE, this becomes a high-risk area to organizations supporting Web-based applications.
- **Software test:** The proliferation of components coupled with dynamic composition creates a tremendous burden on test groups to validate that the software system will remain robust in the face of various component configurations. When can the group be comfortable that a sufficient coverage of

deployable configurations is stable? Note testing activities are separated from Quality Assurance processes, as one may claim that CBSE has a greater potential for delivering quality software out of development, and that point is not debated.

- **Deployment:** This process packages software and physically installs it onto appropriate hardware resources and then configures the system under the constraints presented in the hosted environment. The explosion of distributed and component-based software systems increases system complexity dramatically, and is a focal point of the ideas presented in this chapter.
- **Application support:** Post-deployment processes for supporting deployed production-level software systems. These processes include production system monitoring, post-deployment application tuning, execution failover/rollover processes, and deploying upgrades and patches.
- **Other business issues:** There are also a host of organizational and social issues involved in deploying component-based systems. Customers often must be educated on new ways of receiving software and support. Licensing terms must be stated that define legal concerns, intellectual property, and cost issues for systems composed of components obtained from a wide variety of sources. Application support personnel need to be trained with new skill sets for deploying and maintaining component-based software systems. Strategic planners (executive decision-makers) and tacticians (project managers) must understand the ramifications of CBSE.

These issues are not new to the software engineering community; they are in fact why a large body of software lifecycle and software process literature exists. However, these issues are re-emerging in the CBSE community and will continue to grow due to several influences:

- The availability of off-the-shelf architectures and frameworks, both commercial and open source;
- Design techniques like aspect oriented programming and model driven architecture that continue to separate system functionality into more refined modules;
- The growing popularity of open source computing;
- Innovations in training and education;
- The impact of the Internet and distributed computing;
- Software economic pressures such as an increase in outsourcing, elevated risks (and costs) of failure, and the dot-com bust force software organizations to reconsider reuse and component assembly.

The separation between development and deployment has all but become institutionalized. Development teams and deployment teams meet on organizational charts at a relatively high level. Lines of communication are informal where deployment teams may borrow developers to help resolve their issues. Product vendors for software development also contribute to the separation by aligning tools along the developer and deployer/operator line. For example, developer tools (e.g., IBM-Rational Application Developer) and the deployment operation products (e.g., IBM WebSphere, IBM Tivoli) may not developed by the same organizations. While they are "integrated," the products are created, marketed, and sold separately.

CHARACTERIZING THE COMPONENT PROBLEM

The term software component has a broad meaning in the software community. However, most definitions share some common themes—"encapsulates functionality," "communicates through well-defined interfaces," "provides a simple plug-and-play model," and "allows easy replacement of system behaviors." Few definitions consider all the forms a component will manifest itself throughout the development lifecycle. Cheesman & Daniels (2002) presented a view of components from the complete software lifecycle and discussed components in the following "forms:"

- **Component interface:** Defines a set of features offered; analogous to a CORBA or Java interface
- **Component specification:** A software unit that describes behavior for a component as a set of Component Interfaces
- **Component implementation:** An assembly that implements a Component Specification
- **Installed component:** A specific installation and configuration of a Component Implementation
- **Component object:** A run-time instance of a component with unique state and identity

Java Database Connectivity (JDBC) is a popular software package that will serve as a concrete example. The JDBC Component Specification defines several Component Interfaces (Connection, Statement, ResultSet) and has many vendor Component Implementations, including MySQL, Oracle, and so forth. Figure 1 provides a diagram illustrating the relationship of these three component forms defined by Cheesman & Daniels (2002). A deployer installs the binary representation of the Installed Component as a file mysql.jar on all appropriate nodes requiring connectivity. The Installed Component might require configuration in some property file that must also be deployed with the system. Finally, at run-time there is an instance of that component providing connectivity to the external data source.

Component Dependencies Issues

There can be many hidden dependencies with this simple example that impact a deployment. The first simple dependency is that JDBC refers to classes in the Java library, meaning the component's

Figure 1. Example component interface, specification, and implementation

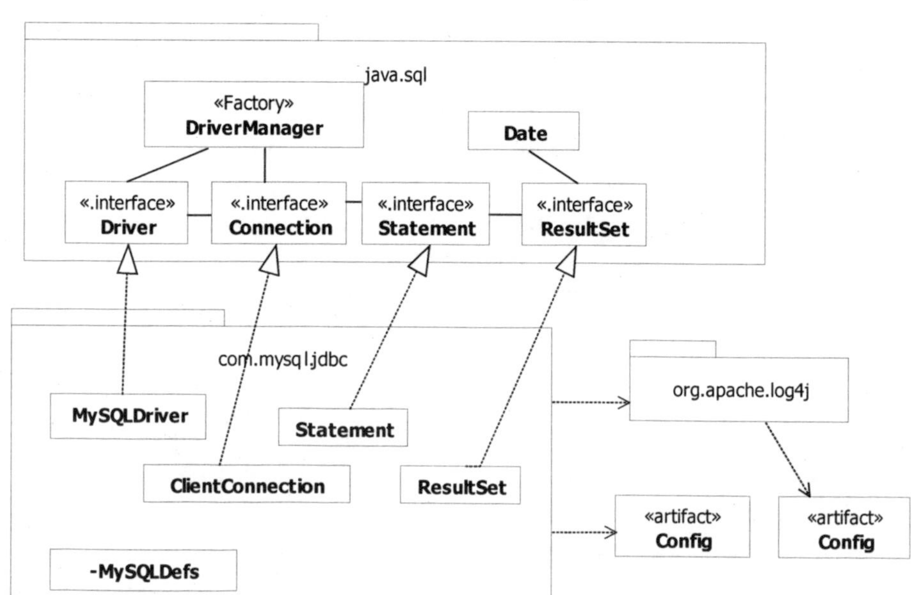

specification, and therefore its implementation, has an implicit dependency on some version of the J2SE. Also, the component implementation (MySQL in this example) obtains configuration information from an external property file so that file must exist and be properly configured at run-time. It also requires a specific version of the log4j component which itself depends on a property file for configuration. Finally, the implementation depends on a database process executing on some node and the configuration file be properly configured to reference this node.

Quality of service is also an issue. Replacing the MySQL implementation with a new one can greatly impact the execution profile for the node running the component. Processing and memory requirements vary widely across implementations.

Lastly, versioning can be an issue. While the MySQL component requires a specific version of the log4j component, a component in another part of the system may require a different version. Separating the executable space of components can be solved in some instances (Java Classloaders provide a nice facility), but the system must have been initially designed and architected to support such a feature.

Developers rarely have issues with hidden dependencies, if they have them at all. They configure their environment early and rarely need to change it. When there are changes, they experience a mild amount of pain reconfiguring. But once completed, the process and lessons learned are typically undocumented and then forgotten. When the pain repeats itself too often, developers will commonly create scripts to speed the reconfigure process and also help them remember it (e.g., homegrown scripts for connecting to different databases, such as a development database or a test database). While undocumented activities like this (a form of duct tape and glue) are effective in development, production deployments have larger and more variable environments, and this approach will not scale for the application support team.

Build-Time vs. Run-Time

The previous section described the many forms a component takes through the software lifecycle. Software components are an independently deployable and enactable portion of a software system. Deploying and enacting a component (the run-time view of a component) can lead to issues. The 4+1 software architecture model (Krutchen, 1995) espoused by the rational unified process (RUP) addresses, at least through notation, the issues of deploying and enacting components. Figure 2 shows the 4+1 model and its related views.

The Logical View provides the system's logical organization defining how functionality is parti-

Figure 2. Rational unified process 4+1 architectural views

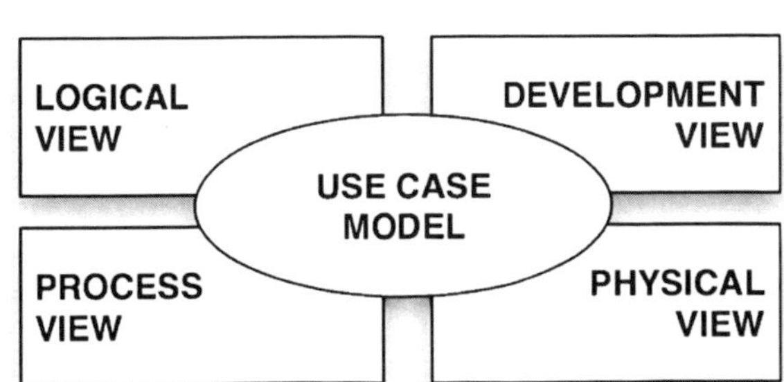

Figure 3. Deployment view of a logical component model

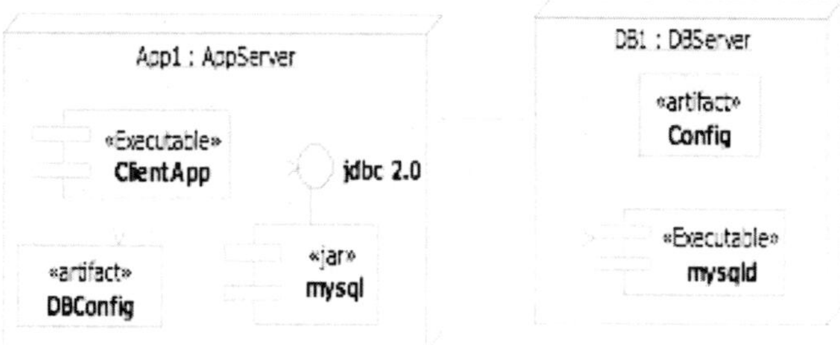

tioned into packages, subsystems, and interfaces, as well as those elements' implementations. UML package and class diagrams are commonly drawn to illustrate this view. The Implementation view describes how the conceptual logical elements map to physical artifacts (source files) and then how those physical elements are built into components (UML's definition of component) and other artifacts required for execution. Makefiles, Ant build descriptors, and other scripts commonly represent this view. The Process View depicts the processes that execute in the system and the Deployment View depicts those processes on physical nodes in which they execute.

The Logical and Implementation views (the *build*-time views) are vital to developers and are typically understood on most successful software projects. However, the Process and Deployment views (the *run*-time views) are not as well considered on most projects. In fact, many projects do not properly consider deployment aspects of the Implementation view. Figure 3 shows an example deployment view of Component Objects at execution time for the database example discussed earlier.

With the proper performance and quality characteristics, Deployment Diagrams can help deployment and operations teams make informed decisions regarding the installation and configuration of a system. Questions like, "do we need a JDBC 2.0 compliant database?" or "will a given node support the computational and memory requirements for this set of components?" or "what property files must I configure?" can be answered with a Deployment View of the system. Many researchers have advocated quality of service information and languages for CBSE. However, it is unfortunate that few of those efforts have made it into practice so deployers and operators are left with few if any characteristics to reason about their systems.

REAL WORLD SCENARIOS

This section presents experiences from actual systems that reveal the pain and cost of deploying highly configurable systems based on components.

Hidden Dependencies

Hidden dependencies are a large problem in understanding and deploying systems. Some motivational examples were presented in the previous section. In this section we elaborate on types of hidden dependencies and present real world scenarios.

Configuration Files

There are many examples of hidden dependencies in configuration files. These files allow developers, testers, deployers, operators—any user of the system—to dynamically configure it. Software components then use these values to configure and tune themselves. While a great asset for developers who configure their system for a rather static development/QA/System test environment in an otherwise relatively fixed deployment, understanding all the tuneable parameters for a large system is a complex problem. Systems place configurable items like IP addresses, filename and paths, JNDI and CORBA names, and a plethora of tuning parameters in these files that need to vary across deployments.

An operator spent days spent reconfiguring files to make a system operational. The system configuration properties worked in the development environment, but the same configuration file in the mock production environment generated many exceptions. When the code read property values, it always assumed the values would be legitimate and performed no error checking. Missing or incorrect property values caused format or null pointer exception. The operators were not Java programmers, so members of the development team spent several hours helping the application support team get the system operational. Ironically, the next release had identical problems with new property values added to the configuration files.

Old habits die hard in any organization, and institutional policies (and policing those policies through reviews), are commonly required to institute changes.

Versions of Components

While component dependencies are not well documented, dependencies between versions are even less documented. Open source tools are a particular challenge, primarily because the user community is the integration testbed. The first user to try a combination of component versions is the one that will reveal any issues.

A developer was experimenting with an open source Model Driven Architecture tool to understand how it could help his organization's software development. He had previously installed the necessary tools—Ant, JBoss, and so forth—but continued to get nondescript errors when he deployed the application. In frustration he reinstalled everything on a different computer and, to his surprise, it ran successfully. He eventually discovered the product was not compatible with his version of JBoss and that those dependencies were not documented anywhere and there was no error indicating this conflict. It was only by accident that he installed a compatible version of JBoss. Unfortunately most businesses cannot rely on luck to resolve component problems.

Deployment problems with component versions can also appear when two components in a system depend on different versions of the same component. Problems arise when the component space is cluttered with multiple versions of the same component.

In an early version of a nameless servlet engine, Web applications shared their loaded component space (classes in jar files) with all other Web applications in the engine, meaning different Web applications could not load different versions of the same component. In order to enable the deployment, operators instantiated multiple instances of the servlet engine, each with its own web server, and used router tricks to route traffic to appropriate machines.

Component version problems like this are difficult to discover in development or even QA. Developers are responsible for their application, but typically not responsible for its operation with other applications. It is not until the deployment team configures the system for a particular site that these problems are typically revealed, then resolved with a little duct tape and glue.

Another common example of versioning components is found in HTML for browsers. Developers and testers must consider the different types and different versions of browsers found in their user base. Browser variants can differ dramatically with variants in HTML, different Javascript language support, supported plugins. The problem is compounded when considering the variable WAP devices and their different form factors.

Plug-n-Play—Plug-n-Poof

One of the major benefits touted by the component community is the ability to drop in alternate implementations of a component transparently to the rest of the system. The ability to provide an alternate implementation, the argument goes, allows for greater deployment flexibility leading to easier maintenance, localization of error, "best-of-breed" economies for components, and so on. However, experience has found components can rarely be interchanged in a production system without potentially serious consequences. In this section we examine how things can go awry.

Dynamic Component Assembly

Dynamic component assembly at run-time is considered a huge win, and significant effort is being spent in the component community constructing formalisms and methodologies for assembly. The existence of well-defined interfaces and configuration and wiring "standards" supposedly allows for the "plug" in plug-n-play.

A content delivery system included a particular version of the open source XSLT processor Xalan. The next deployment included an XSLT upgrade to Caucho's Resin processor to speed rendering requests. By leveraging the TraX API, the software was able to incorporate this change transparently to developers. Some performance enhancements were mentioned in the release notes, but there were no interface changes at the programmatic level. The system built and functional unit tests passed. However, the system and scalability tests failed. Several days of analysis showed the internal implementation of the XSLT processor improved its performance by holding references to stateful internal buffers. While caching these buffers improved performance, it caused dramatic changes to the resource consumption profile for the component in terms of memory and CPU utilization.

While the idea of "plug-n-play" components is attractive, it is not enough for a component to conform to a standard interface. Deployers must understand how the run-time characteristics of alternative components affect the overall behavior of the system. Compile-time interfaces provide encapsulation, but at run-time the entire set of components deployed to a node must compete for the same set of resources. In other words, the resource consumption characteristics of components should not be encapsulated, because the resources consumed are not encapsulated, but instead shared across all of the components of the system.

Binding to Non-Standard Interfaces

Component vendors provide value-added services as a means to draw customers to use their components. These services include robust toolsets (such as IDE integration), well-written documentation, and the inclusion of unit test suites. One common form of value-add service is to include additional APIs outside the APIs defined for a component standard. Customers who use these APIs are (often unwittingly) tied to the vendor implementation, and their software will no longer port to alternative component implementations.

In an open source enterprise portal, the 2.0 implementation had a compile-time dependence on Xerces XML parsing APIs, so when an alternative parser was introduced, the system would not build. The problem was identified and rectified in the next version of the portal, but demonstrates that software development with a component-based philosophy must be built against alternative component implementations to ensure tight coupling to specific extensions has not occurred.

The preceding example is common and troublesome, but the bright side is that such issues can be addressed during development as they are compile-time issues. More serious issues occur downstream during test and deployment if alternative component implementations are provided that build with the rest of the system, but throw run-time errors when certain functionalities are not supported.

Early versions of the Postgres and JTDS SQL Server JDBC drivers threw MethodNotSupportedExceptions for unsupported JDBC 2.0 standard APIs. Although JDBC 2.0 provides methods to check for the support of various features, it is difficult to programmatically check and provide alternative implementations at a method level for each interface that is used, and there is no guarantee that even if you do so you will prevent all possible exception scenarios. So developers using these incomplete database drivers would not reveal their problem until run-time. Hopefully developer unit tests catch these problems, but some may be missed and are hopefully revealed in more formal system testing.

Vendors need to provide value-added services that may take the form of standard API extensions or partial implementation of API standards. The issue this causes for deployers is an uncertainty that alternative implementations may be introduced without rebuilding and retesting the software across all possible configurations, introducing costs instead of saving time and money.

Component Quality of Service

A final variant on plug-n-play situations occurs when the semantics of the component implementation deviate from expected semantics, yet there is no mechanism for expressing the different behavior.

A Web application used Oracle for its database server and was being ported to SQL Server. The application would occasionally deadlock when trying to perform database writes. Analysis showed the developers assumed row-level database locking when accessing objects, yet the version of SQL Server being used provided page-level locking.

Certainly, when testing against the SQL Server database such situations should be uncovered, but how will they be addressed? If they are addressed by rewriting sections of application code to avoid deadlocks, then in fact we have again violated the plug-n-play principle by forcing changes in an application component to deal with the level of service provided by the database component. Furthermore, there is no standard mechanism by which one can express this assumption; it is merely "understood" by developers and implementers alike. It is possible that Ontology-based approaches (for example, the KAON project (Oberle, Eberhart, Staab, & Volz, 2004)) might address this problem, but these solutions have not made it from the lab to the market yet.

The plug-n-play scenarios in this section are certainly preventable, but not without additional effort spent on testing various configurations that include alternative component implementations. It is also counter to a culture that grabs the latest open source implementation without stressing conformance and compliance processes.

Missed Requirements for Operations

Poor planning and requirements also lead to run-time duct tape and glue. While many requirements address the system's functionality and behavior, requirements on the support and operation of the running system are often late or incomplete if they are even considered.

Despite years in planning and development, a large, global telecommunications system did not have an automated strategy for periodically uploading required routing data to its distributed nodes. The interim solution was to manually transfer generated files to the proper nodes scattered around the world until the system had an automated method for getting the proper data onto the proper nodes. Surprisingly this task, which kept the entire communication system functioning, was completely manual. Instead of scripts, operators manually uploaded data files via ftp on a daily basis. Even more surprising, the operators did not have an appreciation for the activity's importance; that failure to upload the files by a specific time, loading the wrong file on the wrong node, or an error in the file's transmission would cause a catastrophic system failure. Fortunately the manual process was quickly replaced. After the system went live, developers were kept on at the deployment site and one of them who knew the possible ramifications recognized the risks with the current process. Scripts were created to automate the process and check the uploaded files for errors.

Situations such as these provide several important lessons. First, deployment requirements are typically not sufficiently expressed even in large complex systems with significant deployment responsibilities. Operational requirements are easy to overlook since most of the "interesting" requirements occur with the system in operation, not in keeping the system operating. Second, the cost of deployment is expensive if it requires developers on site to provide assistance. Finally, scripts are a better, more reliable, duct tape and glue than people.

BUSINESS IMPACTS

The previous section related experiences with the immaturity of deployment technology; unfortunately this immaturity dwarfs in comparison to the understanding of the full range of business impacts in this area. This is a large area in which the community has very little understanding.

Lifecycle Model Improvements

Software process models should be revisited to provide more support for back-end, post-development activities. The community needs to raise deployment to the level of a first class process phase. While most software methodologies have a *deployment* activity or phase, their major activities and discussions center on the ones close to development. The inclusion of deployment is almost a token to show completeness. For example, the rational unified process (RUP) defines workflows for requirements, analysis, design, implementation, test, and deployment. (Jacobson, Booch, & Rumbaugh, 1999) devotes an entire chapter to every workflow in the Rational Unified Process, except deployment.

In addition, the information and the respective representation advocated by process models in the deployment area also need to be revisited, particularly as they relate to components. The UML provides notation support with Component Diagrams and Deployment Diagrams, but that information is not semantically rich enough to address the problems characterized here. While work in the area of component quality of service and contracts will help, components need to provide more deployment information through metadata similar to J2EE Deployment Descriptors. Additional work in component metadata facilities (JSR-175, 2003), languages, and additional quality attributes for deployment will lead to better processes and tools.

The scripts, tools, and "hacks" described above are vital to deployment organizations and to the success of many products. However, process models do not discuss them as a formal activity. Consequently, they are typically created outside the scope of the planned product lifecycle. These activities need to be raised to first class status and considered a business asset. They should be managed projects with a formal requirements activity, regular release cycle, and supporting documentation.

Economics

Research into the cost models for component-based systems remains open. While many tout the cost benefits of CBSE, practical experiences show increased pain and expense on the backend likely not considered before now. While it would be foolish to move back to monolithic systems, those expressing the benefits of components are not considering all the costs. What does the intersection of application types, deployment scenarios, and business models tell us about the complete lifecycle costs of CBSE? Are there more factors that help determine the costs? Can some of the risk be mitigated and if so, how?

CBSE has also changed the nature of deployment and operations, requiring more sophisticated skill sets and training for those positions. Not long ago operators were strong system administrators with networking and scripting skills. Component based systems have become so complex that we now find traditionally trained software developers deploying and monitoring the health of these systems. Now we commonly see workers with M.S. degrees in Computer Science with significant development experience filling these job roles. Both are vital to their system's success. However the cost of an experienced, graduate degreed developer is significantly more than the traditional operator.

Universities need to address these gaps. There is some understanding of the skill sets required, but these need to make their way in to university curricula. Furthermore, the curricular tracks, even when present, are too separated; a more integrated form of learning experience is required. For example, students may receive a Bachelor's degree in Computer Science, Information Systems (CIS/MIS), or Information Technology, but different academic units in different colleges within the university typically deliver these. While an MIS student may be asked to take a programming course or two, this does not expose them to how scalable enterprise Web applications are constructed, nor does it truly expose them to working in teams with their future software engineering counterparts. Interdisciplinary study and integrated hands-on learning experiences must be provided in university programs.

Better Understanding of Product Scope

Lack of planning is perhaps the largest reason we experience pain and expense deploying component-based systems. While the 4+1 Architecture Model (Krutchen, 1995) advocates scenarios in all four views, few systems have formal requirements around deployment and operational scenarios. Most requirements and analysis activities focus primarily on functional requirements and secondarily on non-functional requirements. Requirements are commonly evaluated using the FURPS framework (Grady & Caswell, 1987)—functionality, usability, reliability, performance, and supportability. The supportability requirements discussed here are typically very limited, non-existent, or discovered too late in the process to properly address before deployment. Developers can only plan and design for known requirements.

The business side of a software product organization needs to understand and commit to a component strategy suitable to their market. If platform X and database Y are on the product roadmap, then developers and testers exercise those component combinations on every release. The engineering side has the obligation to communicate the cost of supporting those combinations so the business organization can decide where they belong on the product roadmap and how to charge for them. A salesperson the committing to the product running on some other platform with some other database in a short timeframe unwittingly commits the organization to a component deployment it cannot support.

The organizational commitment to a component roadmap extends to licensing issues as well. An organization must perform due diligence to ensure the acquired component may be used in the ways the provider's licensing model allows. Licensing models may provide unanticipated constraints, such as pricing by CPU versus pricing by concurrent requests versus pricing over an aggregate load on the Web application in a given timeframe (daily, monthly, etc.). These constraints may be difficult to understand, add complexity to a deployment architecture, and lead to unpleasant cost surprises, particularly in hosted environments. Open source components add another dimension to these constraints. Open source components may suffer from poor understanding of the license and specialized variants of standard licensing models. There is a significant difference in intellectual property and commercialization of software distributed under the GNU public license (GPL), lesser GPL (LGPL), and BSD style licenses. Newer dual-licensing models, such as the MySQL and PHP licenses, attempt to promote open source while at the same time protecting commercial interests, but may inadvertently make it more difficult for adopters to understand how the solutions may be repackaged. While the community is fond of the colloquial "open source" term, the proliferation of license variants today makes attempting to combine off-the-shelf components into a solution a harrowing experience.

DISCUSSION AND OPEN DIRECTIONS

Insufficient attention has been paid to component-based deployment and post-deployment processes. We are not alone in this belief. Recent symposia and workshops include tracks dedicated to these issues (including Component Deployment 2002, 2004, and 2005, ICSE Workshop on CBSE 2003, 9th IEEE Conference on Engineering of Computer-based Systems 2002). While certainly welcome, more discussion is given to architecture description languages (ADLs), dynamic component assembly, and QoS/resource optimization than to the tangible pain of application support teams. When deployment is discussed, the focus is on specific techniques in transitioning architectures into deployed systems, and not on a comprehensive discussion of the full range of issues facing application support teams.

This is not to say that there is no discussion in academic and industry research. Crnkovic (2003) presents an overview of CBSE and includes a discussion on deployment issues, though not quite at the same playing field as more evolutionary issues such as component maintenance. Van Ommering (2001) describes four aspects of "independent deployment," yet these aspects really map to an orderly description of versioning issues between components across the full component lifecycle. Weyuker (1998) also tackles a specific dimension of the component problem, noting as long as a decade ago that testing component-based systems may lead to additional complexity and cost. Hoffmann & Neubauer (2005) acknowledge that to discuss practical component-based systems today requires considering platform-specific technologies, and compares platform-independent technologies eODL, UML 2.0, and OMG's

Deployment and Configuration (DnC) configuration (2003 version). Lau & Ukis (2006) identify that too few discussions focus on binding binary components in the deployment phase.

Perhaps the most comprehensive academic study of component deployment issues is provided by (Carzaniga, Fuggetta, Hall, Heimbigner, van der Hoek, & Wolf, 1998). Although dated, this group walks the reader through a complete treatment of deployment from the process perspective and presents a range of challenges for component deployment that exist still today. They then suggest a taxonomy for framing the set of technologies related to this area. This is presented almost exclusively from a process perspective, which is a useful viewpoint, but does not provide a neat conceptual mapping for the set of related methodologies (such as ADLs, notations like UML, etc.). However, to date this is the most comprehensive treatment, and yet it is a technical report nearly a decade old!

Industry "standard" component models such as Java's J2EE model (JCP 2002) and the CORBA Component Model (CCM 2002) provide deployment configuration specifications. The J2EE component model identifies the roles of Application Component Provider, Application Assembler, and Deployer, but does not include in the model an expression of meta-information for determining how to arrange components. Deployment descriptors provide a description of a deployed environment, but are not the means to get there. An assumption is made that tools will be available to each role to facilitate the process; but again it is not clear how the tools would assist in decision making.

The CORBA Component Model specification (OMG 2002) suffers from the same gap. Component assembly and deployment on various host environments is described in the specification, including a discussion of the participants in deployment process. But a sample deployment scenario starts by assuming "The deployment application has a conversation with the user to determine where each component or collocation is to be placed" (p. 6-68). No recommendations on how such as decision is made are provided. The deployment and configuration specification (OMG 2006) addresses this gap by taking making the usual platform independent model (PIM) and platform specific model (PSM) mappings, and defining a UML Profile with specific actors for executing the process. While this abstract model is useful for tool developers, it is not directly useful for application support people.

Vendors do not necessarily facilitate emerging standards in this area as well. They are aware that open, standard deployment environments may sacrifice some value-added capabilities in their tools. Application support managers, whose purchasing power typically falls under the IT division of an organization (instead of the engineering division) will make buy or no-buy decisions based on the robustness of tools for supporting deployment and post-deployment processes. So while standards exist for developers in technology platforms such as J2EE and CORBA, the usability of tools provided by vendors varies widely and may facilitate proprietary extensions.

Unfortunately, the real answers to how scope and importance in these issues are not within our grasp today. Further research must be done, not just on specific tools and experiences, but on broader and more ambitious agendas that define the problem, the range of solutions, and the economic impact of these downstream activities on the health of web-based application and system providers. Here are some directions for the community:

- **Common process models:** RUP's 4+1 architecture model is a start, but merely implies a use case driven process where one evolves from logical to physical (or concrete) deployment models. A more explicit and comprehensive consideration of process models, such as that given in (Carzaniga,

Fuggetta, Hall, Heimbigner, van der Hoek & Wolf 1998), is needed.

- **Common terminology:** A common vocabulary for deployment is required. Too often narrowly defined problems such as versioning or component assembly are described in general deployment terminology.
- **Formal definitions of languages and models:**Liu & Smith(2006) have broken new ground with an attempt to formally characterize software deployment processes and constraints. This work is fresh in the sense is deals with deployment directly, instead of through an extension to an existing notation such as UML or an ADL. While extensions to such languages may eventually be folded in, the direct formal representation of deployment processes will help mature understanding of the field.
- **Deployment evaluation:** Extensions to architecture evaluation techniques to account for deployment— Quality attributes identifying and quantifying deployment characteristics could have the same impact on deployment as quality attributes for software architectures have had for systems design.
- **Economic impact models:** There is a need to model downstream costs versus upstream benefits.This model should incorporate cost models, licensing models, risk assessments, and organizational impacts. Data is needed to shed light on just how expensive this problem is, where the greatest pain lies, and uncover cause-and-effect between upstream design activities versus downstream deployment and post-deployment activities.The impacts on the full range of business activities must be considered.
- **Better education and training:** Universities need to respond to the challenge to teach students that a software system is not complete if it cannot be reliably deployed and re-deployed.

In this chapter we have attempted to energize the community around the practical issues of deployment of component-based systems. We believe this area is undernourished and is fertile ground for the development of models, tools, and methods for both academia and industry. We have presented motivating scenarios based on our practical experience in industry, and outlined some research issues to follow. A survey of the community shows that others are also beginning to address the issues, but much more work is needed for us to both understand the problem space and recommend solutions.

REFERENCES

Carzaniga, A., Fuggetta, A., Hall, R., Heimbigner, D., van der Hoek, A., & Wolf, A. (1998). *A characterization framework for software deployment technologies* (Tech. Rep. CU-CS-857-98). Boulder, CO: University of Colorado, Department of Computer Science.

Cheesman, J. & Daniels, J. (2002). *UML components: A simple process for specifying component-based software.* Reading, MA: Addison-Wesley.

Crnkovic, I. (2003). *Component-based software engineering – New challenges in software development.* Paper presented at the 25th International Conference on Information Technology Interfaces. Cavtat, Croatia.

Grady, R. & Caswell, D.L. (1987). *Software metrics: Establishing a company-wide program.* Upper Saddle River, NJ: Prentice-Hall.

Hoffman, A., & Neubauer, B. (2004). *Deployment and configuration of distributed systems.* Paper presented at the 4th International SDL and MSCWorkshop. Ottawa, Canada.

Jacobson, I., Booch, G. & Rumbaugh, J. (1999). *The unified software development process.* Reading MA: Addison-Wesley.

Krutchen, P. (1995). The 4+1 View Model of software architecture. *IEEE Computer, 12*(6), 42-50.

Java Community Process (2002). *JSR 88: Java EE™ application deployment.* Retrieved on January 21, 2007, from http://jcp.org/en/jsr/detail?id=88

Java Community Process (2003). *JSR 175: A metadata facility for the Java™ programming language.* Retrieved on January 21, 2007, from http://jcp.org/en/jsr/detail?id=175

Lau, K. & Ukis, V. (2006). *Defining and checking deployment contract for software components.* Paper presented at the 9th International SIGSOFT Symposium on Component-based Software Engineering *(CBSE'06).* Stockholm, Sweden.

Liu, Y. & Smith, S. (2006). *A formal framework for component deployment.* Paper presented at the 21st ACM Conference on Object-Oriented Programming, Systems, Languages, and Applications (OOPSLA'06). Portland, OR.

Oberle, D., Eberhart, A., Staab, S., & Volz, R. (2004). Developing and managing software components in an ontology-based application server. In *Proceedings of the 5th International Middleware Conference*, Toronto, Canada: Springer-Verlag.

Object Management Group (2002). *CORBA component model, version 3.0.*

Object Management Group (2006). *Deployment and configuration of component-based distributed applications specification, version 4.0.*

Van Ommering, R. (2001). *Techniques for independent deployment to build product populations.* Paper presented at the 2001 Working IEEE / IFIP Conference on Software Architecture. Amsterdam, The Netherlands.

Weyuker, E. (1998). Testing component-based software: A cautionary tale. *IEEE Software. September/October 1998.*

Chapter VIII
Evolving Web Application Architectures:
From Model 2 to Web 2

David Parsons
Massey University, New Zealand

ABSTRACT

This chapter explores how Web application software architecture has evolved from the simple beginnings of static content, through dynamic content, to adaptive content and the integrated client-server technologies of the Web 2.0. It reviews how various technologies and standards have developed in a repeating cycle of innovation, which tends to fragment the Web environment, followed by standardisation, which enables the wider reach of new technologies. It examines the impact of the Web 2.0, XML, Ajax and mobile Web clients on Web application architectures, and how server side processes can support increasingly rich, diverse and interactive clients. It provides an overview of a server-side Java-based architecture for contemporary Web applications that demonstrates some of the key concepts under discussion. By outlining the various forces that influence architectural decisions, this chapter should help developers to take advantage of the potential of innovative technologies without sacrificing the broad reach of standards based development.

INTRODUCTION

Web applications grew out of the World Wide Web in the 1990s, driven by the need, particularly within e-commerce applications, for dynamic content. Early Web content, limited to static hypertext markup language (HTML) pages, could not support on-line sales, transactions, personalisation, or any of the other features of the Web that we now take for granted. Subsequently, technologies that could build Web pages on the fly, such as common gateway interface (CGI) scripts, active server pages (ASPs) and Java Enterprise Edition components like servlets and JavaServer Pages (JSPs), transformed the landscape of the Web by introducing the Web *application*, rather than just the Web *site*.

The cycle of innovation that saw the change from static to dynamic content in the mid 1990s has continued unabated. The post 'dot com' era has seen Web application design move into a number of new areas. On the one hand we see a rapid evolution of the Web client into the mobile, cross-platform space, with mobile browser-hosted technologies such as XHTML-MP (extensible hypertext markup language—mobile profile, the standard mark-up promoted by the Open Mobile Alliance, a consortium of commercial interests in the mobile communications industry), and micro browsers like Opera Mobile that run on top of a Java Micro Edition environment. On the other hand we see a change on the server side from the 'walled garden' Web application, providing only content from a single source, to open Service-Oriented Architectures that enable disparate services from many sources to be integrated by exchanging data using the extensible markup language (XML). On both client and server we see service driven 'mashups' (re-used, intermingled services) and the programmable Web, aspects sometimes associated with the umbrella term *the Web 2.0*. These changes raise questions about the future of the Web application, for example, can we realistically consider both a rich client and a mobile application that will work across many different types of device? How do extensible markup language (XML) based technologies like Web services and XSLT (extensible stylesheet language transformations) fit in to Web application architecture? How will the Web client evolve with the increasing use of browser hosted applications and Ajax (Asynchronous JavaScript and XML)? Much of the discussion around the Web 2.0 focuses on the social networking aspects of Web based applications, but there is an equally important discussion to be had about the underlying architectures and technologies of applications in the Web 2.0 era. In this chapter we try to address some questions about how Web application architecture continues to evolve, what forces come into play, and how apparently conflicting paths of development may converge into new approaches. Illustrative examples are provided using the Java Enterprise Edition and supporting open source tools.

THE WEB FRAGMENTATION CYCLE

One of the characteristics of the evolution of the Web is what we might term the *Web fragmentation cycle*, which is the effect of technology driven change in two conflicting directions. On the one hand, we have seen that a particular Web technology, for example a particular type of browser or server side application, can drive Web evolution using features specific to that technology. An early example of this was the introduction into Netscape Navigator of HTML tags for presentation (Lie & Bos, 1999). Netscape's development of LiveScript (later JavaScript), and Microsoft's introduction of the XMLHttpRequest into Outlook Web Access (Van Eaton, 2005) can also be seen in this light. On the other hand we see proliferation of Web access tools that increase the reach of the Web across many different types of client. One example of this was the consortium of mobile phone network operators that joined in the WAP (wireless access protocol) Forum in the late 1990s to enable Web access via a wide range of mobile devices by introducing the WAP browser and the wireless markup language (WML).

The effect of technology specific innovation is to narrow the accessibility of Web content to those who have the appropriate technology. While these innovations may increase qualities such as usability and functionality for some, they will exclude those who do not have the right technology. In contrast, the effect of wider reach is to encourage generic technologies for Web access that enable more types of client to access Web-based content, while potentially decreasing usability and functionality due to the need to run on a lowest common denominator platform. The resolution to this dichotomy has traditionally been the introduction of standards on a post hoc

Figure 1. The Web fragmentation cycle

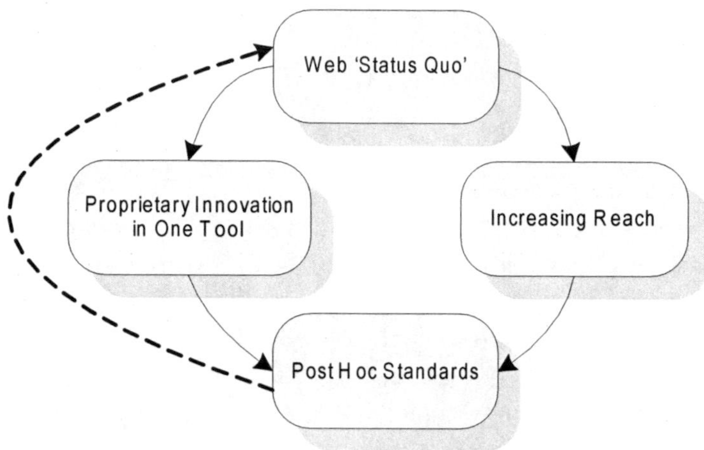

basis, which set an industry-wide benchmark that most products will ultimately comply with. These standards enable the fragmentation caused by innovation to be (at least partially and temporarily) resolved by integrating propriety innovation into a common Web technology platform (Figure 1). Once proprietary innovations have been integrated into a common standard, those technologies enjoy broader implementation and wider reach, and the Web 'status quo' moves forward.

Identifying the Web Fragmentation Cycle in Practice

Of course the fragmentation cycle shown in Figure 1 is an over-simplification, since there are many overlapping cycles related to different Web application technologies. Therefore the concept of the Web status quo is somewhat misleading, since there is never a stable point from which we embark on a new cycle. However it is possible to identify specific instances of this cycle at various points in time and identify the two aspects of innovation and reach. One such example was the level 1 document object model (DOM), which was designed to introduce a standard, cross platform, language independent view of how a document should be accessed and manipulated, but was partly based on the concept of the 'level 0' DOM, which was the de facto standard already implemented in the browsers of the time. Similarly, at around the same time, the HTML specification, including its many contradictions and lack of well-formedness, was engineered from innovations already established by browsers in common use. While the fragmentary, proprietary views of the DOM and HTML were emerging, others were attempting to maintain the original concepts of separation of structure, content and presentation that were originally intended for Web-based documents, for example, by promoting the standardisation and use of style sheets (Lie & Bos, 1999). Over a series of standardisation cycles, that concept has gradually been reasserted, with the development of the extensible hypertext markup language (XHTML) and cascading style sheets (CSS). Unlike HTML, XHTML does not allow the arbitrary mixing of content, structure and presentation, delegating presentation and, to some extent, structural management of Web page content (written in XHTML) to external style sheets (written in CSS).

The Emergence of Ajax

A recent example of the Web fragmentation cycle has been the emphasis in browser hosted tools for providing a richer client experience, including Web service clients for mashup services and, in particular, Ajax (asynchronous JavaScript and XML). Ajax is not a particularly new concept, following on as it does from a longer tradition of client side processing. This began in the mid 1990s with the introduction of JavaScript into Netscape Navigator and Java applets into the HotJava browser, and was followed by dynamic HTML (DHTML) which combined HTML, style sheets and scripts in a way that enabled documents to be 'animated' (dynamically manipulated) in the browser, with a view of the document that has since been formalised by the W3Cs document object model (Le Hégaret, Whitmer, & Wood, 2006). However the significant difference between Ajax and previous approaches is the concept of the "one page Web application," whereby page content is updated asynchronously from the server without the whole page being rebuilt. The two main advantages of this approach are that it enables a more interactive user experience, and that it can reduce the amount of Web traffic required to update a page, though both of these are dependent on careful design. An early example of this approach was Google Suggest, which was able to dynamically populate a search text box with suggestions for search terms as characters were typed into it, providing, of course, that the browser was able to support it. Ajax itself is not a technology but a label, applied by Garrett (2005), to a way of building Web applications that uses the XMLHttpRequest object within client side scripts to seamlessly update Web pages. Garret summarised Ajax as a combination of:

- Standards based presentation using XHTML and CSS
- Dynamic display and interaction using the document object model
- Data interchange and manipulation using XML and extensible stylesheet language transformations (XSLT)
- Asynchronous data retrieval using the XMLHttpRequest
- JavaScript binding everything together

Figure 2 shows the general architecture of Ajax based systems. The key to this architecture is that the Ajax engine mediates between the user interface and the server, processing on the client where possible (using DHTML) and, where necessary, sending asynchronous HTTP requests and receiving XML data (or indeed data in any other suitable format) that it renders in the browser via the DOM. Of course the technologies listed

Figure 2. Ajax architecture

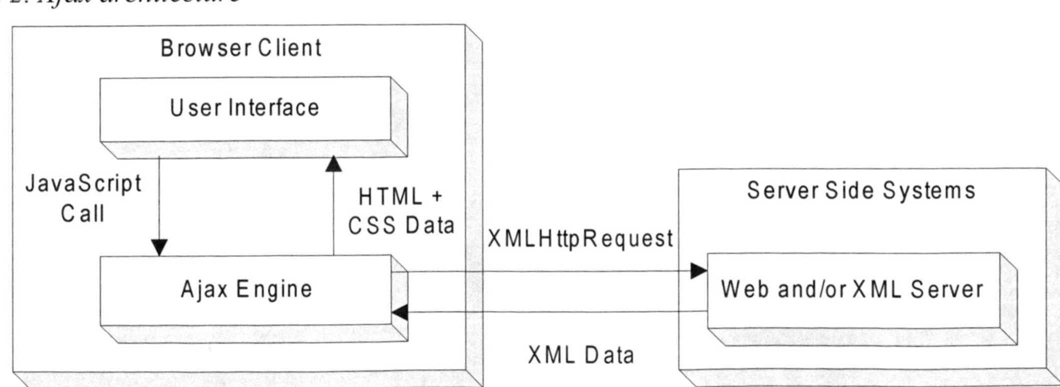

by Garrett are not the only way to provide one-page applications on the Web, since alternative technologies like Flash can be used to similar effect.

As a simple example of how Ajax might be applied, we might consider how it could be used to manage the submission of login data from a Web page. In the usual client server interaction, an HTML form will submit data to the server using a standard action that posts the data to a URI. JavaScript may perform some surface validation on the client (e.g., checking for empty fields or password length), but true validation (checking the user's ID and password against the security domain) is done after the page is submitted to the server. Here for example an HTML 'form' tag submits to a server side URI after invoking a local validation function

```
<form action="/processlogin" method="post"
onsubmit="return validateLoginData(this);">
```

The server response will then be to generate a different client page, either a regenerated login page (if there are errors) or the next logical page in the business process. In contrast an Ajax approach might be to trigger a local Ajax function by an event such as a button being pressed (or tabbing fields, or keys being pressed) rather than submitting a form to a server side URI.

```
<input name="submit» type="button"
onclick="ajaxLogin();" value=»Submit" />
```

The local Ajax function will use an XML-HttpRequest to communicate asynchronously with the server, and trigger a function that processes the return value and interacts with the current page to update it appropriately.

```
...
xhrequest = getXMLHttpRequest();
...
xhrequest.onreadystatechange = processLogin;
xhrequest.open("post", url, true);
xhrequest.send(null);
...
```

Figure 3. Standard multi-page processing compared with the Ajax "one page" approach

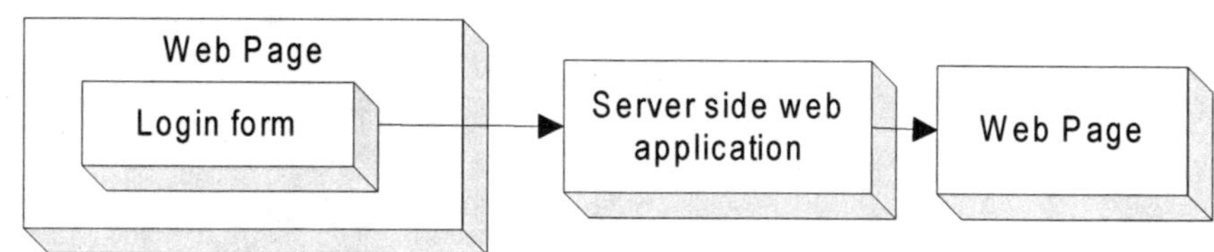

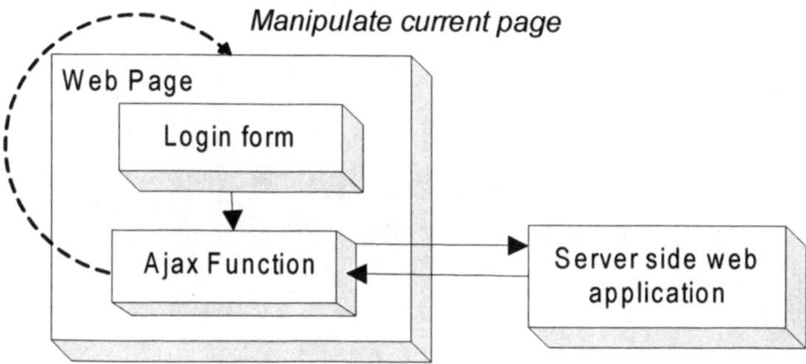

```
function processLogin()
...
// process the server response data and manipulate the current page
```

Figure 3 shows how the two approaches differ in terms of their page flow. The Ajax version is the "one page Web."

Although they can bring significant benefits in the contexts in which they work, the aspect of fragmentation that Ajax and similar technologies bring is that they rely on the programming platform within the Web browser, excluding those Web clients that do not support these technologies. At the same time there has been a separate move towards broader Web access on mobile devices, which is in conflict with the reliance on client side JavaScript or similar tools that are needed to support Ajax. Indeed there is a potential conflict between two of the Web 2.0 patterns, as described by O'Reilly (2005), namely "software above the level of a single device" and "rich user experience." Of course there is no reason why these two patterns should be mutually exclusive, but the Ajax approach to a rich user experience, with the current level of mobile browser technology, is potentially fragmentary. Whilst some rudimentary Ajax implementations may be possible with JavaScript enabled mobile browsers like Opera Mobile, most of the highly interactive, desktop style Ajax application cannot currently be accessed from mobile devices.

Mobile Standardisation: The .mobi Domain

At the same time as the desktop browser was being increasingly leveraged to develop a rich, interactive user experience based on asynchronous server requests, there was a move by the mobile phone network operators and handset manufacturers in the Open Mobile Alliance to increase the reach of the Web by introducing the .mobi top level domain in 2006. The intention of this new domain was to provide a standard way for users to access the mobile Internet by guaranteeing that a site with a .mobi extension was designed for mobile device access. The primary mechanism for this was to use XHTML-MP (extensible hypertext mark-up language—mobile profile), the successor to the WAP Forum's wireless mark-up language (WML) as the standard page mark-up for .mobi domains (the WAP Forum was the precursor to the Open Mobile Alliance). This approach, however, has come in for some severe criticism. Tim Berners Lee, the 'father' of the World Wide Web, commented that:

> *This domain will have a drastically detrimental effect on the Web. By partitioning the HTTP information space into parts designed for access from mobile access and parts designed (presumably) not for such access, an essential property of the Web is destroyed* (Berners Lee, 2004).

However, over time we can expect that this separation between mobile and non-mobile Web sites will fade away as first technology, and then standards, will minimise the differences between desktop and mobile content delivery. Further development of mobile browsers will begin to provide Ajax or alternative rich client functionality on more types of mobile device. We can also expect the technologies that underlie the .mobi domain, such as XHTML-MP, to evolve over time and support such an increase in functionality. After a period dominated by fragmented tools and techniques, we can expect a common set of standards to emerge that will give consistency across the Web, regardless of browser, rendering specialised mobile content unnecessary. Figure 4 shows how Ajax and specialised mobile content can be seen to fit into the Web fragmentation cycle, and that the two trends should resolve into device independent rich content.

Figure 4. The fragmentation between specialised mobile content and Ajax

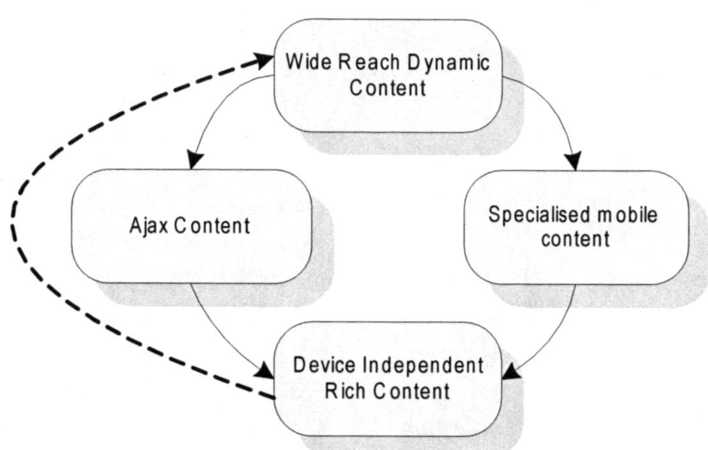

Listing 1. The innerHTML property and DOM methods compared

```
<?xml version="1.0"?>
<!DOCTYPE html PUBLIC "-//W3C//DTD XHTML 1.1//EN"
 "http://www.w3.org/TR/xhtml11/DTD/xhtml11.dtd">
<html xmlns="http://www.w3.org/1999/xhtml" xml:lang="en">
 <head>
  <title>innerHTML and the DOM</title>
 </head>
 <body>
  <div id="section1"></div>
  <div id="section2"></div>
  <script type="text/javascript">

 // add an element containing text using innerHTML
   document.getElementById("section1").innerHTML =
    "<p id='para1'>Section 1 uses innerHTML</p>";

  // add an element containing text using the DOM
  var para2 = document.createElement("p");
  para2.setAttribute("id","para2");
  para2.appendChild(document.createTextNode
   ("Section 2 uses the DOM"));
  document.getElementById("section2").appendChild(para2);

  </script>
 </body>
</html>
```

Innovations That Do Not Become Standards

Although we have so far presented the fragmentation cycle as a repeating pattern of innovation and reach being resolved by standardisation, this is also an over-simplification in the sense that not all innovations are neatly resolved by the development of agreed standards. One example where this process does not appear to be satisfactorily resolved is the common use and implementation of the innerHTML property within many browser DOMs, despite the slim chance of it becoming a standard part of the DOM specification published by the World Wide Web Consortium (W3C). Use of this property is driven largely by its simplicity, since it makes it easy for Web developers to insert dynamic content into a text node of a Web based document. This makes it very useful for developing Ajax applications. Objections to the use of this property are primarily that it can be easily misused, since content, structure and presentation aspects of the document can become muddled together with client side scripts when content is manipulated in this way. Unfortunately the alternative approach, to use the standard APIs of the DOM, is often considerably more complex, compounded by variations in the way that the DOM APIs are actually implemented in different browsers. The simple example in Listing 1 shows an XHTML page that includes a script (written in JavaScript) that adds an element containing text using the innerHTML property, and adds a similar element using the methods of the DOM. Even for such a simple operation, the DOM version is much more complex. Not only is the code itself longer, but the performance of the innerHTML approach is much faster in most situations (Koch, 2006)

Given the conflicting viewpoints in this debate, it is difficult see to how the fragmentation can be easily resolved. Therefore we should acknowledge again that the simplistic view of the fragmentation cycle that sees a clean merging of innovations and standards will not always apply.

WEB 1.0 ARCHITECTURE

So far in this chapter we have focused largely on the evolving client side components of Web application architecture. However it is on the server that the major Web application processes actually take place. Therefore, we also need to consider how Web application architectures have evolved in terms of server side components and interactions. Before exploring how current innovations and standards may affect the future direction of Web application design, we will first review some of the main architectural aspects of established practice. In the spirit of the level 0 DOM, we might perhaps categorise pre-Web 2.0 architectures as "the Web 1.0." In this type of architecture, there are some standard patterns that are commonly used and integrated into popular Java Web application frameworks such as Struts and JavaServer Faces. These patterns include the Model 2 architecture (Seshadri, 1999), which is loosely based on the Model View Controller pattern (Buschmann et al., 1996) and the Template View pattern (Fowler, 2003).

The Server Page Template Model

Established Web application technologies that support dynamic content, like JavaServer Pages (JSPs) and Microsoft's Active Server Pages, use a relatively simple model of a layered Web based architecture, whereby server side application execution can be integrated into presentational mark-up. Dynamic content is based on the transformation of database content into Web based pages, and form based input into database updates. Architectural patterns for this type of Web application are based on simplifications of the Model View Controller pattern, like the JSP

Listing 2. An example of the Template View pattern using a JavaServer Page and the JSTL

```
<?xml version="1.0"?>
<jsp:root xmlns:jsp="http://java.sun.com/JSP/Page"
 xmlns:c="http://java.sun.com/jsp/jstl/core"
 doctype-public="-//W3C//DTD XHTML 1.1//EN"
 doctype-system=" http://www.w3.org/TR/xhtml11/DTD/xhtml11.dtd "
 version="2.1">

 <jsp:useBean id="choice" class="com.webhomecover.beans.QuoteChoice"
scope="session" />
 <jsp:useBean id="contents" class="com.webhomecover.beans.ContentsDetails"
scope="session" />
 <jsp:useBean id="buildings" class="com.webhomecover.beans.BuildingsDetails"
scope="session" />

 <html xmlns="http://www.w3.org/1999/xhtml" xml:lang="en">
  <head>
   <link href="webhomecover.css" rel="stylesheet"
    type="text/css" />
   <title>WebHomeCover Insurance Quote</title>
  </head>
  <body>
   <h1>Here is your insurance quote from WebHomeCover</h1>
   <c:if test="${choice.buildings}">
    <h2>Your Buildings Quote:</h2>
    <jsp:getProperty name="buildings"
     property="formattedInsuranceQuote"/>
   </c:if>
   <c:if test="${choice.contents}">
    <h2>Your Contents Quote:</h2>
    <jsp:getProperty name="contents"
     property="formattedInsuranceQuote"/>
   </c:if>
   <form action="welcome.jsp" method="get">
    <input type="submit" value=
    "Thanks for the quote, now take me back to the home page" />
   </form>
  </body>
 </html>
</jsp:root>
```

Model 2 architecture (Seshadri, 1999). Here, a server side component at the controller layer, the *FrontController* (Fowler, 2003), is responsible for interpreting client requests, delegating to other server side components in the model layer to provide the necessary data objects, and ensuring that the thread of control is ultimately forwarded to a component suitable for rendering the client response. This component is usually a server page, which is able to embed dynamic content into presentational markup using special tags. Fowler (2003) characterises this as the *Template View* approach to page generation, where the server page provides the presentation template and dynamic content is inserted into it at predefined locations, preferably using XHTML compliant tags. The code example in Listing 2 shows a JSP that uses the template view approach. It is a document in XML format that contains a combination of standard mark-up tags, using XHTML (the template, highlighted in bold text), and other tags, specific to the Java server side environment, that generate

Figure 5. The 'Web 1.0' application architecture, based on the Model 2 and Template View patterns

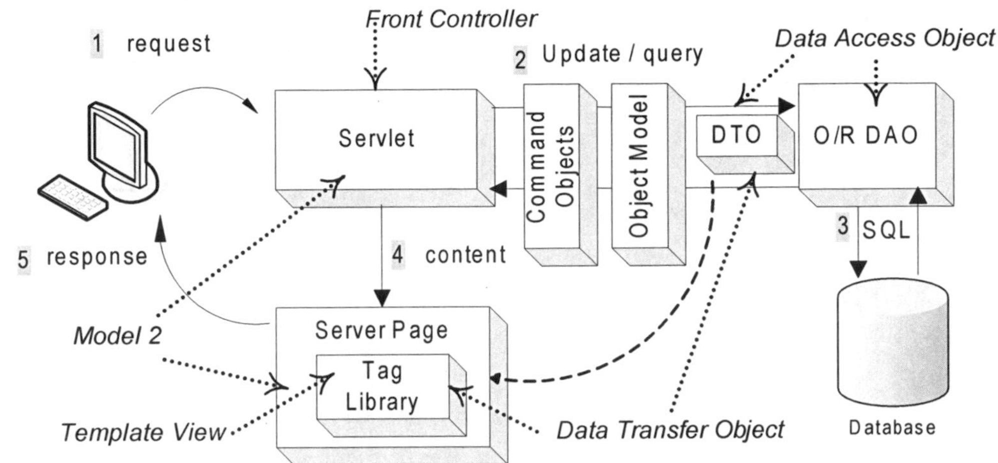

dynamic content. These tags are a combination of standard JSP tags ('jsp:root,' 'jsp:useBean,' 'jsp:getProperty') and the 'if' tag from the JSP Standard Tag Library (JSTL).

Behind the view/controller layer, the mapping between the data store and the model is the responsibility of data access objects (DAOs) that interact with the higher levels of the system via data transfer objects (DTOs) (Alur et al., 2003). In most real world cases the data store is a relational database, requiring some object relational (O/R) mapping, enabling the structured query language (SQL) to be used to read and write persistent data. The DTO components that encapsulate updates or query results can be embedded into server pages using appropriate tag libraries.

Figure 5 illustrates the basic 'Web 1.0' architecture. Initial routing of all hypertext transfer protocol (HTTP) requests from Web clients is handled by the front controller component that receives all the requests, parses their parameters and delegates to the appropriate object model layer components via command objects (1). There are a number of frameworks that encapsulate this design pattern, including Struts and JavaServer Faces. In both cases, the front controller component is a Java servlet. Any interaction between the model and persistent data is done via the DAO/DTO layer (2), which interacts with the underlying data store (3). Content is returned in DTOs that can be processed by the object model layer and/or embedded into server page components using tag libraries (4). These server page components can then generate a dynamic response (5). In "traditional" Web applications, that response has typically been in the form of HTML.

The XML Transformation Model

More recently, XML (extensible markup language) based technologies have become popular in Web application architectures, in no small part due to the limitations of HTML, which is purely a page mark-up syntax and cannot be used to represent data at any level of abstraction, whereas XML can be used to represent semi-structured data that is not restricted to page mark-up (Abiteboul, Buneman, & Suciu, 2000). Certain principles underlie the use of XML in a Web application. Its role is essentially to provide services that cannot be supported using more traditional approaches to Web applications based on representing content directly in HTML. There are four broad categories where XML provides such services (Bosak, 1997):

- Applications that require the Web client to mediate between two or more heterogeneous databases;

Figure 6. The Transform View pattern, where a source document is transformed into a client page using an XSL Transformation

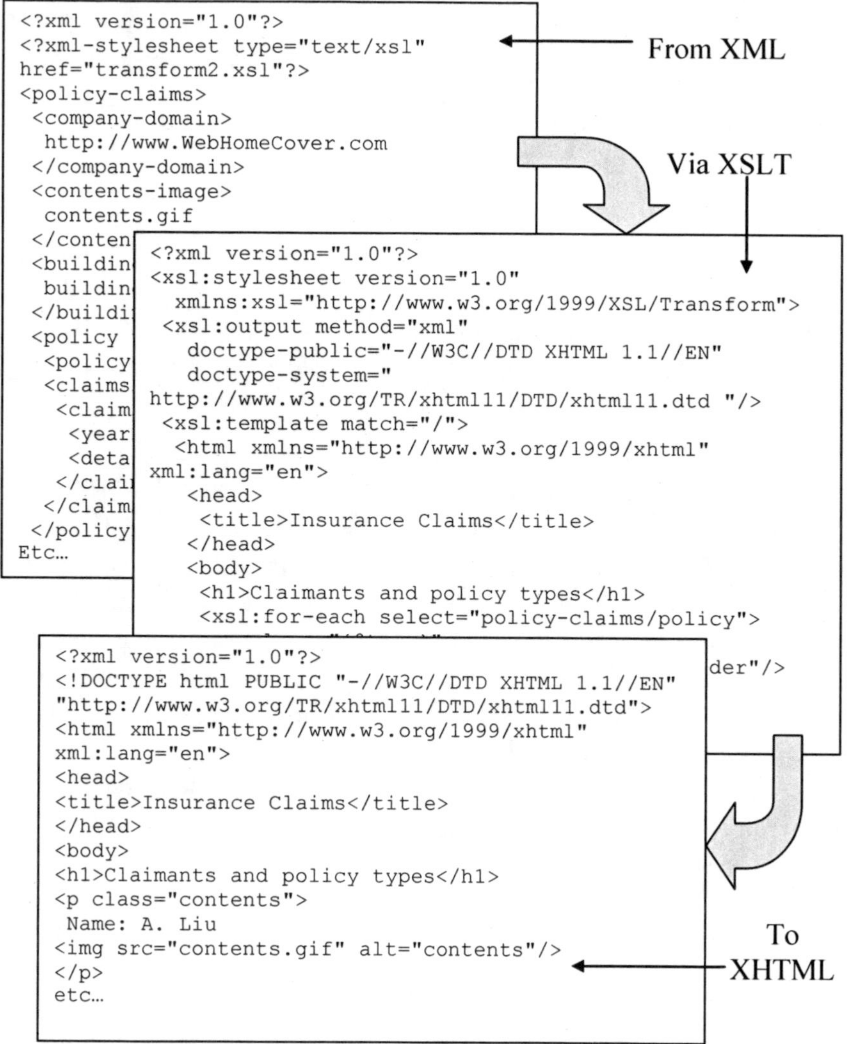

- Applications that attempt to distribute a significant proportion of the processing load from the Web server to the Web client;
- Applications that require the Web client to present different views of the same data to different users;
- Applications in which intelligent Web agents attempt to tailor information discovery to the needs of individual users.

Meeting these requirements means that a Web application that uses XML needs a more complex server side architecture than the simple template model, and data has to be represented in, and translated between, multiple formats, which may include XML, object model and database level representation. Whereas the template view pattern assumes a component based plugging-in of object model components into a server page template, the XML based model requires some kind of transformation to take place from data, taken from some source and represented by XML, into mark-up that can be interpreted by the client. Fowler (2003) characterises this approach as the *Transform View* model where, instead of using a server page to embed object model components into a page template, an extensible stylesheet language (XSL) processor is needed to transform an XML document into a generated client page. In order to make this transformation, the XSL processor applies an XSL transformation (XSLT) stylesheet that defines how the input document is to be transformed.

Figure 6 shows how an XML document is transformed by an XSL Transformation into an XHTML client page. This transformation can take place either on the client or the server, depending on whether the client is able to process the transformation (most desk top browsers are able to do this).

The main problem with executing transformations on the client, even assuming the client is able to perform those transformations, is that they will provide static content, since the assignment of the transforming style sheet will be hard coded into the XML document. Alternatively, we would have to rely on the client providing the necessary scripting and DOM support to perform the dynamic transformation. In contrast, a transformation executed on the server can be applied dynamically and reliably, independent of the client. Listing 3 shows an example of how such transformations can be executed on the server using tags from the JSP Standard Tag Library. Specifically, the 'transform' tag from the JSTL XML library applies an XSL transformation style sheet to an XML document.

A simplified visualisation of the distinction between the template and transform approaches in shown in Figure 7. In the template view, the server page integrates components from the object model into a mark-up based template. The page template is embedded in the server page itself. In the transform approach the XSL processor integrates elements from XML documents into a mark-up based transformation, with the page template embedded in the transformation style sheet. If the transformation is performed on the server, the server page simply triggers the transformation process.

XML in Web Applications

Bosak's (1997) four roles of XML can be supported in various ways by the transform model. Mediation between heterogeneous databases can be seen from more than one perspective. In some

Listing 3. The Transform View pattern implemented on the server using the JSP Standard Tag Library

```
<?xml version="1.0"?>
<jsp:root xmlns:jsp="http://java.sun.com/JSP/Page"
  xmlns:c="http://java.sun.com/jsp/jstl/core"
  xmlns:x="http://java.sun.com/jsp/jstl/xml"
  version="2.0">

 <jsp:directive.page contentType="text/html"/>
 <jsp:output omit-xml-declaration="true" />

  <c:import url="policies.xsl" var="stylesheet"  />
  <c:import url="policies.xml" var="xmldocument"  />
  <x:transform doc="${xmldocument}" xslt="${stylesheet}" />

</jsp:root>
```

Figure 7. The Template View and Transform View patterns compared

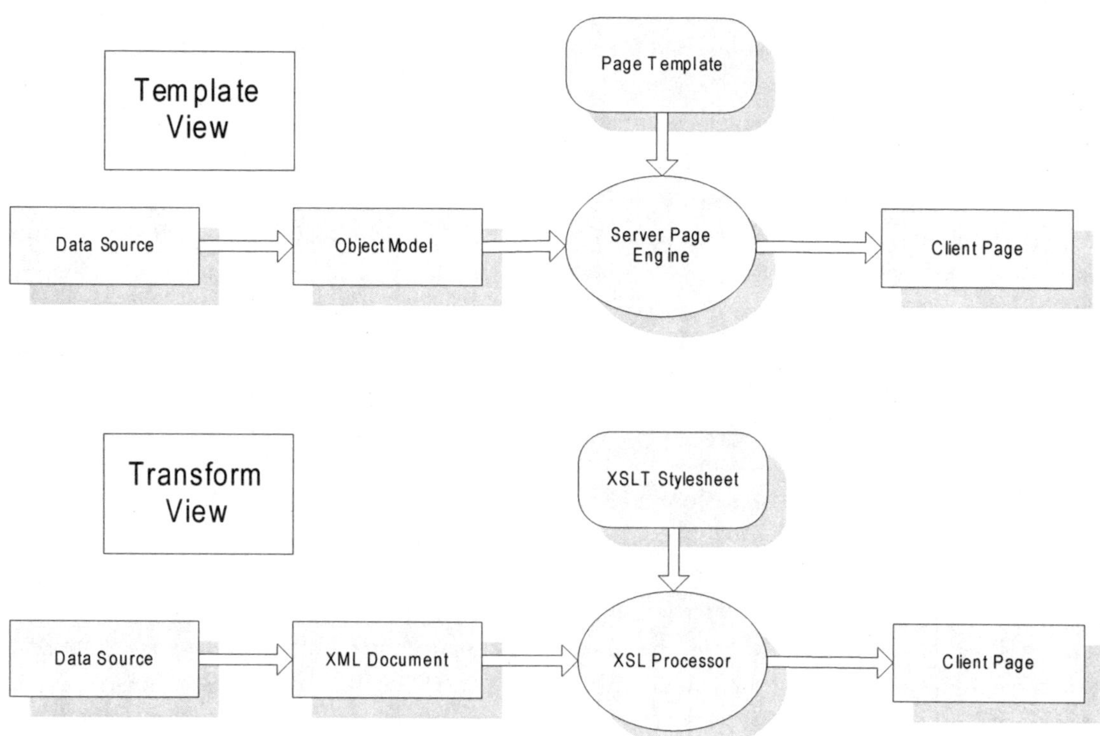

cases, XML Web services can be used to integrate the various data sources of a single organisation or collaborating group of organisations. In this sense XML can be seen as the natural successor to more traditional forms of electronic data interchange (EDI). In other cases, such as those included in the patterns of the Web 2.0, there is the concept of specialised databases being the core item of value in a system, with the opportunity to reuse these databases in new combinations of services. In this type of scenario the databases in question may be distributed over a wide range of providers, with no formal relationship or set of contracts between them. Here, the XML Web services are in the public domain. From any of these perspectives, raw XML data may be the way of communicating between systems, but transformations are necessary to apply queries or filters to the data and transform it to meet the specific requirements of a client system.

There are various ways that XML can be used to offload some of the processing from the server to the client, but again some aspect of transformation is essential. XML documents provide a presentation neutral way of representing the underlying content of an application, but the typical role of the Web client is to render that data in an appropriate way. Tools for this type of transformation, such as XSLT, are therefore an essential component of an XML-centric Web application architecture. This is what underlies the concept of the transform model, since XSLT is able to express how one XML document may be transformed into another document, which may be XML or may use a different syntax such as XHTML. The way that processing can be offloaded from the server to the client ranges in sophistication from the simple separation of XML data, XSL transformations and CSS (cascading style sheets, used to style the presentation of a document) to Ajax applications. In the former case, we rely on the ability of the browser to download, cache, and process by transformation and styling, XML documents. Since the same transformations and style sheets can be used with multiple XML documents from similar data sources (i.e., related but different queries across the same data source) all that is required of the server is to send the XML documents to the client for processing. In the latter case, various Ajax tools can be used to develop rich client processes based on JavaScript and asynchronous XML messaging. This is a more complex approach, since instead of relying on the browser's native ability to process XSLT, we are relying on a JavaScript application to support client side processing and parsing of XML documents. While it is possible to offload some presentational style sheet processing to small devices using this kind of markup, not all mobile clients are able to manipulate XML documents using XSL Transformations or by using client side scripting languages like JavaScript. However, many small devices can support the use of Java Micro Edition to manipulate XML documents on the client, using small footprint parsers such as kXML. Therefore an Ajax-style approach is possible even in the absence of JavaScript.

Where it is not possible to offload the processing necessary for XML transformations onto the client, the server can take responsibility for enabling the Web client to present different views of the same data to different users. One of the useful ways that we can apply XSL transformations of XML data in a Web application is to provide different transformations for different types of client device. For example, the same content, represented in XML, could be transformed into XHTML for desktop browsers, wireless markup language (WML) for 'legacy' mobile phones, XHTML-MP for more recent mobile phone browsers, and so on. The same techniques can be use for personalisation or customisation of the content itself, targeting the user as well as the device

Finally, agents can utilise XML because it provides metadata about Web-based content. This aspect of XML is what supports the concept of the semantic Web, enabling a greater degree of reflection about Web-based information sources on the part of Web applications. Although discussion of the Semantic Web is beyond the scope of this chapter, XML and XML Schema are important foundation technologies that support the higher level aspects of the Semantic Web technology stack such as the resource description framework (Hendler, 2001).

The Limitations of the Transform Model

Whilst transformations, used directly, can have certain advantages such as the ability to leverage the common abilities of browsers to perform XSL transformations on the client, on their own they have a number of drawbacks. First, the syntax of the XSL transformation is complex, and can be difficult to maintain, particularly where the "apply-templates" approach is used to transform the XML documents using pattern matching. Second, using a transformation directly cannot easily be integrated with other server side processes, even if the transformations are actually executed on the server rather than the client. The advantage of the XSLT approach is that a relatively small set of transformations can be used to manage a large number of XML documents in a Web application. However they are not particularly flexible. In contrast, using a tag-based template approach gives us a highly flexible way of integrating both mark-up and programmatic logic. Passani & Transatti (2002) argue that XSL Transformations are a very poor approach to providing adaptive mark-up for different devices, because they mean writing separate transforms for each type of client. In contrast, template based tools such as the wireless abstraction library (WALL) enable us to encapsulate different page generation behind processing tags (Passani & Transatti, 2002). It is possible, however to use parts of XSL combined with tags to gain the benefits of both approaches.

Figure 8. Combining transform and template views

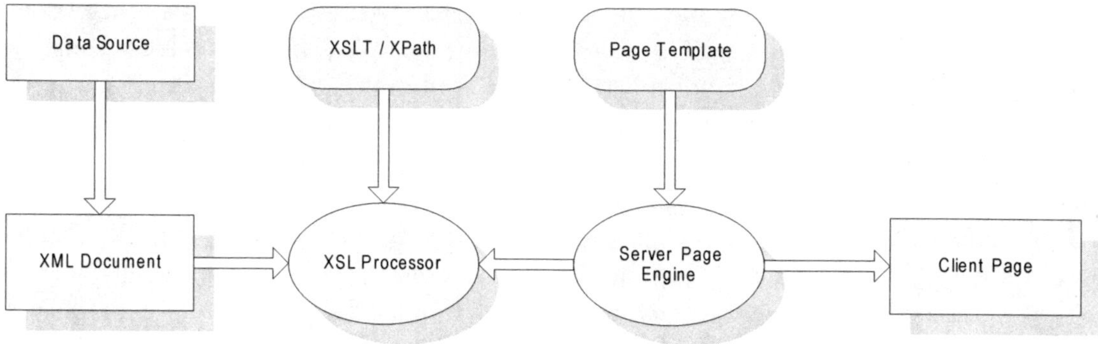

Listing 4. Combining the Template View and Transform View approaches in a JSP

```
<?xml version="1.0"?>
<jsp:root xmlns:jsp="http://java.sun.com/JSP/Page"
   xmlns:c="http://java.sun.com/jsp/jstl/core"
   xmlns:x="http://java.sun.com/jsp/jstl/xml"
   version="2.1">
 <jsp:directive.page contentType="text/html"/>
 <jsp:output omit-xml-declaration="false"
  doctype-root-element="html"
  doctype-public="-//W3C//DTD XHTML 1.1//EN"
  doctype-system="http://www.w3.org/TR/xhtml11/DTD/xhtml11.dtd" />

 <html xmlns="http://www.w3.org/1999/xhtml" xml:lang="en">
  <head>
   <link href="webhomecover.css" rel="stylesheet"
     type="text/css" />
   <title>WebHomeCover Insurance Policies</title>
  </head>
  <body>
   <div>
    <!— main page presentational content here,
       including XPath expressions, e.g.
    —>
    <c:import url="./policies.xml" var="xmldocument"/>
    <x:parse doc="${xmldocument}" var="xml"/>
    <x:out select="$xml/policies/policy/policy-holder"/>

   </div>
  </body>
 </html>
</jsp:root>
```

Rather than transforming documents in a single process, we can combine partial transformations with template based mark-up. This can help us to gain the benefits of XML in Web applications without losing the programmatic flexibility of the template model. However the key is to combine both transformations and templates into a single server page to ensure adequate performance. It is technically possible to generate a server page, marked up using device adaptive tags, from a transformation, and then run that generated page through the server page engine. However this two stage process can be very inefficient. The alterative approach is to integrate partial transforms, for example using XPath filters on the XML data, within the same server page as the adaptive mark-up. This means a single pass through the server page (Figure 8).

The code extract in Listing 4 shows the basic principles of this approach. The template view provides the main structure of the page and generates the XHTML tags that build the presentation. Within this page, however, XPath expressions are applied to an XML document to insert dynamic content. One of the main contrasts between the transform and mixed template and transform approaches is that the responsibility for the XHTML mark-up is in the XSLT document in the transform approach, but in the server page in the combined template/transform approach.

WEB 2.0 GENERATION APPLICATION ARCHITECTURE

In the previous sections we introduced some of the architectural features of Web applications that may use either template or transform approaches, or a combination of both. In this section we describe a Web 2.0 generation architecture for adaptive Web applications that leverages XML as a data representation format within an adaptive architecture, integrates service oriented aspects, and combines both template and transform elements. Perhaps the most important feature of this architecture is the introduction of an XML data layer that integrates

Figure 9. A Web application architecture integrating XML

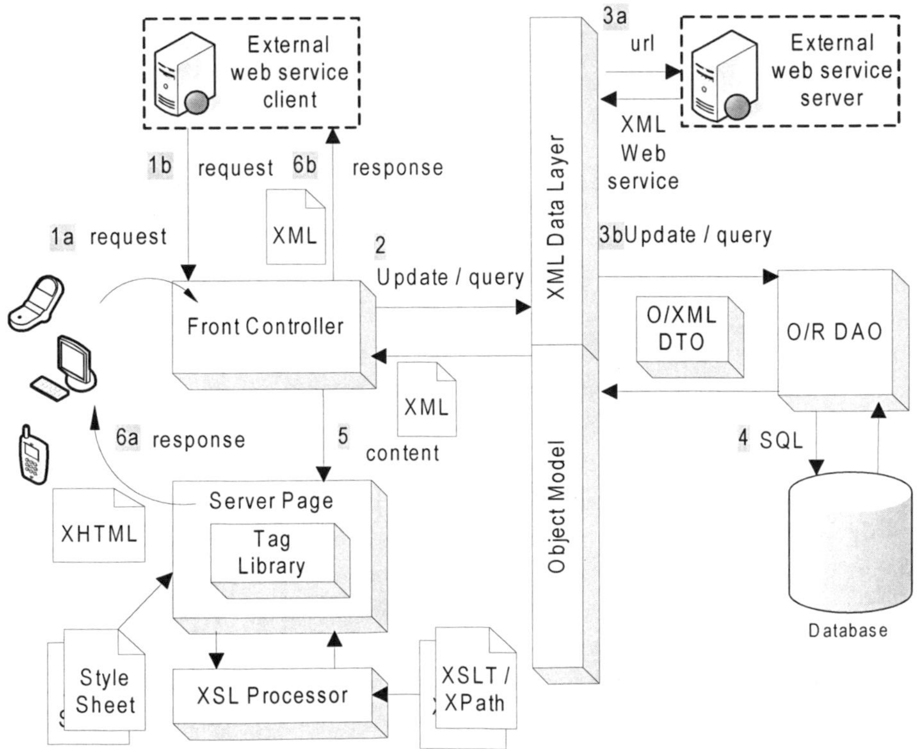

disparate data sinks and sources into a cohesive XML management process at the same level as the object model. This enables a system to manage requests and responses in a consistent way, regardless of whether the client request is for a Web service or presentation mark-up, and whether our data sources are local databases or remote Web services. An outline of the architecture and its core processes is shown in Figure 9.

In this architecture, client requests to a front controller component may come either from presentation oriented devices (1a) or external Web service clients (1b). The request is delegated to a joint XML data and object model layer (2). Depending on the requirements of the request, queries delegated via the XML data layer may either retrieve XML content from one or more external services running on other Web servers (3a), generate XML content from a local database via suitable queries (3b) or perhaps combine multiple data sources. Where content is based on external Web services, content will already be in XML format. Where content is held locally in a database that is not natively XML, some kind of transformation will be required between relational and XML data. Rather than implementing such a direct transformation, which would provide only XML to the controller layer, it is probably wise to maintain the data transfer object (DTO) pattern, but add XML generation capabilities to the DTOs. This approach means that the DTOs can be used both as objects in their own right within the object model of the underlying entities in the system and as sources of XML documents. A layer of object/relational data access objects (DAOs) is required to perform the translation from relational data to object model and create the XML generating DTOs (4). Multiple XML documents may be generated from a model that is created from a single database query, since multiple object level queries can be used against any graph of DTOs. The XML content returned from the XML data layer (5) is transformed via an XSL process utilizing XPath queries to build parts of a document. These XML path queries should be performed via a server page using tag libraries that integrate both XML data retrieval and adaptive markup to ensure encapsulation and reuse. Once the content is transformed it can be transferred to the device in the generated mark-up format for processing by the client. This may be a device oriented mark-up (6a) or a Web service (6b). Of course many mobile devices can both process mark-up and act as end points for Web services. It should be noted that services may take different forms. For example, a server side component may provide XML messaging service for Ajax clients by using the mechanism described above, whereby program components generate XML documents and stream these to JavaScript components on the client. However Web services may take a more heavyweight form, using standards such as Web services description language (WSDL), simple object access protocol (SOAP) and universal description, discovery and integration (UDDI). In this case, generation of XML from programming components will typically be done via tool support to generate the necessary XML documents, stubs and skeletons to enable client/server interoperability. The current generation of Java Web service tools are based on the Java API for XML Web services (JAX-WS).

This type of XML centric architecture has a number of benefits. First, we can gain reuse of both external Web services and local components. Second, we can provide adaptivity that integrates both client device mark-up and Web service end-points. From a business perspective it provides maximum leverage of external services and APIs while gaining maximum potential distribution across all possible client types. Encapsulation of data access and transformation should provide benefits in terms of maintainability and cost. For example, separating out concerns between XML, object models and the database, rather than directly generating XML from the database, can assist in the reuse of legacy data sources and provide added security through additional layering. In terms of the Web fragmentation cycle, the architecture is based on common XML, Java and SQL standards while integrating some aspects of innovative Web 2.0 service based and cross-device architectures.

Integrating Java and XML

So far we have introduced a number of features of Web application architecture relating to either XML or Java, but indicated that a Web application architecture need to be able to support data objects that can convert to and from Java and XML. In this section we discuss some aspects of how Java and XML can be integrated, including some standard tools. In a Java based architecture, the O/XML DTOs should be based on the relevant parts of the JavaBean specification that enable them to be used via tag libraries in JSPs, and be manipulated by the JSP Expression Language. These beans provide both an object model to organise the requested set of data and a way of generating XML documents from one or more beans. This is particularly useful where a client page will include data from multiple tables that have some kind of associative relationship, realised in the application as a related graph of objects. Generating XML from JavaBeans can be done using frameworks or custom code. Geary (2001), for example, provides some guidance on how to implement JavaBeans that can generate XML. In cases where the XML content encapsulates nested elements from multiple domain components, the beans will form a composite pattern (Gamma et al., 1995) with the composite objects organising sub elements and the leaf objects generating element content. Listing 5 shows a simple implementation of this pattern, where the class that implements the 'getXmlElement' method (representing an insurance policy) is associated with a collection

Listing 5. Generating XML content from composite JavaBeans

```
public String getXmlElement(boolean graph)
 {
  StringBuffer element = new StringBuffer();
element.append
("<policy policy-number=\"" + getPolicyNumber() + "\">");
element.append
("<start-date>" + getFormattedDate() + "</start-date>");
element.append
("<annual-premium>" + getAnnualPremium() +
"</annual-premium>");
element.append
("<number-of-claims>" + claims.size() +
"</number-of-claims>");
  element.append("<paid-up>" + getPaidUp() + "</paid-up>");
  if(graph)
  {
   element.append("<claims>");
   Iterator<Claim> claimIter = claims.iterator();
   while(claimIter.hasNext())
   {
    Claim claim = claimIter.next();
    element.append(claim.getXmlElement());
   }
   element.append("</claims>");
  }
  element.append("</policy>");
  return element.toString();
 }
```

of 'Claim' objects that generate their own XML content via a call to a polymorphic 'getXmlElement' method. Each of these methods generates a fragment of a larger XML document.

The link between the JavaBeans and the database will depend on the nature of the database and the transactional requirements of the persistence layer. In some cases a simple manual mapping using JDBC may be adequate, but it is likely that frameworks implementing standards such as the Java Persistence API will be required for industrial strength persistence. How the JavaBeans interact with the XML layer will also depend on the requirements of the application. As we have described, manual solutions are possible, but it may be more appropriate to use tools such as the open source XMLBeans (Apache Software Foundation, 2006) or Java XML binding (JAXB) which is integrated into the standard Java platform from version 6 onwards. The choice here will often depend on whether the application is mainly driven by an object model or by XML documents, since frameworks like XMLBeans and JAXB depend on the data model being derived from XML Schemas, and Java objects being generated subservient to that XML data model. Another important consideration is the generation of Java objects from XML documents. This can be done manually using the Java API for XML processing (JAXP) but is perhaps better performed by tools like JAXB, particularly since more recent versions of JAXB, unlike the original implementation, can convert both from XML to Java and from Java to XML. The partial code example in Listing 6 shows how JAXB can be used to generate XML documents from Java objects via a 'Marshaller,' in this case a collection of 'Policies' objects. The collection of Policies is derived from an ObjectFactory class generated by the JAXB framework.

Once XML documents are generated from the JavaBeans layer, XML processing should be performed via JSPs using tags from libraries such as the JSP standard tag library (JSTL), which includes a dedicated sub-library for XML processing. However there are a number of other XML processing tag libraries available in Java that may be used instead. Listing 7 shows the JSTL being used for a simple XSL transformation—similar to listing 3—but this time, not using a static XML document. A JavaBean is used to generate an XML document which is transformed by a style sheet using JSTL tags. In this code example we assume that the getXmlDocument method is responsible for marshalling the various components of a complete XML document, in contrast to the

Listing 6. Transforming objects into XML documents using JAXB

```
ObjectFactory policiesFactory = new ObjectFactory();
Policies policies = policiesFactory.createPolicies();
JAXBContext ctx = JAXBContext.newInstance(Policies.class);
...
Marshaller m = ctx.createMarshaller();
m.setProperty(Marshaller.JAXB_FORMATTED_OUTPUT, Boolean.TRUE);
m.marshal(policies, System.out);
 try
 {
  OutputStream os = new FileOutputStream (new File("newpolicies.xml"));
  m.marshal(policies, os);
 }
 catch(Exception e)
 {
  e.printStackTrace();
}
```

Listing 7. Using the JSTL tags for processing of JavaBeans that generate XML

```
<?xml version="1.0"?>
<jsp:root xmlns:jsp="http://java.sun.com/JSP/Page"
  xmlns:c="http://java.sun.com/jsp/jstl/core"
  xmlns:x="http://java.sun.com/jsp/jstl/xml"
  version="2.0">

 <jsp:directive.page contentType="text/html"/>
 <jsp:output omit-xml-declaration="true" />

 <c:import url="claims.xsl" var="stylesheet" />
 <x:transform doc="${claim.xmlDocument}" xslt="${stylesheet}"/>

</jsp:root>
```

getXmlElement method we saw in Listing 5, which generated a fragment of a larger document. Using the JSP Expression Language, 'claim.xmlDocument' refers to the getXmlDocument method of a JavaBean stored in the scope of the Web page under the lookup name 'claim.'

Adapting to Client Devices

If an architecture is to be flexible enough to manage multiple types of client browsers, the server side page generation has to be able to adapt to different types of client device, which can be identified by the 'User-Agent' field in the HTTP request header. If an XML transform layer is in place, the underlying content can easily be adapted, but how it is actually adapted is another question. Fortunately, there are a number of Java-based technologies that can assist us in adapting server-side content to the client device without manually interrogating the user-agent field. One example is the wireless abstraction library (WALL), a JSP tag library that builds on the wireless universal resource file (WURFL) (Passani & Trasatti, 2002). WURFL is an XML database that is able to map user agent information to the capabilities of the originating device, while WALL generates device specific mark-up. JSP pages based on the template view model, using WALL syntax, can be integrated with XML elements using the transform view model by applying tags (such as those from the JSTL) that enable XPath queries to be applied to an XML document and the results included in a server page. During the page processing, the WALL tags will be turned into the appropriate mark-up for the client device while the XPath elements will provide the content from the source XML document. Listing 8 shows part of a server page that incorporates WALL tags to generate adaptive mark-up, along with XPath queries using the JSTL. The key difference between the XML processing in Listing 7 and that in Listing 8 is that the former uses the transform view pattern, whereas the latter uses the combined template and transform view pattern outlined in Figure 8.

By using the WALL device aware library we are able to provide a single server page for many different types of device. However we might in addition choose to provide our own customised transform for particular types of device. For example, a transform could be used that would take advantage of the rich client technologies available on the desktop. Many Web 2.0 applications leverage Ajax and Flash, though there are many other possibilities. At the least, we might apply a cascading style sheet (CSS) to manage

Listing 8. Combining the WALL library with the JSTL to generate adaptive mark-up

```
<%@ taglib uri="http://java.sun.com/jstl/xml" prefix="x" %>
<%@ taglib uri="/WEB-INF/tld/wall.tld" prefix="wall" %><wall:document>
<wall:xmlpidtd />
<wall:head>
 <wall:title enforce_title="true">Claim Display</wall:title>
</wall:head>
 <wall:body>
  <wall:block>
  <wall:br />
  <jsp:useBean id="claims" class="webapp.classes.ClaimsBean" />
   <x:parse doc="${claims.xmlDocument}" var="xml"/>
   <x:forEach select="$xml/claims/claim">
    Claim Description: <x:out select="description"/>
    <wall:br/>
    Claim amount: $<x:out select="amount"/>
    <wall:br/>
   </x:forEach>
  </wall:block>
 </wall:body>
</wall:document>
```

the presentation in the browser. Where the client is a service end point, and the system provides content as a Web service, direct XSLT transforms can be used, since they would not be generating presentation mark-up but XML documents suitable for Web service use.

CONCLUSION

In this chapter we have described how Web applications have evolved from static content, through dynamic content based on a server page template model, to contemporary architectures that rely heavily on the transformation of XML documents and increasingly complex client side applications. We have discussed this evolution within a model that shows how Web technologies tend to fragment as a result of innovation, and then find broader reach as a result of standardisation. To develop Web applications in this context we need to be aware both of new technological developments and also how generic standards support is evolving. For example we may wish to develop Ajax applications but also deliver cross platform applications to both desktop and mobile clients. To be innovative while broadening reach as much as possible it is necessary both to work within well supported standards (e.g., parts of the DOM) but also to leverage tools that provide flexibility across platforms, whether it be increasing support for certain technologies on multiple platforms (e.g., JavaScript support in mobile browsers) or tools that adapt themselves to different clients (e.g., tag libraries for adaptive mark-up.) To support the integration of different approaches and technologies, a highly flexible architecture that leverages patterns, layers and XML centric data management is necessary. In this chapter we have described a reference architecture that is based on Java, XML and server side tag libraries that supports the transformation of local or service based content into client-adaptive output formats, using a combination of the template view and transform view design patterns. This architecture takes into account both current thinking on the technologies of Web-based applications and the encapsulation and reuse of standard libraries.

Building Web applications based on this architecture should enable developers to gain maximum flexibility and reach without relying too heavily on proprietary tools.

REFERENCES

Abiteboul, S., Buneman, P., & Suciu, D. (2000). *Data on the Web - From relations to semistructured data and XML*. San Francisco: Morgan Kaufmann.

Alur, D. Crupi, J., & Malks, D. (2003). *Core J2EE patterns: Best practices and design strategies*, 2nd Edition. Upper Saddle River, NJ: Sun Microsystems Press / Prentice Hall.

Apache Software Foundation. (2006). *Apache XMLBeans*. Retrieved on July, 2007, from http://xmlbeans.apache.org/index.html

Berners Lee, T. (2004). *New top level domains .mobi and .xxx considered harmful*. Retrieved on January, 2007, from http://www.w3.org/DesignIssues/TLD

Bosak, J. (1997). *XML, Java, and the future of the Web*. Retrieved on July, 2007, from http://www.ibiblio.org/pub/sun-info/standards/xml/why/xmlapps.htm

Buschmann, F., Meunier, R., Rohnert, H., Sommerlad, P., & Stal, M. (1996). *Pattern-oriented software architecture: A system of patterns*. Chichester: Wiley.

Fowler, M. (2003). *Patterns of enterprise application architecture*. Boston: Addison-Wesley.

Gamma, E., Helm, R., Johnson, R., & Vlissides, J. 1995. *Design patterns: Elements of reusable object-oriented software*. Reading, MA.: Addison-Wesley.

Garrett, J. (2005). *Ajax: A new approach to Web applications*. Retrieved on July, 2007, from http://www.adaptivepath.com/publications/essays/archives/000385.php

Geary, D. (2001). *Advanced JavaServer pages*. Upper Saddle River, NJ: Sun Microsystems Press / Prentice Hall.

Hendler, J. (2001). Agents and the Semantic Web. *IEEE Intelligent Systems, 16*(2), 30-37.

Koch, P. (2006). *ppk on JavaScript*. Berkeley, CA: New Riders.

Le Hégaret, P., Whitmer, R., & Wood, L. (2006). *Document object model (DOM)*. Retrieved on July, 2007, from http://www.w3.org/DOM/Overview

Lie, H. W., & Bos, B. (1999). *Cascading style sheets: Designing for the Web*, 2nd edition. Harlow, England: Addison Wesley Longman.

O'Reilly, T. (2005). *What is Web 2.0: Design patterns and business models for the next generation of software*. Retrieved on July, 2007, from http://www.oreillynet.com/pub/a/oreilly/tim/news/2005/09/30/what-is-web-20.html

Passani, L., & Trasatti, A. (2002). *WURFL*. Retrieved on July, 2007, from http://wurfl.sourceforge.net/

Seshadri, G. (1999). Understanding JavaServer pages model 2 architecture: Exploring the MVC design pattern. *JavaWorld, December*.

Van Eaton, J. (2005). *Outlook Web access - A catalyst for Web evolution*. Retrieved on July, 2007, from http://msexchangeteam.com/archive/2005/06/21/406646.aspx

Chapter IX
Applying Agility to Database Design

Guoqing Wei
FedEx Corporation, USA

Linda Sherrell
University of Memphis, USA

APPLYING AGILITY TO DATABASE DESIGN

Agile methods are flexible, allowing software developers to embrace changes during the software development life cycle. But the introduction of agile practices into software organizations may cause unhealthy tensions between the developers and data professionals. The underlying reason is that when agile methodologies are employed, the two communities use incompatible approaches, namely simple design and iterative development, which are practices associated with all agile methodologies, and big design up front (BDUF), a popular database technique. BDUF is inflexible, as once the database foundation is set, it is difficult to make changes throughout the software development life cycle. This chapter describes a database development method for a Web environment. Using this method, a data professional divides the database into loosely coupled partitions and resolves the above conflicts by applying certain agile practices. The result is that the database development becomes more iterative and incremental . This has the added benefit of supporting rapid application development in a dynamic environment, a fundamental characteristic of most Web applications.

BACKGROUND

Traditional Software Engineering

Traditional software life cycle models, also known as plan-driven or document-driven approaches, treat software development as a linear planning process where the majority of the planning is completed before design. The waterfall model, which consists of five phases, as shown in Figure 1, is the most common example. In the requirements phase, developers aim to determine their

Figure 1. Waterfall model

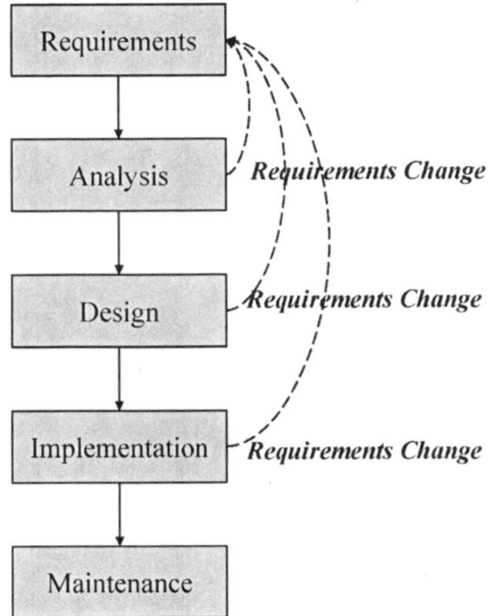

clients' needs based on interviews or questionnaires. Then in the analysis phase, developers analyze and refine the requirements and amend their findings in a software requirements specification (SRS) document, which acts as a contract between the developers and clients. Once the SRS is agreed upon, developers move to the design phase. The implementation phase does not start until completion of the overall design. After the product implementation phase, product delivery to the clients occurs, and finally the software development lifecycle moves to the last phase, post-delivery maintenance (Schach, 2003).

About 40% of software development companies still use the traditional waterfall model (Neil & Laplane, 2003). The document-driven approach is seen as an advantage for large, complex projects such as safety critical software or aerospace projects where there may be multiple contractors. The associated standards with their predictability help to mitigate personnel turnover and allow management to better compare projects (Boehm & Turner, 2004).

However, there are several drawbacks to the waterfall model. In particular, when the requirements change, the developers have to stop the current analysis, design or implementation, return to the requirements phase, and renegotiate the SRS. This process may also involve a change control board. The development can not continue until the requirements have been finalized again; this consumes time and resources. Another drawback is that clients do not see the product until the development organization finishes the implementation and releases the product. Furthermore the different background knowledge between developers and clients can also lead to alternative understandings of the SRS with the final product often not meeting the client's expectations.

To help counteract the limitations of the waterfall model, the rapid prototyping software life cycle model may be employed. In particular, this model allows feedback from the customer during requirements analysis, so there is a better likelihood that the final product will meet the customer's needs. Software developers may even decide to use an incremental model in order to gain feedback from the customer multiple times in the software life cycle. Note that this model still requires extensive planning of the overall architectural design before increments are implemented (Schach, 2003).

Although rapid prototyping and incremental development allow for more feedback from the customer than the waterfall model, plan-driven approaches are still more appropriate when the requirements are mostly stable (Boehm & Turner, 2004). In modern software development, it is almost impossible to ask stakeholders to give all of the requirements at the beginning of the project and expect them not to make changes later on. Projects with dynamic requirements due to rapidly changing markets are more commonplace. Especially in the Internet-based economy, clients not only require the flexibility to change or add requirements during the software development lifecycle, but they also want quick product delivery.

In summary, plan-driven methodologies are especially problematic for Web applications for three main reasons. First, most Web applications require rapid application development. Second, Web applications are highly dynamic in nature, and third, feedback from the customer should not wait until final delivery of the product.

Modern Agile Software Engineering

In order to better satisfy the demands of software development with a dynamic nature, agile software engineering approaches were invented. Some of the agile methodologies currently in use include Scrum (1986), Schwaber & Beedle (2002), Crystal (1996), Cockburn (2005), Feature Driven Development (1995), Palmer & Felsing (2002), eXtreme Programming (1996), and Beck (2005). All the agile methodologies share a set of common goals, which were formalized in the Agile Manifesto in 2001. The following quote clearly outlines the agile objectives in relationship to traditional software development.

Individuals and interactions over processes and tools.
Working software over comprehensive documentation.
Customer collaboration over contract negotiation.
Responding to change over following a plan.

That is, while there is value in the items on the right, we value the items on the left more (Boehm & Turner, 2004, p. 195).

In addition, a truly agile method must be "iterative (several cycles) and incremental (not deliver the product all at once)" (Boehm & Turner, 2004, p. 17)

Many software developers or companies have begun to use agile methodologies (Lindvall, et al., 2002; Sanja, 2005). For example, Sabre Airline Solutions Corporation has advocated using and adapting XP throughout the organization for many years. To date, there are over 30 teams, with more than 300 developers that have successfully used XP to develop projects (Layman, Williams, and Cunningham, 2004). At Microsoft, Scrum has been accepted and used in different development teams (Schwaber, 2003; Taft, 2005).

How Agile Methodologies Resolve Traditional Limitations

As mentioned above, there are two major limitations to the waterfall model—the inability to facilitate changing requirements and the release of the product to the customer near the end of the software life cycle. Agile methodologies solve these problems with their incremental and iterative development philosophy. To illustrate, we use the life cycle model of extreme programming (XP) (Figure 2). Projects are divided into a number of increments, and each increment includes user stories that are prioritized by the customer. User stories consist of one or two sentences on 3" X 5" index cards. The implementation (see inside the dashed line frame) includes a number of iterations. Each iteration typically lasting about two weeks, implements one piece of the product. During the implementation, customers can revise

Figure 2. eXtreme programming process

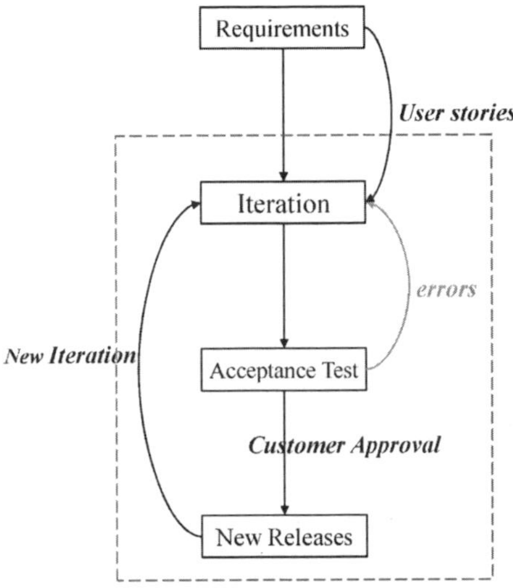

their requirements whenever they think it is necessary. Based on the changes, software developers re-factor the code to satisfy the customers. At the end of each iteration the developers invite the customers to perform acceptance testing to make sure this piece of the product meets their needs. The incremental and iterative development limits the effects of the requirements change since the change only affects the current piece of the project. Even in the worst case, a piece of the code can easily be completely redone (Beck, 2005). In addition, stakeholders gain a concrete feeling about the current product during acceptance testing at the end of each iteration thereby resolving the second limitation of the waterfall model.

One of the original practices of XP is the metaphor, an expression or an analogy used to foster project understanding during development. Throughout the remainder of this chapter, the authors will use some metaphors to give the reader a flavor of this practice. For more information about the 13 XP practices as described by Jeffries (2001) and the authors' experiences in a previous agile project, the interested reader may refer to (Mills, et. al., 2006).

IS DATABASE DEVELOPMENT READY FOR AGILE METHODS?

As previously mentioned, compared to conventional approaches, agile software development focuses on working code and customer collaboration as opposed to contract negotiation, and responding to change other than following a plan (Boehm & Turner, 2004). This is exactly what many project stakeholders need in today's rapidly changing marketplace; however, in the data professional community, *common wisdom dictates that the entire data model be carefully designed up front and protected from change thereafter* (Harriman, Leo, & Hodegetts, 2004, p. 100). The reason for this mandate is that the data model is usually the bottom layer of the application architecture, as shown in Figure 3.

The test code, applications, input data file, and output data file all rely on the database. Therefore, normally database refactoring costs more than code refactoring. For example, if the requirement causes the data's column value to change from 10 to 10.0, in order to preserve informational semantics along with behavioral semantics, the associated variable type must be changed from

Figure 3. Application architecture

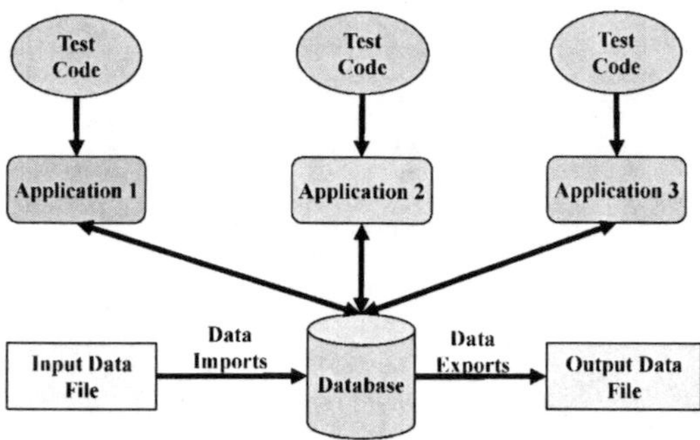

Figure 4. Big design up front process

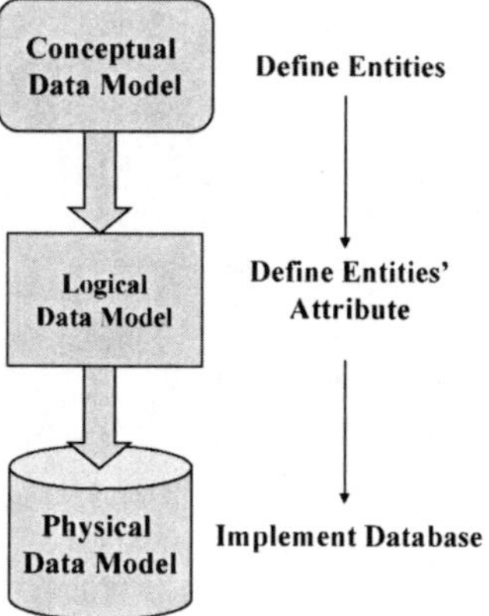

integer to double or float in the application and testing code. Especially when there are multiple applications that rely on one database, the propagation of changes affect each of the other components and their effects are widespread. The situation is even worse when the development team structure has the software developers responsible for the design and implementation of the GUI and the corresponding application, while the data professionals focus on building the data model. In particular, refactoring the database will also require refactoring by the software developer.

Big Design Up Front

Based on the above reasons, database development is still a linear and serial process. The leading development approach for databases is big design up front (BDUF), as illustrated in Figure 4. In the beginning of development, the data professionals gather, analyze and finalize all of the requirements from the clients. Then the data professionals design the conceptual data model based on the requirements. They define a database prototype with all the entities and their relationships. Next the data professionals define the attributes and the corresponding data type for each entity. In other words, the conceptual model is transformed to the logical data model. At this stage, the data professionals must determine the relationships between entities. Finally, based on the logical data model, the physical data model will be implemented by the data definition language (DDL) and stored on the hard disk. Also at this time, each attribute's data constraints will be defined.

BDUF works perfectly with the conventional software engineering approach. The software developers work with data professionals together acquiring and analyzing the requirements in the requirements and analysis phase. In the design phase, the domain model and data model are designed by each community respectively. Then in the implementation phase, software developers use programming languages such as Java or C++ to pass data to the database, and data professionals use the data manipulation language (DML) to store the data and update the database.

However, apparently BDUF is not suitable for agile approaches. When the software developers start the first iteration, there is no database for them because data professionals are still waiting to finalize the whole set of requirements.

Related Literature

Ambler (2003) points out that it is possible to adapt agile methodologies to make database development more flexible. He lists all the possible practices that could be suitable for agile database development, such as applying agile modeling (Ambler, 2004), using test-driven development, and taking advantage of database encapsulation techniques and development tools.

Harriman, Leo, and Hodegetts (2004) report on how they switched from BDUF to a "*just in time design*" approach for database development. In the beginning of their project, these authors used the BDUF approach to implement the database based on the initial requirements and then selected agile practices to develop the software application. However, they soon experienced difficulties. Because of the lack of complete details in the initial requirements, they designed and implemented entities and attributes in the anticipation of future needs. Nevertheless, as the requirements kept changing, the early anticipation designs became even more speculative, and the authors were not sure whether it was worth it to implement the designs. Therefore, in order to get rid of the distraction from speculative designs, they eliminated the speculative attributes and entities before coding an iteration, and implemented the database to just fit the current requirements.

Morien (2005) explains the "*focal entities prototyping approach.*" Instead of fully implementing a database based on the initial requirements, he uses the Entity Modeling technique, which focuses on identifying the entities and their relationships; then, he transfers this information to the physical database model. After developing a corresponding part of the application, delivery of this partial system and database occurs to allow the client to validate. The development cycle is repeated so that database development becomes iterative and incremental.

Initial Experience with Agile Practices

Based on the related literature above, the authors attempted to apply both the agile philosophy and its practices to database development. Although in the beginning, they discovered database development can be more flexible when agile practices are utilized, later on they noted difficulties in fully applying these practices. Note that agile methods recommend performing the simplest design for an iteration. One reason is that in a highly dynamic environment, the requirements could change at anytime. Therefore, any anticipated design for the future situation could become useless. As an example, in the beginning of a project, we can model a laptop as a laptop entity with attributes such as serial number, manufacturer, size, color and all its components. However, if in the final requirements, it turns out that the laptop entity only needs to store its serial number, then the original model of the laptop was only speculative, and it must model back as an attribute within other entities. This change will impact all of the relationships that were associated with the laptop entity, and propagate up to the correlated domain models and the related code. Therefore, it is important to apply the rule of thumb YAGNI ("you are not going to need it") (Beck, 2005). In other words, only include the necessary features. However, with too simple a design, it is hard to make a project's development move forward. For example, if we originally define the laptop as the attribute, "serial number," but later on the requirements indicate that the laptop should be modeled as an entity, then the database model needs to change and the impact for the development will occur again. With agile development, simple design may result in code that smells bad (Beck, 2005), so developers need to be ready to re-factor the code. Similarly, in agile database development the database can also smell bad. No matter whether we use "just in time" or "focal entities prototyping," as long as requirements changes affect the data model structure, we are not able to avoid database re-factoring.

So applying agility in database development, but at the same time reducing database refactoring as much as possible becomes the major challenge. As mentioned above, the essential idea of agile methods is iterative and incremental development. Also in order to reduce database refactoring, data professionals must do enough up front analysis in database development as Morein (2005, p. 105) points out,

As with any system development, some up-front analysis must be undertaken. How much analysis, and how complete that analysis must be, is the major point of difference between the Heavyweight Models of development, and the alternative Lightweight / Evolutionary development models.

Based on these findings, the authors decided to include up front analysis in agile database development to balance the two techniques. To the authors' knowledge, there is no literature or guidance on how to balance or adjust up-front design with agile practices in a database development environment. Although Morein claims that some of up-front analysis is necessary in agile development, he does not have any further discussion. According to the authors' hypothesis, if the data professionals can have a piece of a high stable database ready for an iteration using agile development, when the software developers start coding the application, the developers can focus on writing the application code, and not be distracted by the database revision. If the data professionals can continue to deliver a high stable database piece to the software developers, the entire development process can become iterative and incremental. So now the question is how to design this database piece? The authors have learned that dividing the initial database model into loosely coupled partitions greatly helps to achieve this objective.

To test their hypothesis about combining agile and database practices, the authors used their new approach in two projects—an online inventory system (OIS) and the marketing simulation study manager (MSSM), both of which are described in the next section. OIS was built using PHP and MySQL, but not using object-oriented and model-view-controller (MVC) design pattern technologies. MSSM was built with JSP, Java, MySQL and the MVC framework, Struts, and the object-relational mapping framework, Hibernate.

CASE STUDIES

Online Inventory System

Initial Requirements

The online inventory system is developed for a company (called XYZ in this chapter) to record the number of filters the company receives, cleans, and ships to their customers. The system can automatically generate online reports for managers of XYZ and their customers, who own the filters, to track filter status based on the time period or the shipments.

Every employee and customer must register an account with the appropriate security level in the system. Each account can only access or modify the current or lower security level's data. For instance, a received level employee account can only access the subsystem that is responsible for receiving the filter, but not the section to clean the filter, while the filter cleaner account can review how many filters were received for cleaning. Similarly, for the customers, for example, a manager of customer A's branch in Memphis only can access the filter data in the Memphis branch, but a president of customer A can access all the filter data throughout the country.

When the dirty filters first arrive at company XYZ, the filter receivers scan the filter barcode (filterID) into the system. If the filter is not in the system, the application must remind the user to register the filter first. Only a registered filter can go on to the clean process and XYZ sends back only cleaned filters to its customers. A filter can be marked as bad if it fails to pass the post test after cleaning. Bad filters must be destroyed, and this information must be recorded in the system. The system monitors each filter's status by its barcode, which is scanned in each step; the process provides insurance that there is no invalid operation.

There are two different report levels for both XYZ and its customers. The high level reports display the overview for the filters, such as how many filters are received, cleaned or returned based on the time period or shipment. The low level reports contain detailed descriptions for each filter, such as, the filter attribute (barcode, size, color, manufacturer), filter status (received, cleaned, destroyed, warehousing, shipped), and when and how the filter was cleaned, who cleaned it and so on.

Database Development

In the online inventory system, based on up front analysis for the initial requirements, the authors divided the initial database model into four partitions: User Account; Filter Receive; Filter Cleaning; and Filter Shipping. The User Account partition only includes the entities that store the information related to the Web user, such as the username, password, contact information, and associated customer with his/her company information. The Receive partition has the entities that store all filter's attributes, such as the size, color, manufacturer, and the receive status. The Cleaning partition stores the filter cleaning data such as remove dusty weight, post test record, cleaned status and so on. The last partition stores the shipping information such as the date and time for shipping and the track number. As shown in Figure 5, these four partitions are loosely coupled. There is only one foreign key (customerID) among the partitions. The foreign key customerID is a reference to the customer that owns the filter. If we know a filterID in the Receive partition, we can use its associated customerID to retrieve the customer information in the User Account partition. Similarly the filterID can be used to join the entities among the Receive, Cleaning and Shipping partitions to retrieve the filter information and its corresponding status.

Figure 5. Online inventory system database partitions

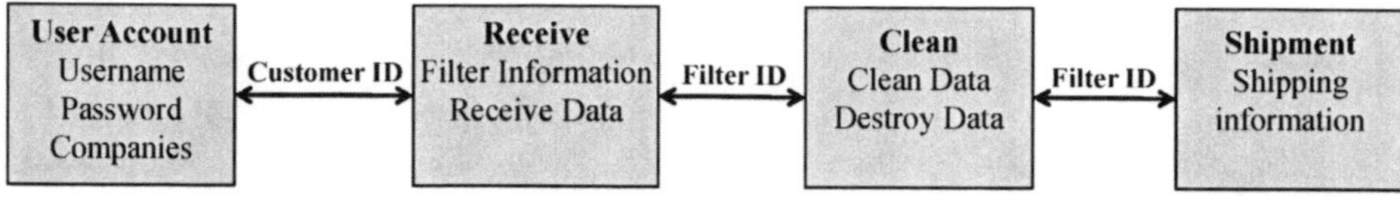

Once the partitions are defined, the next step is to conduct up front analysis for the partition that will be used for the current development iteration. In the dynamic environment, it is difficult for the clients to specify all the requirements in the beginning, but it is possible to capture most of the requirements for the design in a single partition of the database. If there are difficulties for clients to specify the requirements, as with agile practices, you can always develop a quick prototype to facilitate requirements elicitation. Rapid prototyping is especially helpful when there are not enough requirements to start the development. Once the analysis is completed, we can implement both the logic data model and the physical data model for the partition.

Notice partitioning the database is not based on the task load of an agile iteration. In agile development, an iteration is defined by its time period (e.g., two weeks in eXtreme Programming), and each iteration can have various tasks. The basic idea behind partitioning the database is to locate and to group together the entities that are highly coupled, eliminating entities exhibiting no or low coupling. So it is normal if the partition "over fits" the current development iteration by containing entities that are not for the current application development. For example, to implement an account management system for Web application, it may take more than one iteration, but all the entities of the account management partition should be designed and implemented before the first iteration starts.

Marketing Simulation Study Manager

Initial Requirements

The marketing simulation study manager (MSSM) is a Web-based application with a database. The application allows researchers (investigators) to predefine, or generate their own marketing research Web sites (research studies) or to modify existing studies through Web manipulation features. Each of these research Web sites can simulate a set of marketing experiments. Inside each experiment, there will be a number of questions for the respondent to answer about one product. Furthermore, based on whether the experiment will be visited by a control group or one of an experimental group of respondents, the site may or may not have media files (stimuli), which could be videos, audios or pictures for that product. The researchers will compare and find out the difference between these answers based on whether the respondents experience the stimuli, and then conclude how these stimuli could affect the other products in the market.

Each customer (respondent) of a research study Website must register an account, if he or she wants to participate in an experiment. Respondents do not know if they are members of a control group or an experimental group. Group assignments are made manually by the investigator or automatically by the software application. Each group can visit only one experiment per study. As the respondents go through their experiments, the answers from each respondent are saved in the database. Finally, the MSSM resulting data will be transferred from the database to the researchers' analytical tools (e.g., SAS).

Database Development

The database development approach has two steps. The first step is to analyze the initial requirements and to define the database structure, dividing it into a number of loosely coupled partitions. Second, based on the software development requests, the data professionals choose a partition, and using the up-front approach, analyze, define, and implement entities within that partition.

In the MSSM project, based on the initial requirements, the authors split the databasc into three components. The first component stores the Web user's information, such as the administrator, investigators, and respondents. The second part stores the necessary data for investigators to create their research or study Web sites: for instance, the number of experiments that will be in a study; what are the questions for the respondent in each experiment; what are the stimuli that will be displayed in each experiment, and so on. Finally, the third partition of the database is a collection of tables that store the respondents' answers or the study results of each Website. As shown in Figure 6, these three partitions exhibit low coupling. In other words, no matter how much the requirements relating to the Web user are revised, the changes will not affect the data model design for the study partition. As long as one can acquire the identification id of the study from the investigator or respondent, he or she can determine which study belongs to which investigator and which respondents are registered to that study. Also through the study id, the study data can be provided to the investigator. Similarly, one only needs the question ids of a study to reference the questions from the result partition; then it can be determined which answer is corresponding to which question in which study.

In an iteration that implements the features to upload the stimuli and input the Web page text information, questionnaires and multiple choice answers for the study, the software developers only need the stimuli, text information and questionnaires' entities for that iteration. But again if these entities are highly coupled with other entities (e.g., the study entity), then all related entities must be in the same database partition as well. After the definition of each partition of the database is complete, the developer can start using agile practices to gather enough requirements for the partition and implement the physical data model.

Compared to the work of Harriman, Leo, and Hodegetts (2004) and Morein (2005), the main difference here is that before the start of an iteration, there is a fully implemented database partition for the software developers. Although, in a dynamic environment, the database piece could be refactored due to a requirements revision, our approach is better than the "*just in time design*" or the "*focal entities prototyping*" approach. Database revisions are more likely when teams are composed of software developers who take on both the roles of developer and data professional, because of the technologies, culture, visions and priority differences between the software devel-

Figure 6. MSSM database partitions

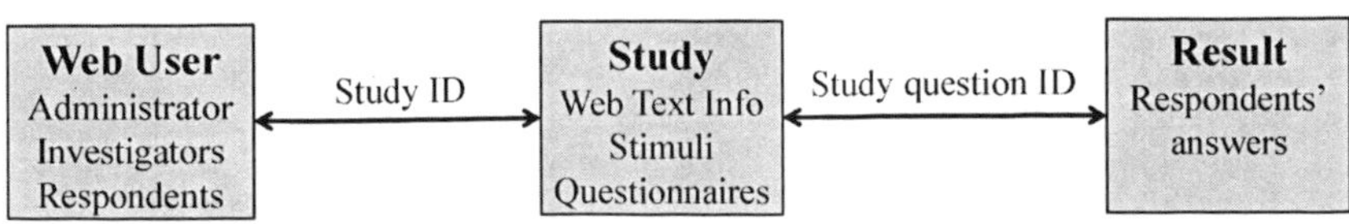

opers and data professionals. In fact, when the software developer and data professional is the same individual, the database maintenance may become unwieldy because the underlying paradigms required for each role are different. When developers implement the database, usually the implementation is based on their perspective as opposed to that of the data professional. In order to resolve this issue the software developers and data professionals should be responsible for the application and database development respectively based on a design agreement (Ambler, 2003). Furthermore, applying some techniques such as database encapsulation will help to resolve the technology differences between the software developer and the data professional. For more information on the difference in roles between the software developer and the data professional, please refer to Ambler (2003).

The Object-Relational Impedance Mismatch

In MSSM, we are using both object-oriented and relational technologies, because they both are leading technologies in programming and database environments, and both are widely used by most organizations. But it is clear that the match between these two technologies is not perfect. The problem is called the object-relational impedance mismatch (Ambler, 2003). So far there is no perfect solution for this problem, since to adapt either, the object-oriented or the relational database technologies to complement the other, the corresponding site will violate the rules of one of the technologies. One author even describes this problem as the Vietnam War in computer science (Newward, 2006). The history behind

Figure 7. Object-relational impedance mismatch

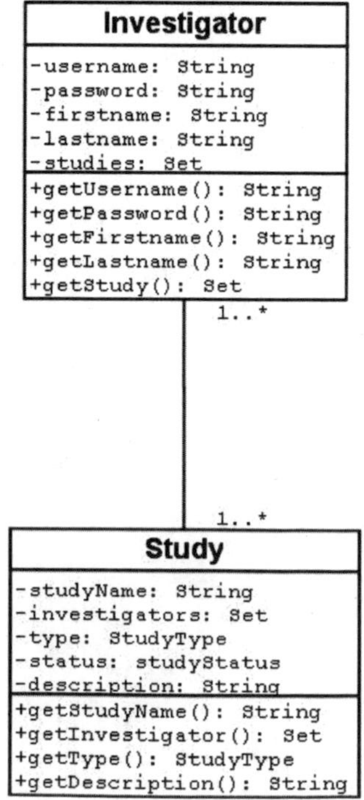

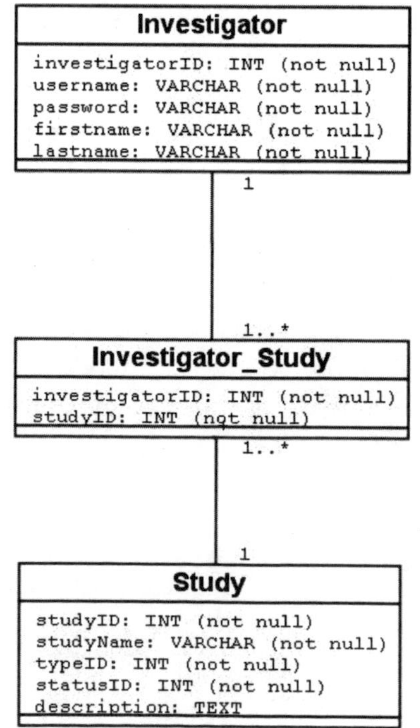

this mismatch involves the different technical and cultural backgrounds between the software and data communities.

According to the documentation of Hibernate (2006), the attribute of a class is called a property, whereas the attribute of a database entity is called an attribute. A predominant impedance mismatch between these two technological features appears in Figure 7. Note that the information in the figure is from the MSSM case study. On the left, is a UML diagram containing Investigator and Study classes. Similarly, the UML diagram for the Investigator and Study entities is on the right. The diagrams look analogous, as both Investigator and Study classes and associated entities have similar properties and attributes. The only major difference is that there is one more associative table, "Investigator Study," between the Investigator and Study entities in the entity model. The reason for this additional table in the MSSM database is that an Investigator needs to be able to create one or more studies in MSSM. Note that a study may involve a group of investigators. Therefore the relationship between the Investigator and the Study entities is a many-to-many relationship. Since relational databases do not support this type of relationship, typically an associative table will be introduced between the entities with multiplicity greater than one. Since object technology naturally supports the many-to-many relationship, there is no need for an association class between the respective classes (Ambler, 2003).

Database Encapsulation Strategies

Ambler suggests that an agile database administrator (DBA) should have the ability to work between both the object and data worlds and between the agile and traditional worlds. And furthermore, he recommends using the database encapsulation strategies to overcome the object-relational mismatch. (Ambler, 2003)

In agile database development, ideally the database system should act as a black box. In other words, the database should hide both the details of the implementation and the physical schema from the software developers. However, the database does provide the services for the software developers to create, read, update, and delete (CRUD) their business objects persistently (Ambler, 2003).

The advantage of database encapsulation is obvious. It dramatically reduces the coupling between the object schema and the data schema. Usually the database encapsulation layer lies between the business object and the data model layers, and all the business objects need to talk to just the database encapsulation layer instead of communicating with the database.

Ambler lists three database encapsulation approaches: (1) Implement the data access object (DAO) layer; (2) Use the transparent persistence technique; (3) Use enterprise service. In MSSM, we selected option two, because it has a big advantage over the other two options. In particular, the SQL can automatically be generated by the persistence framework on the fly. Therefore, the usage of the transparent persistence technique can reduce the code significantly by a factor of four to five when compared to the conventional Java database connectivity (JDBC) call-level interfaces (Barry & Stanienda, 1998).

Hibernate

Hibernate is the transparent persistence framework used in MSSM. In traditional JDBC call-level interfaces, the developers are required to write a large amount of code to store and retrieve business objects to and from the database. This coding job is usually complex, tedious and costly (Baure & King, 2004). However, Hibernate can resolve this coding issue well, as it can provide a batch of highly flexible methods to persistence objects in the relational database. For more information about how to use Hibernate, please refer to the Hibernate documentation (2006) or Peak and Heudecker (2005).

Figure 8. Architecture of hibernate framework

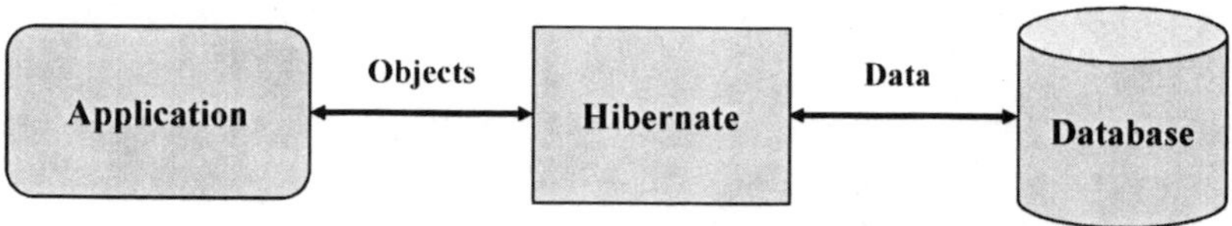

Figure 8 (adapted from (Ambler, 2003)) shows the layout among the business objects, Hibernate, and the database. Because the Hibernate layer lies between the application and the database, it helps separate the objects schema and the data schema. The advantage of Hibernate is it provides a facility to map the business objects to the relational database and automates the loading and saving of objects based on the configuration of the mapping files (Peak & Heudecker, 2005).

For instance, in Figure 7 there are both Investigator and Study classes and entities. Hibernate uses a mapping file for each of these classes, respectively mapping all the properties from a class to its entity or entities in the database.

To store the Investigator and Study objects in the database, software developers simply need to write the code below:

```
Investigator investigator = new Investigator ( );
Study study = new study( );
investigator.setPropertiesValue(propertiesValue);
study.setPropertiesValue(propertiesValue);
Session.save(investigator);
Session.save(study);
Session.getTransaction.commit( );
```

Similarly, to retrieve an Investigator or one Study object from the database, the following code is appropriate

```
Session.load(Investigator.class, investigatorId);
Session.load(Study.class, studyId);
```

Hibernate will generate the SQL statements dynamically and finish saving or loading the objects to and from the database. Furthermore to resolve the object-relational impedance mismatch, Hibernate provides a sophisticated idea. Recall that the mismatch is the additional associative table caused by the many-to-many relationship between the Investigator and Study entities. In Hibernate, data professionals just need to specify this relationship in both the Investigator and Study mapping files. When the objects are mapped to the database, Hibernate will take care of the mismatch. Software programmers can write their object-oriented code as usual. For example when assigning an Investigator to a Study, programmers just need to use a Java collection, (e.g., List, Set or Map) to add the Investigator object to the collection property in the Study class:

```
study.getInvestigator( ).add(Investigator);
```

Whenever this line of code is running, Hibernate will automatically generate the SQL statement to insert a record into the associative table with a pair of correlated identifies: investigatorId and studyId.

Decoupling, Decoupling, Decoupling

Although the persistence framework facilitates good design, the encapsulation strategies are equally as important in Web applications. As Fowler (2001, p. 96) points out,

Any code that does anything with a user interface should only involve user interface code…it should not manipulate the information other than to format it for display.

A layered Approach

The (Web) application component in Figure 8 may be further refined into four different Web logic layers—input logic, application logic, business logic, and presentation (Knight & Dai, 2002). The code for the input logic should handle the HTTP requests from Web users (such as validate the input data integrity and format) and store the user input data for the application controller. The application logic code determines the correct action to respond to the Web users' requests, (e.g., talk to business objects to access the database, or direct forward to another Web page and display it to the user.) Business logic is responsible for accessing the database, storing the data passed from the Web page, and retrieving the data for presentation. Presentation refers to the Web interface or Web pages implemented in HTML, PHP or JSP that will display in the Web users' browser.

It may seem that dividing a project into these four layers will make the architecture ponderous, thereby difficult to develop and maintain. But the layers do not have this effect. In the real world, the Web interfaces are usually designed and implemented by Web page designers. Just as software programmers do not need to know SQL, page designers do not need to know the programming language (e.g., Java). High coupling of the Web interface with other layers will mix the HTML and application code. One drawback of the mix is that page designers are very easily distracted and annoyed by application code. It is similar for the application developers; they do not want to see the HTML tags in their Java code (Davis, 2001). Therefore separating the Web layers helps each community focus on their respective jobs. A second problem with mixing HTML and Java is that, many code debugging and configuration technologies are not supported in the server page; so decoupling the Web interface from the Java code facilitates the debugging activities for the software programmer. Finally, decoupling means that when Web maintenance is performed, Web page designers and software programmers do not need go through the other community's code. Furthermore if a layer needs to be changed, it will not affect other layers, so Web maintenance is much easier.

Model-View-Controller and Struts

The ideas of Knight and Dai actually are based on the model-view-controller (MVC) design pattern (Davis, 2001). In the Web development environment, the Model represents the business objects in the business layer. The View is the Web interface layer and the controller layer includes both the input and the application logic. Therefore separating a Web application into three logics areas

Figure 9. MSSM architecture

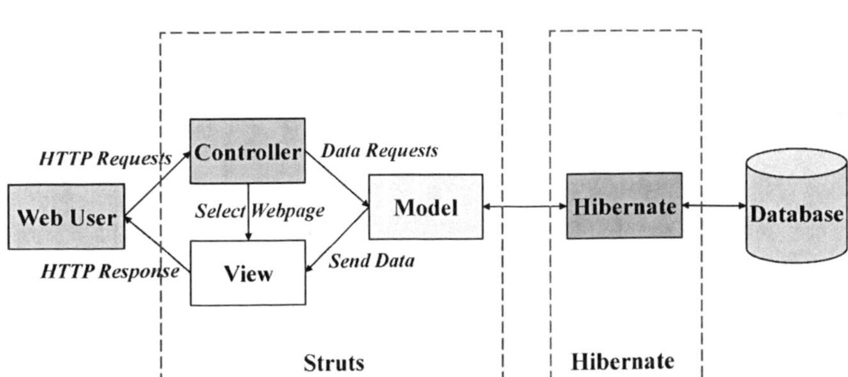

is equal to applying the MVC design pattern to Web development.

However, implementing the MVC design pattern from scratch is time consuming; a better way is to use an existing MVC framework. Since the authors of the MSSM project are using JSP and Java as implementation languages, the preferred framework is Struts. Struts is an open-source framework and fully implements the MVC design pattern. Although several other frameworks are excellent, "*none of these frameworks offer the broad-range of features and functionality for which Struts is known*" (Siggelkow, 2005, p. viii). The authors' objective is to achieve an architecture with the minimal amount of coupling. See Figure 9 for an overall view and clarification of the MSSM architecture. For more information about Struts, and how to use it, please refer to Siggelkow (2005) or the Struts User Guide (2005).

EXPERIENCE REPORT

Since agile methodologies advocate simple design and short release times, there was initially some concern that the amount of time spent on database analysis would slow down development. However, using another metaphor, just as sharpening the chef's knife will not slow down how fast he cuts through foods, some up-front analysis makes the database much more stable and robust. When the application implementation is ongoing, there is no speculation about entities or attributes in the database. Therefore, the up front analysis significantly reduces the chance of database refactoring, allowing the software developers to focus on application development and code revision.

A valuable lesson the authors learned during the development of MSSM is in Internet-based projects, requirements can be classified into three basic groups:(1) Web interface; (2) Web application; and (3) database design and implementation. So by focusing on the gathering of requirements in the third group first, data professionals can start the database design early and, in the mean time, the software developers can still gather their requirements for the application design. The parallel activities can reduce the time for agile database development.

Database Refactoring

In a highly dynamic development environment, refactoring can not be avoided. This section describes the authors' experiences from refactoring the database caused by the requirements changes in MSSM. The associated implementation for a requirements revision requires at least 7 steps, as shown below:

1. Revise the entities tables by inserting (deleting) the attributes.
2. Change the Hibernate mapping file for the affected tables.
3. Add (delete) the corresponding properties to the associate business objects.
4. Revise the Java code in the Struts action class (the application logic code) that needs to access those business objects.
5. Revise related Struts Form Beans (the input logic).
6. Revise the remainder of the related code due to the above changes (e.g., testing code)
7. Add the input field for the attributes in the corresponding Web pages.

The authors have discovered that the cost of changes in agile database development is still much higher than the cost of revisions to application code. More research needs to be undertaken to see if it is possible to further reduce the complexity and cost of database refactoring.

Based on the authors' experiences, having a partition database in place prior to each iteration is absolutely necessary in agile database development. Although the partitions do not facilitate database refactoring, they can dramatically reduce the chance of refactoring. In other words, in software development projects, database refactoring should be limited to requirements revision, not speculative design. Another big advantage of our approach is since the database is incrementally and iteratively developed, when the database refactoring happens, it affects only the current database partition. Compared to the BDUF, our approach is able to avoid the refactoring of the entire database.

Database Performance

The object-relational mapping technique provided by Hibernate greatly benefits the application development. Hibernate users do not need to write as many SQL statements to create, read, update and delete database operations. Although there are many advantages of the object-relational mapping approach, there is a metaphor that we apply to the database world, namely, "there is no free lunch." The reason for this claim is Hibernate always needs to generate extra SQL statements when inserting a new data object into the database if that object was mapped as a child table.

In a relational database, a foreign key of a child table references a tuple (record or instance) of its parent table. In the object-relational mapping, a tuple usually represents an instance, which is an object of the class. For example, in MSSM, each Study object associates to one StudyType, and the typeID references an instance of Study type in the StudyType table. A Studytype can include one or more than one Study. So the Study table is the child table and the StudyType is the parent table. The Study and StudyType tables exhibit a one-to-many relationship. In the object-relational mapping, the typeID attribute of Study maps to the studyType class. But the drawback of this mapping is obvious. Without using the object-relational mapping, the system creates a new Study that needs to persist in the database. The corresponding SQL follows:

```
INSERT INTO Study VAULES (typeID, StudyAttribute-value1, StudyAttributevalue2 ...)
```

According to Hibernate Reference Documentation version 3.1.1 (2006), when using the object-relational mapping, Hibernate will generate two SQL statements. The extra SQL is used to retrieve a StudyType instance for the type property of the Study object, because in the business model, the type property is an instance of the StudyType class. So before Hibernate can insert the Study into the database, it has to fill the type property for that Study object.

The one-to-many relationship is the most common relationship in the relational database. A child table can also have more than one foreign key reference to a different parent table. Furthermore, an Internet-based system needs to handle many users accessing the database simultaneously. Therefore generating extra SQL statements to perform the database operation could slow down the database performance. For more information about other object-relational mapping data retrieval concerns, please refer to Neward (2006).

CONCLUSION

Applying agile practices and iterative and incremental development to the database arena makes the database development better suited to agile software engineering processes. Moreover, the whole application development process can be dynamic in nature, thereby more readily satisfying the clients' needs. However, due to the high cost of database refactoring, data professionals should balance agility practices and up front analysis during development. In particular, by introducing some up front analysis, where a database partition is in place prior to all agile development iterations, data professionals can effectively provide the necessary balance. Also using encapsulation techniques can decouple the database from the application as well as the layers inside the application. Decoupling allows each development community to focus on its own project responsibilities, and significantly improves the development process.

Furthermore, decoupling reduces the difficulty of maintenance.

Future Trends

Test-driven development (TDD) or test-first development is a characteristic of some of the agile methods (e.g., XP) As the name suggests, test-first development means software developers design and implement the test cases before they write the application code. The approach is a good way to facilitate the design process, and it can guarantee 100 percent test coverage of the application code (Ambler, 2003; Beck, 2005). According to our experiences, test-driven database development (TDDD) is not as easy as test-driven application development. Most test-driven approaches heavily rely on third party tools. For software applications, the xUnit family is the most popular test-driven framework. However, in the database world, finding tool support for test-driven database development is the big challenge. The open source tool, DBunit, is still an emerging technology (Ambler, 2003). Two commercial tools found in (Ambler, 2007) are available only for database internal testing. The tools are Qute, a Quest unit testing engine for Oracle, and the testing tool in Visual Studio Team System for an SQL server. In summary, it still remains to be seen how much test-driven development will be used in the future for database development.

Although the object-relational mapping (ORM) technique may be considered as one of the methods to solve the object relational impedance mismatch problem, it does have a disadvantage when compared to the data access object (DAO) technique. ORM generates the queries on the fly, which makes the query tuning much harder to do than with the DAO. In other words, the data retrieval mechanism of ORM limits its database performance. Future research should address ORM's performance issues.

Another focus for future study is the scalability of our database partition approach to large, complex projects. Just as the metaphor says "there is no silver bullet;" agile methods are more suitable in the dynamic development environment, but plan-driven approaches have more advantages for large, complex projects when the requirements are mostly stable. Therefore, further combinations of plan-driven and agile methods may be required to handle database modeling, performance, and robustness in large database project environments.

ACKNOWLEDGMENT

The authors would like to thank Mr. David Mills and Dr. Dan Sherrell. Mills is a graduate student in the Department of Computer Science at the University of Memphis and served as Wei's agile partner in the MSSM project. His focus in the project was on the Web site design and implementation. Dr. Dan Sherrell, who is the client for MSSM, is a professor in the Department of Marketing and Supply Chain Management at the University of Memphis.

REFERENCES

Ambler, S. W. (2003). *Agile database techniques.* Indianopolis, IN: Wiley.

Ambler, S. W. (2004). *The object primer: Agile modeling-driven development with UML 2.0*, Cambridge University Press.

Ambler, S.W. (2007). Test-driven development of relational database. *IEEE Software*, *24*(3), 37-43.

Barry, D., & Stanienda, T. (1998). Solving the Java object storage problem. *IEEE Computer, 31*(11), 22-40.

Baure, C., and King, G. (2004). *Hibernate in Action.* Greenwich, CT: Manning Publications.

Beck, K. (2005). *Extreme programming explained.* Upper Saddle River, NJ: Pearson Education.

Boehm, B., & Turner, R. (2004). *Balancing agility and discipline.* Boston: Pearson Education.

Cavangess, C. (2004). *Programming Jakarta struts.* Sebastopol, CA: O'Reilly.

Cockburn, A. (2005). *Crystal clear, A human-powered methodology for small teams.* Boston: Addison-Wesley.

Davis, M. (2001, February) *Struts, an open-source MVC implementation.* Retrieved on January 3, 2007, from http://www-128.ibm.com/developerworks/java/library/j-struts/

Fowler, M. (2001). Separating user interface code, *IEEE Software*, *18*(2), 96-97.

Harriman, A., Leo, M., & Hodegetts, P. (2004) Emergent database design: Liberating database development with agile practices. In *Proceedings of the Agile Development Conference*, (pp. 100-105).

Hibernate (2006). *Hibernate reference documentation version 3.1.1.* Retrieved on January 2, 2007, from http://www.hibernate.org/5.html

Jeffries, R. (2001, November). What is eXtreme programming? *XP Magazine.* [Electronic version] Retrieved on January, 2006, from http://www.xprogramming.com/xpmag/whatisxp.htm

Knight, A., & Dai, N (2002). Objects and the Web. *IEEE Software*, *19*(2), 51-59.

Layman, L., Williams, L., & Cunningham, L. (2004). Exploring eXtreme programming in context: An industrial case study. In *Proceedings of the Agile Development Conference* (pp. 32-41).

Lindvall, M., Basili, V., Boehm, B., Costa, P., Dangle, K., Shull, F., Tesoriero, R., Williams, L., & Zelkowitz, M. (2002). Empirical findings in agile methods. In *Proceedings of XP/Agile Universe 2002*, (pp. 197-207).

Mills, D. L., Sherrell, L. B., Boydstun, J., & Wei, G. (2006). Experiences using agile software development for a shopping simulation. In *Proccedings of IEEE Southeast Con 2006* (pp. 285-290).

Morein, R. (2005). Agile development of the database—a focal entity prototyping approach. In *Proceedings of the Agile Development Conference,* (pp. 103-110).

Neill, C., & Laplane, P. (2003). Requirements engineering: The state of the practice, *IEEE Software*, *20*(6), 40-45.

Newward, T. (2006). *The Vietnam of computer science.* Retrieved on May 28, 2007, from http://blogs.tedneward.com/2006/06/26/The+Vietnam+Of+Computer+Science.aspx

Palmer, S. R., & Felsing, J. M. (2002). *A practical guide to feature-driven development.* Upper Saddle River, NJ: Prentice-Hall.

Peak, P., & Heudecker, N. (2005). *Hibernate quickly.* Greenwich, CT: Manning Publications.

Sanja, A. (2005). *Overview of agile management project perfect.* White Paper. Retrieved on January 2, 2007 from http://www.projectperfect.com.au/info_agile_programming.php

Schach, S. R. (2003). *Object-oriented & classical software engineering.* New York: McGraw-Hill.

Schwaber, K. (2003). *Agile project management with scrum.* Microsoft Press.

Schwaber, K., & Beedle, M. (2002). *Agile software development with scrum.* Upper Saddle River, NJ: Prentice Hall.

Siggelkow, B. (2005). *Jakarta struts cookbook.* Sebastopol, CA: O'Reilly.

Steiner, D. H., & Palmer, D. W. (2004). *Extreme software engineering.* Upper Saddle River, NJ: Pearson Education.

Struts (2005). *The Struts user guide.* Retrieved on January 2, 2007, from http://struts.apache.org/struts-doc-1.2.9/userGuide/index.html

Taft, D. K. (2005, November 11). Microsoft lauds 'Scrum' method for software projects. *eWeek.* [electronic version] Retrieved on January 3, 2007, from http://www.eweek.com/article2/0,1895,1885883,00.asp

Chapter X
Automatic Creation of GUI's for Web-Based ERP Systems

Jorge Marx Gómez
Universität Oldenburg, Germany

Daniel Lübke
Leibniz Universität Hannover, Germany

ABSTRACT

Service-oriented architecture (SOA) is an emerging architectural style for developing and structuring business applications, especially enterprise resource planning (ERP) systems. SOA applications are composed of small, independent, and network-accessible software components, named services. The service composition is normally based on the enterprise's business processes. However, current composition standards like BPEL have no ability to interact with users. Therefore, we propose a mechanism for including user interaction descriptions into the composition and extending the composition platform for generating user interfaces. In our case study, a federated ERP (FERP) system, this mechanism has been implemented in a prototype based on yet another workflow language (YAWL) dynamically generating Web pages for accessing the ERP system. Because every aspect including the user interfaces can be manipulated through the service composition, such systems are highly flexible yet maintainable.

INTRODUCTION

Enterprise resource planning (ERP) systems have become the most critical system in the IT infrastructure for most enterprises. However, most ERP software packages still follow a monolithic design. In recent years the software design of such systems tends to move to a new emerging architectural style—service-oriented architectures (SOA). Within an SOA, functionality is realized by small, fine-grained, independent, and network-accessible components called services. These services can be composed along the business processes of an enterprise. Since the compositions should be on a very high abstraction level, the resulting system can be changed easily.

An SOA is defined in this chapter as an enterprise-wide distributed software architecture for business applications that consists of services as its elementary software components. Those services are composed according to given business processes, and linked to the processes at run-time. SOA's main design goals are flexibility and maintainability in regard to changes affecting these business processes.

SOA is currently most often realized by utilizing Web service standards. Web services are based on XML. For calling Web services the SOAP protocol (Gudgin, Hadley, et al., 2003) has been defined. Composition of Web services is defined in the business process execution language (BPEL) (Andrews, Curbera, et al., 2003). While BPEL aims to support the alignment of Web services with business processes, it lacks one critical component of business processes—interaction with and involvement of responsible users. Recently, BPEL4People (Kloppmann, Koenig, et al., 2005)—an extension for BPEL has been proposed. However, BPEL4People treats end-users merely as just other services. The developers and system integrators still have to develop matching user interfaces in traditional programming languages. Development of user interfaces requires much effort and runs contrary to the suggested flexibility of SOA.

This problem becomes even more dominant for smaller enterprises, which need to be very flexible in the market for maintaining their competitive edge but do not have many resources that they can devote to their IT infrastructure.

In this chapter we present a solution on how to make user interfaces for ERP systems more flexible by incorporating parts of user interface models into the service composition. Thereby, the user interface can be easily modeled and updated without having programming skills.

Within the course of this chapter, the differences between business processes and service compositions are explained first. Afterwards, our concept for annotating user interfaces is presented. In the following section the generation algorithm is presented. Then a case study, namely the federated ERP system, is presented, which heavily utilizes the generation facilities described before. Finally, a conclusion and an outlook are given.

BUSINESS PROCESSES AND SERVICE COMPOSITIONS

Business processes define how activities in an enterprise should be performed. Additionally, they define who is responsible for performing these activities and which resources are affected. Business processes are designed from the management point of view. Based upon their definition, activities within enterprises can be improved, and supporting IT systems can be designed. There are many notations available for modeling business processes, among them event-driven process chains (EPCs) (Mendling & Nüttgens, 2005) and the Business Process Modeling Notation (BPMN) (White, 2006). Taking an order as an example for an administrative business process is illustrated in EPC notation in Figure 1.

Service Compositions have a similar structure compared to business processes (Henkel, Zdravkovic, et al., 2004). However, they offer a technical view of the IT landscape of an enterprise. They define the control-flow between different services, related error-handling, and data management. While the control-flow should optimally match the one of the underlying business processes, technical constraints often require slight differences. For example, it is possible that certain activities cannot be performed in parallel due to constraints of the software. For maintaining a SOA it is therefore a critical task to synchronize changes of the business process with the service composition.

However, this is not the only difference between business processes and service composi-

Figure 1 Sample order process

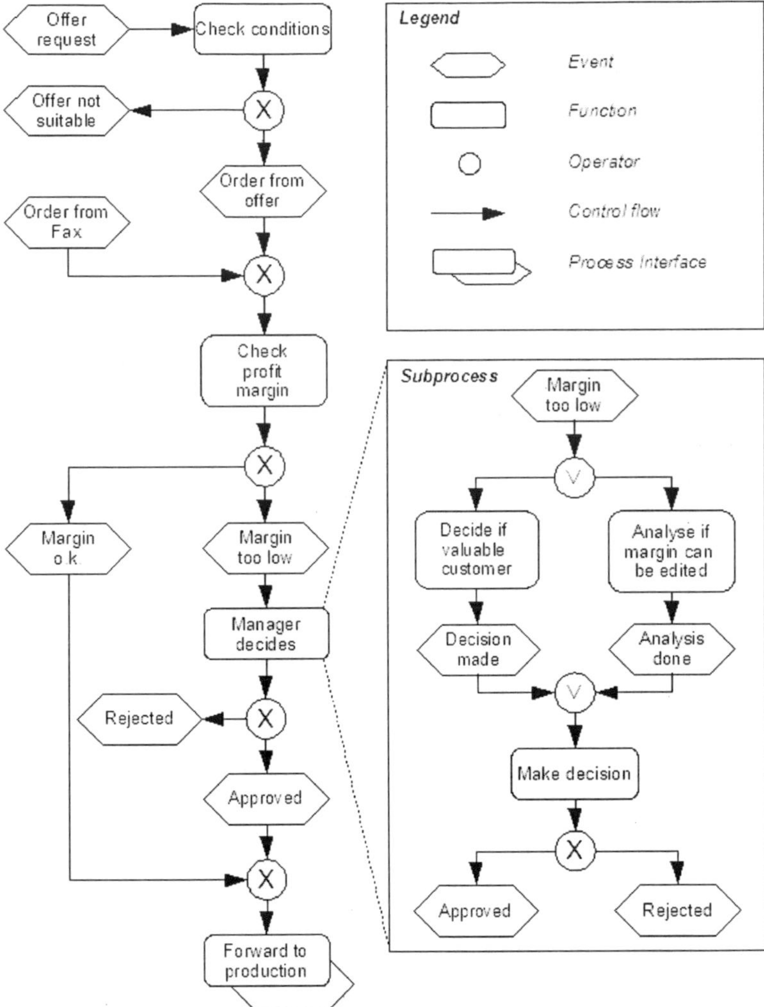

tions. Currently, the most often used service composition language is BPEL, which focuses on system-to-system interaction. Consequently, only fully automated business processes without human interaction can be modeled satisfyingly in BPEL. As a consequence, BPEL has often been criticized as being not suitable for building complex SOA systems. The general structure of the control-flow for BPEL is shown in Figure 2—the clients have to call the BPEL processes that return some data. Then, the clients can show user interfaces and invoke the BPEL process again.

As a reaction to the missing user interaction, BPEL4People (Agrawal, Amend, et al., 2007) has been proposed, which integrates users into the BPEL composition.

However, it treats human users merely as different services, which can send XML to and receive it from BPEL compositions. During development, user interfaces need to be developed, which offer matching user interfaces which can handle the corresponding XML data. While the BPEL4People composition can better reflect the control-flow of the original business process

Figure 2. Control-flow partition for user interactive processes using BPEL

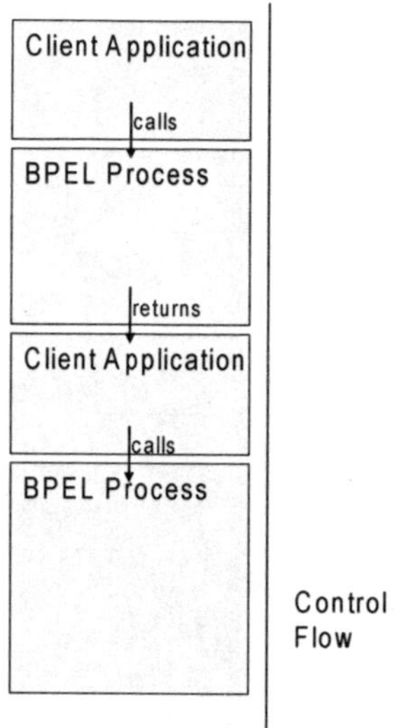

Figure 3. Control-flow partition for user interactive processes using BPEL4People

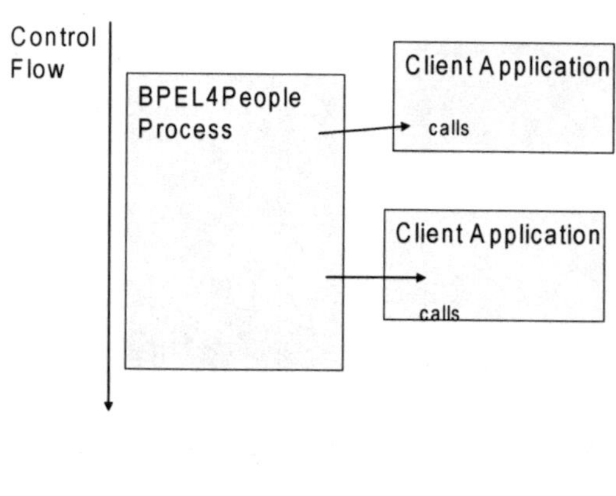

(Lübke, 2007) (see Figure 3 for the control-flow of the BPEL4People solution), the development and maintenance effort still remains very high. In the following section, a concept for modeling the user interface on the same abstraction level as the composition is presented which addresses these problems.

USER INTERFACE DESCRIPTIONS IN SERVICE COMPOSITIONS

Some approaches for modeling user interfaces exist. For example, TERESA (Mori, Paternò, et al., 2004) and an approach for supporting business processes with user interfaces by Sliski et al. (Sliski, Billmers, et al., 2001) have been proposed. These are very heavyweight user interface models that allow modeling of rich user interfaces. However, creating such models is a time-consuming task and hindered the adoption of these approaches in practice. The same has happened with all heavy-weight user interface modeling approaches. Furthermore, such approaches are not integrated into service compositions but require new models.

The main goal of our approach is to make an SOA—and consequently the applications built upon it—more flexible by managing the user interfaces completely with the service composition. Not only the control-flow of the composition resembles more closely the one of the business process, but furthermore, the user interfaces do not need to be developed using traditional languages. Therefore, service compositions can be easier modified. As a result, the deploying organization can change its business processes more often and with less effort and associated costs.

To achieve this goal, the user interface must be modeled within the service composition and consequently on the same abstraction level. For some time, research has been done in the area of model-based user interfaces (MB-UI) (Lübke, 2007), which has yielded many results. Many approaches are based upon so-called Task Models (Silva, 2000). These task models describe the activities a user should be able to perform. However, many models from this area have been too complex to be useful in practice.

Table 1. Annotations for business processes and service compositions

Name	Description	Notation
Selection	The user selects data from a collection of possible choices.	Select production plant
Edit	The user edits some information object from the data model attached to the business function.	Edit purchase order
Control	The user explicitly invokes some action. This is used to model navigational decisions.	Perform action on order
Visualize	Data is presented to the user.	Visualize difference

Figure 4. Annotated part of the sample process

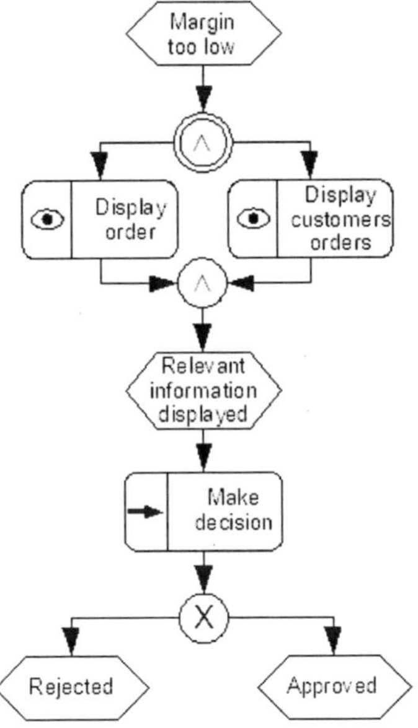

Since task models resemble business processes on a very low level (Trætteberg, 1999), they can be integrated well into service compositions, which shall resemble the business process control-flow. Our approach uses activities which already exist in all business process modeling languages and service compositions. The activities are decomposed using sub-processes. The sub-process is treated as a task model and annotated. To achieve this, following annotations are introduced (Lübke, Lüecke, et al., 2006), which together resemble a small task model—Selection, Edit, Control, and Visualize. This task model can be practically used due its small size. The annotations are illustrated in Table 1.

Additionally, each user-performed activity is assigned a typed variable. The variable type is used to generate a matching viewer and editor, and to store the results delivered by the user. Variable Types can be simple types and aggregate types, like lists.

Using this method, it is possible to annotate business processes and service compositions alike. For instance, a part of the example process with user interface annotations is shown in Figure 4.

Using these annotated processes or compositions our algorithm can generate user interfaces at run-time and pre-generate them at design-time. The algorithm is presented in the following section.

GENERATION OF USER INTERFACES

Algorithm

The generation algorithm for generating the matching user interfaces works by recursively going through all data types required for the concerned variables. Each data type is traversed separately. If for the main data type a matching editor or viewer (depending on the task) is available, it is used for the user interface. If no such editor or viewer is available in the system, a screen area is reserved that is populated with the sub-types. For each sub-type an editor or viewer is looked up. If none is available, the structure is traversed further. The recursion ends at a minimum if a primitive data type is detected, for which a viewers and editors are available in the system at a minimum. Primitive data types are string, integer, Boolean, date and so on. The generation of a customer editor is shown in Figure 5.

The whole generation of a sample process to a matching user interface is shown in Figure 6; for each activity in the task model a matching viewer or editor is created as discussed before. The generated components are aggregated on a single form as shown in the sub-window on the right hand side of Figure 6. The client surrounding this sub-window provides the management of tasks. Users can initiate new business processes and manage their tasks by using the standard functionality of the client application.

After the user has entered all data and has submitted them, the client sends the data back to the service composition which can proceed and call a service or initiate another user interaction.

The presented algorithm is independent from the implementation. Target user interfaces can be for example traditional clients and Web interfaces. Additionally, the interface is described on such a high abstraction level, that interfaces can be generated for different devices with different screen sizes. For example, a PDA can have a registry of components, which is completely different from a desktop computer. Therefore, this simple annotation of business processes can be used to accommodate different devices—only matching generators need to be implemented once.

Figure 5. Generation of a screen mask (b) from a data type (a)

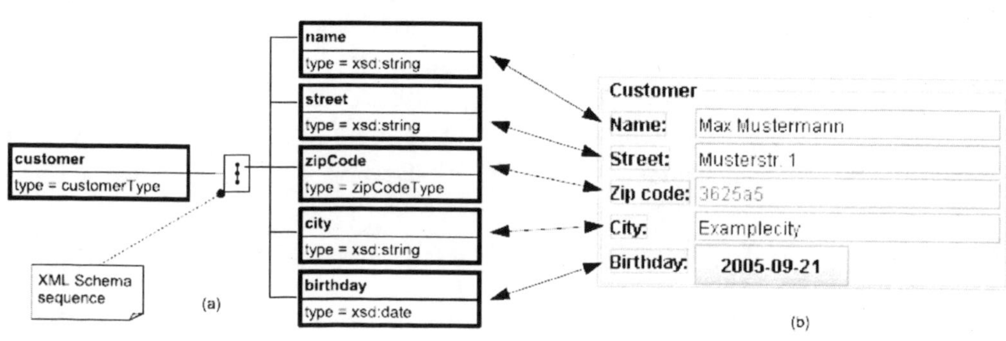

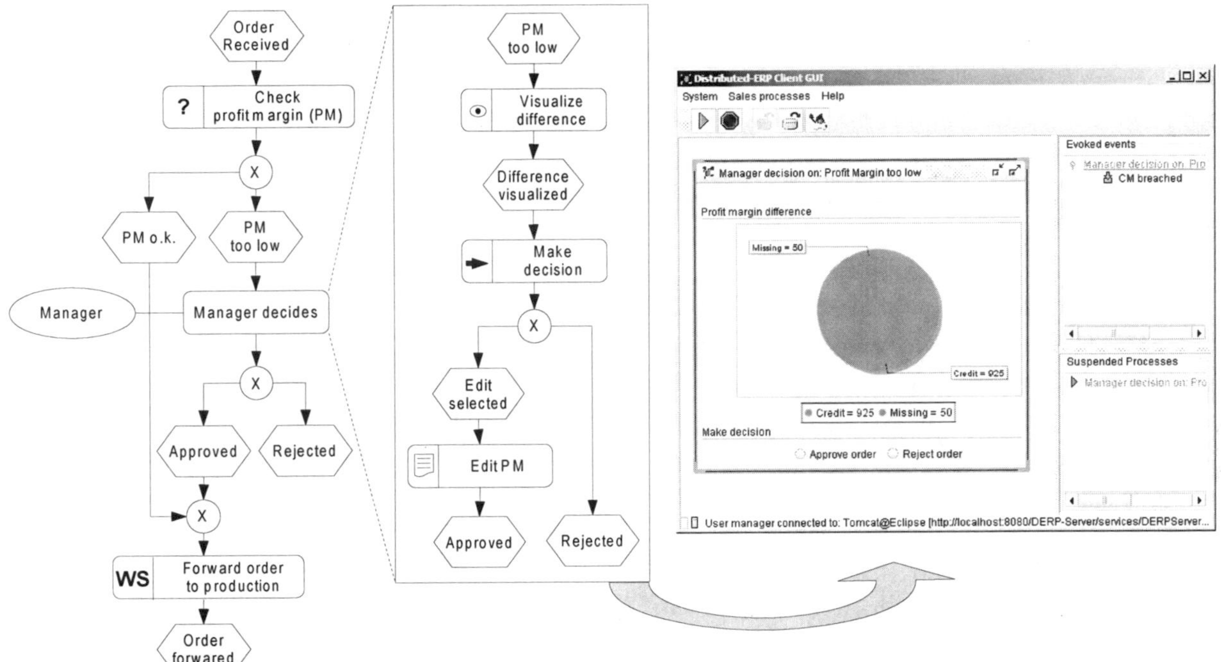

Figure 6. Generation of a whole user interface from a business process

Furthermore, the editor and viewer registry allows flexibility for advanced user interface design when needed. Advanced components can be stored in the registry for their corresponding data types. Although they need to be implemented in traditional languages, they can be reused by all business processes, still saving some effort.

Run-Time Generation

The presented algorithm can be used for different purposes in the development process. This section will explain the usage scenario for using run-time generation.

In run-time generation, the composition places a user interaction request until the user logs into the system. The request contains the task model, the data type description, and the data itself. When the user proceeds with the user interaction, the client downloads the task model and its accompanying data and generates the matching user interface on the fly. Therefore, the editor and viewer registry must be available for each client.

Run-time generation allows easy maintenance of the service compositions, since the composition can change without the need to change the client if no additional editors and viewers are needed. Furthermore, the service composition designer does not need to have technical knowledge in programming user interfaces. However, the client's generator is the limiting factor. Since the task model itself is very simple, advanced forms cannot be modeled.

The run-time generation is very suited for developing initial software versions and prototype applications which can be changed on the fly during interviews and discussion. Additionally, it is well-suited within projects, which have the ability to deliver a sufficient number of editors and viewers to their customers. Therefore, the customers can modify the services without the need to develop software themselves.

Design-Time Generation

If the user interfaces need major modifying, run-time generation may be too limited. Either, the generators and the task models need to be improved, thereby making them more complex, or the user interfaces need to be reworked. If the task model would become to complex to handle, it can be better to generate user interface code from the annotated business process or service composition. This approach works equally for both model types, in contrast to run-time generation which works only on service compositions. The generated classes can be modified by developers. The main logic can be generated, which addresses messaging and data serialization and deserialization. However, the customized user interfaces need to be maintained separately and cannot be controlled by the service composition alone. Changes in the service composition may require additional rework on the client side. Furthermore, each client update needs to be rolled out. Since the source code has changed, a new binary must be installed at each user device.

CASE STUDY: FEDERATED ERP-SYSTEMS

Motivation

Classical ERP systems offered by major software companies provide a big area of functionality. However they are very expensive. There are many low-cost ERP solutions on the market, which aim at small and medium enterprises (SME), but their functional range is often not adequate for the targeted audience. This problem is described as the ERP Dilemma for SMEs. A study conducted in our area (631 SMEs and 27 enterprises with more than 250 employees have been interviewed) has found that SME face similar requirements (see Figure 7) compared to major enterprises. However, due to their limited financial resources, SMEs are not able to afford classical ERP systems offering the required functionality.

The costs of the adoption of classical ERP systems are not limited to the acquisition of the software. A very important factor of those systems is the high hardware demands. In many cases it is necessary to allocate high-end systems. Furthermore, lots of external service providers are required, which install, introduce and customize the software [5].

Because SMEs are achieving their competitiveness due to the focus on differentiation strategies associated with a high level of flexibility and innovative ability, their business processes are modified more often, which requires high-priced customization of their ERP software. One opportunity for decreasing the costs is the application service providing (ASP), whereat the whole ERP application is offered by an external service provider, and the actual enterprise accesses their own system via the Internet. In this case the costs for the utilization can be paid as a bill, depending on the used volume of the software or in a way of a monthly payment.

However, concerning former experiences, this strategy is not well accepted in practice, because enterprises are not willing to save their valuable data on third-party providers' systems (Walsh, 2003). Thus a solution is necessary, which supports local data management, is inexpensive in adoption and maintenance, and offers an easy adjustment for changing business processes. This is the intention for the development of federated ERP systems, which offer their functionality in a shared architecture (Brehm & Gómez, 2006). Inside the enterprise the FERP-client-software is used, which offers basic functionality and data management. All further functionality is available as Web services on different peers inside the FERP-Network. Thus, the enterprise only pays for the functionality, which it finally uses as Web service. Because of the minor complexity of the client software the cost for installation and maintenance is decreased. Furthermore, it is much easier for providers of ERP components to enter the market, because they don't have to offer an entire ERP system anymore, but can rather concentrate on developing separate and not-yet available functionality. The architecture of a federated ERP system as illustrated in Figure 8 is based on researches considering the FERP approach and represents the fundament for the prototypical development of the workflow- and GUI components. The FERP client is deployed at the particular enterprise, which uses the federated ERP system. It acts as the client, because it presents one peer of the whole FERP network. Considering the enterprise, this component presents a kind of ERP server [8], which provides the entire functionality for the end user, who does not notice the communication with the Web services.

Figure 7. Percent of operational software in particular divisions (in %)

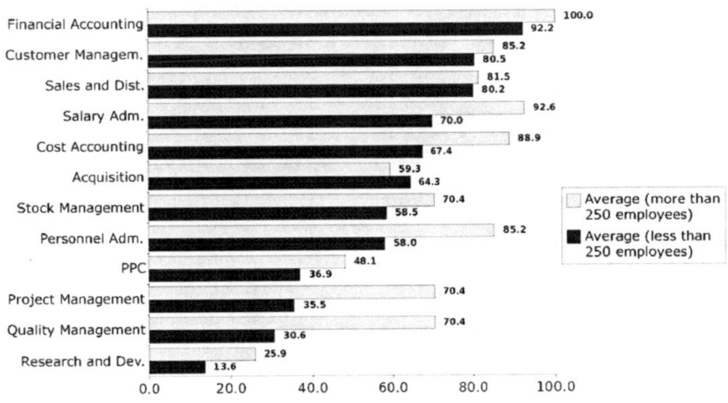

Figure 8. FERP system architecture

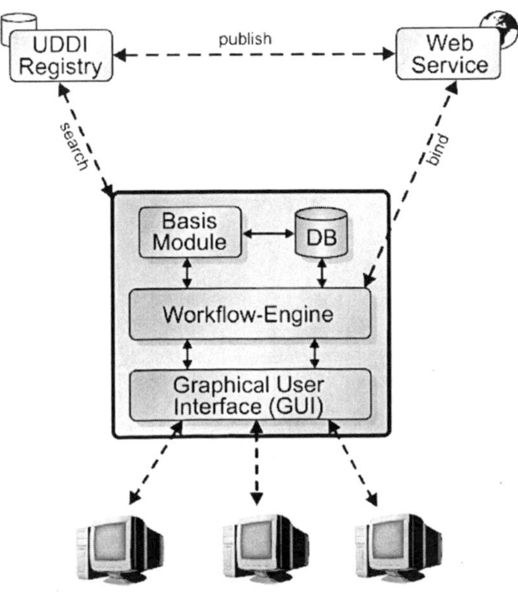

The function of the GUI component is to realize the communication between the workflow-engine and the end-user who runs the system. If interaction with the user is required during workflow processes, the GUI generates the needed views. The user accesses the system using a common Web browser via HTTP. The GUI is realized using a JSP- and Servlet-Technology running on a Java application server. The workflow engine's main function is to take care of the execution of the stored workflows, which represent the business processes within the enterprise. The engine controls the sequence of activities considering the terms and conditions, which are defined by the workflow designer.

Prototype Implementation

The prototype was developed mainly to gain experiences with the following components: the workflow engine; the Web service interface; the process parser; and the generation of graphical user interfaces. Modules including basic functionality and the connection to the enterprises database are not realized yet. It is assumed that these components are realizable via Web services later. The core of the application is the workflow engine, which connects all the involved components. The engine is responsible for realizing the business processes' workflow and the handling requests for functionality as well as reacting on incoming events. For generating the required user interfaces during run-time, it is necessary to supplement existing models of existing workflows and composition languages Web services. There are two possible approaches for this purpose. On the one hand, it would be possible to add additional information to the Web services, which specify the required GUI elements. Principally, the existing constructs and standards would be adequate for that. On the other hand, the second approach for integrating GUI information into the process flow is about to expand the business process definitions. In this case, the process designer (workflow designer) is the one, who adds the required information to the process definitions. An overview about the planned structure of the prototype is shown in Figure 9.

The prototype architecture is based on the model view controller (MVC) pattern. Thus, there is loose interconnection between those three parts, so that changes in one of these parts will hopefully not require modification of the others. For realizing the MVC concept, the Struts Framework is used. Figure 10 shows the MVC structure and its specification regarding the FERP prototype.

Figure 9. Raw architecture of the FERP-prototype

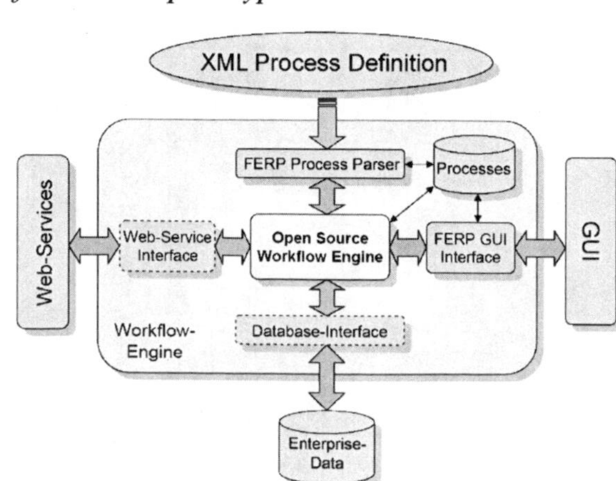

Figure 10. Architecture of the FERP prototype

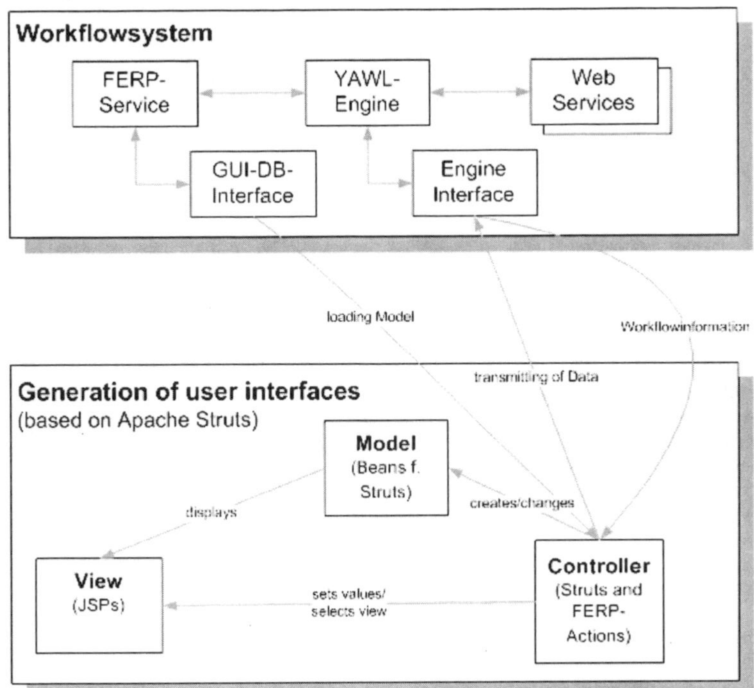

The view component is realized by Java Server Pages for generating HTML code, which can be interpreted by every Web browser. The design of the optical characteristics is implemented via cascading styles sheets (CSS), whereby the GUI pages will be more flexible for customizing. The management of the workflow data is part of the workflow component.

The core is represented by the used YAWL engine and its possibility to include Web services. If a GUI task is activated during a workflow, the YAWL service, called "FERP Service," will be connected, which refers the required information from the workflow engine. This data are latched via the GUI-DB interface in a database, until a user is going to login to complete the task. Therefore, the data is read out from the database and transferred into Struts business objects. The engine interface provides interfaces wherewith workflow information can be retrieved out of the workflow engine. Furthermore, this interface returns input by a user back to the YAWL engine.

CONCLUSION AND OUTLOOK

SOA promises to make IT infrastructure more flexible. However, existing standards are only made to address the flexible and easy arrangement of services to offer the logic required within a business process. The matching user interfaces are neglected. Within this chapter we proposed annotations to complement existing business process and service composition notations. Using these annotations user interfaces can be generated without needing further human intervention.

SMEs do not have much IT knowledge in-house. Therefore, changing their IT systems is costly. This is especially cumbersome when changes are needed for accommodating business process changes which can be vital. In the presented Federated ERP system, the user interfaces are generated using the presented method. This allows even non-programmers to make changes to the business process and change the application itself. All of this can be done without resorting to traditional programming languages. The Federated ERP system is based on the YAWL workflow language and realizes the user interface as a Web front-end. The user interface generator works completely on the server-side. As such, no client software updates are needed when new business processes are rolled out. The FERP system has been implemented as a prototype and first example business processes have been implemented. This prototype will be used to further evaluate the usability and suitability of the user interface generation. Such evaluation has already been done for smaller systems (like a student theses management system) and led to usability improvements of the generator component.

REFERENCES

Agrawal, A., M. Amend, et al. (2007). *WS-BPEL extension for people (BPEL4People), Version 1.0.*

Andrews, T., F. Curbera, et al. (2003). *Business process execution language for Web services Version 1.1.*

Brehm, N., & J. M. Gómez (2006). *Distribution of ERP system components and security considerations.* Paper presented in the 17th IRMA International Conference - Managing Modern Organizations with Information Technology, Washington, USA.

da Silva, P. P. (2001). *User interface declarative models and development environments: A survey.* (LNCS 1946, 207-226).

Gudgin, M., M. Hadley, et al., (2003). *SOAP version 1.2 part 1: Messaging framework.* World Wide Web Consortium.

Henkel, M., J. Zdravkovic, et al. (2004). *Service-based processes: Design for business and technology.* ICSOC '04: Proceedings of the 2nd international conference on Service oriented computing.

Kloppmann, M., D. Koenig, et al., (2005). *WS-BPEL extension for people—BPEL4People* (A Joint White Paper by IBM and SAP), IBM/SAP.

Lübke, D. (2007). *User interface design styles in SOA applications.* Paper presented at the 8th Annual Global Information Technology Management Associaton World Conference, Napoli, Italy.

Lübke, D., T. Lüecke, et al., (2006). *Model-driven development of business applications using event-driven process chains.* GITMA 2006, Orlando Florida.

Mendling, J., & Nüttgens, M. (2005). EPC markup language (EPML) - An XML-based interchange format for event-driven process chains (EPC). *Information Systems and e-Business Management (ISeB) 4*(3), 245-263.

Mori, G., F. Paternò, et al., (2004). Design and development of multidevice user interfaces through multiple logical descriptions. *IEEE Trans. Software Eng., 30*(8), 507-520.

Sliski, T. J., M. P. Billmers, et al., (2001). *An architecture for flexible, evolvable process-driven user-guidance environments.* Paper presented at the8th European software engineering conference held jointly with 9th ACM SIGSOFT international symposium on Foundations of software engineering, Vienna, Austria.

Trætteberg, H. (1999). *Modelling work: Workflow and task modelling.* CADUI.

Walsh, K. R. (2003). Analyzing the application ASP concept: Technologies, economies, and strategies. *Communications of the ACM, 46*(8), 103-107.

White, S. A. (2006). *Business process modeling notation specification, object management group standard.*

Chapter XI
Prototyping in Web Development

Clif Kussmaul
Clif Kussmaul, Elegance Technologies, Inc., USA & Muhlenberg College, USA

Roger Jack
Elegance Technologies, Inc., USA

ABSTRACT

This chapter addresses issues, alternatives, and best practices for prototyping in Web development. The chapter's primary objective is to provide a clear and concise overview of key concepts and best practices for practitioners and students, as well as other audiences. The chapter focuses on graphical user interface (UI) prototyping for Web development, but many of the principles apply to non-UI prototyping and other sorts of software development. First, we introduce and motivate the chapter, and review the major objectives, benefits and risks, and classifications of prototypes. Second, we describe the major approaches to prototyping. Finally, we conclude with future trends and a summary of best practices.

Introduction

In software development, prototyping is the process of creating a preliminary version for evaluation, before investing more resources. The prototyping process can be divided into four key steps (see Figure 1); Floyd (1984) uses similar steps, but different terminology. First, we *prioritize* the objectives and scope of the prototype, so we can focus on critical aspects of the prototype, and avoid aspects that are not immediately relevant. This is discussed in the section titled "Objectives and Scope of Prototyping," Second, we *create* the prototype, which typically is much easier than constructing the final system. A set of approaches is described in the section titled

Figure 1. Key steps in prototyping

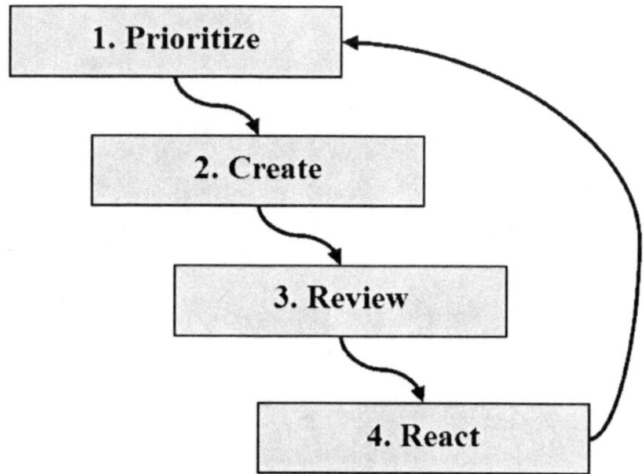

"Creating Prototypes." Third, we *review* the prototype to understand what works well, what could be improved, and what new issues have been identified. Several evaluation methods are described in the section titled "Reviewing and Reacting." Fourth, we *react* to the prototype and determine what to do next. In some cases there may be nothing more to learn from the prototype, and it can be discarded, or archived for future reference. More often, the evaluation will lead to more iterations through these four key steps.

This chapter provides a concise overview of key concepts and best practices for practitioners and students, as well as other audiences. The chapter focuses on graphical user interface (UI) prototyping for Web development, but many of the principles apply to non-UI prototyping. Similarly, prototyping issues and approaches apply to many sorts of software development, but the chapter emphasizes Web applications, based on a variety of Web development projects the authors have been involved with in the last ten years. We provide a multidisciplinary perspective, since effective prototyping involves a variety of disciplines, including business, psychology, and software engineering. The chapter also draws on our experiences developing a tool to help design, simulate, document, and review prototypes for Web applications and other software systems.

BACKGROUND

In this chapter, "development team" is the group of analysts, designers, software developers, and other people responsible for building a Web-based application, or "system," which may be divided into numerous "components." "Stakeholders" are other people or organizations with a significant interest in the project, including upper management, other business functions such as sales, and (of course) the intended end users of the system.

Web development, like other software development, involves a wide variety of activities. These activities can be grouped into four broad categories that span the system's life cycle. First, the team must *analyze* the problem to understand users, their needs, important tasks, and other relevant requirements. Second, the team must *design* the system or component to fulfill these requirements. Third, the team must *build* the system or components; this includes activities such as coding, integration, and testing. Finally, the team must *maintain* the working system as long as it is in use, by correcting defects and adding enhancements. In "waterfall" development (e.g., Royce, 1970), activities are generally performed in stages that parallel the categories described above; the requirements are analyzed, then the entire system is designed and built, and so forth. However, a pure waterfall approach is not feasible in many situations, since work in one stage may reveal omissions or errors in an earlier stage, and because requirements may change during development. Thus, some organizations prefer "iterative" or "spiral" development models, which consist of multiple cycles, each of which contains activities from all four categories. "Agile" development processes such as Scrum may contain many cycles, of a few weeks each (e.g., Cockburn, 2002; Highsmith, 2002).

Web development differs from other types of software development in some important ways. For example, Ginige and Murugesan (2001), Ginige (2002), and Murugesan and Ginige (2006) identify several challenges in Web systems, including the continual growth of requirements, the continual change in information content, a diverse set of stakeholders, and the need for multidisciplinary development teams. They describe a hierarchy of Web-based software, ranging from static Web pages through database-driven Web sites to complex Web systems, and propose a set of processes to help address these challenges. Kappel, et al. (2003) describe a similar but larger set of challenges, which are categorized by their focus on product and content, usage context, system development, or system evolution. Chen and Heath (2005) also identify challenges, including usability design, content maintenance, scalability, security, fast deployment, and competing architectures, platforms, and tools. Similarly, Mendes, Mosley, and Counsell (2006) identify 12 areas of difference, and emphasize the wider range of people involved, the wide range and rapid change of technologies, and the diversity of potential users, as compared to more traditional software systems.

Some of these differences have particular implications for prototyping. First, Web applications often serve a diverse and evolving set of users, and users can easily switch to alternatives (Jeenicke, Bleek, & Klishewski, 2003). As a result, Web applications need to place special emphasis on user requirements and expectations. Prototyping can help development teams understand how users will approach a Web application and what they will try to do with it (Preece, Rogers, & Sharp, 2002). Second, Web applications often require multidisciplinary development teams; prototypes provide a simple, visual way for such teams to communicate. For example, Holter (2006) states that "the root of most Web development problems is not technical but rather it is the failure to communicate technical information non-technically" (p25). Third, Web applications evolve continually in response to user needs and changing technologies, and because it is relatively easy to deploy new versions to a server. Prototyping makes it easier to explore possibilities and quickly identify the most promising.

OBJECTIVES AND SCOPE OF PROTOTYPING

Why Is Prototyping Important?

Prototyping is essentially a way to reduce risk, by exploring specific aspects of the system before making final decisions. Prototyping can be used in almost any aspect of development; for example, a Web development team might use prototypes to evaluate the effectiveness of an open source library, or to compare the performance of different database configurations.

This chapter focuses on UI prototyping. Roughly 50% of application code is devoted to UIs (Myers & Rosson, 1992). Furthermore, roughly 50% of software projects experience significant cost overruns, and inaccurate estimates are often caused by UI issues (Lederer & Prasad, 1992; Standish Group, 1994). Poor UIs can make products difficult or impossible to use (Norman, 2002). For example, a survey found that 63% of new mobile phones returned as defective were not broken; rather, users did not understand how to use the phones (Williams, 2006). Conversely, Nielsen (1993) cites multiple examples of significant cost savings from improved UIs, and customers will often pay a significant premium for high-tech products that are also high-style, for example, attractive and easy to use (Cagen & Vogel, 2002). However, developing effective UIs presents challenges throughout the development process. System requirements can be difficult to define accurately. As with any software system, there are usually multiple stakeholders with competing priorities. In contrast with other types of software development, Web applications often target users who are widely dispersed, use the application briefly or sporadically, can readily switch to alternatives, and have evolving expectations. It can be difficult to design and implement a Web application in the face of continual improvements in network bandwidth, storage space, processing speed, and other technologies. Developers and users often think quite differently about the application architecture; Hohmann (2003) refers to these as "tarchitecture" (technical views for developers) and "marketecture" (marketing views for customers). Too often, the UI is not a priority until late in the development process, although it is usually one of the first things noticed by users. Effective prototyping can help to address all of these challenges.

When and How Can Prototyping Be Used?

Prototyping can be used throughout the development process. Floyd (1984) identifies three broad classes of prototyping, although the boundaries are necessarily vague. First, *exploratory prototyping* helps to clarify the system's broad scope and direction, by identifying missing requirements and considering multiple approaches. Thus, prototyping is useful in the early stages of ideation, product planning, and market research. Prototyping is also useful during requirements analysis and system design, since it can help to identify additional requirements, and evaluate the effectiveness of various implementation models. Second, *experimental prototyping* evaluates the suitability of a proposed solution before committing to full scale implementation. Third, *evolutionary prototyping* explores the consequences of incremental changes to an existing system. The relative position of each class in the system life cycle is illustrated in Figure 2.

Benefits and Risks

Prototyping provides many benefits. Prototyping encourages creative trial and error. Prototyping

Figure 2. Prototyping across the system life cycle

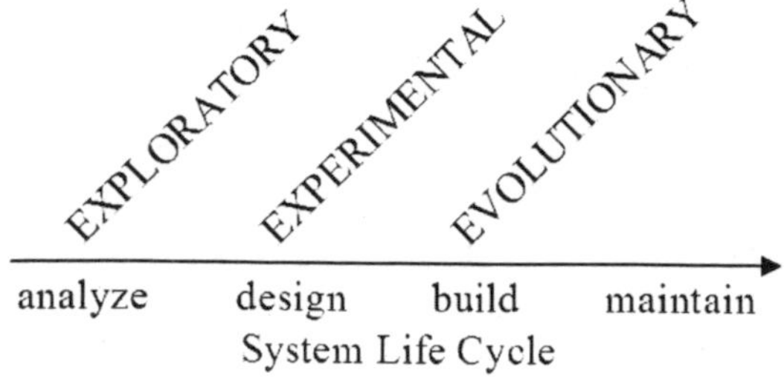

makes it possible to explore more options for the UI, starting early and continuing throughout the development process. Prototyping improves communication with project stakeholders, particularly non-technical stakeholders who find experimenting with a prototype to be easier and more inviting than reading technical documents. Thus, prototyping helps the development team develop more accurate and complete requirements and designs for the UI. Better requirements lead to fewer and less severe defects in the system. Fewer defects lead to reduced development and maintenance costs, and more satisfied users. There is a rule of thumb that the cost to fix a defect increases exponentially with the time between occurrence and detection of a defect (Boehm, 1976). In general, prototyping helps organizations make better decisions faster, which usually saves time, money, and other resources.

Prototyping also presents some risks. Prototyping tools make it easy for developers to become immersed in construction activities when they should really be focused on analysis or design, which should be user-focused (Cooper, 1994). Stakeholders can become too attached to a particular prototype, and be reluctant to consider other alternatives, or focus on details that are not yet relevant. For example, when reviewing a site navigation prototype, they may focus instead on color choices and fonts. At the same time, a more polished prototype may lead some stakeholders to assume that the entire system is nearly complete. Thus, development teams should clearly articulate the issues addressed in a particular prototype. At the other extreme, it is possible to fall into "analysis paralysis," where the team becomes overwhelmed by the range of possible issues and directions, and is unable to make necessary decisions.

Taxonomies

Prototypes can be classified using several taxonomies, including scope, fidelity, longevity, and the prototyping technique.

Scope or *size* describes how much of the system is represented in the prototype. Figure 3 illustrates related terms using a graphical site map, with pages in grey connected by links. A *shallow* prototype shows a few key elements of the system, while a *deep* prototype provides much greater detail. Similarly, a *broad* prototype addresses many aspects of the system, while a

Figure 3. Prototyping terminology

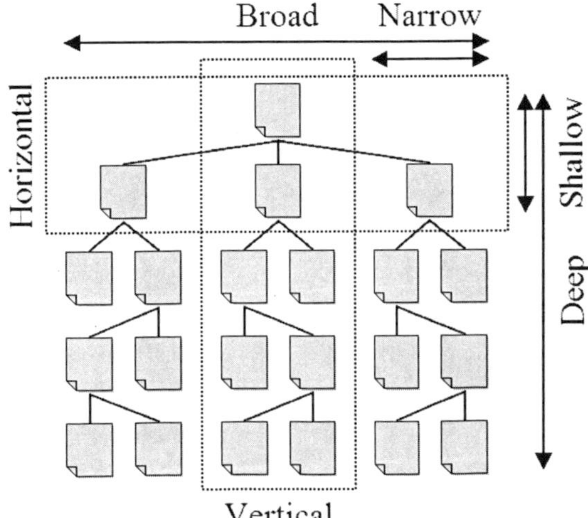

Figure 4. Visual and interaction fidelity

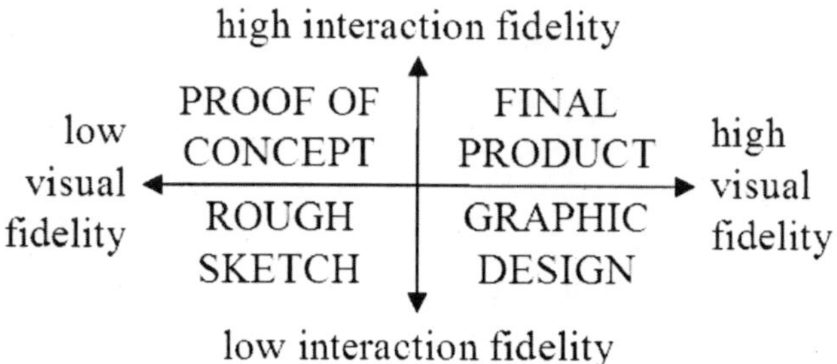

narrow prototype focuses on particular elements of interest. Basic navigation for a new system will usually start out with a broad and shallow (or *horizontal*) prototype, and then add depth as needed. Conversely, changes or new features for an existing system are often explored by deep and narrow (or *vertical*) prototypes.

Fidelity describes how closely a prototype's appearance and behavior match the final system. *Low fidelity* prototypes may be very rough, and are often created quickly with pen and paper or basic drawing software. *High fidelity* prototypes match the final look and feel more closely, are created using specialized prototyping tools or normal developer tools, and require more time to create. Rudd, Stern, and Isensee (1996) argue that designers should use both high and low fidelity techniques, since both have advantages and disadvantages. It can be useful to distinguish two sorts of fidelity. *Visual fidelity* focuses on the visual appearance of the prototype, while *interaction fidelity* focuses on the extent to which the prototype behaves like the final system. Figure 4 illustrates the relationship between visual and interaction fidelity. Thus, graphic designers might create images with high visual fidelity, but low interaction fidelity, while software developers might create proof of concept software with high interaction fidelity but low visual fidelity.

Longevity describes how long the prototype is used. *Throwaway* prototypes are used briefly and then discarded, while *evolutionary* prototypes are intended to gradually develop into the final system. Throwaway prototypes can explore concepts and possible directions quickly, since developers can simplify or even omit non-essential features and processes. However, any useful functionality in the prototype may need to be recreated in the final system. This apparent duplication of effort can be difficult to justify to non-technical stakeholders who see a system that appears to be nearly complete. In many cases, however, the final system will benefit from revisiting these issues. To quote Brooks (1999), "plan to throw one away; you will, anyhow" (p. 116). Conversely, evolutionary prototypes require more initial effort, or more rework. This can reduce the total development effort, but can also commit the development team to decisions that were made prematurely, but are difficult to change. Often, the best approach is to use specialized prototyping tools and techniques for the initial, throwaway prototypes, and then to switch to evolutionary prototyping once the team is confident about key decisions and directions.

Other taxonomies have been proposed. For example, Baumer, Bischofberger, Lichter, and Zullighoven (1996) identify four types of prototypes (presentation, functional, breadboard, and pilot systems) and four tool categories (Hypercard-like, interface builders, fourth generation systems, and application frameworks).

CREATING PROTOTYPES

Prototyping techniques can be divided into three broad categories—paper prototyping, mock prototyping, and code prototyping.

Paper Prototyping

Paper prototyping (Rettig, 1994; Snyder, 2003) uses paper, and pens or pencils instead of software. Typically, each screen, major section, or variation is represented by a sheet of paper, on which key UI elements are sketched. Dialogs, tabs, and menus are represented by smaller pieces of paper (e.g. index cards or Post-It® notes) which can be added and removed as needed. Paper prototyping focuses on the key interactions and UI elements, and makes no attempt to include or describe details. When a stakeholder reviews the prototype, a member of the development team responds to each action by presenting the appropriate piece of paper. If necessary, the prototype can be modified dynamically by adding features to existing pages or creating new pages. Thus, paper prototypes are usually quick, inexpensive, flexible, and can be created by non-experts. Snyder (2003) provides a detailed and highly recommended introduction to paper prototyping. Arnowitz, Arent, and Berger (2007) describe paper prototyping and other paper-based techniques, including card sorting prototyping, wireframe prototyping, and storyboard prototyping.

For example, Figure 5 shows a paper prototype of a Web page where users can renew library books. The symbols at the top represent standard browser controls. The page contains a two field form where the user can type their ID and password and then click on a button to show a table of their books. The user can select specific books, and then click on a button to renew.

Paper prototyping has advantages and disadvantages. It requires no special tools and little training, although experience is certainly helpful. Paper prototypes can be created and modified quickly; in fact, it is easy and common to modify them while they are being tested. They are necessarily throwaway, although this is rarely a concern because they can be created so quickly. Paper prototyping is a good way to explore interaction design and the general layout of screens, but is not intended to capture details such as data validation and business logic. These requirements and design decisions may be captured in separate documents, or may not even be written down. Thus, a mem-

Figure 5. Paper prototype

ber of the development team who understands the intended functionality must be available to help a stakeholder review a paper prototype, and as a result the prototype may behave differently at different times. Furthermore, any prototyping technique also risks omitting essential details through oversimplification, causing problems that must be corrected later.

Mock Prototyping

Appropriate tools can be used to develop *mock prototypes* with better fidelity than paper prototypes, but more efficiently than coding. Such tools enable users to quickly lay out the UI, usually by dragging and dropping elements from a menu. Users can also modify the size, shape, and color of each UI element, and reuse them. Although some tools can simulate the sequence of screens, the flow of data between screens, or other functionality, in most cases users will need to document business logic, either using the tool or in separate documentation. General purpose tools, such as drawing software (e.g., Adobe Illustrator and Photoshop, Dia, or Microsoft Visio), presentation software (e.g., Apple Keynote, Microsoft PowerPoint, or OpenOffice Impress), or word processing software (e.g. Microsoft Word or OpenOffice Writer) are already familiar to most computer users, making it easy to get started. For example, Nam and Gill (2000) and Verhoeven (2003) describe prototyping techniques using Microsoft PowerPoint, and Arnowitz, Arent, and Berger (2007) describe techniques using a variety of applications, including Adobe Acrobat, and Microsoft Word, PowerPoint, Excel, and Visio. HTML editors (e.g., Mozilla Composer, or NVu) can be useful for Web applications as well as other sorts of user interfaces. Dedicated tools (e.g., Axure RP, Irise Studio, or Lucid Spec) are designed specifically for prototyping, and range in price from a few hundred dollars to tens of thousands of dollars. Novice users may find some tools difficult or intimidating, but experienced users can be very productive. Dedicated tools may offer powerful functionality, such as templates for common screens and components, usability testing, or code generation.

For example, Figure 6 shows a mock prototype of the book renewal page, with the same general organization as the paper prototype. The tool (Lucid Spec) provides features to facilitate prototyping. To the left of the prototype is a panel to organize screens in the current document, and a set of controls that can be used in the current screen. To the right of the prototype is a panel for

Figure 6. Mock prototype

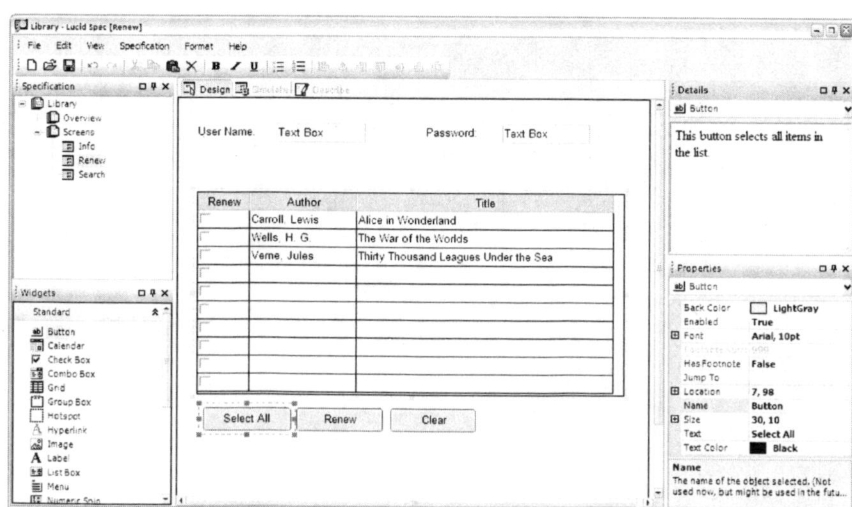

annotating the screen or specific controls, and a panel to view and adjust their properties.

Mock prototyping has advantages and disadvantages. The prototypes are usually either throwaway, or must be maintained as design documents, although some tools may generate HTML or other representations that can be used later in the development process. General purpose tools are familiar to many people, reducing the cost and the learning curve, although they may or may not provide needed functionality, depending on the goal of the prototype. Conversely, special purpose tools are specifically designed for prototyping, but development teams must purchase them and learn to use them effectively. Both general and special purpose tools often present a tradeoff between ease of use and complexity; tools that are easy for novices to use may provide limited functionality, while more powerful tools may be too complex for some users. Both types of tools can also be used by non-developers, making it easier for other stakeholders to create, edit, or review prototypes. Prototyping tools make it easy to adjust details that may not be relevant for a particular prototype, which can waste resources. Thus, Holter (2006) recommends the use of "greyscreen" prototypes that focus on functionality rather than what the Web application will look like.

Code Prototyping

Code prototypes are created using the same development tools, techniques, and even processes that would be used for the final system. For Web development, these might include Web authoring tools such as Adobe Dreamweaver and Microsoft Office FrontPage, as well as integrated development environments such as Eclipse, NetBeans, or Microsoft Visual Studio. Compared to the final system, however, a code prototype has lower fidelity, and the scope is restricted to specific issues of interest. Thus, a code prototype might include little or no error checking for user input, use dummy data instead of a real database, and use simple but inefficient algorithms. It is possible to develop code prototypes using different technologies than the final system. For example, a large, complex Web site might be prototyped

Figure 7. Code prototype

using PHP or Python although the final implementation uses Java.

Figure 7 shows a code prototype of the book renewal page, in a regular browser.

Code prototyping has advantages and disadvantages. It uses tools and techniques that are already familiar to the development team, which reduces the required skill set. While other prototyping approaches may restrict what can be created, code prototyping can use any features or capabilities that could be used in the final system, and can become evolutionary prototypes. Thus, there is a risk of adding unnecessary detail to a code prototype. For example, developers may be tempted to add error checking or graphic design elements (such as colors, fonts, or images) to a prototype intended to explore basic screen layout. Similarly, code prototyping is more resource intensive than other approaches. As a result, development team may be reluctant to give up or throw away code prototypes, and so incidental or erroneous choices may become architectural decisions that will be even more difficult to change later.

Comparing Approaches

The three approaches to prototyping are distinguished by the tools and techniques employed, but also differ in the resource investment required, and the fidelity of the resulting prototype. These relationships are illustrated in Figure 8.

There is evidence that both high and low fidelity prototyping are equally good at finding usability issues (Walker, Takayama, & Landay, 2002), although they emphasize different sorts of problems (Virzi, Sokolov, & Karis, 1996; Nielsen, 1990). Similarly, Bailey and Konstan (2003) compared three approaches (paper, an experimental low-fidelity tool, and a commercial high-fidelity tool) and found that different designers and clients preferred different tools for different tasks; no one approach was best in all situations.

Using Multiple Approaches

Given the advantages and disadvantages of the different approaches, it is often most effective to use multiple techniques in different parts of the development process. Newman, Lin, Hong, and Landay (2003) observed and interviewed Web designers, and found that they switch between lower and higher fidelity representations, using a range of tools suited for each task. When a system or feature is first being explored, paper prototyping is often most useful, since it is fast and simple, making it easy to explore a variety of possibilities. At this stage, it is easy for developers to spend too much time using other techniques. As the requirements or design alternatives are

Figure 8. Comparing effort and fidelity of prototyping methods

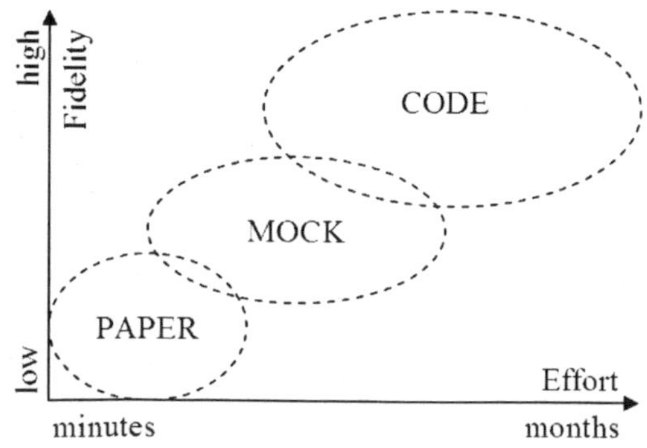

Figure 9. Model parameters

Prototyping Approach	Paper	Mock	Code
Defects Findable	33%	33%	33%
Cumulative Defects Findable	33%	67%	100%
Relative Prototype Cost Per Pass	10	50	250
Relative Defect Removal Cost	10	50	250
Defect Identification Rate Per Pass	50%	50%	50%
Defect Cost Multiplier Per Pass	1.50	1.50	1.50

Figure 10. Model results for different combinations of prototyping

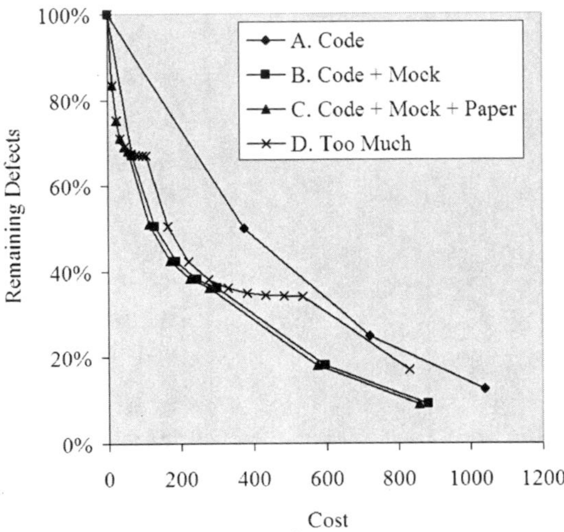

better understood, mock prototyping provides more detail for the more promising approaches. Code prototyping can be reserved for issues that can not be explored using other approaches, or for evolutionary prototyping that will iterate toward the final system.

The benefits and risks of using paper, mock and code prototyping can be illustrated using a spreadsheet model of the relative cost of finding and fixing defects, and the percentage of total defects found. The model makes some simplifying assumptions. First, the prototyping process consists of multiple passes, each using one approach (paper, mock, or code). Second, each approach finds a certain percentage (33%) of the total defects, since some defects can only be found using higher fidelity approaches, but each approach can find defects that could not be found in the previous approaches. Third, there is a fixed cost to create prototypes and remove defects, which increases by a factor of five between approaches. Fourth, each pass finds an equal percentage of the remaining defects (50%). Fifth, the cost of fixing a defect increases by a factor of 1.5 for each pass that is remaining undetected. The model parameters are summarized in Figure 9.

Figure 10 illustrates the results of the model for four scenarios, which invest similar prototyping costs with differing results. Scenario A uses three passes of code prototyping. Scenario B uses five mock passes, then two code passes. Mock prototyping finds a subset of defects much more quickly; this results in fewer remaining defects as well as lower total cost, since finding defects sooner is less expensive. Scenario C uses five paper passes, then four mock passes, then two code passes, for roughly the same total prototyping effort. Paper prototyping finds some defects very quickly, providing a further head start for the mock and code passes. Finally, scenario D shows that each approach reaches a point of diminishing returns, when it has found most of the remaining defects. Thus, too much prototyping may fail to continue reducing the number of defects, and may even increase the overall cost. (These results are qualitative, not quantitative, but could be calibrated for specific organizations and projects.)

REVIEWING AND REACTING

Developers can learn a great deal from the process of creating a prototype, even if no one else ever sees it. However, prototypes are usually reviewed with other stakeholders. Like a prototype, a review should have clearly defined objectives. Reviews can identify requirements that were missing, or imperfectly understood. Reviews can also help to identify features that work well, features that need to be improved, or significant problems with a design. Reviews typically involve several distinct roles, illustrated in Figure 11.

First is the *user*, who should be selected based on the objectives. For example, the user might be a typical intended user of the Web application, with little or no special knowledge. The user might be a stakeholder with insight into customer needs and preferences, such as someone in sales, support, documentation, or quality assurance. The user might be a developer with more specialized knowledge. In each case, the user will bring a different set of experiences and expectations, which can affect the outcome of the review.

The second role is the *facilitator*. This is usually a member of the development team, or perhaps a usability specialist. The facilitator has two main objectives. First, he/she watches and listens to the user, to try to understand what the user is thinking and trying to do. The facilitator may also ask questions, although inappropriate questions may bias the user's responses. Second, the facilitator tries to keep the review focused on the defined objectives.

The third role is the *computer*. For a code or mock prototype, the computer is a physical device, although for a mock prototype the facilitator or an assistant may need to supplement the prototype's functionality with explanations. For paper prototypes, the computer is a person who presents the appropriate sheets of paper based on the user's actions and comments. This person may even modify or create new screens during the review session, to address problems or explore new possibilities.

Figure 11. Roles in reviewing prototypes

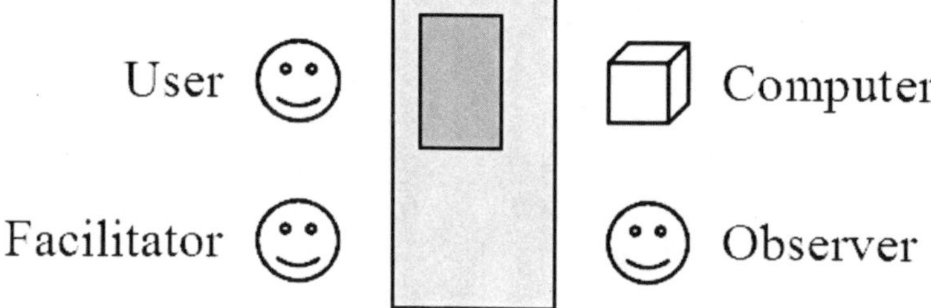

The fourth role is the *observer*, who also watches and listens to the user and takes notes where applicable. The observer may notice things the facilitator does not, since the facilitator is focused on the user and ensuring that the review addresses its objectives.

It can be helpful to make a video recording of the entire review, including the user, facilitator, and reviewer, although some users may find this distracting or intimidating. There is a broad literature (e.g., Kuniavsky, 2003; Rubin, 1994) on usability analysis and observational techniques, such as surveys, focus groups, and usability tests.

After reviewing a prototype with one or more stakeholders, the development team should evaluate the results, and decide how to proceed. For example, if the prototype is for a proposed new feature and the user finds it intuitive and useful, the developers may be ready to implement it. Alternatively, the review may identify potential problems or new possibilities that warrant further prototyping.

FUTURE TRENDS

The advantages of prototyping are well established, and prototyping will be used increasingly by a variety of organizations. Prototyping can be used throughout the Web development life cycle, and can be used as part of almost any development methodology, though it is particularly well suited to iterative or agile processes. In fact, Schrage (2000) argues that prototyping and other simulation-based activities are an essential component of innovative product development at world-class companies.

The use of prototyping in Web development projects has implications for executives and managers. For a modest investment, prototyping yields clear benefits to a variety of project stakeholders. Although prototyping can benefit from specialized experience and complex tools, some concepts and skills should be familiar throughout the organization. For example, basic experience with paper prototyping will benefit anyone who interacts with potential customers or is involved in product development. At the same time, it is important to clearly define the objectives for each prototyping effort, to minimize risks such as analysis paralysis or premature commitment to an inappropriate design.

Prototyping also has implications for students and educators. The basic concepts can be learned quickly and applied to a wide variety of situations. In particular, prototyping makes it possible to explore a wider variety of options in a shorter timeframe, which makes it very useful for providing realistic experiences in a compressed academic timeframe. It also demonstrates the value of iterative development processes, and of considering multiple approaches before committing to a specific direction. Several authors have reported on ways to help students learn about prototyping (e.g., Karat & Dayton, 1995; Frank, Naugler, & Traina, 2005; Kussmaul & Jack, 2006).

Dedicated tools are evolving rapidly to support the growing interest in prototyping. Increasingly, these tools will target specific stakeholders (e.g., salespeople, developers, or end users), technologies (e.g., Web, desktop, or embedded applications), and even application domains.

Web authoring tools and other development tools will increasingly support prototyping as well.

CONCLUSION

Prototyping is easy to do, but it is also easy to do poorly. The following best practices can help maximize the benefits of prototyping. First, clearly define the objectives for a prototype, and communicate those objectives to stakeholders who will develop and review the prototype. Second, focus on the most common or important tasks and goals; an old rule of thumb suggests that users spend 80% of their time on 20% of the application, so focus on that 20%. Third, match the prototyping technique to the issues you want to address, and the stage of development. Understand how much fidelity is needed, and resist the urge to add unneeded detail, which wastes resources and can distract people evaluating the prototype. Fourth, learn as much as possible from prototype reviews; for example, see if users can figure out how to use the prototype without detailed instructions. Finally, use prototyping iteratively; start with simple, throwaway prototypes, and refine gradually as you understand the problem and the solution.

The value of prototyping can be summarized in two key ideas. First, prototyping enables Web development teams to perform multiple small experiments and learn from the results. Tom Peters puts this succinctly as "test fast, fail fast, adjust fast." Second, prototyping improves communication among project stakeholders, who may have a wide variety of perspectives and backgrounds. To quote DeMarco and Lister (1999, p. 4), "the major problems of our work are not so much technological as sociological." Prototypes and prototyping tools are not silver bullets. The real value is in the prototyping process, not the resulting product, which is reflected in our preference for using "prototype" as a verb rather than a noun.

REFERENCES

Arnowitz, J., Arent, M., & Berger, N. (2007). *Effective prototyping for software makers*. San Francisco: Morgan Kaufmann Publishers.

Bailey, B. P., & Konstan, J. A. (2003). Are informal tools better? Comparing DEMAIS, pencil and paper, and Authorware for early multimedia design. In *Proceedings of the SIGCHI Conference on Human Factors in Computing Systems* (pp 313-320). New York: ACM Press.

Baumer, D., Bischofberger, W., Lichter, H., & Zullighoven, H. (1996). User interface prototyping-concepts, tools, and experience. In *Proceedings of the 18th International Conference on Software Engineering (ICSE'96)* (pp 532-541). IEEE Computer Society.

Boehm, B. (1976). Software engineering. *IEEE Transactions on Computers*, 25, 1226-1241.

Brooks, F. P. (1995). *The mythical man-month: Essays on software engineering*. Boston: Addison Wesley.

Cagen, J. & Vogel, C. M. (2002). *Creating breakthrough products: Innovation from product planning to program approval*. Upper Saddle River, NJ: Prentice Hall.

Chen, J. Q., & Heath, R. D. (2005). Web application development methodologies. in W. Suh (Ed), *Web engineering: Principles and techniques*. Hershey, PA: Idea Group Publishing.

Cockburn, A. (2003). *Agile software development*. Boston: Addison Wesley Professional.

Cooper, A. (1994). The perils of prototyping. *Visual Basic Programmer's Journal*, August September 1994, 1.

DeMarco, T. & Lister, T. (1999). *Peopleware: Productive projects and teams* (2nd ed). New York: Dorset House.

Floyd, C. (1984). A systematic look at prototyping. In Budde, R., Kuhlenkamp, K., Mathiassen, L. & Zullighoven, H., *Approaches to prototyping*, (pp 1-18). Berlin: Springer Verlag.

Frank, C. E., Naugler, D., & Traina, M. (2005). Teaching user interface prototyping. *Journal of Computing Sciences in Colleges*, *20*(6), 66-73. Consortium for Computing Sciences in Colleges.

Ginige, A. & Murugesan, S. (2001). Web engineering: An introduction. *IEEE Multimedia*, *8*(1), 14-18.

Ginige, A. (2002). Web engineering: Managing the complexity of Web systems development. In *Proceedings of the 14th International Conference on Software Engineering and Knowledge Engineering*, (pp 721-729). New York: ACM Press.

Highsmith, J. (2002). *Agile software development ecosystems*. Boston: Addison Wesley Professional.

Hohmann, L. (2003). *Beyond software architecture: Creating and sustaining winning solutions*. Boston: Addison-Wesley Professional.

Holter, E. (2006), *Client vs developer wars*. Chapel Hill, NC: Newfangled Web Factory.

Jeenicke, M., Bleek, W.-G., & Klisechewski, R. (2003). Revealing Web user requirements through e-prototyping. In *Proceedings of the 15th International Conference on Software Engineering and Knowledge Engineering (SEKE 03)*. Skokie, IL: Knowledge Systems Institute.

Kappel, G., Pröll, B., Reich, S., & Retschitzegger, W. (2003). An introduction to Web engineering. In G. Kappel, B. Pröll, S. Reich, & W. Retschitzegger (Eds), *Web engineering: The discipline of systematic development of Web applications*. Hoboken, NJ: John Wiley & Sons.

Karat, J., & Dayton, T. (1995). Practical education for improving software usability. *Conference on Human Factors in Computing Systems* (pp 162-169). New York: ACM Press.

Kuniavsky, M. (2003). *Observing the user experience*. San Francisco: Morgan Kaufmann Publishers.

Kussmaul, C., & Jack, R. (2006). User interface prototyping: Tips & techniques. *Journal of Computing Sciences in Colleges*, *21*(6), 188-190. Consortium for Computing Sciences in Colleges.

Lederer, A. L. & Prasad, J. (1992). Nine management guidelines for better estimating. *Communications of the ACM*, *35*(2), 51-59.

Mendes, E., Mosley, N., & Counsell, S. (2006). The need for Web engineering: An introduction. In E. Mendes & N. Mosley (Eds.), *Web engineering*. Berlin: Springer Verlag.

Murugesan, S. & Ginige, A. (2005). Web engineering: Introduction and perspectives. In Suh, W. (Ed.), *Web engineering: Principles and techniques* (pp 1-30). Hershey, PA: Idea Group Publishing.

Myers, B. A. & Rosson, M. B. (1992). Survey on user interface programming. In *Proceedings of the ACM CHI'92 Conference* (pp 195-202). New York: ACM Press.

Nam, T. J. & Gill, S. (2000). An effective prototyping method for delivering interaction design in industrial design education. In *Proceedings of the IDATER Conference*, 2000.

Newman, M. W., Lin, J., Hong, J. I., & Landay, J. A. (2003). DENIM: An informal Web site design tool inspired by observations of practice. *Human-Computer Interaction*, *18*(3), 259-324.

Nielsen, J. (1990). Paper versus computer implementations as mockup scenarios for heuristic evaluation. In *Proceedings of the IFIP Third International Conference on Human Computer Interaction* (pp 315-320). Amsterdam: North-Holland.

Nielsen, J. (1993). *Usability engineering*. San Francisco: Morgan Kaufmann Publishers.

Norman, D. (2002). *The design of everyday things*. New York: Basic Books.

Preece, J., Rogers, Y., & Sharp, H. (2002). *Interaction Design: Beyond Human-Computer Interaction*. Hoboken, NJ: John Wiley & Sons.

Rettig, M. (1994). Prototyping for tiny fingers. *Communications of the ACM*, *37*(4), 21-27.

Royce, W. (1970). Managing the development of large software systems. In *Proceedings of IEEE WESCON* (pp 1-9). New York: The Institute of Electrical and Electronics Engineers.

Rubin, J. (1994). *Handbook of usability testing.* Hoboken, NJ: John Wiley & Sons.

Rudd, J. and Isensee, S. (1994). Twenty-two tips for a happier, healthier, prototype. *Interactions, 1*(1), 35-40.

Rudd, J., Stern, K., & Isensee, S. (1996). Low vs. high-fidelity prototyping debate. *Interactions, 3*(1), 76-85.

Schrage, M. (2000). *Serious play: How the world's best companies simulate to innovate.* Boston: Harvard Business School Press.

Snyder, C. (2003). *Paper prototyping: The fast and easy way to design and refine user interfaces.* San Francisco: Morgan Kaufmann Publishers.

Standish Group (1994). *The CHAOS report.* Retrieved January 1, 2007 from http://www.standishgroup.com/sample_research/chaos_1994_1.php

Verhoeven, J. (2003). Prototyping with PowerPoint. Retrieved on January 1, 2007, from http://www.jansfreeware.com/articles/misc-prototyping.html

Virzi, R. A., Sokolov, J. L., & Karis, D. (1996). Usability problem identification using both low- and high-fidelity prototypes. In *Proceedings of the SIGCHI Conference on Human Factors in Computing Systems* (pp 236-243). New York: ACM Press.

Walker, M., Takayama, L., & Landay, J. (2002). High-fidelity or low-fidelity, paper or computer medium? Choosing attributes when testing Web prototypes. In *Proceedings of the Human Factors and Ergonomics Society 46th Annual Meeting,* pp. 661-665. Santa Monica, CA: Human Factors and Ergonomics Society.

Williams, C. (2006). Smart phones, stupid punters? *The Register,* July 13, 2006.

Chapter XII
Testing for Web Applications

David L. Mills
University of Memphis, USA

ABSTRACT

This chapter explores the concepts and challenges behind testing Web applications, and explores the latest testing techniques and best practices. As our reliance on the Internet grows, the quality and reliability of online resources become critical. Unfortunately, significant research shows that the current approaches to modern Web development are woefully inadequate. It is important that there are processes in place and best practices established to ensure that the development of Web applications can take place with an assurance of quality. In addition to offering an initiation to some of the modern testing methods and tools, the authors hope to motivate readers to consider testing as a multi-purpose tool to be used throughout all stages of development.

INTRODUCTION

Since its inception the Internet has been expanding in size, pervasiveness, and functionality with astonishing speed. It has become a fundamental tool in business, government, and education, not to mention our personal lives. In order to support growing demands, the Web has evolved from static pages of information to complex applications with functionality equivalent to that of modern software. As we grow more dependent on online resources, their quality and reliability become critically important (Hendrickson, 2002). Unfortunately, research suggests that practices to ensure these qualities are generally weak or missing from Web development projects.

This chapter aims to introduce the fundamental concepts of testing Web applications, and to provide insight about current best practices and resources. Not only will this chapter inform readers about the challenges of Web testing and the tools/techniques available, but it should also provide motivation for software developers to integrate a sound testing solution into their software development life cycle model(s).

Concepts in Conventional Testing

Software testing is a method of *quality assurance* involving the analysis of software to evaluate features and ensure that the requirements are met. The tasks involved in testing are aimed at verification and validation (IEEE, 1990). The goal of *verification* is to make sure the product conforms to specifications. It is the process behind the question "are we building the product right?" Some sample verification tasks include inspections, reviews, and testing. Alternatively, *validation* is aimed at making sure that product build meets the needs of the user. It is the process behind the question "are we building the right product?" Evaluations and customer testing are typical validation tasks.

Testing techniques can be categorized as either black-box or white-box. *Black-box* testing takes place from an external view of the system, meaning that it is carried out without access to the code or knowledge about the internals of the program. The focus of such testing is on the functionality of the program. *White-box* testing, which is also called *structural* or *glass-box* testing, is performed with an internal view of the system. Tests are generated using the code as a guide (Myers, 2004).

There are various types of tests, each having a different focus and taking place at a different level. *Unit testing* targets individual components of a program. This usually refers to individual methods and/or classes (IEEE, 1990; Myers, 2004). Unit tests evaluate modules independently of one another. *Integration testing* is performed on combined components of the system, and focuses on their interaction. *System testing* is a form of black-box testing in which the system as a whole is tested. These tests can target functional or non-functional specifications.

Functional testing involves test cases that were created based on functional specifications, as opposed to *non-functional* testing, which is aimed at features such as speed and ease of use. Some common non-functional tests include performance testing, stress testing, and usability testing. *Performance testing* is evaluating the system compliance with performance specifications. *Stress testing* evaluates the behavior system when it is pushed with a heavy load to the limits of performance. *Usability testing* assesses the ability of users to learn to operate a system. This type of testing is often performed by specialists that observe humans interacting with the system.

Acceptance (aka Customer) testing is a form of black-box testing that defines the criteria that a system must meet to be acceptable for delivery, that is, a way of validating the requirements. This is a powerful and beneficial tool. It enables both the customer and developer to ascertain what work is complete and what work remains at any point, and it is way for the developer to validate that the customer is satisfied (Jeffries, et. al., 2001). Since the acceptance tests serve as a testable version of the critical requirements, when all the acceptance tests for a system pass, the system can be considered complete. Acceptance tests rarely involve code at the specification level, but they should be clear and understandable to both the customer and the developer. Collaboration is acceptable, even encouraged, but it is critical that the customer is always directly involved in writing the test cases. To facilitate their communication, acceptance tests are usually provided in the form of spreadsheets, tables or scripts that can be executed directly or as automated tests.

Regression testing is the re-execution of previous tests created for a system or component. They are typically run after significant changes are made, in order to ensure that intended functionality was not lost. Many types of tests can be included in the regression test suite, including unit, integration, system, and acceptance tests; though only a select set of tests is usually included as it can be impractical to continually re-run all tests (IEEE, 1990).

Testing often requires the development of resources that will not be included in the final product in order to enable testing to take place. Such resources are called *scaffolding*. These simulate the functions of missing components; some common examples include *stubs* and *mock objects*.

DEVELOPMENT OF WEB APPLICATIONS

The World Wide Web (WWW) was originally created to enable people to share information. Early Web sites were just static pages of textual information with hyperlinks to one another. Over time the WWW has evolved new functionality and complexity, becoming an environment for applications ranging from the simple static information pages to powerful, commercial sites with features equivalent to conventional programs (Mendes, 2005). In this chapter, we reserve the term Web application for the latter, more powerful type of online resource—with characteristics of both Web sites and conventional software for delivery on the WWW.

Yet, despite the vast advances in the technology the history of Web development is littered with significant issues and expensive failures. A survey published in 2000 by the Cutter Consortium (Deshpande, 2002; Mendes, 2005), identified the following problems:

- 63% of projects exceeded the budget
- 79% of projects were behind schedule
- 53% of delivered systems did not have required functionality
- 84% of delivered systems did not meet business needs
- 52% of deliverables were of poor quality

Such failures are typically the result of weak development and management processes. This seems to be confirmed by studies suggesting that approaches to Web development applications have generally been informal at best—lacking any well-defined defined process or quality control (Ginige, 2002; Mendes, 2005).

Web Engineering

Web development is a relatively young field. The issues facing it are analogous to those faced by the conventional software development community during the software crisis in the 1970's, when large-scale projects were all too often delivered late, over budget, and/or with defects. This was the driving factor for the development of models, processes, and methodologies—the birth of software engineering (Ginige, 2002).

In a similar fashion, Web engineering has emerged as a new discipline. Ginige (2002) defines it as "the application of scientific and mathematical principles to practical ends such as the design, manufacture, and operation of efficient and economical structures, machines, processes, and systems." There is an obvious need for discipline and structure, but does this represent a new field or is it a subset of convention software engineering?

While most agree that there is significant overlap, supporters claim that the Web environment poses unique concepts and problems that require a new and distinct approach (Ginige, 2002). Though an extensive comparison of the two disciplines is outside the scope of this chapter, a look at the potential differences offers valuable information which can help guide an approach to testing. A summary of the comparisons adapted from Ginige (2002), Mendes (2005), and Stout (2001) follows.

Characteristics and technologies. The architectures of traditional applications are typically simpler than those of Web applications, which are usually distributed cross-platform systems.

Also, Web applications are comprised of more distinct components and technologies (such HTML/XML, multimedia, databases, servlets, and server-side pages) than traditional applications.

Releases and maintenance. Maintenance cycles for Web applications are measured in days or hours, rather than weeks for traditional software. Also, releases do not occur as batched events—small updates are made frequently, underscoring the iterative and evolutionary nature of Web development.

Team composition and skills. The people involved in the development process for Web applications are often more diverse in terms of skills and disciplines. While both Web and conventional development incorporate requirements engineering, software engineering, and testing; Web projects often require additional expertise from marketing, art/hypermedia, graphic design, and network management to name a few.

Dynamic nature. The Web itself is by nature a dynamic entity defined by shifting technologies, changing content, and infrastructure that is in a state of constant flux. Modern Web applications have evolved from pages with static information to complex, dynamically-generated pages.

Interactivity. Developers have very limited control over the user's actions. For example, users can close the browser or hit the back button at any time, and no notification may be sent to the application.

User environments. Rather than being limited to a group or company, the community of users may include people anywhere in the world and few assumptions can be made about user's hardware or software environment.

Protected content. With the potential for global exposure, there are special concerns about the content provided in Web applications. Items that are protected by copyrighted must often be protected against downloading and personal information about users or must be protected as there may be ethical or legal issues surrounding such items.

Agile Software Development

The Web is perhaps the best illustration of the challenge of applying traditional engineering principles and practices complicated by the need to remain flexible with increasingly dynamic requirements. New approaches have come about that make traditional processes such as testing more fluid and easier to integrate, and many of these techniques have been made more popular or more effective through the emergence of agile software development over the last decade.

The needs of Web applications are uniquely compatible with the emphasis of agile approaches on quality and speed (Hendrickson, 2002). Agile approaches offer an alternative to traditional methodologies, which are typically heavy processes that are driven by documentation (Mills, et al., 2006). Unlike plan-driven techniques, agile practices embrace change and promote a strong relationship between the developer and the customer (Boehm & Turner, 2005); To be agile is to be 'alert,' 'nimble,' and 'responsive,' all of which are important characteristics of the agile paradigm.

There are a number of agile methodologies available, varying in the practices they employ as well as the degree of discipline, but all of them subscribe to a set of common goals. In 2001, these goals were formalized in a document called the Agile Manifesto. Part of this declaration appears below.

Individuals and interactions over processes and tools.
Working software over comprehensive documentation.
Customer collaboration over contract negotiation.
Responding to change over following a plan.

That is, while there is value in the items on the right, we value the items on the left more (Boehm, 2005, pp. 195).

The Manifesto juxtaposes agile and traditional objectives, but it is important to note that agile approaches do not abandon ideals such as documentation or discipline. Rather, they consider the value of interactions, working software, and responsiveness to change. Extreme programming (XP), one of the more well-recognized agile methodologies, identifies four values that capture the spirit of the agile community: communication, simplicity, feedback, and courage (Boehm, 2005). In XP, these are achieved through practices such as short iterations, simple design, pair programming, test-driven development, and continuous integration.

TESTING WEB APPLICATIONS

Although Web applications are relatively young, their growing importance in critical e-commerce applications means that Web development activities must be followed in a manner that ensures their quality. One of the most important means of accomplishing this is testing. However, as Web applications continue to grow larger and more complex, it is vital that standards and practices are developed to meet the demands for developing quality Web applications in an efficient manner.

Considerations and Techniques

The goals for testing conventional applications are the same as those for Web applications, but additional elements such as *availability, reliability, performance, scalability,* and *security* are important considerations in the measure of quality for a Web application. This section includes testing areas and techniques for consideration adapted from Di Lucca (2002), Myers (2004), Sneed (2004), Stout (2001), Wu (2004), and Xu (2005).

Static techniques focus on elements that do not change, and therefore do not require execution. Such tests are usually run on a page that has already loaded into a browser (Stout 2001). The public exposure of Web applications makes it vital to ensure that presented elements are evaluated for quality. For many businesses, the Web site is the face of the company; poor impressions can have substantially impact your customers and cause them to question your product. Considerations should include:

- Correctness of spelling and grammar
- Accuracy of data
- Functionality of hyperlinks and images
- Aesthetics and consistency of visual/presentation elements

Figure 1. Sample flow of execution for a simple login

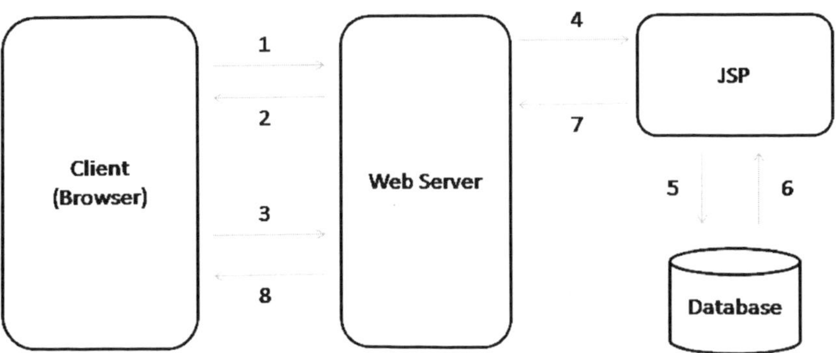

Alternatively, *dynamic* techniques focus on testing the functional and non-functional aspects of the application. These techniques involve the execution of code and interactivity with the server (Ricca, 2001; Stout, 2001). The figure below shows a simple example of the way a Web server transaction occurs; the sample below is a model of a user login.

In Figure 1, the flow of execution proceeds as follows:

1. Login page requested
2. Login page returned
3. Login form submitted
4. JSP (server page) called to handle submission
5. Server page queries the database for validation
6. Database returns validation results to the server page
7. After processing, the server page generates the HTML
8. Based on the validation, the appropriate page is returned

Server-side page tests. Within most technologies used for modern Web applications such as servlets, JSP (Java server pages), and ASP (active server pages)—a great deal of execution occurs at the server. The server-side software can include not only traditional elements such as classes/objects, but Web-specific components such as servlets and scripted server-pages. Conventional practices such as unit testing, integration testing, and regression testing should be performed on server-side components.

Client-side/browser-page tests. For many applications, some execution will also take place on the client end. Typical examples of such items include applets, scripts (controlling interaction, field validation, presentation, etc.), and visual effects. These items can be dynamically generated, which introduces further complexity. However, it is important to tests such code to ensure both correctness (that the logic is accurate) and compatibility (that it will run in varied environments).

Transaction testing. Actions on the Web take place in transactions, beginning with a request on the side of the client and ultimately coming back as a response from the server. The example in Figure 1 demonstrates the transaction(s) for a login. It is important to test the flow of end-to-end transactions to evaluate the transfer of control, data consistency, and data formatting.

Non-functional testing. As mentioned at the beginning of this section, there are many components that contribute to the quality of Web applications. Tests should be written to ensure that each of these characteristics meets pre-determined specifications. An additional non-functional consideration is configuration/environment. Test plans should include tests for common user environments, or those that are most important to you. Important characteristics include connection type and speed, platform/operating system, browser software/versions, additional applications/plug-ins, and security settings.

Best Practices

Process management and planning. It is well-accepted and understood that it is critical to have a process in place for all but the most trivial projects. While there is a great deal of debate over whether conventional techniques are appropriate for Web applications, practices from both traditional and agile approaches should be considered. No approach is universally applicable, even across Web development projects. Projects benefit most by combining appropriate practices (Boehm & Turner, 2005). Regardless of the process and practices used, testing must be an integral part at all times, beginning with the requirements. At each step and decision, testability should be a stated goal.

Automated testing tools. It is important that tests are run early and often, and this is made easier and more efficient by the use of automated testing tools. Such tools can help with the tasks of generating, executing, and managing tests (Stout, 2001). Automated testing is recommended as part of many agile methodologies and in test-driven development. A balance is important however, as writing and executing tests can get costly, especially for large and/or complex systems.

Testing environments. To enable effective testing, it is critical to have a separate environment available. Furthermore, it is critical that the test environment is identical to and independent from the production environment (Stout, 2001). This allows development to proceed without any potential effects on the availability or stability of the production application. An even better approach, as suggested by Stout (2001) and Myers (2004), is to have three identical and independent environments—one for on-going development, another for testing, and finally the production environment.

Test-driven development (TDD). According to Ron Jeffries, the goal of test-driven development is "clean code that works" (Beck, 2003). This sounds like a reasonable goal for any project, but what exactly is involved in adopting this technique? TDD is more than just vigorously incorporating tests in a project; rather it is quite literally the practice of putting testing in the driver's seat of development. In fact, the name itself implies that this is a technique for development as a whole, not for testing (Ambler, 2004).

Like an agile methodology, TDD is an evolutionary approach. The process begins by writing a test, which will initially fail, and then writing code to support the intended functionality (Ambler, 2004). TDD is driven by two basic rules: (1) write code only when a test fails; and, (2) get rid of duplication as you come across it. These rules are the foundation for the TDD building cycle described by the TDD mantra of "*red, green, refactor*" (Beck, 2003).

During the *red* step, the developer writes a test to cover a small piece of functionality (for a new program or to be added to an existing program). This test will of course fail (i.e., return a red bar), since the functionality is not yet in place. During the *green* step, the developer writes just enough code to get the test to pass (i.e., return green). Finally, during the *refactor* step, the developer cleans up the new and existing code to simplify and reduce duplication.

TDD has been gaining increasing support, and adopters are reporting numerous benefits from its use. Building software using TDD tends towards *simplicity*. More specifically, since only enough code is developed to pass the tests, the design is kept inherently simple. There should never be any unnecessary objects in the design, since everything that exists after refactoring should be in use (Beck, 2005). By using TDD, developers apply the agile concepts "Keep It Simple Stupid" (KISS) and "You Aren't Going To Need It" (YAGNI).

TDD forces developers to write automatically *testable code*. Since the tests are written first, the objects are built with a user in mind. This changes the focus by shifting attention to the interface and not the implementation. Also, code tends to be more cohesive and less coupled. Since the objects serve both the application and the tests, the objects will likely be developed with a clean and simple interface (Beck, 2005). Developing the objects with multiple clients in mind also tends to make the code more reusable (Hieatt, 2002).

In addition to driving design decisions, TDD simplifies the process of moving forward. Each step consists of writing a small test, writing code to pass it, and then refactoring. Developers should never be faced with planning and implementing a large chunk of functionality. All aspects of functionality are broken down into small testable pieces, making it much easier to identify what to do next, giving them the *confidence* to implement another cycle. Of course, sometimes it may be difficult to identify what to choose as the next piece; it will take experience to become efficient. However, it is very important to follow the TDD cycle, and not to code until you have a test in place first. There is a rule in TDD "If you can't write a test for what you are about to code, then you shouldn't even be thinking about coding" (Georgy & Williams, 2003, p.1).

In each short TDD development cycle, a small amount of value/functionality is added and each test in a growing suite is passing. Also, the gap between making a decision and getting valuable *feedback* is greatly minimized. This process of consistently adding functionality, coupled with the frequent reinforcement of passing tests fosters courage, confidence, and trust within developers and teams (Georgy & Williams, 2003).

Tools

General Purpose Frameworks

The *xUnit* family is a collection of testing frameworks based on an original design by Kent Beck. They aid in automated testing by allowing a developer to easily write tests using assertions to define expected conditions or outcomes. Implementations have been written for many of the most popular languages available today. The table below contains a partial listing of popular xUnit implementations.

JUnit is a general purpose automated testing framework for Java. It is also is the most familiar and widely-used member of the xUnit family of testing packages, gaining acceptance and popularity in industry and academia. It has also been extremely popular with adopters of test-driven development and agile methodologies such as eXtreme Programming. This success is due, at least in part, to its simplicity and flexibility. Also, a great deal of documentation and online resources exists for JUnit, making it easy for those just getting involved.

Fortunately, using xUnit tools encourages the separation of business logic into its own layer, which could be tested outside of a Web application. Consider the example of a calculator on the Web; the functionality (adding, subtracting, etc.) can be separated into a calculator class, enabling the use of an xUnit tool.

At some point, however, it will be necessary to get access to Web features that are inaccessible by JUnit alone. A number of tools are available to overcome the special challenges of testing on the Web. A few of these tools are described below.

Mock objects. Mock objects are a form of scaffolding, which simplify testing when the code under evaluation depends on another module or resource that does not exist. Open source packages such as *EasyMock* or *DynaMock* provide such functionality, and easily integrate with other testing resources.

HTTP/servlet testing. When Web-specific elements of a program need to be tested, JUnit alone cannot provide all the necessary tools. *HTTPUnit* is an open-source testing framework capable of emulating a browser, submitting forms, working with JavaScript, using cookies, and ana-

Table 1. Partial listing of popular xUnit implementations

Implementation	Language
JUnit	Java
csUnit	.NET
NUnit	.NET
PHPUnit	PHP
jsUnit	JavaScript
Test::Unit	Ruby
PyUnit	Python

lyzing Web pages. ServletUnit is an extension that adds the ability to test applications using servlet technology.

In-container testing. For some testing, it will be necessary to have access to servlet container resources that are unavailable with mock object approaches. Cactus is an extension to JUnit that enables server-side testing in a real servlet environment.

Acceptance and functional testing. A number of tools are available to enable functional testing of Web applications. Two of the most popular are *JWebUnit* and *Selenium*. These tools provide capabilities for evaluating Websites, including support for GUI elements and executable content, such as JavaScript.

CONCLUSION

The development of Web applications is complex and presents new challenges that are not addressed by conventional software development. As they continue to gain importance and complexity, it will become increasingly vital that there are measures in place to ensure their quality and reliability. This chapter presented some of the current challenges in Web development, as well as testing considerations, tools, and best practices. A successful test strategy takes all aspects of quality into consideration. While future technology will most likely offer solutions for many of the issues being experienced today, they will almost certainly present new ones as well.

REFERENCES

Ambler, S. W. (2004) *The Object Primer: Agile Model-Driven Development with UML 2.0.* Cambridge University Press.

Beck, K. (2001). Aim, fire. *IEEE Software, 18*(5), 87-89.

Beck, K. (2003). *Test-driven development: By example.* The Addison-Wesley signature series. Boston: Addison-Wesley.

Beck, K., & Andres, C. (2004). *Extreme programming explained: Embrace change.* Boston: Addison-Wesley.

Boehm, B. W., & Turner, R. (2003). *Balancing agility and discipline: A guide for the perplexed.* Addison-Wesley Professional.

Deshpande, Y., S. Murugesan, et al. (2002). WEB ENGINEERING. *Journal of Web Engineering, 1*(1), 003-017.

Di Lucca, G. A., A. R. Fasolino, et al. (2002). Testing Web applications. Software Maintenance, 2002. In *Proceedings.* International Conference on.

Ginige, A. (2002). Web engineering: Managing the complexity of Web systems development. In *Proceedings of the 14th international conference on Software engineering and knowledge engineering,* (pp. 721-729).

Ginige, A., & Murugesan, S. (2001). Web engineering: An introduction. *Multimedia IEEE, 8*(1), 14-18.

Hendrickson, E., & Fowler, M. (2002). The software engineering of internet software. *Software IEEE, 19*(2), 23-24.

Hieatt, E., & Mee, R. (2002). Going faster: Testing the Web application. *Software, IEEE, 19*(2), 60-65.

IEEE (1990). *IEEE Standard 610.12-1990, IEEE Standard Glossary of Software Engineering Terminology.*

Janzen, D., & Saiedian, H. (2005). Test-driven development concepts, taxonomy, and future direction. *Computer, 38*(9), 43-50.

Jeffries, R., Anderson, A., & Hendrickson, C. (2001). *Extreme programming installed.* The XP series. Boston: Addison-Wesley.

Mendes, E., Mosley, N. (2005). *Web Engineering.* Berlin, Germany: Springer.

Mills, D., L. Sherrell, et al. (2006). Experiences using agile software development for a marketing simulation. SoutheastCon, 2006. In *Proceedings of the IEEE*: 285-290.

Myers, G. J., Badgett, T., Thomas, T. M., & Sandler, C. (2004). *The art of software testing.* Hoboken, N.J.: John Wiley & Sons.

Ricca, F., & P. Tonella (2001). Analysis and testing of Web applications. In *Proceedings of the International Conference on Software Engineering,* (pp. 25-34).

Sneed, H. M., A. GmbH, et al. (2004). Testing a Web application. Web Site Evolution, 2004. WSE 2004. In *Proceedings of the Sixth IEEE International Workshop,* (pp. 3-10).

Stout, G. A. (2001). Testing a Website: Best practices. Retrieved on from whitepaper on www.reveregroup. com

Wu, Y., J. Offutt, et al. (2004). Modeling and testing of dynamic aspects of Web applications. Submitted for publication: 04-01.

Xu, L., B. Xu, et al. (2005). Testing Web applications focusing on their specialties. *ACM SIGSOFT Software Engineering Notes, 30*(1).

Chapter XIII
Outsourcing Issues in Web Development

Clif Kussmaul
Clif Kussmaul, Elegance Technologies, Inc., USA & Muhlenberg College, USA

Roger Jack
Elegance Technologies, Inc., USA

ABSTRACT

This chapter addresses issues, alternatives, and best practices that apply when outsourcing Web development. The chapter's primary objective is to provide a concise overview of key concepts and best practices for practitioners and students, as well as other audiences. First, we introduce and motivate the chapter, provide background, and present three key ideas that are expanded and developed in the two subsequent sections. The first describes four steps to help executives and upper management address strategic issues and decisions in outsourcing. The second describes four more steps to help managers, team leaders, and development teams address more tactical issues. We conclude with future trends and implications, and a summary of the best practices.

INTRODUCTION

Outsourcing is the use of external companies to perform services, rather than using internal staff. According to a 2003 survey of chief information officers (CIOs), 70% of companies outsource some IT function or application (CIO Insight, 2003). **Offshoring** is the use of staff in other countries, and is often associated with India, China, and the former Soviet Union. Forrester Research estimates that 277,000 computer jobs and a similar number of management and operations jobs in the United

States will move offshore by 2010 (Engardio, 2003). An ACM Task Force (Aspray, Mayadas, & Vardi, 2006) estimates that each year 2-3% of IT jobs in the United States are offshored, but that many more IT jobs are lost and created each year. Furthermore, the task force notes that it is very difficult to predict such changes accurately, particularly since many of the studies are produced or supported by organizations with a vested interest in the outcomes (e.g., American Electronics Association, 2004; Behravesh and Klein, 2004; Farrell, 2003; Vogel & Connelly, 2005).

This chapter's primary objective is to provide a concise overview of key concepts and best practices for practitioners and students, as well as other audiences. The chapter emphasizes Web applications and includes examples from outsourced Web development projects the authors have led, overseen, and reviewed. However, many of these approaches and issues apply to many sorts of outsourced development work.

To outsource Web development more effectively, organizations should focus on three key ideas, which we expand and develop below. First, **structure projects in short iterations**, so that there are multiple opportunities to find and correct problems; in words attributed to Tom Peters, "test fast, fail fast, adjust fast." Second, recognize that "the major problems of our work are not so much *technological* as *sociological*" (DeMarco & Lister, 1999, p. 4, original emphasis); in other words, **the major benefits and risks are tied to people and relationships** rather than to tools. Third, **match the structure of the development team to the desired structure of the system**. This relationship, first described by Conway (1968), is particularly important in outsourcing relationships, where it can be more difficult to restructure the project.

Throughout the chapter, we use examples distilled from a variety of outsourced Web development projects the authors have led, overseen, or reviewed. SPC ("Software Product Company") had invested significant resources to develop a powerful and flexible software engine to perform data processing operations. SPC planned to use the software engine in a series of Web-based applications, customized to specific domains, so that customers could take advantage of the engine's capabilities without the expertise and effort needed to configure it. Thus, the Web-based applications have complex databases and interactions with the software engine. SPC decided to consider outsourcing the development of these Web-based applications, for reasons discussed below.

The following section provides essential background. We then describe four steps to help executives and upper management address strategic issues and decisions in outsourcing. Next, we describe four more steps to help managers, team leaders, and development teams address more tactical issues. We conclude with future trends and implications, and a summary of the best practices. We provide a multidisciplinary perspective, since outsourcing involves a variety of disciplines, including business, software development, psychology, and sociology. Outsourcing, especially offshore outsourcing, can be controversial; we present a balanced view of the benefits and risks, based on experience.

BACKGROUND

In this chapter, **client** is the organization that is deciding if and how to use outsourcing in a Web development project, **provider** is the organization performing the outsourced work, **joint team** includes everyone at the client and provider responsible for building the project, and **stakeholder** is anyone with a significant interest in the project, including upper management, other business functions such as sales, and, of course, target end users.

Distributed development is when the team is at multiple locations, and **global** development is when the team is in multiple countries. Outsourcing and offshoring have inspired other related terms. **In sourcing** occurs when internal staff perform previously outsourced services. **Near shoring** is the use of developers in nearby countries, often to reduce travel costs and time zone differences. For example, US companies might use staff in Canada, Mexico, or the Caribbean. **Rural sourcing** is the use of staff in lower cost areas of the same country.

Web Development

Web development differs from other types of software development in some important ways. For example, Ginige and Murugesan (2001), Ginige (2002), and Murugesan and Ginige (2006) identify several challenges in Web systems, including the continual growth of requirements, the continual change in information content, a diverse set of stakeholders, and the need for multidisciplinary development teams. They also describe a hierarchy of Web-based software, ranging from static Web pages through database-driven Web sites to complex Web systems, and propose a set of processes to help address these challenges. Kappel, et al. (2003) describe a similar but larger set of challenges, which are categorized by their focus on product and content, usage context, system development, or system evolution. Chen and Heath (2005) also identify a challenges, including usability design, content maintenance, scalability, security, fast deployment, and competing architectures, platforms, and tools. Similarly, Mendes, Mosley, and Counsell (2006) identify twelve areas of difference, and emphasize the wider range of people involved, the wide range and rapid change of technologies, and the diversity of potential users, as compared to more traditional software systems.

Some of these differences have particular implications for outsourcing. For example, requirement changes must be communicated across organizational boundaries, and may require project costs or schedules to be renegotiated. The range of necessary skills (e.g., analysis, information architecture, graphic design, software design, coding, content development, testing, and usability) can be assembled quickly through outsourcing, but using multiple providers can compound communication problems. Outsourcing can provide "just in time" access to new tools and technologies; however, over a period of years this can result in a diverse set of applications for client is unable to maintain.

Figure 1. Four strategic steps

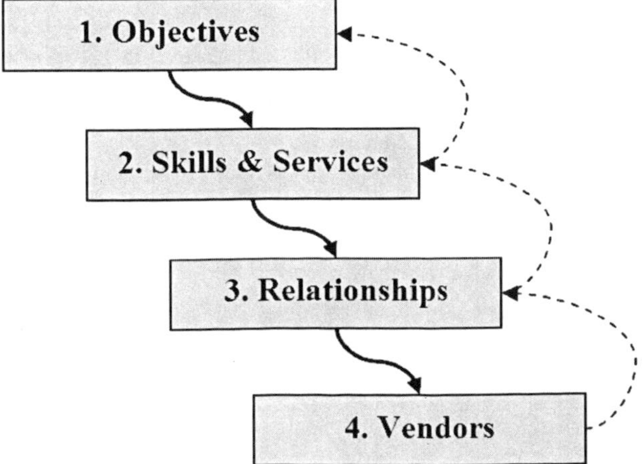

STRATEGIC ISSUES

Four key steps will help executives and upper management realize the benefits of outsourcing and minimize the risks. These steps are illustrated in Figure 1.

First, identify the objectives for outsourcing and whether these objectives are realistic. Second, select the specific skills and services to be outsourced. Third, determine the appropriate type of relationship. Fourth, select a provider or providers. Other authors describe similar progressions (e.g., Lacity & Willcocks, 2000). Note that these steps are not entirely sequential; later steps can feed back to affect earlier steps, just as in the traditional waterfall model for software development. For example, final provider selection may affect the nature of the relationship, or the specific skills and services that are outsourced. Each step is described in more detail below, including relevant background, application in the case study, and a checklist of key questions.

Objectives

The first strategic step is to identify the objectives for outsourcing, and assess whether these objectives are realistic. Outsourcing is done for a variety of reasons. A survey asking CIOs to list their top three reasons for outsourcing IT applications (CIO Insight, 2003) found that the four most commonly cited reasons were lack of resources, lack of expertise, cost effectiveness, and speed of development, as summarized in Figure 2.

Many IT projects run into problems; for example, a Standish Group survey of application development projects estimates that 28% "succeeded", 49% were "challenged", and 23% "failed" (Johnson, et al., 2001). Similarly, 49.8% of companies report problems with outsourcing (CIO Insight, 2003) including poor quality, missed deadlines, unexpectedly high costs, poor domain knowledge, and insufficient value, as summarized in Figure 3.

Clearly, the benefits of outsourcing often parallel the common problems. This should not be surprising, since these are key factors in software development. They are discussed in more detail below, along with the implications for Web development.

First, outsourcing can enable organizations to **increase and decrease project resources**, especially the number of people. This is particularly useful in Web development and product development, where the need for designers, programmers, testers, and other roles can vary dramatically over time.

Second, outsourcing can take advantage of **specialized skills or facilities**. These resources may be expensive or unavailable locally, or may be needed only briefly or intermittently. For example, companies might choose to outsource their entire

Figure 2. Common reasons for outsourcing

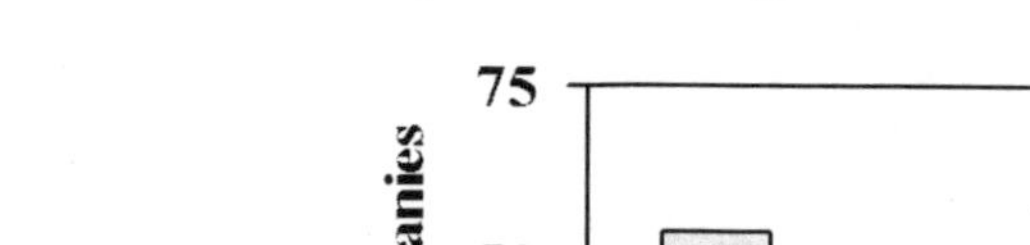

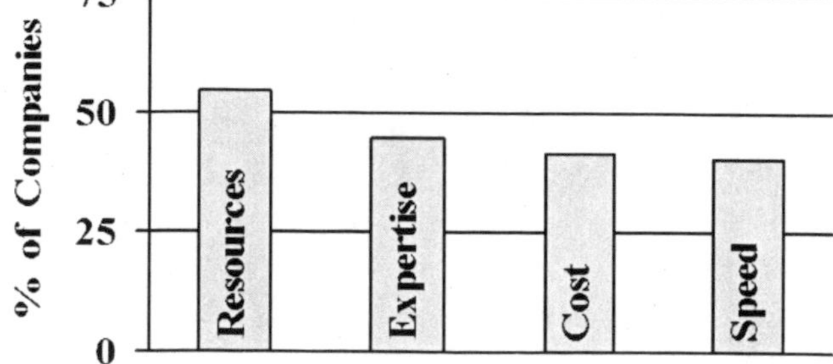

Figure 3. Common problems with outsourcing

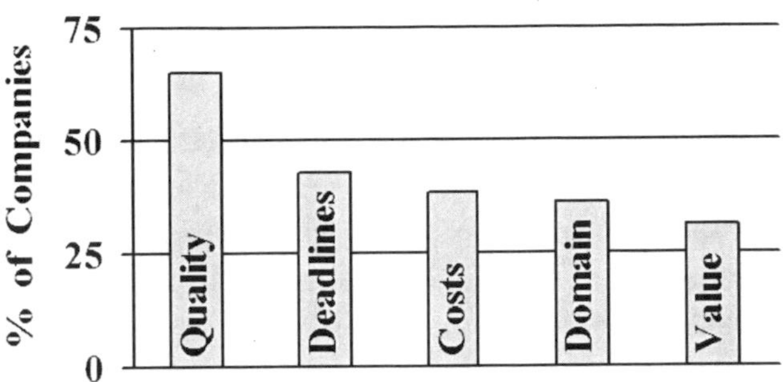

Web presence, if they do not have or otherwise need in-house expertise in Web development. Outsourcing for skills or facilities can result in higher quality, although outsourcing can also lead to quality problems, particularly if the provider misrepresents capabilities or the client does not monitor the project carefully enough.

Third, outsourcing can **reduce costs**. Lower cost is the most commonly cited (59.8%) advantage of offshoring, although 73.5% of CIOs surveyed feel that outsourcing is overrated as a cost-cutting strategy (CIO Insight, 2003). Although hourly rates in countries like India and China can be less than 10% of those in US (Dignan, 2003), most IT organizations save 15%-25% during the first year and up to 40% by the third year (Davison, 2003).

Fourth, outsourcing can **speed development and reduce time to market**, by providing specific skills just when they are needed. For example, making major changes to an organizational Web site might require a long time for a small Web development team, which might otherwise be underutilized. A provider could complete such changes more quickly using a larger team, and then assign them to work for other clients. Under the right circumstances, onshore staff can focus on activities such as testing and customer interaction, pass a set of requirements to an offshore team at the end of the day, and have the resulting changes implemented the next morning. This round-the-clock development can be very effective. In long projects, it also discourages staff from working excessive hours, which can lead to burnout and other morale problems.

Clients should consider whether their objectives are reasonable, since outsourcing is not a silver bullet. It is easy to assume that paying $15 per hour will dramatically lower total development costs, but most organizations realize more modest savings. Similarly, a larger joint team may be able to deliver a system more quickly, but only within limits; "adding manpower to a late software project makes it later" (Brooks, 1995, p. 25).

Clients should also consider how the objectives could be assessed; often, planning such assessment will help determine whether the objectives are reasonable. For example, to assess the cost savings from an outsourced project, a client will need to estimate internal project costs. Developing such estimates will help to uncover potential risks, and may even provide a model for tracking the outsourced project. Financial objectives are relatively easy to assess, but assessing other objectives may be more difficult, and depend on the specific processes used in the project.

For SPC, one objective for outsourcing was to reduce time to market and respond quickly to changes in the market. Furthermore, the number and intensity of Web application projects was likely to vary over time, and diverting SPC's development team to work on the Web applications would adversely affect the software engine. Thus, a second objective was to be able to adjust the size and skill sets of the team during the course of the project. In addition, outsourcing would reduce direct salary costs, and infrastructure costs, such as office space and equipment. Outsourcing had few indirect costs, since SPC already had offices in multiple locations and encouraged telecommuting. SPC considered these objectives to be reasonable based on experience with previous software projects; some of the objectives were also assessed using techniques described below.

In summary, organizations should determine:

a. What are our main objectives for outsourcing?
b. Are our objectives reasonable and attainable?
c. How will we assess our objectives to know if they are achieved?

Skills and Services

After identifying objectives, the second strategic step is to determine the specific skills and services that should be outsourced in order to reach those objectives. A wide variety of skills and services can be outsourced, but clients should consider the potential benefits (described above) of each, as well as the relative drawbacks, which are influenced by several factors, described below.

It is harder to outsource work that involves core competencies or key intellectual property. First, the information or skills must be shared or transferred to the provider, and this process must be repeated with each new provider. Second, there is the risk that the provider will share such information or skills with competitors, or use it to develop competing products. This is a particular concern for smaller clients dealing with offshore providers, since legal proceedings can be difficult and expensive. Conversely, there is less risk in outsourcing work that is only indirectly related to the organization's main business. Thus, technology organizations are less likely to outsource IT functions than are organizations in other sectors. For example, a company that uses a Web site to sell physical products is more likely to outsource Web development than a company whose core offering is a Web-based service.

It is easier to outsource widely available and interchangeable skills and services, and new providers often start with such offerings. If a particular client-provider relationship ends, both parties are more likely to find new partners. For example, it will be easier to outsource systems based on the LAMP (Linux, Apache, MySQL, and PHP) stack (Kunze, 1998) than on less popular technologies or technologies that are used together less frequently. However, clients often select and switch providers for such work based on price. Thus, most providers continually work to develop more specialized or higher level offerings, which clients will value for factors other than price. For example, a provider that provides testing services may add development services, and then add analysis and design services. These general trends, and some of the associated risks, are described by Christensen (1997).

It is easier to outsource work that can be clearly defined, easily evaluated, and is unlikely to change. Vague, poorly understood, or rapidly changing requirements are more difficult, since they generally involve more complex communication and more flexible agreements between the client and provider. Similarly, it is easier to outsource work that can be clearly separated from work that will remain in-house, and more difficult to outsource work that overlaps or must be integrated with in-house work.

Many of the efficiencies of outsourcing require a minimum size. Often, it is best to start by outsourcing a small pilot project, in order to minimize risk, and then expand the scope of the project once the basic approach is validated.

For SPC, the software engine was a core competence, and should be developed in-house. Furthermore, the engine's complexity would make it difficult to outsource work to providers who did not understand the overall architecture and its implications. SPC sought to outsource analysis, design, and development of the Web applications, however, because they used widely available skills and tools, addressed rapidly changing market conditions and shifting requirements imposed by industry leaders, customers, and competitors.

In summary, organizations should determine:

a. What work could we outsource?
b. What are the benefits and risks of outsourcing this work?
c. What will we do if problems occur?
d. How will we specify and assess the work that needs to be done?

Relationships

After determining required skills and services, the third strategic step is to determine the most appropriate type of provider relationship, which will likely depend on the objectives, and the services and skills required. For example, if the system has precise and easily defined requirements, it may be relatively simple to transmit them to an offshore team for implementation, and wait for a final delivery. On the other hand, a much closer relationship may be appropriate for projects that are mission critical, have incomplete or changing requirements, or that have multiple connections to other parts of the organization.

Kishore, et al. (2003) describe a framework that analyzes outsourcing relationships along two dimensions. **Dependence** is the degree to which ownership and control is transferred to the provider. **Strategic impact** is the degree to which the outsourcing relationship affects competitive positioning and long-term strategy. Together, these dimensions categorize outsourcing relationships into four quadrants, illustrated in Figure 4.

Support relationships are used selectively to simplify internal operations, such as payroll processing, basic IT functions, or custodial services. They are easily monitored based on outcomes, and typically have low setup and switching costs, so it is easy to change providers. **Alignment**

Figure 4. Outsourcing relationships

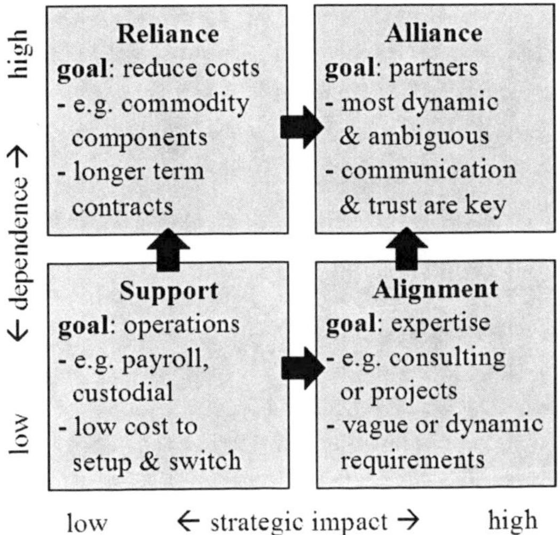

relationships are also used selectively, but usually for more strategic purposes, such as expert consulting, or specific projects. Requirements are typically more vague and dynamic, making it more difficult to measure outcomes. **Reliance** relationships are usually used to reduce costs, on a broader scale and for longer time periods than support relationships. There are typically well-defined processes and outcomes, and thus easy to monitor, though it is more difficult to switch providers. **Alliance** relationships are strategic partnerships that usually develop from other types of relationships. They are the most uncertain and dynamic relationships, and require significant trust and communication, and often shared goals or incentives.

Kishore, et al. conclude that "outsourcing should be considered more as a management of relationship with services providers rather than as a simple subcontract for IS commodities" and that "the most important factor affecting success of outsourcing appears to be a mutual understanding between clients and their service providers" (p. 92). This is certainly the case with outsourcing product development, which can have strategic consequences.

Outsourcing relationships also have financial characteristics, which can take on a variety of forms. At one extreme are projects where the client and provider agree on a fixed price for the entire project. Although fixed price projects may appear to limit risk, they actually present several challenges. First, the project scope may be unclear or incomplete. As a result, the provider must build in a margin to cover the unexpected. Second, fixed price projects set up a conflict between the client and provider. Once a fixed price is set, the client has an incentive to add functionality to the project, while the provider has an incentive to reduce functionality and minimize the total effort. At the other extreme are projects where the provider bills the client for time and materials, usually on an hourly basis. In such projects, a provider can do whatever the client requires; however, the client may have difficulty estimating or managing the total cost of the project. An intermediate approach is to develop a joint estimate of the entire project, but to structure the project as a series of time-limited iterations, each with an agreed price range. In each iteration, the provider attempts to deliver as much functionality as possible within the range. After each iteration, the client and provider revise the joint estimate, and agree on priorities for the next iteration. This example illustrates a more general principle that will be discussed further below—a client and provider should structure their relationship so that their incentives are aligned whenever possible. Multiple negotiations tend to encourage cooperation (e.g., Axelrod, 1984).

Finally, outsourcing relationships should consider the organizational cultures of the client and the provider. Tightly coupled relationships are more difficult when the cultures are different, or when one organization has reasons to distrust the other. For example, it is difficult for providers to work closely with client staff who fear that they will lose their jobs or status to outsourcing.

Such relationships are easier to develop when the organizations have similar cultures, locations, languages, and time zones, and when there is already experience with distributed teams.

The relationship between SPC and its outsourcing provider evolved over time. Initially, it was a support relationship, where an hourly consultant worked with SPC to study and evaluate specific application domains. It then evolved to an alignment relationship, where the consultant helped define requirements and a high-level architecture for a specific Web application. Once a proof-of-concept was completed, an onsite liaison and a small offshore development team began to design, implement, and test. The relationship then shifted toward reliance as the offshore team expanded, the joint team developed confidence, the market opportunities became clearer, and the scope of the overall project increased. Finally, the relationship evolved toward an alliance, where

SPC and its provider worked together to identify future directions. As the first Web application matured, the offshore team was reduced and SPC took over ongoing system maintenance. Thus, the relationship shifted back toward alignment. Several factors contributed to this successful evolution. First, SPC and its provider were already accustomed to distributed teams. Second, SPC's developers were confident that they would not be replaced by the provider, but that the provider was enabling SPC to address new opportunities. Finally, the project was divided into time-limited iterations, as described above, which encouraged cooperation.

In summary, organizations should determine:

a. How strategic is the relationship to our core business?
b. How dependent will we be on the provider?
c. How can we make this a win-win relationship?
d. How conducive to outsourcing is our organizational culture?

Providers

After determining the appropriate relationship, the fourth step is to select a provider. The best provider usually depends on the skills, services, and relationships needed, as well as a variety of other factors.

The provider's capabilities are clearly important. However, it can be difficult to assess the capabilities of a large group, particularly at a remote location. General reputation in an industry can be helpful, although there can be great variety within an organization; DeMarco and Lister (1999) argue that productivity varies by a factor of ten both between and within organizations. Certifications can demonstrate some capabilities, or satisfy external regulations (e.g., in pharmaceutical industry). It is usually easier to evaluate teams and individuals. Clients should insist on talking to references in other organizations. Clients should also insist on interviewing their primary contacts with a vendor, to assess both technical ability and how well each contact will interact with the client organization. Clients should ask about the provider's turnover rate, and what accommodations will be made when someone leaves the project. Although people change jobs for a variety of reasons, such changes can have adverse effects. In a strong job market, it can be difficult to keep the best people, and providers also have an incentive to move their best people to newer, larger projects when they can.

Location is also a key factor. Clients should decide whether all, some, or none of the provider's staff should work onsite. It can be easier to interact and develop trust with onsite staff, but they also require space and possibly other support. If the provider has offices in the same geographic area, it is easier to have staff work onsite for extended periods; otherwise, staff may need to travel frequently or have temporary housing. Offsite, near shore, or offshore staff often have lower hourly rates, but are more difficult to interact with, particularly if problems arise. One common approach is to have a few provider staff onsite, either initially or throughout the project, to serve as liaisons for staff in other locations.

Size is another key factor. A small provider may not be able to provide everything a client needs while a large provider may have too many other priorities. Many clients prefer to deal with a provider that is roughly the same size, as a way of balancing these concerns.

Cost is clearly an important factor, but as noted above outsourcing is often overrated as a strategy to reduce costs. Too much emphasis on a low hourly rate may result in poor quality and/or support.

It is relatively easy to assess factors such as location, size, cost, and some capabilities. Thus, they can be useful to filter an initial list of potential providers. Cultural and individual factors are more difficult to evaluate, even for the most promising potential providers. Some of these factors are described in more detail in the next section.

SPC selected its provider for several reasons. The most important factors were the technical and interpersonal abilities of the provider's onsite staff. In addition, the provider was located within driving distance of SPC's development center and primary sales office. Finally, the provider could manage an offshore team for SPC, which would speed development and reduce costs.

In summary, organizations should determine:

a. What capabilities do we need, and how do we evaluate them?
b. Where do we want the provider's staff to work? (onsite, offsite, offshore, etc)
c. What size of provider do we prefer?
d. How important is cost?
e. Who will be our primary contacts? What skills and traits should they have?
f. What other characteristics are important, and how do we evaluate them?

TACTICAL ISSUES

Once a client decides to outsource Web development activities to a provider, managers, team leaders, and developers face other challenges. Battin, et al. (2001) describe five risk categories for global software development projects, including communication, coordination, geography, team identity and cohesiveness, and culture. Four more steps will help managers, team leaders, and development teams work effectively together and respond to common problems. These steps are illustrated in Figure 5.

First, determine responsibilities. Second, determine methodologies and processes. Third, manage and monitor the project. Fourth, develop trust and collaboration. Note that these steps are not entirely sequential; later steps can feed back to affect earlier steps, just as in the traditional waterfall model for software development. For example, as collaboration improves during the course of the project, the processes may become less structured and more flexible. This in turn may enable the client and provider to shift or share responsibilities. Furthermore, these steps should be a mutual responsibility of the client and provider in an outsourcing project. The client

Figure 5. Four tactical steps

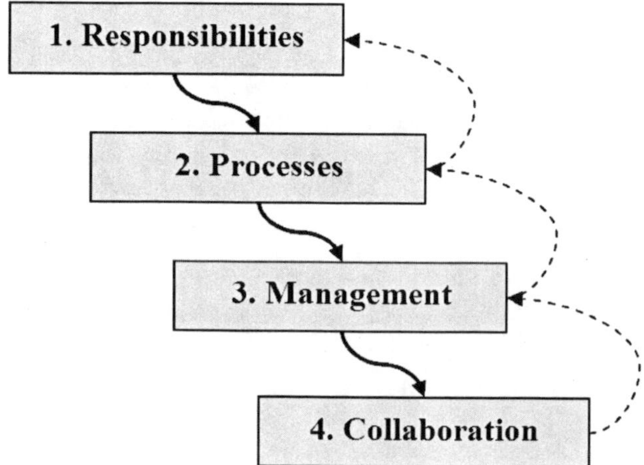

may have a better sense of what is needed, but the provider may have more experience making such relationships work effectively. Each step is described in more detail below, including relevant background, application in the case study, and a checklist of best practices.

Responsibilities

First, **determine responsibilities** in the project. Just as a development team must clearly define the interfaces between system components, clients and providers should work together to define responsibilities, interfaces, and supporting processes between organizations, key individuals, and teams.

As in any technical design, joint teams should seek to minimize coupling (relationships between components), which reduces the amount of communication between groups. Joint teams should also seek to maximize cohesion (the degree to which a single component has a single purpose), which makes it easier for each group to focus on particular objectives. As noted above, the structure of the joint team should parallel the structure of the Web system they are developing.

Carmel and Agarwal (2001) describe the importance of reducing intensive collaboration across organizational boundaries. Battin, et al. (2001) emphasize the importance of assigning teams to subsystems for the entire lifecycle, so that each subsystem is supported and enhanced by the same people who designed and constructed it. Mockus and Weiss (2001) define an analysis process to identify appropriate components and boundaries for distributed development.

Joint teams should focus on the boundaries between groups, and let each group allocate responsibilities within that group. These boundaries are likely to shift as the outsourcing relationship evolves. Bhat, Gupta, and Murthy (2006) identify characteristics of outsourced projects that affect requirements analysis; all of them involve boundaries or differences between the client and vendor. It is often beneficial to have one person or a small group, act as a liaison (and bottleneck) between the groups, to provide conceptual integrity and prevent miscommunication. Strong liaisons are often identified as key success factors for outsourced or offshored projects (e.g., Battin, et al., 2001; Cusick & Prasad, 2006).

For SPC, the provider's staff used their experience with outsourcing and offshoring development to help SPC define responsibilities and interfaces to meet key objectives. The project consisted of a client team and a provider team, each with a team leader. In addition, one person was designated as the technical lead for the project, and coordinated activities between the two teams.

Since the engine was designed to access databases, the joint team decided to use a database as the primary interface between the engine and the Web applications, in order to reduce coupling and minimize the impact on the application of changes in the software engine. By increasing the cohesion of the Web application, the provider team could focus on it without detailed knowledge of the software engine. At the same time, the client team could focus on configuring the engine and developing supporting utilities, without detailed knowledge of the Web application.

In summary, clients and providers should work together to:

a. Minimize coupling between groups and maximize cohesion within groups;
b. Concentrate on interfaces and boundaries, not internal operations;
c. Focus on win-win scenarios with common goals and ongoing negotiations.

Methodologies & Processes

After determining responsibilities, the second step is to determine the methodologies and processes for the project. This can be challenging, since it

Figure 6. Software development methodologies

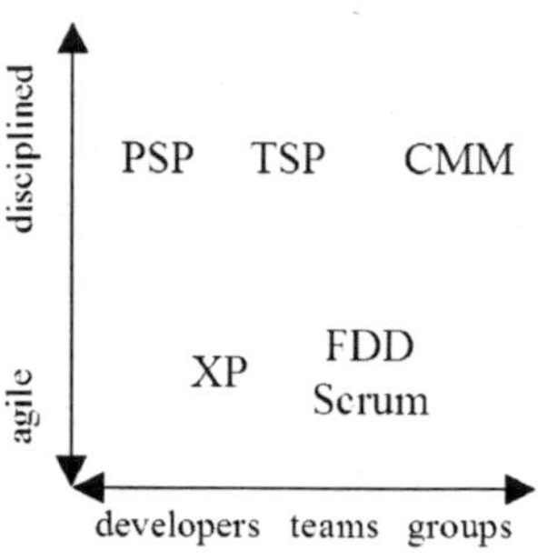

may require coordination between different development cultures. Concentrate on the processes at the interfaces between organizations, rather than trying to define all of the processes for everyone involved in the project. It is quite feasible, and at times even preferable, for the in-house developers to be using one methodology, and the out-sourced or offshore teams to be using different methodologies, since they may have different cultures and requirements.

Currently, software development methodologies can be divided into two broad categories, referred to as "disciplined" methods and "agile" methods (Boehm & Turner, 2003). Note that both of these names are misleading; some "disciplined" methods adapt quickly to change, and some "agile" methods are formal and prescriptive. Methodologies also address different levels of the organization; some focus on day-to-day developer activities, while others focus on project and team management or higher level organizational structures, as shown in Figure 6.

Disciplined approaches treat software development as a manufacturing process, which can be defined, measured, and optimized using well-known techniques. These approaches are often associated with quality initiatives such as ISO 9000 (e.g., Patterson, 1994) or the Capability Maturity Model (CMM®) (e.g., Paulk, et al., 1994). CMM-style disciplined processes can be particularly effective for large projects, or for highly repeatable projects. CMM is used by many offshore development centers, particularly in India; offshore companies represent roughly 74% of CMM-4 organizations and roughly 84% of CMM-5 organizations (Software Engineering Institute, 2003). Disciplined approaches have also been developed for individuals (e.g., Humphrey, 1994; Humphrey, 1996) and for teams (e.g., Humphrey, 1999).

Agile approaches treat software development as a specialized craft, performed by master craftspeople with supporting staff. Agile approaches share a particular set of values defined in the Agile Manifesto (Beck, Beedle, van Bennekum, Cockburn, et al., 2001):

- Individuals and interactions over processes and tools;
- Working software over comprehensive documentation;
- Customer collaboration over contract negotiation;
- Responding to change over following a plan.

Typically, agile methodologies have multiple iterations, each of a few weeks, which integrate analysis, design, implementation, and testing activities (e.g., Cockburn, 2002; Highsmith, 2002). These and other agile principles help a project to react to dynamic conditions, such as poorly understood requirements and aggressive schedules. Although agile methodologies often assume that all team members are in one location, they have also been adapted to include distributed or outsourcing teams (Fowler, 2006; Jensen & Zilmer, 2003; Kussmaul, Jack, & Sponsler, 2004; Paasivaara & Lassenius, 2004; Ramesh, Cao, Mohan, & Xu, 2006; Simons, 2002). In fact, Forrester Research concludes that there are unique benefits to combining agile methodologies and offshore outsourcing (Moore & Barnett, 2004).

There is growing recognition that both agile and disciplined approaches have advantages, and that carefully designed combinations can be very effective (e.g., Anderson, 2004; Boehm & Turner, 2003; Paulk, 2001). Furthermore, industries, markets, projects, and development teams can vary dramatically. Thus, organizations should understand the benefits and risks of multiple methodologies, and identify and adapt them to suit the situation, instead of trying to make the situation fit a methodology.

Methodologies have been developed specifically for Web development, and for global Web development (e.g., Cusick & Prasad, 2006). Unfortunately, Web systems are often developed with informal or completely ad hoc methodologies. For example, Taylor, McWilliam, Forsyth, and Wade (2002) surveyed 25 organizations, and found relatively little use of design techniques, layout standards, documentation, or testing procedures.

Informal or ad hoc processes may be sufficient for small, collocated teams, although they would likely benefit from articulating and reflecting on their processes. Processes are more important for larger teams, especially distributed or joint teams that include client and provider staff. However, as the team's size and diversity increase, it becomes more difficult to define processes for the entire team, since different parts of the team will have different objectives, training, and cultures (organizational, disciplinary, or regional). Thus, it may be more effective to have high-level processes to direct the overall project, but to allow or encourage different parts of the joint team to work in different ways.

As noted above, if the provider and client tasks are clearly separated, and the requirements are stable, it may be feasible to develop a detailed specification and receive a single delivery at the end of the project. However, such assumptions are often not true in Web development. Instead, some of the requirements are vague or likely to change, and there is often overlap between client and provider responsibilities. In such projects, it is essential to have multiple iterations with frequent deliveries, so that everyone can see what is actually working, find and fix problems, provide timely feedback, build confidence and trust. This applies throughout the system life cycle, to items such as requirements, designs, GUI mockups, software, tests, and documentation. Within each iteration changes should be kept to a minimum, since there is more risk of miscommunication with outsourced or distributed teams. However, the project direction can change dramatically between iterations. This is particularly useful in Web development projects, for several reasons. First, after each iteration, the updated system can be deployed relatively easily, since users access it via a browser rather than installing software. Second, Web applications are more likely to change in response to evolving user expectations, competitors, and changing technology; thus it is very valuable to be able to periodically redirect the project.

SPC's development group was relatively small, co-located, and used relatively informal processes. In contrast, the provider's offshore team was in a much larger development center and had years of experience with disciplined CMMI-5 processes. The provider's onsite staff had experience with both agile and disciplined development processes. The Web applications involved an evolving software engine, an unfamiliar application domain, shifting external requirements, and a rapidly changing market environment; an agile approach seemed most appropriate in such circumstances, although a certain amount of discipline and formality was needed to coordinate the multiple locations and organizations. The high-level development process was inspired by SCRUM and feature driven development (Highsmith, 2002; Palmer & Felsing, 2002; Schwaber & Beedle, 2001). For each iteration, provider staff in the US worked with SPC product management to identify a list of high-level requirements. The provider staff then worked with the offshore team to develop more detailed requirements, which were then reviewed with the SPC product manager. The provider's offshore team used CMMI processes, while the in-house processes gradually became more formal as the team grew and the system matured. The groups tended to work separately early in each iteration, and interact more frequently later in the iteration as they integrated and tested individual components.

In summary, clients and providers should work together to:

a. Identify and adapt processes to fit the project, not the reverse;
b. Encourage diversity; different parts of the project can be done in different ways;
c. Manage the project pace, especially the rate of change;
d. Schedule frequent deliveries throughout the project.

Project Management

Regardless of the methodology and processes used, organizations need continual visibility into the status and progress of any outsourced activities. Project management and monitoring can be more challenging with outsourcing, since "managing by walking around" is more difficult when people are at multiple locations and in different organizations. Conversely, extra effort to plan and coordinate with an experienced provider can lead to better visibility for everyone concerned.

Most importantly, client and provider managers should focus on creating and maintaining "win-win" relationships in the joint team. If everyone in the joint team feels that they are working together toward a common goal, there will be fewer problems, and those problems will be easier to resolve. Conversely, if the client and provider teams feel that they are competing or in conflict, it will be nearly impossible to keep the project on track.

Defining requirements requires particular attention for several reasons. Requirements definition is often one of the first joint activities in an outsourcing project, before both parties have developed effective communications and a comfortable working relationship. Outsourcing projects generally require more detailed and precise requirements than do in-house projects. Finally, incorrect or incomplete requirements can lead to significant problems later in the project. Roughly 20% of project defects originate in requirements (Jones, 2000, p. 104), and the cost of fixing an error increases exponentially with the time between when it was introduced and when it was detected and corrected (Boehm, 1981, p. 40). Organizations with considerable in-house experience developing detailed requirements may have no difficulty delivering them to a provider. In many cases, however, it is much more effective to use the provider's staff onsite to help develop a set of requirements for the outsourcing development team, particularly if the team is offshore. Defining requirements more carefully, and having both in-house and outsourcing team review them, can help both organizations identify and resolve problems sooner and more efficiently. Grünbacher (2003) describes some of the challenges in adapting traditional requirements engineering for Web applications, including the multidisciplinary teams, lack of developer experience, unavailability of stakeholders, fixed schedules, volatility of requirements, unpredictable operating conditions, the impact of legacy systems, and several aspects of quality. Grünbacher recommends that Web developers work to understand the system context, involve stakeholders, use iterative requirements definition, and focus on architecture. Escalona and Koch (2004) review the roles and uses of requirements analysis in ten Web application development methodologies, focusing on the types of requirements addressed, the techniques utilized, and the degree of detail. They note that most of the methodologies focus more on eliciting and specifying requirements, and less on validating them. They also note that most of the methodologies focus more on the steps in of a process, and less on the set of techniques or the results of the process.

Written documentation can be a particular challenge; teams or their managers must decide which documents to create, and for how long to maintain them. Brooks defines the **documentary hypothesis**: "a small number of documents become the critical pivots around which every project's management revolves" (Brooks, 1995, p. 107). However, this set of documents may vary from project to project, and even during the course of a project. Frequently, some of these documents are tables or diagrams. For example, it might be a list of key use cases, an architecture diagram, a detailed data model, or a Gantt chart. Often, the documents show the current status and expected completion date of system components.

Other traditional project management activities are still important in outsourced projects. Too often, clients assume that outsourced activities do not require supervision, which leads to problems. It is still important to manage near-term priorities and stakeholder expectations, to set and maintain a sustainable pace for the joint team, and to monitor costs and other resources.

Although people and interactions have the biggest impact on projects, appropriate tools can certainly contribute to project success. Focusing on a few key tools enables everyone to become proficient. Choosing flexible tools allows them to evolve with the project and its processes, rather than committing the joint team to a particular approach. New tools should be added cautiously, to avoid disruption. Three categories of tools are particularly valuable for outsourced Web development. First, tools designed to help archive project information, including source code control systems, and ticket tracking systems for tasks, defects, and requested enhancements. Second, automation tools to assist in developing, building, testing, or deploying the system. Third, tools to facilitate communication (described in more detail below). In each category, there are a variety of commercial and open source options.

The Web application project also contributed to the quality of SPC's software engine. The in-house QA team focused primarily on the GUI used to configure the engine. Developing the Web applications enabled in-house and outsourced developers to identify defects and potential enhancements. Because the offshore team was less familiar with the software engine, they identified additional issues. Similarly, the in-house development and QA groups identified issues and potential enhancements in the outsourced Web application. Every new product has defects, but the development teams were able to identify and resolve many of them before they were found by customers.

In summary, clients and providers should work together to:

a. Focus on win-win interactions with common goals, and frequent interactions;
b. Define requirements carefully throughout the project, since change is inevitable;
c. Identify and maintain the key project documents;
d. Monitor near-term priorities, sustainable pace, and resources;
e. Use tools to provide leverage.

Trust and Collaboration

Trust and collaboration are essential to any Web engineering effort, since there are typically numerous stakeholders: managers and executives, analysts, system architects, developers, graphic designers, technical writers, and end users, to name a few. Too little can lead to misunderstandings, errors and omissions, but too much can use valuable time and energy. These issues are particularly important in outsourcing relationships, where staff can be spread across multiple locations, time zones, and cultures (e.g., Lacity & Willcocks, 2000). There is an extensive literature on virtual teams (e.g., Duarte & Snyder, 2006; Katzenbach & Smith, 1993; Martins, Gilson, & Maynard, 2004; Pinsonneault & Caya, 2005; Powell, Piccoli, & Ives, 2004; Sundstrom and Associates, 1998).

It can be difficult to determine and maintain the appropriate levels of trust and collaboration. Alignment and alliance relationships generally require more trust than support or reliance relationships. Within a project, alignment of stakeholder goals can cause individual relationships to shift (Lacity & Willcocks, 2000). Trust generally increases as joint teams gain experience, but some initial trust is necessary for them to start working effectively. Lander, Purvis, McCray, and Leigh (2004) studied trust building mechanisms in an outsourced IT project. They surveyed fourteen participants, including management, client team members, provider team members, and users. They found that the most important trust building mechanisms were communication, followed by identification with the joint team and predictability. However, several mechanisms were perceived quite differently by different stakeholders. For example, client team members ranked communication lower than the other groups, while users ranked predictability lower than the other groups. Herbsleb and Mockus (2003) found that distributed teams had very little informal, unstructured communication, and as a result the developers knew much less about the state of the project, what other developers were doing, and who had expertise in particular areas. They also suggest strategies to encourage such communication and knowledge.

Developing trust and effective communication practices early in the project is a priority, for several reasons. First, initial expectations and habits are likely to persist, and be difficult to change. Second, the joint team is often smaller at the beginning of the project, making it easier to work through problems and develop appropriate patterns. Third, key initial decisions about priorities, system architecture, responsibilities, and processes may be difficult or impossible to change later in the project. Thus, it is particularly important to have frequent deliverables and interactions near the start of a project.

If the joint team includes members from different organizations, national cultures, or language backgrounds, seemingly minor differences can lead to problems. Language is a particular challenge, and a key factor in India's success as an outsourcing destination for the US. Thus, communication can be improved by developing a common project vocabulary of terms, documents, diagrams, and metaphors. Carmel and Agarwal (2001) describe the importance of reducing cultural distance by using liaisons and focusing on language skills.

The mechanics of communication are also important; joint teams need to match activities to appropriate communication methods, since each has distinct benefits and risks. Synchronous communication, such as face-to-face meetings, online chats, and teleconferences, is ideal for quick status meetings, brainstorming sessions, and reviews. Instant messaging can be more effective than phone conversations, particularly if people have trouble understanding accents. Face to face meetings or video conferences are most valuable near the beginning of the project; as the joint team gains experience and trust, they become less necessary. Asynchronous communication, such as email, mailing lists, and shared documents, can provide a persistent record of discussions and decisions, and doesn't require participants to be available at the same time.

Finally, joint teams need to be sensitive to cultural differences between small groups, larger organizations, and geographic regions. Some cultures respect seniority and expect top-down direction, while others value initiative and entrepreneurship, even if it challenges existing structures. Asking questions is seen as a sign of interest in some cultures but as a challenge to authority in other cultures.

Since key SPC staff were in multiple locations across the US, email, instant messaging, teleconferences, and Web conferences were already part of the SPC culture, which helped ease the transition into an outsourcing relationship. Every 1-2 months there was a high-level status meeting with the leadership of SPC. During the project, there was continual email and instant messaging between SPC and the provider's US staff, and face-to-face meetings, teleconferences, or Web conferences several times a week. The provider's onsite staff were trained and experienced with using offshore resources and at working with customers. The provider's development teams usually had short online chats once or twice a day (when both US and offshore staff were awake) to discuss current status and problems.

In summary, clients and providers should work together to:

a. Ease into relationships—start small, and develop useful bottlenecks;
b. Develop a common vocabulary of terms, and diagrams, and metaphors;
c. Use appropriate communication channels, both synchronous and asynchronous;
d. Be sensitive to cultural differences.

FUTURE TRENDS

In the future, globalization and other competitive factors will continue to push organizations to operate more effectively and efficiently. For many organizations, this will include appropriate use of outsourcing and offshoring. As organizations and individuals continue to gain experience with distributed development models, they will find ways to maximize benefits and minimize risks. This is particularly true of Web development, since for many organizations Web presence is an important asset but not a core competence. Thus, it makes sense to outsource Web development to specialist organizations.

The increased use of outsourcing has implications for students and teachers. Students are likely to need more than technical knowledge and abilities to function effectively. They need to be able to form effective relationships with a variety of other people, and apply their combined expertise and abilities to understand incompletely defined problems and develop solutions. Levy and Murnane (2005) refer to these skills as complex communication and problem-solving. In particular, students need experience working in virtual teams and other structures typical of outsourced relationships. Thus, educators should find ways to help students obtain such experiences. Klappholz and Bernstein (2001) describe Live-Through Case Histories, which force students to live through specific situations to learn specific lessons about software engineering. Student experiences might involve interactions with other departments on campus as well as with other institutions. There are multiple studies of projects involving student teams spanning diverse institutions (e.g., Jarvenpaa, Knoll, & Leidner, 1998; Kaiser, Tuller, & McKowen, 2000).

Researchers should continue to identify and document best practices for outsourcing and other distributed teams, particularly in situations where relationships are likely to evolve. As best practices are understood, they can be organized into methodologies and tools, which can be adapted to different situations.

CONCLUSION

This chapter has reviewed the state of Web application outsourcing, and provided a concise overview of important concepts, key questions, and best practices, which are summarized in

Figure 7. Summary of key questions and best practices

Objectives	What are our main objectives for outsourcing?
	Are our objectives reasonable and attainable?
	How will we assess our objectives to know if they are achieved?
Skills & Services	What work could we outsource?
	What are the benefits and risks of outsourcing this work?
	What will we do if problems occur?
	How will we specify and assess the work that needs to be done?
Relationships	How strategic is the relationship to our core business?
	How dependent will we be on the provider?
	How can we make this a win-win relationship?
	How conducive to outsourcing is our organizational culture?
Providers	What capabilities do we need, and how do we evaluate them?
	Where do we want the provider's staff to work? (onsite, offsite, offshore, etc.)
	What size of provider do we prefer?
	How important is cost?
	Who will be our primary liaisons? What skills and traits should they have?
	What other characteristics are important, and how do we evaluate them?
Responsibilities	Minimize coupling between groups and maximize cohesion within groups.
	Concentrate on interfaces and boundaries, not internal operations.
	Focus on win-win scenarios with common goals and ongoing negotiations.
Methodologies & Processes	Identify and adapt processes to fit the project, not the reverse.
	Encourage diversity; different parts of the project can be done in different ways.
	Manage the project pace, especially the rate of change.
	Schedule frequent deliveries throughout the project.
Project Management	Focus on win-win interactions with common goals, and frequent interactions.
	Define requirements carefully throughout the project, since change is inevitable.
	Identify and maintain the key project documents.
	Monitor near-term priorities, sustainable pace, and resources.
	Use tools to provide leverage.
Trust & Collaboration	Ease into relationships—start small, and develop useful bottlenecks.
	Develop a common vocabulary of terms, and diagrams, and metaphors.
	Use appropriate communication channels, both synchronous and asynchronous.
	Be sensitive to cultural differences.

Figure 7. However, many of these issues and approaches also apply to other sorts of outsourced development work.

When outsourcing Web development, organizations should focus on three key ideas. First, structure projects in short iterations, so that there are multiple opportunities to find and correct problems. This is particularly true early in outsourcing relationships, when stakeholders are less sure of their roles and relationships. Second, recognize that the major benefits and risks are tied to people and relationships rather than to tools. If individuals are committed to the project and team, they will find ways to utilize available tools to solve their problems. Outsourcing projects often depend on a few key liaisons to facilitate communication and prevent misunderstandings. Third, match the structure of the development team to the desired structure of the system.

REFERENCES

American Electronics Association. (2004). *Offshore outsourcing in an increasingly competitive and rapidly changing world: A high-tech perspective.* Retrieved on January 1, 2007, from http://www.aeanet.org/publications/IDMK_AeA_Offshore_Outsourcing.asp

Anderson, D. (2004). *Agile management for software engineering: Applying the theory of constraints for business results.* Upper Saddle River, NJ: Prentice Hall PTR.

Aspray, W., Mayadas, F., & Vardi, M. Y. (Eds.) (2006). *Globalization and offshoring of software: A report of the ACM Job Migration Task Force.* New York: Association for Computing Machinery. Retrieved on January 1, 2007, from http://www.acm.org/globalizationreport

Axelrod, R. (1984). *The evolution of cooperation.* New York: Basic Books.

Battin, R. D., Crocker, R., Kreidler, J., & Subramanian, K. (2001). Leveraging resources in global software development. *IEEE Computer, 18*(2), 70-77.

Beck, K., Beedle, M, van Bennekum, A, Cockburn, A. et al. (2001). *Manifesto for agile software development.* Retrieved on January 1, 2007, from http://www.agilemanifesto.org

Behravesh, N. & Klein, L. (2004). *The comprehensive impact of offshore IT software and services outsourcing on the US economy and the IT industry.* Report prepared by Global Insight for the Information Technology Association of America, Arlington, VA.

Bhat, J., Gupta, M., & Murthy, S. (2006). Overcoming requirements engineering challenges: Lessons from offshore outsourcing. *IEEE Software, 23*(5), 38-44.

Boehm, B. (1981). *Software engineering economics.* Upper Saddle River, NJ: Prentice Hall PTR.

Boehm, B., & Turner, R. (2003). *Balancing agility and discipline: A guide for the perplexed.* Boston: Addison Wesley Professional.

Brooks, F. (1995). *The mythical man-month.* Boston: Addison Wesley Professional.

Carmel, E., & Agarwal, R. (2001). Tactical approaches for alleviating distance in global software development. *IEEE Software, 18*(2), 22-29.

Chen, J. Q., & Heath, R. D. (2005). Web application development methodologies. In W. Suh (Ed.), *Web engineering: Principles and techniques.* Hershey, PA: Idea Group Publishing.

Christensen, C. M. (1997). *The innovator's dilemma: When new technologies cause great firms to fail.* Boston: Harvard Business School Press.

CIO Insight (2003). Outsourcing: How well are you managing your partners?. *CIO Insight, 1*(33), 75-85.

Cockburn, A. (2003). *Agile software development.* Boston: Addison Wesley Professional.

Conway, M. E. (1968). How do committees invent? *Datamation, 14*(4), 28-31.

Cusick, J., & Prasad, A. (2006). A practical management and engineering approach to offshore collaboration. *IEEE Software, 23*(5), 20-29.

Davison, D. (2003). *Top 10 risks of offshore outsourcing.* META Group Research.

DeMarco, T., & Lister, T. (1999). *Peopleware: Productive projects and teams.* New York: Dorset House.

Dignan, L. (2003). Leaping, then looking. *Baseline, 1*(22), 17-29.

Duarte, D. L., & Snyder, N. T. (2006). *Mastering virtual teams.* Hoboken, NJ: Jossey-Bass.

Engardio, P., Bernstein, A., Kripalani, M., Balfour, F., Grow, B., & Greene, J. (2003). The new global job shift. *Business Week, February,* 36-42,44,46.

Escalona, M. J., & Koch, N. Requirements engineering for Web applications—A comparative study. *Journal of Web Engineering, 2*(3), 193-212.

Farrell, D. (2003). *Offshoring: Is it a win-win game?* McKinsey Global Institute, Aug 2003.

Fowler, M. (2006). *Using an agile software development process with offshore development.* Retrieved on January 1, 2007, from http://www.martinfowler.com/articles/agileOffshore.html

Ginige, A., & Murugesan, S. (2001). Web engineering: An introduction. *IEEE Multimedia, 8*(1), 14-18.

Ginige, A. (2002). Web engineering: Managing the complexity of Web systems development. In *Proceedings of the 14th International Conference on Software Engineering and Knowledge Engineering,* (pp 721-729). New York: ACM Press.

Grűnbacher, P. (2003). Requirements engineering for Web applications. In G. Kappel, B. Pröll, S. Reich, & W. Retschitzegger (Eds.), *Web engineering: The discipline of systematic development of Web applications.* Hoboken, NJ: John Wiley & Sons.

Herbsleb, J.D., & Mockus, A. (2003). An empirical study of speed and communication in globally distributed software development. *IEEE Transactions on Software Engineering, 29*(6), 1-14.

Highsmith, J. (2002). *Agile software development ecosystems.* Boston: Addison Wesley Professional.

Humphrey, W. S. (1994). *A discipline for software engineering.* Boston: Addison Wesley Professional.

Humphrey, W. S. (1996). *Introduction to the personal software process.* Boston: Addison Wesley Professional.

Humphrey, W. S. (1999). *Introduction to the team software process.* Boston: Addison Wesley Professional.

Jarvenpaa, S. L., Knoll, K., & Leidner, D. E. (1998). Is anybody out there? Antecedents of trust in global virtual teams. *Journal of Management Information Systems, 14*(4), 29-48.

Jensen, B., & Zilmer, A. (2003). Cross-continent development using Scrum and XP. *Extreme programming and agile processes in software engineering.* Paper presented at the 4th International Conference. Berlin: Springer.

Johnson, J., Boucher, K. D., Conners, K., & Robinson, J. (2001). Collaborating on project success. *Software Magazine, February/March.*

Jones, T. C. (2000). *Software assessments, benchmarks, and best practices.* Boston: Addison Wesley Professional.

Kaiser, P. R., Tuller, W. L., & McKowen, D. (2000). Student team projects by internet. *Business Communication Quarterly*, *63*(4), 75-82.

Kappel, G., Pröll, B., Reich, S., & Retschitzegger, W. (2003). An introduction to Web engineering. In G. Kappel, B. Pröll, S. Reich, & W. Retschitzegger (Eds.), *Web engineering: The discipline of systematic development of Web applications.* Hoboken, NJ: John Wiley & Sons.

Katzenbach, J. R., & Smith, D. K. (1993). *The wisdom of teams: Creating the high performance organization.* Boston: Harvard Business School Press.

Kishore, R., Rao, H. R., Nam, K., Rajagopalan, S., & Chaudhury, A. (2003). A relationship perspective on IT outsourcing. *Communications of the ACM, 46*(12), 87-92.

Klappholz, D. & Bernstein, L. (2001) Getting software engineering into our guts. *Crosstalk: The Journal of Defense Software Engineering, 14*(7). Retrieved on January 1, 2007, from http://www.stsc.hill.af.mil/crosstalk/2001/07/bernstein.html

Kunze, M. (1998) Let there be light: LAMP : Freeware Web publishing system with database support. *c't*, Dec 1998, 230.

Kussmaul, C., Jack, R., & Sponsler, B. (2004). Outsourcing and offshoring with agility: A case study. *Extreme programming and agile methods—XP / Agile Universe*, (pp 147-154). Berlin: Springer.

Lacity, M.C., & Willcocks, L.P. (2000). Relationships in IT outsourcing: A stakeholder perspective. In R. W. Zmud (Ed.), *Framing the domains of IT management: Projecting the future through the past.* Cincinnati, OH: Pinnaflex Educational Resources, Inc.

Lander, M. C., Purvis, R. L., McCray, G. E., & Leigh, W. (2004). Trust-building mechanisms utilized in outsourced IS development projects: A case study. *Information and Management, 41*(4), 509-528.

Levy, F., & Murnane, R. J. (2005). *The new division of labor: How computers are creating the next job market.* Princeton, NJ: Princeton University Press.

Martins, L. L., Gilson, L. L., & Maynard, M. T. (2004). Virtual teams: What do we know and where do we go from here? *Journal of Management, 30*(6), 805-835.

Mendes, E., Mosley, N., & Counsell, S. (2006). The need for Web engineering: An introduction. In E. Mendes & N. Mosley (Eds.), *Web engineering.* Berlin: Springer Verlag.

Mockus, A., & Weiss, D.M. (2001). Globalization by chunking: A quantitative approach. *IEEE Software*, *18*(2), 30-37.

Moore, S., & Barnett, L. (2004). *Offshore outsourcing and agile development.* Forrester Research, Inc.

Murugesan, S., & Ginige, A. (2005). Web engineering: Introduction and perspectives. In W. Suh (Ed.), *Web engineering: Principles and techniques* (pp. 1-30). Hershey, PA: Idea Group Publishing.

Paasivaara, M., & Lassenius, C. (2004). Using interactive and incremental processes in global software development. In *Proceedings of the International Conference on Software Engineering (ICSE) Third International Workshop on Global Software Development*, (pp. 24-28).

Palmer, S. R., Felsing, J. M. (2002). *A practical guide to feature-driven development.* Upper Saddle River, NJ: Prentice Hall PTR.

Patterson, J. (1994). *ISO 9000: Worldwide quality standard.* Stamford, CT: Crisp Learning.

Paulk, M. C., Weber, C. V., Curtis, B., Chrissis, M. B., et al. (1994). *The capability maturity model: Guidelines for improving the doftware process.* Boston: Addison Wesley Professional.

Paulk, M. C. (2001). Extreme programming from a CMM perspective. *IEEE Software, 18*(6), 19-26.

Pinsonneault, A., & Caya, O. (2005). Virtual teams: What we know, what we don't know. *International Journal of e-Collaboration, 1*(3), 1-16.

Powell, A., Piccoli, G., & Ives, B. (2004). Virtual teams: A review of current literature and directions for future research. *ACM SIGMIS Database, 35*(1), 6-36.

Ramesh, B., Cao, L., Mohan, K., & Xu, P. (2006). Can distributed software development be agile? *Communications of the ACM, 49* (10), 41-46.

Schwaber, K., & Beedle, M. (2001). *Agile software development with SCRUM.* Upper Saddle River, NJ: Prentice Hall PTR.

Simons, M. (2002). Internationally agile. *InformIT,* March 15, 2002.

Software Engineering Institute (2003). *Process maturity profile: Software CMM® -CBA IPI and SPA appraisal results.*

Sundstrom, E. & Associates (1998). *Supporting work team effectiveness: Best management practices for fostering high performance.* San Francisco: Jossey-Bass.

Taylor, M. J., McWilliam, J., Forsyth, H., & Wade, S. (2002). Methodologies and Website development: A survey of practice. *Information and Software Technology, 44*, 381-391.

Vogel, D. A., & Connelly, J. E. (2005). Best practices for dealing with offshore software development. *Handbook of Business Strategy.* Bradford, UK: Emerald Group Publishing Limited.

Chapter XIV
Engineering Wireless Mobile Applications

Qusay H. Mahmoud
University of Guelph, Canada

Zakaria Maamar
Zayed University, UAE

ABSTRACT

Conventional desktop software applications are usually designed, built, and tested on a platform similar to the one on which they will be deployed and run. Wireless mobile application development, on the other hand, is more challenging because applications are developed on one platform (like UNIX or Windows) and deployed on a totally different platform like a cellular phone. While wireless applications can be much smaller than conventional desktop applications, developers should think in small terms of the devices on which the applications will run and the environment in which they will operate instead of the amount of code to be written. This chapter presents a systematic approach to engineering wireless applications and offers practical guidelines for testing them. What is unique about this approach is that it takes into account the special features of the new medium (mobile devices and wireless networks), the operational environment, and the multiplicity of user backgrounds; all of which pose new challenges to wireless application development.

INTRODUCTION

The general mobile computing model in a wireless environment consists of two distinct sets of entities (Figure 1): Mobile Clients (MCs) and fixed hosts. Some of the fixed hosts, called mobile support stations (MSSs), are enhanced with wireless interfaces. An MSS can communicate with the MCs within its radio coverage area called wireless cell. An MC can communicate with a fixed

Figure 1. Mobile computing model

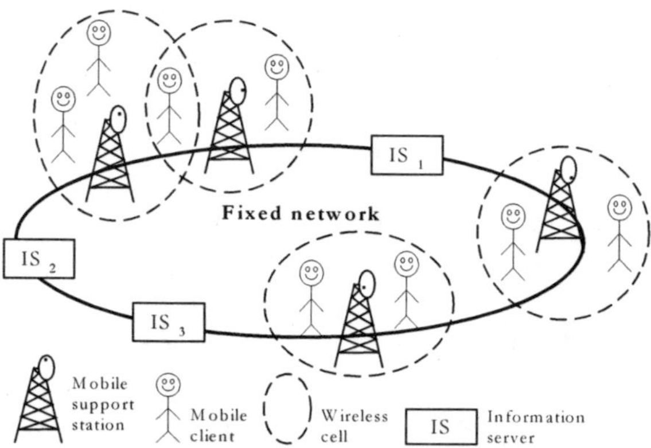

host/server via an MSS over a wireless channel. The wireless channel is logically separated into two sub-channels: an uplink channel and a downlink channel. The uplink channel is used by MCs to submit queries to the server via an MSS, whereas the downlink channel is used by MSSs to disseminate information or to forward the responses from the server to a target client. Each cell has an identifier (CID) for identification purposes. A CID is periodically broadcasted to all the MCs residing in a corresponding cell.

A wireless mobile application is defined as a software application, a wireless service or a mobile service that can be either pushed to users' handheld wireless devices or downloaded and installed, over the air, on these devices.[1] Such applications must work within the daunting constraints of the devices themselves:

- **Memory:** Wireless devices such as cellular phones and two-way pagers have limited amounts of memory, obliging developers to consider memory management most carefully when designing application objects.
- **Processing power:** Wireless devices also have limited processing power (16-bit processors are typical).
- **Input:** Input capabilities are limited. Most cell phones provide only a one-hand keypad with twelve buttons: the ten numerals, an asterisk (*), and a pound sign (#).
- **Screen:** The display might be as small as 96 pixels wide by 54 pixels high and 1 bit deep (black and white). The amount of information that can be squeezed into such a tight screen is severely limited.

In addition, the wireless environment imposes further constraints: (1) wireless networks are unreliable and expensive, and bandwidth is low; (2) they tend to experience more network errors than wired networks; and (3) the very mobility of wireless devices increases the risk that a connection will be lost or degraded. In order to design and build reliable wireless applications, designers need to keep these constraints in mind and ask themselves, what impact do wireless devices with limited resources have on application design?

The motivation for this chapter is provided in part by the above characteristics that form some of the foundations for pervasive computing environments. Such characteristics pose several challenges in designing wireless mobile applications for mobile computing. This chapter provides a detailed treatment of the impact of these characteristics on engineering wireless mobile applications and presents a systematic approach for designing them. In addition, it offers practical design techniques for wireless application design and development.

WIRELESS APPLICATIONS

Wireless applications can be classified into two streams (Beaulieu, 2002; Burkhardt, Henn, Hepper, Rintdorff, & Schack, 2002):

1. **Browser-based:** Applications developed using a markup language. This is similar to the current desktop browser model where the device is equipped with a browser. The wireless application protocol or WAP (http://www.openmobile alliance.org) follows this approach (Open Mobile Alliance, 2005).
2. **Native applications:** Compiled applications where the device has a runtime environment to execute applications. Highly interactive wireless applications are only possible with the latter model. Interactive applications, such as mobile computer games, are a good example. Such applications can be developed using the fast growing Java 2 Micro Edition (J2ME) platform (http://www.java. sun.com/j2me), and they are known as MIDlets.

Another stream is the hybrid application model that aims at incorporating the best aspects of both streams above. The browser is used to allow the user to enter URLs to download native applications from remote servers, and the runtime environment is used to let these applications run on the device.

WAP Might be Dead, but What Did We Learn?

WAP and J2ME MIDP solve similar problems but each can learn a couple of things from the other. There are special features that are available in WAP but not in MIDP and *vice versa.* These features are summarized as follows:

- MIDP provides a low-level graphics APIs that enable the programmer to have control over every pixel of the device's display. This is important for entertainment applications (such as games) in a wireless environment.
- MIDP is the way to go for games. The nature of MIDlets (they exist on the device until they are explicitly removed) allows users to run them even when the server becomes unavailable (support for disconnected operations).
- WML provides tags and possible presentation attributes, but it doesn't define an interaction model. For example, WML defines a SELECT tag for providing a list. Some WAP-enabled devices interpret the SELECT tag as a popup menu list while others interpret it as a menu that can be used for navigation. Therefore, there is no standard interaction model defined for this element. If a developer uses it, the application may run well on some devices and poorly on others. MIDlets, on the other hand, provide a clearly defined standard for interaction using commands.

A Micro Browser is Needed

MIDlets combine excellent online and off-line capabilities that are useful for the wireless environment, which suffers from low bandwidth and network disconnection. Integrating WAP and MIDP opens up possibilities for new wireless applications and over the air distribution models. Therefore, WAP and MIDP shouldn't be viewed as competing but rather as complementing technologies. In order to facilitate downloading wireless applications over the air, there is a need for some kind of an environment on the handset that allows the user to enter a URL for a MIDlet Suite, for

Figure 2. Combining WAP and J2ME

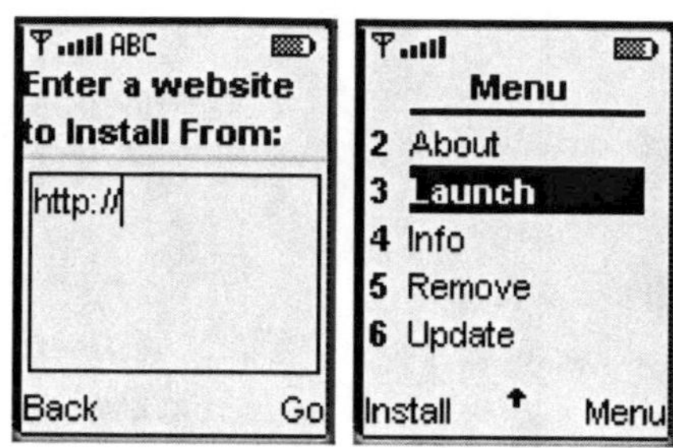

example. This environment could very well be a WAP browser as shown in Figure 2.

Similar to Java Applets that are integrated into HTML, MIDlets can be integrated into a WML or an XHTML page. Such a page can then be called from a WAP browser, and the embedded MIDlet gets downloaded and installed on the device. In order to enable this, a WAP browser is needed on the device. Another alternative approach for over-the-air provisioning is the use of a short message service (SMS) which has been done by Siemens where the installation of MIDlets is accomplished by sending a corresponding SMS. If the SMS contains a URL to a Java application descriptor (JAD) file specifying a MIDlet Suite, then the recipient can install the application simply by confirming the SMS.

DESIGN CHALLENGES AND POSSIBLE SOLUTIONS

In this chapter, we are more concerned with native interactive applications that can be developed using the J2ME platform or a similar technology. J2ME-based wireless applications can be classified into local (stand-alone) and network applications. Local applications perform all their operations on a handheld wireless device and need no access to external data sources through a wireless network. Examples include calculators and single-player games. Network applications, on the other hand, consist of some components running on a wireless device and others running on a network, and thus depend on access to external resources. An example would be an e-mail application with a client residing on a wireless phone interacting with a simple mail transfer protocol (SMTP) server to send/receive e-mail messages. A major difference between local and networked applications is in the way they are tested. Local applications are easier to test than network applications. For example, a calculator application can run on a wireless device even when it is not connected to any network, but an e-mail client will not work without a connection to e-mail servers.

Challenges

The constraints discussed earlier pose several crucial challenges, which must be faced in order for wireless applications to function correctly in the target environment.

- **Transmission errors:** Messages sent over wireless links are exposed to interference (and varying delays) that can alter the content received by the user, the target device, or the server. Applications must be prepared to handle these problems. Transmission errors may occur at any point in a wireless transaction and at any point during the sending or receiving of a message. They can occur after a request has been initiated, in the middle of the transaction, or after a reply has been sent. While wireless network protocols may be able to detect and correct some errors, error-handling strategies that address all kinds of transmission errors that are likely to occur are still needed.

- **Message latency:** Message latency, or the time it takes to deliver a message, is primarily affected by the nature of each system that handles the message, and by the processing time needed and delays that may occur at each node from origin to destination. Message latency should be taken into account and users of wireless applications should be kept informed of processing delays. It is especially important to remember that a message may be delivered to a user long after the time it is sent. A long delay might be due to coverage problems or transmission errors, or the user's device might be switched off or have a dead battery. Some systems keep trying, at different times, to transmit the message until it is delivered. Other systems store the message then deliver it when the device is reconnected to the network. Therefore, it is important to design applications that avoid sending stale information, or at least to make sure that users are aware that it is not up-to-date. Imagine the possible consequences of sending a stock quote that is three days old without warning the user!
- **Security:** Any information transmitted over wireless links is subject to interception. Some of that information could be sensitive, like credit card numbers and other personal information. The solution needed really depends on the level of sensitivity. To provide a complete end-to-end security solution, you must implement it on both ends, the client and the server, and assure yourself that intermediary systems are secure as well.

Possible Solutions

Here are some practical hints useful to consider when developing mobile applications.

- **Understand the environment.** Do some research upfront. As with developing any other software application, we must understand the needs of the potential users and the requirements imposed by all networks and systems the service will rely on.
- **Choose an appropriate architecture.** The architecture of the mobile application is very important. No optimization techniques will make up for an ill-considered architecture. The two most important design goals should be to minimize the amount of data transmitted over the wireless link, and to anticipate errors and handle them intelligently.
- **Partition the application.** Think carefully when deciding which operations should be performed on the server and which on the handheld device. Downloadable wireless applications allow locating much of an application's functionality of the device; it can retrieve data from the server efficiently, then perform calculations and display information locally. This approach can dramatically reduce costly interaction over the wireless link, but it is feasible only if the device can handle the processing that the application needs to perform.
- **Use compact data representation.** Data can be represented in many forms, some more compact than others. Consider the available representations and select the one that requires fewer bits to be transmitted. For example, numbers will usually be much more compact if transmitted in binary rather than string forms.
- **Manage message latency.** In some applications, it may be possible to do other work while a message is being processed. If the delay is appreciable—and especially

if the information is likely to go stale — it is important to keep the user informed of progress. Design the user interface of your applications to handle message latency appropriately.

- **Simplify the interface.** Keep the application's interface simple enough that the user seldom needs to refer to a user manual to perform a task. To do so: reduce the amount of information displayed on the device; make input sequences concise so the user can accomplish tasks with the minimum number of button clicks; and offer the user selection lists.

AD-HOC DEVELOPMENT PROCESS

An ad-hoc development process for wireless applications comprises three steps:

1. Write the application. Several integrated development environments (IDEs) are available for developing Java-based wireless applications, for example, Sun's J2ME Wireless Toolkit, and Metrowerks CodeWarrior.
2. Test the application in an emulation environment. Once the application compiles nicely, it can be tested in an emulator.
3. Download the application to a physical device and test it. Once the application's performance is satisfactory on one or more emulators, it can be downloaded to a real device and tested there. If it is a network application, it is tested on a live wireless network to ensure that its performance is acceptable.

It is clear that many important software engineering activities are missing from this ad-hoc development process. For example, there is no formal requirements analysis phase, and so following an ad-hoc development process may lead to building a product different from the one customers want. Also, testing an application without knowing its requirements is not an easy task. In addition, issues related to the operating environment such as network bandwidth should be considered during the design so that the performance of the application will be satisfactory.

WIRELESS SOFTWARE ENGINEERING

While wireless application development might appear to have less need for the coordination that a process provides, aspects of development, testing, evaluation, deployment, and maintenance of a wireless application have to be integrated in the design process throughout the full development life cycle. We have put forward a systematic approach to developing wireless applications, which is compatible with the rational unified process or RUP (Jacobsen, Booch, & Rumbaugh, 2000) in the sense that it is iterative and responsibility-driven. We have developed this systematic approach based on our experience designing and building wireless applications. We recognized that the development of a wireless application is not a one-shot task, and testing wireless applications is more challenging than testing conventional desktop software applications; therefore, an ad-hoc development process cannot be used.

Development Activities

Our software engineering approach to wireless application development consists of a set of manageable activities that, if followed properly, leads to reliable and maintainable wireless applications. The activities of our approach are shown in Figure 3.

Planning. This iterative process begins with a planning phase, which is an activity that identifies the objectives of the wireless application and specifies the scope of the first increment. In addition, the costs of the overall project are

Figure 3. Wireless application development activities

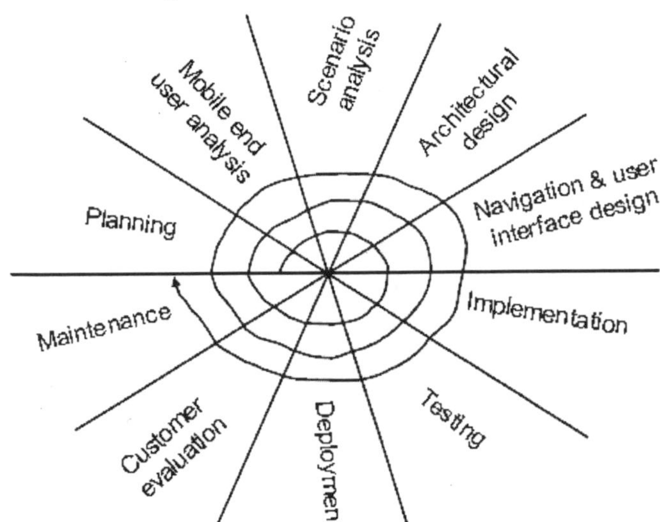

estimated, the risks are evaluated, and a tentative schedule is set.

Mobile user analysis. First, we must understand the audience of the application and the environment in which it will operate. As an example, if the application is a wireless network-aware application such as a multi-player game, the study will include the users of the application and how they plan to use it. The output at the end of this phase is a wireless application plan document that serves as the mobile end-user requirement.

Scenario analysis. This phase is similar to conventional software requirements analysis, and therefore concepts and principles of requirements analysis can be applied here (Pressman, 2005). In this phase, the mobile end user, an interaction designer, and a developer sit together to come up with a complete scenario analysis model that takes into account the following types of scenario analysis:

- **Screen and interaction analysis:** The basic unit of interaction between the user and the mobile device is the screen, which is an object that encapsulates device-specific graphic user input. Therefore, the content to be displayed on the screen is identified. Content may include text fields, menus, lists, and graphics. Interaction analysis specifies how the user interacts with the application. In order to find out how the user will interact with the application, UML (Booch et al., 2000) use cases are developed.
- **Usage analysis:** The use case model developed during screen and interaction analysis is mainly related to how users interact with the application through the screen. The whole functionality of the application should be described with use cases.
- **Environment analysis:** The environment in which the application will operate should be described in detail. This includes the different wireless networks and back-end systems used. In addition, target mobile devices such as cellular phones and PDAs on which the application will run should be described in detail.

The output of this phase is an information analysis model document produced by the interaction designer and developer that outlines the functional requirements of the application and the constraints of the environment. This document is reviewed by developers and other stakeholders and modified as required.

Architectural design. This phase is concerned with the overall architecture (or structure) of the wireless application. Architecture is very important for any application, and no optimization techniques will make up for an ill-considered architecture. Design patterns can be used in this phase to reuse experience in order to come up with an extensible, high-performance architecture. Some of the most important design goals should be to minimize the amount of data transmitted over the wireless link, and to anticipate errors and handle them intelligently. Other design and architecture issues include:

- **Application partitioning.** Designers need to think carefully when deciding which operations should be performed on the server and which on the wireless device. J2ME allows designers to locate much of an application's functionality on the device; it can retrieve data from the server efficiently, then perform calculations and display information locally. This approach can dramatically reduce costly interaction over the wireless link, but it is feasible only if the device can handle the processing your application needs to perform.
- **Message latency.** In some applications, it may be possible to do other work while a message is being processed. If the delay is appreciable—and especially if the information is likely to go stale — it is important to keep the user informed of progress.

The outcome of the architectural design phase is a design document that details the system architecture.

Navigation and user interface design. Once the application architecture has been established and its components identified, the interaction designer prepares screen mockups and navigation paths that show how the user moves from one screen to another to access services. Figure 4 shows a simple example where the user will have to login before she is able to check her messages.

The user interface is the face of the application to users. A poorly designed user-interface will scare the user away, and a well-designed user interface will give a good first impression and improves the user's perception of the services offered by the application. The user interface must be well-structured and easy to use. Here are some guidelines that can help in designing simple yet effective user interfaces for mobile devices with tiny screens.

- Keep the application's interface simple enough that the user seldom needs to refer to a user manual to perform a task.
- Reduce the amount of information displayed on the device.
- Make input sequences concise so the user can accomplish tasks with the minimum number of button clicks.

Figure 4. Screen mockups

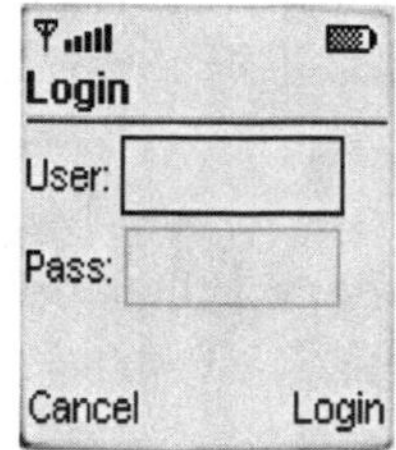

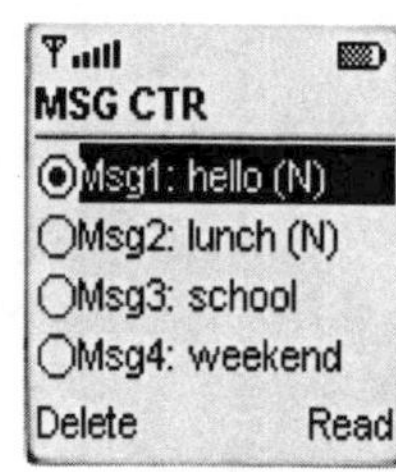

- Offer the user selection lists.
- Do not depend on any specific screen size.

The output of this phase is a user manual that describes the screen mockups and the navigational paths.

Implementation. In this phase development tools are used to implement the wireless application. There are several tools available for building wireless applications such as Sun's J2ME Wireless Toolkit. We would recommend using a tool that allows installing the application in various emulation environments. Conventional implementation strategies and techniques such as coding standards and code reviews can be used in this phase.

Testing. Software testing is a systemic process to find differences between the expected behavior of the system specified in the software requirements document and its observed behavior. In other words, it is an activity for finding errors in the software system and fixing them so users can be confident that they can depend on the software. Errors in software are generally introduced by people involved in software development (including analysts, architects, designers, programmers, and the testers themselves). Examples of errors include mismatch between requirements and implementation.

Many developers view the subject of software testing as "not fashionable," and, as a result, too few of them really understand the job software testers do. Testing is an iterative process and should start from the beginning of the project. Software developers need to get used to the idea of designing software with testing in mind. Some of the new software development methodologies such as eXtreme Programming (XP) (Beck, 1999) stress incremental development and testing. XP is ideally suited for some types of applications, depending on their size, scope, and nature. User interface design, for example, benefits highly from rapid prototyping and testing usability with actual users.

Wireless applications, like all other types of software, must be tested to ensure functionality and usability under all working conditions. Testing is even more important in the wireless world because working conditions vary a lot more than they do for most software. For example, wireless applications are developed on high-end desktop machines but deployed on handheld wireless devices with very different characteristics.

One way to make testing simple is to design applications with testing in mind. Organizing the system in a certain way can make it much easier to test. Another implication is that the system must have enough functionality and enough output information to distinguish among the system's different functional features. In our approach, and similar to many others, the system's functional requirements (features that the system must provide) are described using the Unified Modeling Language (Booch et al., 2000) to create a use-case model, then detailing the use cases in a consistent written form. Documenting the various uses of the system in this way simplifies the task of testing the system by allowing the tester to generate test scenarios from the use cases. The scenarios represent all expected paths users will traverse when they use the features that the system must provide.

Deployment. Deploying and running applications in an emulation environment is a very good way to test the logic and flow of your application generally, but you will not be certain it will satisfy users until you test it on a real physical device connected to a wireless network. Your application's performance may be stunning in the emulator, which has all the processing power and memory of your desktop machine at its command, but will it perform well on the handheld device, with its limited memory and processing power, low bandwidth, and other constraints? In this phase, the application is deployed on a live network and evaluated.

Customer evaluation. Once the application has been deployed, it is ready to be downloaded by users for evaluation and usage. In this phase, users start using the deployed application and report any problems they may experience to the service provider.

Maintenance. Software maintenance encompasses four activities: error correction, adaptation, enhancement, and reengineering (Pressman, 2005). The application will evolve over time as errors are fixed and customers request new features. In this phase, users report errors to and request new features from the service provider, and developers fix errors and enhance the application.

TESTING ISSUES AND TESTING ACTIVITIES

The wide variety of mobile devices such as wireless phones and PDAs results in each device running a different implementation of the J2ME environment. Varying display sizes add to the complexity of the testing process. In addition, some vendors provide proprietary API extensions. As an example, some J2ME vendors may support only the HTTP protocol, which the MIDP 1.0 specification requires, while others support TCP sockets and UDP datagrams, which are optional. Here are some guidelines for testing wireless applications.

Implementation validation. Ensuring that the application does what it is supposed to be is an iterative process that you must go through during the implementation phase of the project. Part of the validation process can be done in an emulation environment such as the J2ME Wireless Toolkit (Sun Microsystems, 2005), which provides several phone skills and standard input mechanisms. The toolkit's emulation environment does not support all devices and platform extensions, but it allows for the application to look appealing and to offer a user-friendly interface on a wide range of devices. Once the application has been tested on an emulator, you can move on to the next step and test it on a real device, and in a live network.

Usability testing. In usability testing, the focus is on the external interface and the relationships among the screens of the application. As an example, consider an e-mail application that supports entry and validation of a user name and password, enables the user to read, compose, and send messages, and allows maintenance of related settings, using the screens shown in Figure 3, among others.

In this example, start the test at the Login window. Enter a user name and a password and press the soft button labeled Login. Enter a valid user name and password. The application should display the main menu. Does it? The main menu should display a SignOut button. Does it? Press the SignOut button. Does the application return to the Login screen? Write yourself a note to raise the question, "Why does the user 'log' in but 'sign' out?" Now enter an invalid user name or password. The program should display a meaningful message box with an OK button. Does it? Press the OK button. Does the application return to the Login screen?

You need to test the GUI navigation of the entire system, making notes about usability along the way. If, for example, the user must traverse several screens to perform a function that's likely to be very popular, you may wish to consider moving that particular function up the screen layers. Some of the questions you should ask during usability testing include: is the navigation depth (the number of screens the user must go through) appropriate for each particular function, does the application minimize text entry (painful on a wireless phone) or should it provide more selection menus, can screens of all supported devices display the content without truncating it, and if you expect to deploy the application on foreign devices, does it support international character sets?

Network performance testing. The goal of this type of testing is to verify that the application performs well in the hardest of conditions (for example, when the battery is low or the phone is passing through a tunnel). Testing performance in an emulated wireless network is very important. The drawback with testing in a live wireless network is that so many factors affect the performance of the network itself that you cannot repeat the exact test scenarios. In an emulated network environment, it is easy to record the result of a test and repeat it later, after you have modified the application, to verify that the performance of the application has improved.

Server-Side Testing. It is very likely that wireless applications communicate with server-side applications. If your application communicates with servers you control, you have a free hand to test both ends of the application. If it communicates with servers beyond your control (such as quotes.yahoo.com), you just need to find the prerequisites of use and make the best of them. You can test server-side applications that communicate over HTTP connections using testing frameworks such as HttpUnit (http://httpunit.sourceforge.net), and measure a Web site's performance using httperf (http://citeseer.nj.nec.com/mosberger98httperf.html), a tool designed for measuring the performance of Web servers.

Testing Checklists

Here we provide checklists that are useful when testing your application, in both emulation and live environments. These checklists include tests that are usually performed by certification programs offered by Nokia and Motorola (Motorola Application Certification Program).

Navigation checklist. Here are some items to check for when testing the navigational paths of wireless applications:

- **Successful startup and exit:** Verify that your application starts up properly and its entry point is consistent. Also make sure that the application exits properly.
- **Application name:** Make sure your application displays a name in the title bar.
- **Keep the user informed:** If your application does not start up within a few seconds, it should alert the user. For large applications, it is a good idea to have a progress bar.
- **Readable text:** Ensure that all kinds of content are readable on both grayscale and color devices. Also make sure the text does not contain any misspelled words.
- **Repainting screens:** Verify that screens are properly painted and that the application does not cause unnecessary screen repaints.
- **Soft buttons:** Verify that the functionality of soft buttons is consistent throughout the application. Verify that the whole layout of screens and buttons is consistent.
- **Screen navigation:** Verify that the most commonly used screens are easily accessible.
- **Portability:** Verify that the application will have the same friendly user interface on all devices it is likely to be deployed on.

Network checklist. Some of the items that should be inspected when testing wireless applications are:

- **Sending/Receiving data:** For network-aware applications, verify that the application sends and receives data properly.
- **Name resolution:** Ensure that the application resolves IP addresses correctly, and sends and receives data properly.
- **Sensitive data:** When transmitting sensitive data over the network, verify that the data is being masked or encrypted.
- **Error handling:** Make sure that error messages concerning network error conditions (such as no network coverage) are displayed properly, and that when an error message box is dismissed, the application regains control.
- **Interruptions:** Verify that, when the device receives system alerts, SMS messages, and so on while the application is running, messages are properly displayed. Also make sure

that when the message box is dismissed the application continues to function properly.

PROVISIONING WIRELESS APPLICATIONS

Developers usually build, test, and evaluate an application on a platform similar to the one on which it will be deployed and ran. Development of wireless applications is more challenging because they typically are developed on one platform (such as Solaris or MS Windows) but deployed on a totally different one (such as a cell phone or PDA). One consequence is that, while emulators enable developers to do some of their testing on the development platform, ultimately they must test and evaluate the application in the very different environment of a live wireless network.

Wireless applications fall into two broad categories:

- **Local applications** perform all their operations on a handheld wireless device and need no access to external data sources through a wireless network. Examples include calculators and single-player games.
- **Network applications** consist of some components running on a wireless device and others running on a network, and thus depend on access to external resources. An example would be an e-mail application, with a client residing on a wireless phone that interacts with an SMTP server to send messages.

Although these two types of applications are different, they are deployed in the same way. The big difference shows up later: Local applications are easier to test than network applications. For example, a calculator application can run on a wireless phone even when it is not connected to any network, but an e-mail client won't work without a connection to the SMTP server that actually transmits the messages.

Over the Air Provisioning

For some time, wireless portals in Europe such as Midletcentral have allowed customers to download applications directly to their phones, over the air. Over-the-air provisioning of wireless applications (OTA) is finally available in

Figure 5. Over-the-air provisioning

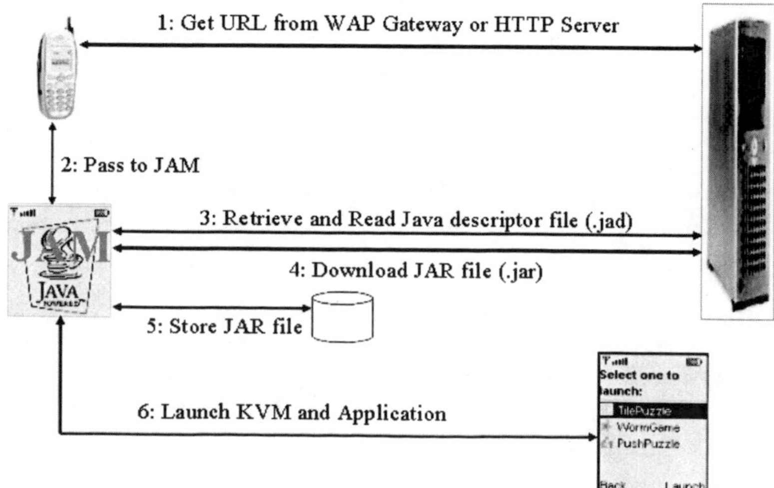

North America. Nextel customers, for example, can download network-aware wireless applications without an update data cable. OTA is the deployment of wireless Java applications (*MIDlet suites*) from the Internet to wireless devices over a wireless network. Users need not connect their devices to the desktop with a data cable or visit a service center to install or upgrade software. To take advantage of OTA, you must equip your handheld device with a mechanism to discover MIDlet suites available for download, using the device's browser (such as a WAP browser) or a resident application written specifically to identify downloadable MIDlet suites. The process of downloading MIDlets over the air is illustrated in Figure 5.

RELATED WORK

The explosive growth of the wireless mobile application market raises new engineering challenges (Morisio & Oivo, 2003); what is the impact of the wireless Internet on engineering wireless mobile applications for the new wireless infrastructure and wireless handheld devices? Due to the limited experience with wireless technologies and developing wireless applications, little work has been in the area of wireless software engineering. We found a special issue in the *IEEE Transactions on Software Engineering* on "Software Engineering for the Wireless Internet" (Morisio & Oivo, 2003). However, out of the six papers accepted in the special issue only two papers deal with the development process. Ocampo, Boggio, Munch, and Palladino (2003) provided an initial reference process for developing wireless Internet applications, which does not differ significantly from traditional iterative process models but includes domain-specific guidance on the level of engineering processes. Satoh (2003) developed a framework for building and testing networked applications for mobile computing. The framework is aimed to emulate the physical mobility of portable computing devices through the logical mobility of applications designed to run on them; an agent-based emulator is used to perform application-level emulation of its target device.

More recently, Chen (2004) proposed a methodology to help enterprises develop business strategies and architectures for mobile applications. It is an attempt to formulate a life cycle approach to assisting enterprises in planning and developing enterprise-wide mobile strategies and applications. This methodology is more concerned with business strategies rather than technical details, and thus it is targeted at managers rather than developers. And finally, Nikkanen (2004) presented the development work of a browser-agnostic mobile e-mail application. It reports on experiences porting a legacy WAP product to a new XHTML-based browser application and offers guidelines for developing mobile applications.

Our work is different in the sense that we provide a detailed treatment of the impact of the characteristics of mobile devices and the wireless environment on engineering wireless mobile applications; we discuss the challenges and offer practical solutions for developing mobile applications. We present a systematic approach for designing wireless mobile application. Our approach is iterative just like in Ocampo et al. (2003), but differs in the sense that our process has more focus on requirements elicitation and more importantly scenario analysis. We do not provide a testing framework, but our testing strategy and checklist is more practical than using just an emulated environment. Finally, unlike the work reported in Chen (2004), our methodology is targeted at developers and researchers rather than managers. And, unlike the work in Nikkanen (2004), our guidelines and systematic approach is not limited to WAP-based applications, but can be applied to engineering any wireless application.

CONCLUSION AND FUTURE WORK

As the wireless Internet becomes a reality and software developers become comfortable with the methods and processes required to build software, we recognize that the methods developed for conventional systems are not optimal for wireless applications. In particular, wireless application development doesn't always fit into the development model originated to cope with conventional large software systems. Most wireless application systems will be smaller than any medium-size project; however, a software development method can be just as critical to a small software project success as it is to that of a large one. In this chapter, we have presented and discussed a systematic approach to wireless application development, and presented practical guidelines for testing wireless applications. The proposed approach takes into account the special features of the wireless environment. We have successfully used the approach presented to develop various wireless applications ranging from a stock portfolio management application to a mobile agent platform for mobile devices (Mahmoud, 2002). Our future work includes evaluating the effectiveness of the proposed methodology, documenting wireless software design patterns, and building tools to automate the task of testing wireless applications.

There are several interesting research problems in the emerging area of wireless mobile applications and services. Some of these research issues include: novel mobile services in the area of m-commerce and health care; security and privacy issues; mobile agents for mobile services; discovery and interaction of mobile services; enabling roaming of applications and profiles between different wireless standards; and location-aware and context-aware mobile services. We are currently addressing some of these research problems, and research results will be presented in future articles.

ACKNOWLEDGMENT

The authors would like to thank the anonymous reviewers for the many helpful suggestions for improving this chapter. The first author was supported in part by the Natural Sciences and Engineering Research Council of Canada (NSERC) Discovery Grant No. 045635.

REFERENCES

Beaulieu, M. (2002). *Wireless Internet applications and architecture*. Boston: Addison-Wesley.

Beck, K. (1999). *Extreme programming explained: Embrace change*. Addison-Wesley.

Booch, G., Rumbaugh, J., & Jacobsen, I. (2000). *The Unified Modeling Language user guide*. Boston: Addison-Wesley.

Burkhardt, J, Henn, H., Hepper, S., Rintdorff, K., & Schack, T. (2002). *Pervasive computing technology and architecture of mobile Internet applications*. London: Addison-Wesley.

Chen, M. (2004). A methodology for building mobile computing applications. *International Journal of Electronic Business, 2*(3), 229-243.

Jacobsen, I., Booch, G., & Rumbaugh, J. (2000). *The unified software development process*. Boston: Addison-Wesley.

Httperf. Retrieved January 13, 2005, from http://www.hpl.hp.com/research/linux/httperf

HttpUnit. Retrieved January 13, 2005, from http://httpunit.sourceforge.net

Mahmoud, Q. (2002). MobiAgent: An agent-based approach to the wireless Internet. *Journal of Internet Computing, special issue on Wireless Internet, 3*(2), 157-162.

Morisio, M., & Oivo, M. (2003). Software engineering for the wireless Internet [Guest Editor's Introduction]. *IEEE Transactions on Software Engineering, 29*(12), 1057-1058.

Motorola Application Certification Program. (n.d.). Retrieved February 10, 2005, from http://qpqa.com/motorola/iden

Nikkanen, M. (2004). User-centered development of a browser-agnostic mobile e-mail application. In *Proceedings of the Third Nordic Conference on Human-Computer Interaction*, Tampere, Finland (pp. 53-56). New York: ACM Press.

Ocampo, A., Boggio, D., Munch, J., & Palladino, G. (2003). Towards a reference process for developing wireless Internet services. *IEEE Transactions on Software Engineering, 29*(12), 1122-1134.

Open Mobile Alliance. (2005). Retrieved from March 15, 2005, http://www.openmobilealliance.org

Pressman, R. S. (2005). *Software engineering: A practitioner's approach* (6th ed.). New York: McGraw Hill.

Satoh, I. (2003). A testing framework for mobile computing software. *IEEE Transactions on Software Engineering, 29*(12), 1112-1121.

Sun Microsystems J2ME. (2005). Retrieved from http://java.sun.com/j2me

Sun Microsystems J2ME Wireless Toolkit. (2005). Retrieved from http://java.sun.com/products/j2mewtoolkit

ENDNOTE

1 We use the terms wireless application and mobile application interchangeably throughout this article.

This work was previously published in International Journal of Information Technology and Web Engineering, Vol. 1, Issue 1, edited by G. Alkhatib and D. Rine, pp. 59-75, copyright 2006 by IGI Publishing, formerly known as Idea Group Publishing (an imprint of IGI Global).

Chapter XV
Project Management and Web Software Engineering

Daniel M. Brandon
Christian Brothers University, USA

ABSTRACT

The process involved with the development of Web applications is significantly different from the process of developing applications on older traditional platforms. This is a difference not only in technologies but in the overall business process and associated methodology, in other words the project management. Web applications generally manage content and not just data, and many Web applications are document centric versus data centric. In addition, there are many more people involved in the definition, design, development, testing, usage, and approval for Web applications. The pace of business life is much quicker today than in the past, thus Web applications need to be deployed much quicker than earlier IT applications and they need to be engineered for more flexibility and adaptability in terms of changing requirements, changing content, changing presentation, and perhaps mass user customization. In addition, security concerns are more prevalent in Web applications since the end users are outside as well as inside the corporate virtual perimeter. Web applications can serve a global audience and thus there are more diverse stakeholders for these applications. Issues such as language, culture, time zones, weights and measures, currency, and international standards need to be considered. This chapter examines the project management issues involved with Web applications.

INTRODUCTION

In this chapter we will discuss the key issues involved with project management for developing Web applications. These issues revolve around the fact that Web applications have many and possibly globally diverse stakeholders operating in a "security challenged world" whose expectations are for "better/cheaper/faster" software. We start out by discussing the project management discipline, and then the relationship between this discipline and the software engineering discipline. Then we consider how both of these disciplines combined can successfully address the opportunities, issues, and problems of modern Web application development including the internationalization and security aspects.

THE PROJECT MANAGEMENT DISCIPLINE

A number of professional organizations have developed around the world to foster the project management discipline (Morris, 2001). These organizations have recognized that there is a distinct skill set necessary for successful project managers, and the organizations are devoted to assisting their members develop, improve, and keep current with these skills [Boyatzis, 1982; Caupin, 1998]. The Project Management Institute (PMI) is the largest of these organizations in the world, but other major international organizations are the Association for Project Management (APM), British Standard Institute (BSI), Engineering Advancement Association (ENAA) of Japan, Australian Institute of Project Management, and the International Project Management Association (IPMA).

Each of these organizations has developed a set of project management standards as has the ISO (International Organization for Standards) with its ISO 10006 "Guide to Quality in Project Management." For comparative purposes the size in words of these various project management standards are (Crawford, 2004):

PMBOK – 56,000
APM BoK – 13,000
IPMA ICB – 10,000
ENAA P2M – 36,000

The APM has developed a Body of Knowledge (BoK) of Project Management Competencies. The APM Body of Knowledge identifies 40 key competencies grouped as:

- **Project management:** Covering the key elements that differentiate projects from general management;
- **Organizations and people:** Detailing the main qualitative skills of a Project Manager;
- **Techniques and procedures:** Details the quantitative methods;
- **General management:** Covers industry specific concepts.

The APM Body of Knowledge also provides a focal point for many of the programs run by the APM including their Certification Program which assesses a person's competence in managing a project; the Course Accreditation Program which reviews training courses run by both commercial private training companies and higher education institutes, and the Project Management Capability Test which assesses a person's knowledge in the APM Body of Knowledge. The British Standards Institute publishes the Guide to Project Management (BS6079). This Standard has been adopted by both the British government and industry and establishes commonly accepted terminology. The stated objectives of BS 6079 are to provide guidance to:

- **General managers:** To enable them to provide proper support for project managers and their teams;
- **Project managers:** To improve their ability to manage their projects;
- **Project support staff:** To help them understand project management issues and solutions thereto;
- **Educators and trainers:** To help them understand the project management environment and the context in which project management methods are deployed.

The International Project Management Association is a federation of national project management associations of several European countries and it publishes the IPMA competency baseline (ICB) in English, French, and German; the first publication being in 1998. While the content is similar to the APM BoK, the organization is different and is termed the "Sunflower." The IPMA encourages each national organization to form its own competency baselines called "National Competency Baselines" (NCB). There are now about 30 countries with such NCB's throughout Europe and also Egypt, India, and China.

The Engineering Advancement Association (ENAA) of Japan has also issued a project management body of knowledge called P2M (A Guidebook of Project and Program Management for Enterprise Innovation). Their PM standard is different from that of the PMI or APM, and is based on how project management can be used to increase business value for an organization and promote innovation. This P2M was a multi-year joint effort between the Japanese Project Management Forum (JPMF) and the Japanese Ministry of Economy, Trade and Industry (METI); it was supported by both Japanese industry and government with a very significant contribution from academic research. The Japanese P2M is based on a "tower structure" which according to the Japanese Standard Committee is focused on the alignment of project management to the business units instead of the European and North American approach which is dedicated to the management of a single project. The four areas of certification in the Japanese program are Objectives, Strategy, Value Management, and Finance.

The largest project management organization is the Project Management Institute (PMI) with over 250,000 members in over 125 countries. The PMI Web site (www.pmi.org) records over 10 million visitors per year. Founded in 1969, PMI establishes project management standards, provides seminars and other vehicles for professional growth, promotes educational programs, funds and encourages research, and provides professional certification that many of the world's organizations desire for their project personnel. PMI produces a number of publications including the Project Management Journal, Project Management Quarterly, PM Network, and PMI Today. In this chapter we will closely follow the PMI standards, however other and broader standards and concepts found in the other major project management organizations will also be included.

PMI established its first body of knowledge in 1976 which finally became "A Guide to the Project Management Body of Knowledge" (PMBOK) around 1987. It was revised several times with major releases in 1996, 2000, and 2005; there are about 1.5 million copies of all PMBOK versions

Figure 1. PMBOK components (Brandon, 2006)

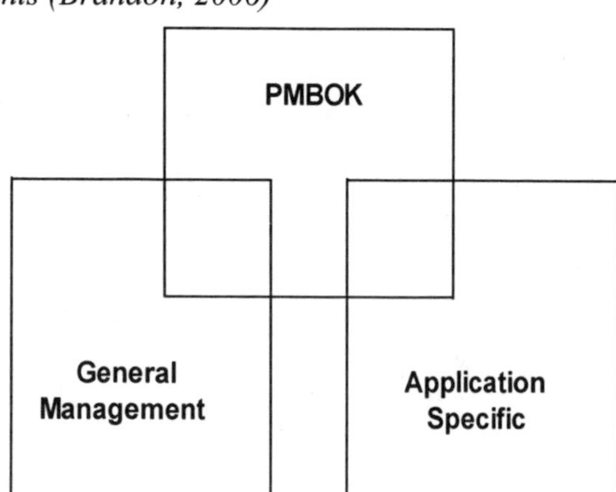

Figure 2. Process representation (Brandon, 2006)

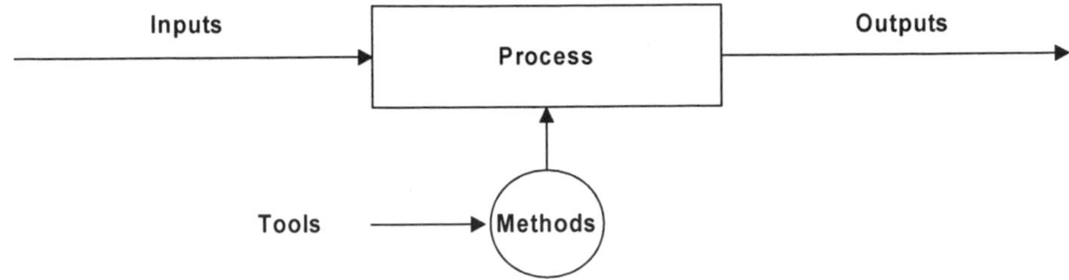

Figure 3. Process group interaction

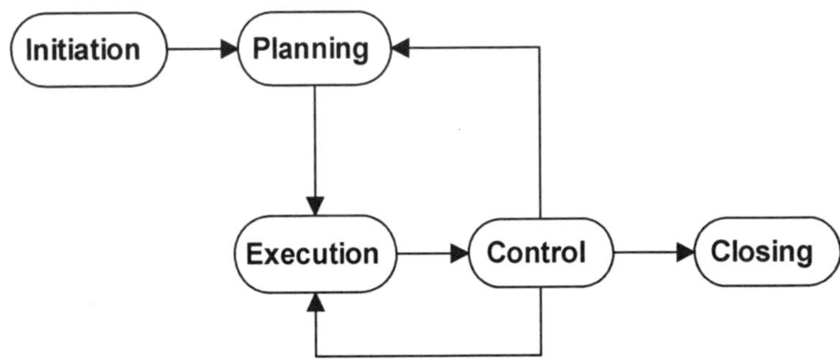

in circulation. The PMBOK embodies generally accepted "best of practice" procedures, methods, and general tools which are derived by a structured consensus of its vast membership. The PMBOK is approved as an American National Standard (ANS) by the American National Standards Institute (ANSI). There is some overlap of project management best practices with practices of general management and also practices specific to various fields as illustrated in Figure 1.

The content of the PMBOK is organized into processes, and 37 "Key Processes" are defined. Each process is defined as procedures which receive various inputs and produce various outputs using various methods and perhaps with the assistance of some general tools. Methods may be management techniques, mathematical techniques, statistical techniques, etc. The tools are typically some type of software. This is shown pictorially in Figure 2.

These processes are grouped into 5 "Process Groups" which relate to how project work is managed, "what to do when;" PMI's 5 Process Groups are:

- Initiation
- Planning
- Execution
- Control
- Closing

They are further sub-divided into 9 "Knowledge Areas." The Knowledge Areas are:

- Integration management [3 processes]
- Scope management [5 processes]
- Time management [5 processes]
- Cost management [4 processes]
- Quality management [3 processes]
- Human resource management [3 processes]
- Communication management [4 processes]
- Risk management [4 processes]
- Procurement management [6 processes]

Figure 4. PMBOK process groups versus. knowledge areas (Brandon, 2006)

PMI Process Groups and Knowledge Areas					
	Initiation	**Planning**	**Executing**	**Controlling**	**Closing**
Integration		Project Plan Development	Project Plan Execution	Overall Change Control	
Scope	Initiation	Scope Planning	Scope Verification	Scope Change Control	Scope Verification
		Scope Definition			
Time		Activity Definition		Schedule Control	
		Activity Sequencing			
		Activity Duration Estimation			
		Schedule Development			
Cost		Resource Planning		Cost Control	
		Cost Estimating			
		Cost Budgeting			
Quality		Quality Planning	Quality Assurance	Quality Control	
Human Resources		Organizational Planning	Staff Acquisition	Team Development	
Communications		Communications Planning	Information Distribution	Performance Reporting	Administrative Closure
Risk	Risk Identification	Risk Identification		Risk Response Control	
		Risk Quantification			
		Risk Response Development			
Procurement		Procurement Planning	Solicitation	Contract Administration	Contract Closeout
		Solicitation Planning	Source Selection		
			Contract Administration		

Figure 3 shows the general sequencing of the process groups in the timeline for a project. However, in practice there may be considerable overlap; that is all initiation processes will not complete before all the planning processes begin. The output from one process in a process group is typically the input to another process either in the same process group or the next one in this sequence.

The overall organization of the PMBOK and the relationship between the processes, process groups, and knowledge areas is shown in Figure 4.

In practice, large projects are typically broken down into phases, and the organization of a project into phases is discipline specific and typically follows some type of development methodology. Each project phase has a beginning and an end, and the five process groups are a part of each. Deliverables from one phase are typically inputs to the next phase; and there may be phase overlap

Figure 5. Project phasing (Brandon, 2006)

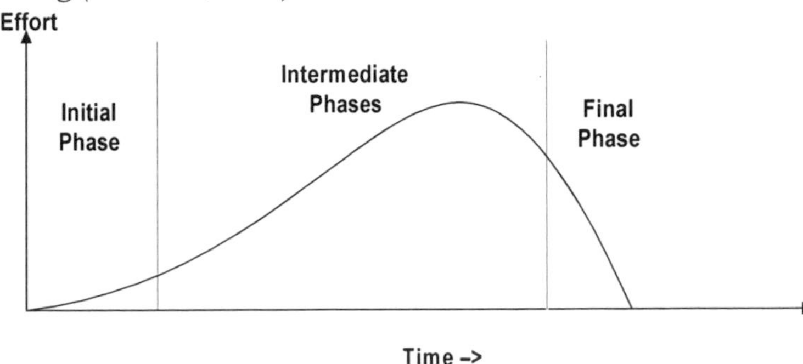

Figure 6. Process groups in each phase

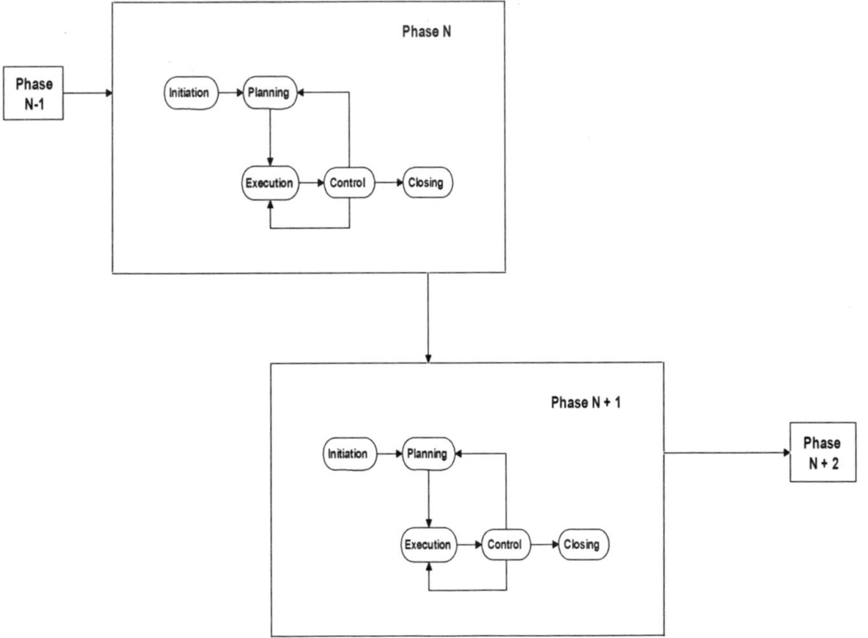

in some methodologies. Figure 5 shows how effort is generally distributed across project phases.

For IT projects a typical phasing might be:

- Feasibility, Proposal, Business Plan
- Requirements Specification
- Design
  - Overall design
  - External (user interaction) specifications
  - Detail design
  - "As designed" internal documentation
  - Test Plans
  - Deployment and integration specifications
- Implementation
  - Coding
  - Unit Testing
  - Module and Feature Testing
  - User documentation
  - "As built" internal documentation

- Installation
  - Conversion
  - Training
  - Network and site preparation
  - Hardware install and integration
  - Software install and integration
  - Integration and acceptance testing
  - Parallel operation
- Operation and Maintenance (O & M)

As stated earlier, for phased projects, the five process groups occur in each project phase. This is illustrated in Figure 6. For some industries and in some methodologies there may be some overlap in project phases. In IT there may also be some project phase overlap, and this depends upon the type of methodology adopted and upon the type on contracting arrangement; both of these aspects are discussed in more detail later in this chapter.

PMI's summarization of key activities by process group are:

- Key Initiation Activities
  - Project feasibility (high level ROI approximation)
  - High level planning
  - Project charter document (memo, letter)
- Key Planning Activities (order important)
  - Develop scope statement
  - Assemble project team
  - Develop work breakdown structure (WBS)
  - Finalize project team
  - Do network type diagram (showing activity dependencies)
  - Estimate cost and time, find the "critical path"
  - Determine overall schedule and budget
  - Procurement plan
  - Quality plan
  - Identify risks, quantify them, develop risk responses
  - Other plans: change control plan, communications plan, management plan
  - Overall project plan
  - Project Plan approval
  - "Kickoff meeting"
- Key execution Activities
  - Execute the project plan
  - Complete work packets (activities)
  - Information distribution
  - Quality assurance
  - Team development
  - Scope verification
  - Progress meetings
- Key Control Activities
  - Overall change control
  - Performance reporting
  - Scope control
  - Quality control
  - Risk response control
  - Schedule control
  - Cost control
  - Manage by exception to the project plan
- Key Closing Activities
  - Procurement audits & contract(s) close out
  - Product verification
  - Formal acceptance
  - Lessons learned documentation
  - Update all project records
  - Archive records
  - Release team

PMI has established a certification program (since 1984) for the project management discipline, and there are several levels of certification. The highest certification level is called a "PMP" (Project Management Professional). The requirements for an individual to be awarded that certification level include:

- 4500 Hours of documented project management experience over 3- 6 years;
- BS/BA Degree and at least 35 contact hours in PM training;
- Passing a very comprehensive 4 hour exam on the PMBOK;
- Adherence to the PMI professional code of ethics.

There are over 200,000 certified PMP's in the world. Those who have been granted PMP

certification must demonstrate an ongoing professional commitment to the field of project management by satisfying PMI's Continuing Certification Requirements Program to retain their PMP status. In 1999, PMI became the first organization in the world to have its Certification Program attain International Organization for Standardization (ISO) 9001 recognition. The IPMA also has certification processes with four levels of certification: practitioner, professional, manager, and director. ENAA's P2M has three levels of certification: architect, manager, specialist (Crawford, 2004).

The PMP certification (or equivalent certification from another international organization) is now being required (or highly recommended) by many large corporations for one to become a project manager in their organization. Additionally many companies expect vendors to provide certified project managers for contracted work (Brandon, 2006).

SOFTWARE ENGINEERING AND PROJECT MANAGEMENT

The term "software engineering" was coined by Bauer who was a principal organizer of the 1968 NATO conference on that subject (Bauer, 1972). His definition of software engineering was "the establishment and use of sound engineering principles in order to economically obtain software that is reliable and works on real machines." The IEEE definition is "the application of a systematic, disciplined, quantifiable approach to the development, operation, and maintenance of software (IEEE Std 610-1990). The modern Webopaedia definition is:

Figure 7. Software engineering versus project management (Brandon, 2006)

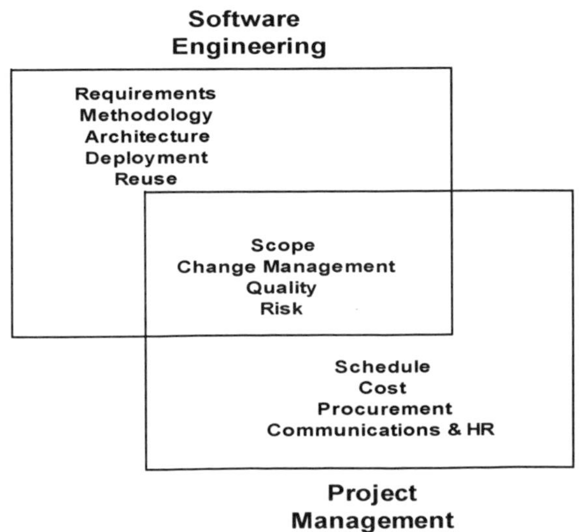

Software engineering is the computer science discipline concerned with developing large computer applications. Software engineering covers not only the technical aspects of building software systems, but also management issues, such as directing programming teams, scheduling, and budgeting.

The disciplines of project management and software engineering overlap considerably as illustrated in Figure 7. The Institute of Electrical and Electronics Engineers (IEEE) software standard 1490-2003 provides for the adoption of PMI Standard (PMBOK).

In the IT industry there is no one methodology, architecture, or set of standards. In other industries, there are typically established codes, frameworks, patterns, methods, and tools that are almost always used when one builds something in one of those industries. In home building there are county building codes, frameworks for house patterns (ranch, colonial, Tudor, contemporary, etc.), subdivision guidelines and limitations, standard methods, and tools of the trades involved. Not so in IT; there are a number of standards, frameworks, architectures, tools, and methodologies to choose from and they all are rapidly changing and evolving.

TRADITIONAL SOFTWARE DEVELOPMENT METHODOLOGY

According to Webster's dictionary, *methodology* is "a system of methods." This author's definition for methodology is "organized know how." The most common and established methodology used in building and/or integrating IT systems has been casually called the "waterfall method" and formally referred to as the "software development lifecycle methodology" (SDLC). This notion and term was first applied to IT systems by Royce (Royce, 1970). The steps in this classical methodology are:

- Definition
- Specification (Requirements)
- Design

Figure 8. SDLC waterfall (Brandon, 2006)

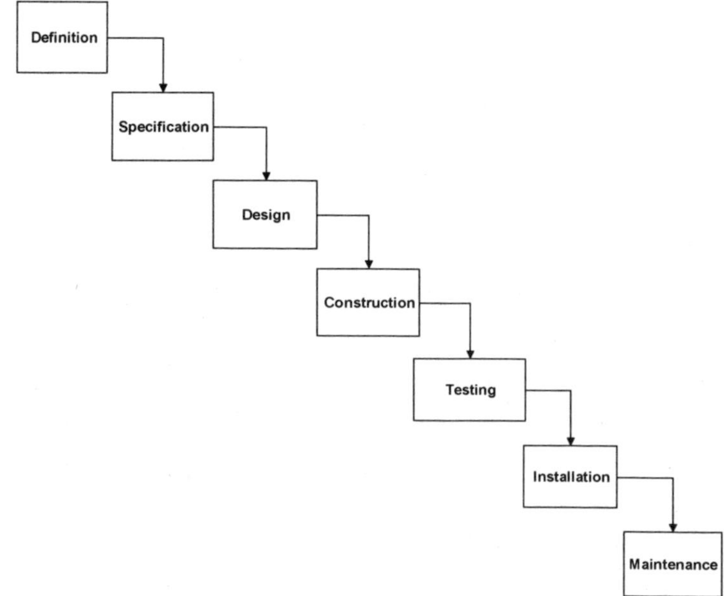

- Construction (Programming and unit testing)
- Testing (system and integration)
- Installation
- Operation and Maintenance

This is illustrated in Figure 8. In theory these steps are not supposed to overlap or to iterate. Today there are many software development methodologies, and some of these newer methodologies are variations of or alternatives to the basic waterfall approach.

For IT projects there is a correlation between the PMI process groups (and their outputs) and the chosen software engineering methodology (and its outputs); Figure 9 is an example of a correlation between PMI process groups and the classical software development life cycle (SDLC) of a single phase project.

One hears many comments about the classical waterfall software development lifecycle:

The software development lifecycle is the cornerstone of development!
The lifecycle is out of date!
Get good people and the lifecycle will manage itself!
There must be better and faster ways to build IT systems!

The SDLC goals are to:

- Do it right the first time!
- Meet customer's stated requirements!
- Have a timely completion!
- Complete within cost constraints!
- Build system with the necessary quality!

The *definition step* involves making a clear statement of goals, identifying why and how the proposed system will be better/cheaper/faster than the system it is replacing, and usually an overall/rough cost-benefit analysis. This phase is typified by frequent customer interaction, elimination of arbitrary constraints, negotiation and compromise on scope (features) versus time and cost, statement of assumptions, rough time and cost estimates, a rough project plan, and a signed "go-ahead" (i.e., the project charter).

The *specification step* involves a complete statement of scope ("requirements"), "use case scenarios," preparation of preliminary user manual (external design specifications), detailed project plan (including work breakdown structure [WBS]), specification of needed resources, refined estimate of time and cost, refined cost-benefit analysis, and signed approval of requirements and user manual by stakeholders. However, in practice the user's manual is rarely written at this stage. The reason the user's manual (or at least a draft) should be done at this step is so that some other dependent activities can began such as: test planning and test scripts, making training plans and materials, and other dependent tasks like internal or external marketing.

The *design step* involves resolution of critical technical issues, selection of architecture and platform(s), adoption of standards, assignment of staff, completion of external design (user interface design), design of critical data structures & database, internal design of algorithms and processes, "Requirements Traceability Matrix," preliminary test scripts, final time and cost estimates, and a final cost-benefit analysis. Often this step is divided into two steps; analysis (or overall design) and design (or detailed design).

The *construction step* involves the implementation of the design (i.e., via coding), unit testing, systems integration, draft internal documentation, and the completion of test scripts.

The *testing step* involves full scale integration and system testing, completion of user documentation, completion of training material, adoption of formal change control procedures, completion of the internal documentation, completion of installation manual and "roll-out" or "phase-in" plan.

The *installation step* involves product roll-out, end user training, producing "lessons learned"

Figure 9. PMBOK and SDLC (Brandon, 2006)

PMBOK		SDLC	
Process Group	**Outputs**	**Stage**	**Outputs**
Initiation	Business Plan		
	Project Charter		
Plan	Overall Plan		
	Management Plan		
	Scope Statement	Definition	Project Plan:
			Communications Plan
			Change Management Plan
		Requirements	Requirements Document
	WBS Document		
	Network Diagram		
	Schedule		
	Resource Plan		
	Cost Plan		
	Procurement Plan		
	Quality Plan		
	Risk Plan		
Execute/Control	Performance Reports	Analysis	Overall Design Documents:
	Stage Gate Reviews		Use Cases
			Preliminary Users Manual
			Test Plan
		Design	Detail Design Documents:
			Menu/Navigation Design
			Screen Designs and Storyboards
			Report Designs
			Database Design
			Algorithms Design
			Prototypes
		Construction	Development Objects:
			Commented Code
			Test Scripts
			Help Screens
		Testing	Test Results Documents
			User Manual
			Training Material
		Installation	Install Documents
Closing	Project Close Out		
	Contract Close Out		
	Lessons Learned		

documentation, and defining procedures for handling: operations, user support and configuration management.

The *maintenance step* involves following and revising procedures for: problem resolution & problem escalation; operations; backup and security; configuration control; and quality/performance monitoring.

At the end of each step there is usually a formal meeting where a document is produced for the culmination of effort in that step. This document is reviewed by project management, the performing organization line management, and the benefiting organization (customer). If any of these stakeholders are not satisfied with the results of that step, the project can be terminated or the step repeated; the project will not proceed unless the stakeholders have given approval to move forward at each step. So in theory this should result in a product being produced that satisfies the initial requirements and the stakeholders.

So what can, and often does go wrong for IT applications:

User requirements were misunderstood, incomplete, not fully documented, not fully implemented

Requirements have changed

Documentation is "unusable"

System is difficult to use

Training is ineffective

Capacity or performance problems are present

Audit and integrity problems are present

"Bugs" and other quality issues are present

Standards were not followed

Estimation of workload was poor

Project was managed poorly

Budget was exceeded

Not completed on time

As well as the above list of potential problems, many feel the classical waterfall approach is too slow for modern Web-based applications.

DEPLOYMENT ACELERATION FOR WEB APPLICATIONS

The classical waterfall methodology can be slow in getting a software product "to market" due to the extensive and formal stakeholder review at the end of each step and the lack of overlap. This is particularly troublesome for Web applications due to the increased number and diversity of stakeholders. Also the waterfall method becomes unstable if the initial requirements are significantly in error or if they change much, as if usually the case with Web applications. Another problem today involves the quickly changing business landscape due to new technologies and global competition. From the time a business problem is analyzed and a solution built, the "shape" of the original problem has changed significantly; thus the developed solution no longer matches the original problem. This is illustrated in Figure 10. Today's IT projects, and particularly larger and Web based projects, are getting harder to complete successfully as evidenced by project success rates (Standish Group, 2004). Projects over $10 million have success rates of only 2%, projects between $3 and $10 million have success rates from 11% to 23%, and projects under $3 million have success rates from 33% to 46%.

For these reasons, a number of variations and alternatives have been suggested and tried with varying degrees of success. Many of these approaches take larger IT projects and break them down into smaller more manageable pieces. However there is no single best "silver bullet" approach (Jones, 1994). Some of the SDLC alternatives, to list a few, include: Yourdon Structured Design, Ward/Mellor, Stradis, Spectrum, SDM/70, LBMS, Information Engineering, IBM AD/Cycle, Gane & Sarson Structured Analysis, DeMarco Structured Analysis, Anderson Method/1, Bachman, Agile and XP, Rational Unified Process (RUP), and Clean Room.

The **overlap or free-flow method** allows any task to proceed as long as its dependent tasks are

Figure 10, Changing shape of IT problems (Brandon, 2006)

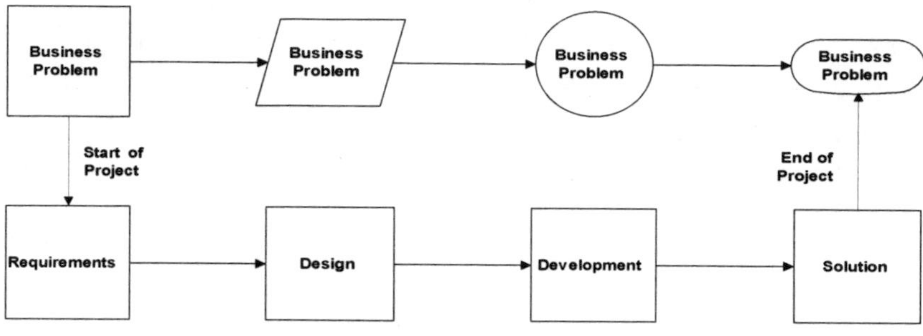

Figure 11. Overlap method (Brandon, 2006)

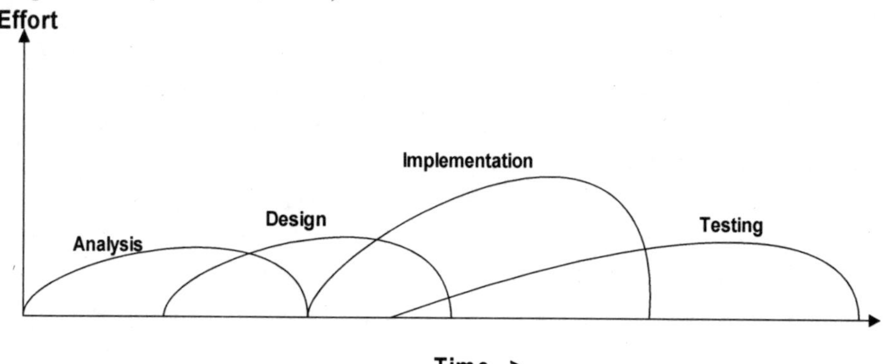

completed. Here the basic waterfall steps may have considerable overlap. This is illustrated in Figure 11. For example, even though the design of the total system is not completed (nor documented and approved), the implementation of those components whose design is completed may begin. This "overlap" is built into the dependency relationships in the project work breakdown structure (WBS). One is betting that the total design (such as may be manifested in UML design drawings) will be approved. The concept is similar to "optimistic record-locking" in an interactive database application. This is also a good technique on contracts where incentives are available for early completion. Obviously risks are greater with larger projects and for projects where requirements can change significantly.

Evolutionary development begins with only the user requirements that are very well understood and builds a first version. Often that first version is just a prototype. Analysis, design, implementation, and testing are done in a free flow overlapping manner without any formal review of documents. This first version is then taken back to the customer for review and definition of further requirements. Next the second version is built, and then taken back to the customer. This process continues until the customer is satisfied with a version and no further extensions are required. This is illustrated in Figure 12. Documentation, training, acceptance testing and other project completion activities are done at that point at which all (or most) of the customer's requirements have finally been included. This methodology is much faster than a waterfall approach and also

Figure 12. Evolutionary development (Brandon, 2006)

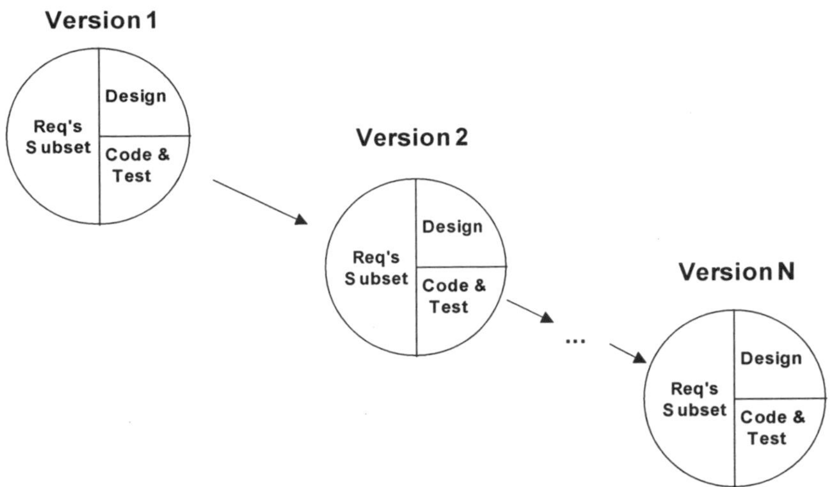

somewhat quicker than the free-flow method. However management visibility may be limited since there is little intermediate documentation produced. Also in a contracted environment (external or internal), a fixed price contract could not be used since the overall scope is not initially determined and priced. For a contracting environment, either a "cost plus" or "time and material" contract would have to be used. Also, internal design is often poor for evolutionary development since the entire scope is not visible from the beginning, and continually changing a system leads to a design which is less adaptable and harder to maintain. If the system is designed and built in a fully "Object Oriented" manner this problem can be minimized.

Incremental development begins with a determination of all the requirements but only in a rough outline form. Next those requirements are prioritized normally based upon those features that are most important from a business perspective. Since time is spent up front looking at all requirements a more appropriate overall platform, architecture, and design can be selected. This is particularly important for security requirements, since security cannot be an afterthought. Good security has to be built into the total product (and the methodology of constructing it), not "bolted on" afterwards.

After the initial requirements phase, development proceeds as in the evolutionary method. This is illustrated in Figure 13. Each increment typically represents a product portion that can be placed into service. Incremental development is not as quick as evolutionary development, but attempts to avoid the design problems caused by not knowing all the major requirements initially. However it suffers from the same contract type issues as the evolutionary method. Another potential problem is that the increments are based on the priorities of the requirements, and sometimes priorities may significantly change during the time of developing the increments. Both the rational unified process and extreme programming as well as several other new methodologies use this technique and these are discussed later.

Bounding box development is similar to incremental development except that each increment is not based on a certain scope (requirements subset) but instead it is based on a measure of effort. If the effort put into an increment is constrained by calendar time then the term "**timebox**" is commonly used. This is illustrated in Figure 14. For contracted development (external or internal), the

Figure 13. Incremental development (Brandon, 2006)

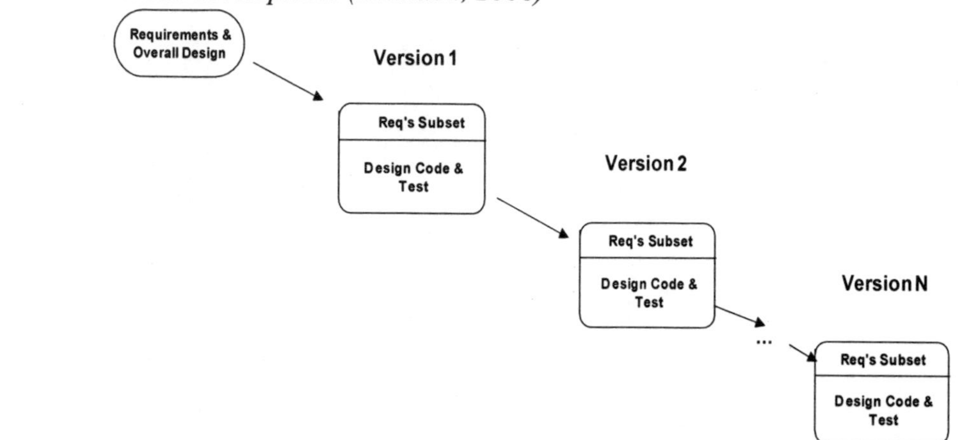

Figure 14. Bounding box development (Brandon, 2006)

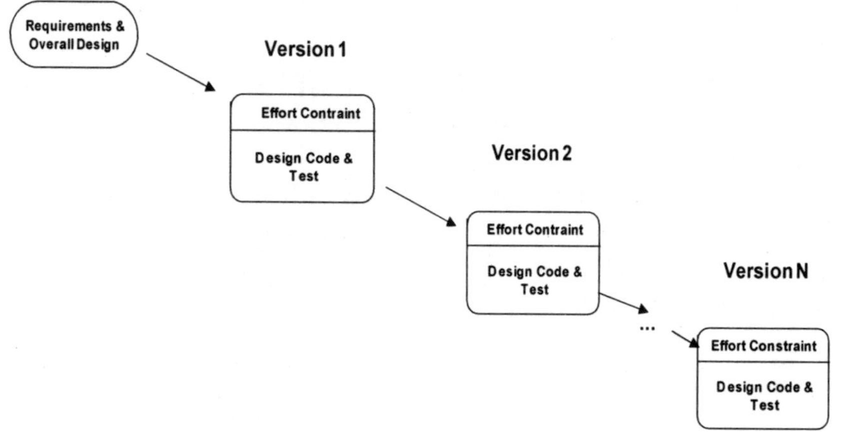

increments are usually based on a dollar (budget) amount. Thus this method does not have the contracting disadvantage that evolutionary or iterative methods have, but customers must be willing to contract for portions of the total system. However, since the amount of scope that will actually be completed in each increment is not known, each increment may not represent a product portion that can be placed into service.

There is no one best methodology. An organization must select the methodology that is most appropriate for the type and size of IT project at hand, the nature of the customer and stakeholders, the contracting environment, and the resources involved both human and financial (Brandon, 2006).

Combinations of these methodologies can also be used such as in the rational unified process which combines incremental and free-flow techniques. Figure 15 shows the primary advantage and disadvantage of each of the methodologies.

The **rational unified process (RUP)** is based upon both the incremental methodology and the free-flow methodology discussed earlier. Instead of attempting to address all of a project's requirements, RUP produces software iteratively that addresses a compromised but known feature

Figure 15. Methodology comparison (Brandon, 2006)

Method	Advantage	Disadvantage	Best For
Waterfall	Sound Development, High Quality	Slowest Method	Fixed Price Contract
Free-Flow	Faster Development	Risky for Unstable Requirements	Incentive Contracts
Evolutionary	Quick Development for Small Applications	Design Problems	Smaller Systems
Incremental	Quick Production of Partial Products	Contract Issues	Rapid Phased Deployment
Bounding Box	Quicker Development within Budget	Partial Products Uncertain	Budget Bound Organization

set and evolves the project over time (Jacobson, 1999; Kruchten, 1998). The difference between evolutionary methods and RUP, though, is that one identifies the requirements for the entire system, but only details the top 20% or so of architecturally significant cases during a single increment. This enables the determination of an appropriate architecture and design which will accommodate the remaining 80% of the requirements without compromising the integrity and quality of the total system. This is particularly important for security requirements, and a "plug-in" to the standard RUP is available called CLASP (Comprehensive Lightweight Application Security Process) which provides a structured way to address security issues in the software development process.

RUP specifies different roles for project participants. Before an architect ever gets involved, an analyst is building use cases and evaluating and prioritizing them with the customer. Before the coders begin implementation, architects work with analysts to identify the architecture which best satisfies requirements and constraints. UML tools are used to build a consistent model from requirements to detail design. RUP uses the free-flow methodology also in that there is considerable overlap in activities of different roles.

RUP is a phased approach that defines four distinct phases:

- **Inception:** Understanding the need, understanding the proposed system to address the need, making the business case for the proposed solution;
- **Elaboration:** Selecting the architecture and developing the project plan;
- **Construction:** Design, coding, integrating, and testing;
- **Transition:** Iinstalling the product on the target platform(s) and getting the user take ownership of the system?

The key to RUP is iteration, both within each of the above phases and also with the incremental production of version. Each iteration within a phase ends in a deliverable, and each increment results in a working product version. RUP defines static workflows, core workflows (business models, requirements, analysis/design, testing, and deployment) and support workflows (change management, project management, environment, and tools). However each of these static workflows is not associated with any one phase, and some degree of each type of workflow goes on within each phase.

The transition from phase to phase is not separated by a "stage gate" and management control is not done by placing dates upon the phase boundaries. Management control is only done upon iterations. The project plan contains a list of proposed iterations (which is likely to change), and each iteration has an estimate (which is also likely to change). The proposed iterations are not assigned due dates, but decision points are set up in time (usually based upon weeks). At each decision point, a decision is made in regard to adding/removing resources, adding/removing iterations form the next release (version) of the product, or killing/holding the project. These decisions are based upon progress, cost, and/or earned value metrics. Thus a key part of the project plan is how risks will be managed; it is a plan of contingencies, as opposed to just a plan of activities.

RUP is not suitable for all IT projects. It is complex and difficult to quantify in a contracting arrangement. However for internal projects that are large and risky, and where quick deployment of partial products is necessary, it may be an appropriate choice.

Agile programming (AP) is a name given to a growing number of "lightweight methodologies" with names like Crystal (Cockburn, 2001), Scrum (Schwaber, 2001), Adaptive (Highsmith, 2000), Feature-Driven Development (Palmer, 2002), Dynamic Systems Development Method [DSDM] (Stapleton, 1997), and Extreme Programming (Beck, 1999). During the 1990's there was such a need to quickly build new IT systems to take advantage of new technologies like Web applications and e-commerce as well as the need to address the Y2K problem, that IT organizations began exploring these lightweight techniques. Lightweight methodologies do away with much of the SDLC process overhead that slows down developers, such as detailed formal requirements definitions and extensive documentation. Some feel that these new development approaches are based on the premise that if you hire competent programmers who always know what they are doing, then any problems they encounter are organizational and communications ones; and those are what the agile approach focuses on. However for managers to make this assumption may be as bad as programmers assuming that all mangers know what they are doing.

Although the various agile methods differ, they have some things in common. Most use an incremental free-flow approach as does RUP. The common intent is to "be agile," so one should embrace change and produce software that is adaptable; thus most of these methods call for the use of object oriented languages. Another common feature is a lot of contact time with users. Still another key focus is a focus on people not process thus emphasizing team morale building. Most AP methods have some core principles including:

- Use a simple design (the old military KISS principle);
- Design as you go, and keep "refactoring" the code;
- Take small incremental steps (when changing or adding code, take the smallest step you can, then test again; one or two weeks is typical);
- Stick to one task at a time (do not add code to accomplish two things at the same time);
- Heavy use of IDE and RAD tools;
- Use only the techniques that really work for you.

Now, to many developers these may seem basic and obvious, but we know many programmers who never follow any of these principles. Some programmers, instead of modifying a module or class for a small change or addition, will spend a great deal of time writing the entire module from scratch. The advantages of these AP principles include:

- Faster reaction to changes in requirements
- The overall design remains simple
- Coding can begin earlier
- By refactoring the code, the most important parts get the most attention (one doesn't invest time into changing what doesn't need to be changed)

- Code in progress is always stable
- Refactoring is one of the core concepts and this is a new word for "cleaning up the code." More formally, refactoring is improving the design and maintainability of code in small incremental steps confined to areas of current interest. One problem with refactoring is that when a programmer comes under pressure to finish faster, the refactoring work may not be done.

AP is relatively new, so the jury is still out on the success and applicability of these methods. It is felt that AP is suitable for small projects and small teams; whether it has practical application for larger environments is still in question (Brandon, 2006).

Extreme programming (XP) is a software development approach initially for small teams on risk-prone projects with unstable requirements (Beck, 1999). Kent Beck, a programmer and project leader, developed XP while working on a long-term project to rewrite Chrysler's payroll application. XP is a form of AP based on a lightweight methodology. XP however, differs from most other agile approaches by being much more prescriptive. Like AP, XP is an incremental method with free-flow. XP advocates say the methodology (creating user scenarios and performing upfront feature testing) allows them to develop and deliver code more quickly with fewer bugs. XP is built around rapid iterations, an emphasis on code writing, and working closely with end users. The 12 basic practices of XP are:

- Customers define requirements via use case scenarios ("stories")
- Teams release small increments into production early
- Teams use standard names and descriptions
- Simple object oriented coding is used
- Designers write automated unit test scripts before coding
- Refactoring is used extensively
- Programmers work in pairs
- Programmers have collective ownership of all code
- Teams integrate and check code back into repositories frequently (no longer than one day)
- Developers work only 40 hour weeks
- User representative(s) remain on-site
- Programmers follow strict coding standards

Although XP in different forms has been used for a few years, many IT organizations have been reluctant to try it. A major issue is that some XP principles contradict long standing IT policies. For example, XP specifies "pair programming," in which two programmers sit side by side working at a single workstation. Pair programming seems inefficient, but studies have shown that it is no less efficient that traditional programming and usually results in fewer code defects (Williams, 2000). Fewer defects eventually mean quicker delivery. However, not all programmers want or are suited for pair programming. Very good programmers should not always be encumbered with a sidekick. Many programmers like the solace and that is one of the reasons they stay programmers. Often programmers consider themselves masters of the trade, and if you have two master chefs in the kitchen there is going to be conflicts. Another problem with XP (like all AP) is its application in contract environments. Still another problematic issue for XP is that all code is generally open for programming pairs to check out and alter. This can open up the team to integrity and security issues. As discussed earlier in this book, internal security is becoming a prime concern for IT organizations for both internally developed code and particularly outsourced programming. Further XP does not address downstream SDLC issues such as training, user documentation, and training.

XP also requires the benefiting organization to take a very active role in the development process even to the extent that users are asked to write tests which will prove requested functions work properly before they are coded. In XP, customers may write needed scenarios or features on index cards for example (one scenario per card). Using and reusing a number of index cards is far cheaper and faster than writing, editing, and reviewing a large formal requirements document. The developers estimate how long it will take to build that feature. Based on the estimates, the customer prioritizes the features. Next the customer writes the test, and the developers write code which will successfully pass the test. Testing is normally automated, and "test harnesses" organize test scripts that related to particular functional areas. However, since testing is limited to "acceptance" type testing, full multi-level testing is seldom performed. This may lead to problems with unanticipated inputs, scalability problems, and security problems. This is discussed further in a later chapter of this book on quality management.

Since XP requires constant communication between the benefiting and performing organizations (as well as among the developers), and since communication time and traffic increases in proportion to the square of the number of communicating parties, XP is not suited to large teams (Beck advises limiting project teams to no more than 12 developers working in pairs). As with JAD and AP methods in general, a customer may not be able to commit his resources to that much involvement.

Thus XP has a number of specific advantages and a number of specific disadvantages. The issue is a hot debate topic in the IT world. It is "extreme;" on the one side it is thought of as a great breakthrough and on the other side it is akin to "letting the inmates run the institution." So XP is not for all IT organizations. It is like "extreme sports;" it is great recreation for my 14 year old nephew, but not for his 38 year old dad (Brandon, 2006).

Component based software engineering (CBSE) is a development philosophy that utilizes existing software modules to build application systems. Any of the aforementioned methodologies may be utilized, but the requirements specification stage here may be longer due to preparation of procurement documentation. CBSE is a formalized system of reuse at a high level; formalized in the sense of a business approach rather than at the software architecture level. CBSE is based on having a cadre of reusable modules or programs and some framework for integrating these modules together. CBSE can be applied at several levels of granularity. At the highest level is the COTS (commercial off-the-shelf software) approach where commercial programs are purchased and integrated together through a data exchange mechanism.

A recently evolving approach to CBSE using the internet is called "Web services." Here different services are provided by different vendors in real time on their servers generally at a per usage price. This new computing architecture is formally called SOA or Service Oriented Architectures (Hall, 2003). Web services are based upon modern open standards; unfortunately some of these standards (SOAP, WSDL, UDDI, etc.) do not have adequate security built into them yet. Web services architecture uses SOAP (simple object access protocol) as a "lightweight" remote method invocation process. Older more complex protocols for distributed object services are Microsoft's DCOM (distributed component object model), Java's RMI (Remote Method Invocation), and OMG's CORBA (common object request broker architecture); RMI and CORBA are more secure than SOAP. Central repositories (registries) catalog which services are available and where using the UDDI (universal description discovery and integration) protocol. Providers list the usage specifications of their services via the WSDL (Web services description language) protocol. Web service applications can be created in a number of languages with most being written in Java, PHP, or Visual Basic-Net. Some think that the "service–oriented architecture" may become the core paradigm for software applications and integration in the future" (Eisenberg, 2004).

The advantages of the CBSE approach are speed of assembly and possibly short term cost. The disadvantages are that no strategic advantage is derived from the resulting product (nothing is proprietary, anyone can do it), compromise of requirements to meet capabilities of available components, vendor dependencies, possible performance issues, and security problems (Brandon, 2006).

WEB APPLICATION STAKEHOLDERS

The identification and management of a project's stakeholders is vital to the complete success of a project. Often well planned and properly executed still fail due to a lack of or inappropriate relationships between the project manager and various stakeholders. The first step in good stakeholder management is the complete identification of all of the stakeholders. Cleland defines stakeholders as (Cleland, 1998):

Stakeholders are people or groups that have, or believe they have, legitimate claims against the substantive aspects of the project. A stake is an interest or share or claim in a project; it can range from informal interest in the undertaking, at one extreme, to a legal claim of ownership at the other extreme.

Stakeholders are associated with both the performing organization and the benefiting organization; these are called "internal" stakeholders. The performing and benefiting organizations may or may not be part of the same company, and there may be a formal or informal contract situation. In addition there are usually "external" stakeholders that are not part of these two groups. Traditionally the key stakeholders have been individuals or other groups closely associated with either the benefiting or performing organizations. "However, long-run changes in the social, political, and economic environment of projects have meant that this is no longer necessarily the case for a number of reasons." (Winch, 2004) So, today there may also be a number of key external stakeholders in terms of various social, political, or economic interests.

Internal stakeholders related to the *benefiting organization* might include:

Project sponsor(s)
Business Owners or stockholders
Customer line management
Customer's users of the IT system
Customer's IT group
Customer's accounting group
Customer's business units affected
Customer's other employees
Customer's customers
Customer's contractors and vendors
Customer's financiers

Internal stakeholders related to the *performing organization* might include:

Project Manager
Business Owners or stockholders
Project Team
Performing organization's line management
Performing organization's IT group
Performing organization's accounting group
Performing organization's other employees
Performing organization's customers
Performing organization's contractors and vendors
Performing organization's financiers

External stakeholders may be concerned individuals, companies, or associations and these can be termed "private" interests; or they may be local, state, federal, or international govern-

Figure 16.

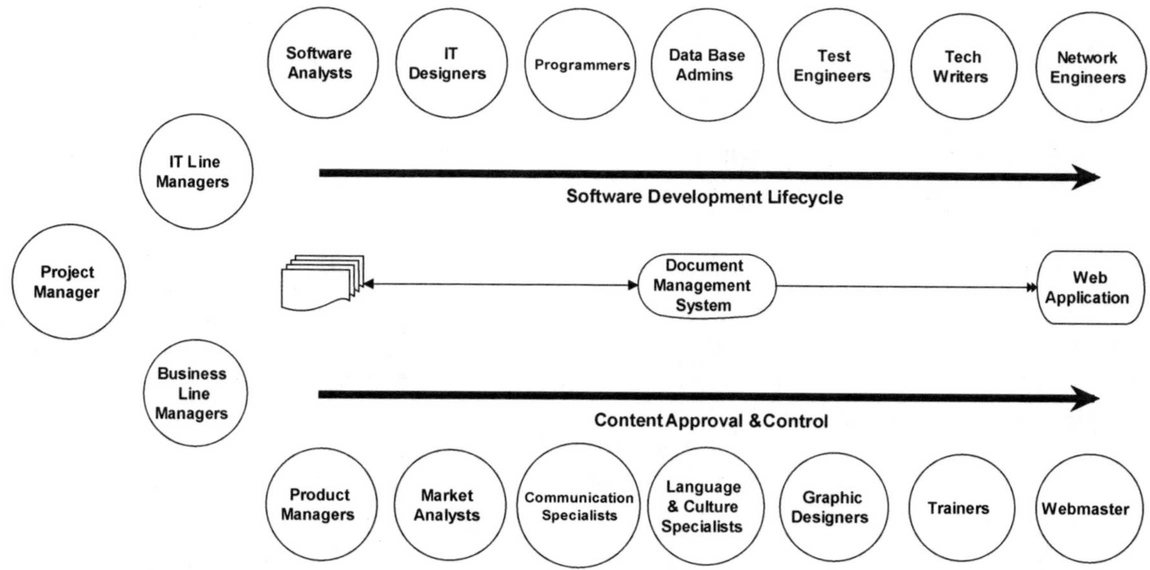

ment bodies and these can be termed "public" (Winch, 2004).

External private stakeholders might include concerned individuals, trade associations, environmental and conservationists associations, neighborhood association, and the like. External public stakeholders might include local governments, state governments, federal regulatory agencies, federal governments, or international agencies.

Since Web applications generally manage content and not just data, and many of them are document centric versus data centric. In addition, there are many more stakeholders involved in the definition, design, development, testing, approval, and usage for Web applications as opposed to traditional applications. This is illustrated in Figure 16.

Once the stakeholders have been identified, then each must be analyzed carefully in order to determine the manner in which that stakeholder should be managed. Stakeholders that are more concerned about a project will have to be managed differently than those that are less concerned; similarly stakeholders that are more powerful will have to be managed differently than those that are less powerful. Each stakeholder or group of stakeholders may have a different type of stake in the project including: monetary, job security, position, influence, reputation, or convenience (time). The interests of all stakeholders in one group may not be the same, if the result of the successful completion of the project may alter the "balance of power" in that organization. For example, within the benefiting organization different stakeholders may be for, against, or neutral about the project and this may depend upon the way the final project manifests itself or the way it is implement in the benefiting organization. In Web projects, which involve a number of corporate business units, there is very much corporate change including work flow, policies and procedures, management structure, and so forth; for this reason, different stakeholders within the same group (i.e., the benefiting organization) may have varying interest in the success of such a large project.

Figure 17 shows a "power versus concern" graph which can be used to visually assign stakeholders into one of these four categories.

Figure 17. Stakeholder classification (Brandon, 2006)

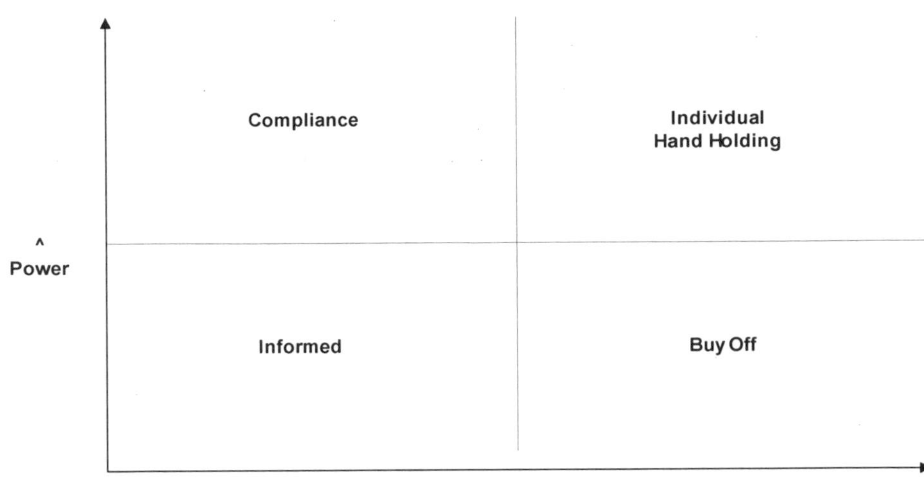

Powerful and highly concerned stakeholders will have to have to be actively managed throughout the project; these are the people who will need their "hands held." For each of these individuals or groups the project manager must ascertain:

- What is the biggest thing this stakeholder has to gain if the project succeeds (fails)?
- What is the biggest thing this stakeholder has to lose if the project fails (succeeds)?

Next the project manager (or line management) must try to best align this person with the project and get early "buy-in;" and then check on this alignment during all phases of the project. Stakeholders that are powerful but not too concerned usually have some standards or other "manifestation of principle" that they wish to be followed. An example here is regulatory agencies. This group of stakeholders is best managed by becoming very knowledgeable and clear on the principles and standards involved and then complying with same. Stakeholders that are not powerful but quite concerned need to be aligned with the project; this is the so-called "buy-off" group. This done first by finding out why they are so concerned, and then seeing if they can be aligned with the project initially; they do not need constant attention as does the powerful and very concerned group. Stakeholders that are not powerful and not too concerned simply need to be kept informed on a regular basis as a group.

In terms of power and influence over project affairs, there are four types of power in most organizations and communities of stakeholders:

1. Power due to position (or rank)
2. Power due to control over resources (like budget)
3. Power due to unique expertise
4. Power due to politics and/or charisma

Some authors define other types of power like physical power, which may have relevance on a sport team, but typically not in a project management environment. Other types of power defined by some are "position" sub-types of power as:

- **Reward:** Power based on being able to give an employee something he/she wants; does not necessarily mean money, may be a position/role on the team, or may be a "positive evaluation" supplied by the PM to a functional line manager

- **Referent:** Power bestowed from a higher point in the organization
- **Penalty:** Power based on being able to penalize someone (may be more than monetary)

A stakeholder may fall into more than one of these categories. Figure 18 shows one method to perform this stakeholder analysis including an analysis of power versus concern and type of power. In this spreadsheet each stakeholder is listed and characterized by checking off columns.

After all the stakeholders have been identified and analyzed, a "stakeholder action plan" should be developed. This plan will explain how each stakeholder is to be managed and who (within the project team, line management, or project management office, is responsible for carrying out each item in the action plan for that stakeholder (Brandon, 2006).

This stakeholder management plan can be a separate plan or be part of the project "communications plan." As part of the stakeholder management plan, potential claims by a stakeholder or other potential unfavorable actions should also be identified and included in the project risk management plan. As well as correlating with the communications plan and risk management plan, the stakeholder management plan should also map to the project quality plan. Finally, access to the stakeholder plan must be properly controlled, in regard to which stakeholders can view such a plan.

GLOBAL DEPLOYMENT FOR WEB APPLICATIONS

According to Computerworld: "Globalization is the marketing and selling of a product outside a company's home country." To successfully do that on the Internet, a company needs to *localize*—make its Web site linguistically, culturally, and in all other ways accessible to customers outside its home territory (Brandon, 2002). "Ever since the end of the Cold War, the world has been rushing toward ever-higher levels of national convergence, with capital markets, business regulation, trade policies, and the like becoming similar (Moschella, 1999). The value of cross-border mergers grew sixfold from 1991 to 1998 from U.S. $85 billion to $558 billion. The world has not witnessed such a dramatic change in business since the Industrial Revolution. (Korper, 2000) More than 95% of world population lives outside of the U.S., and for most countries the majority of their potential market for goods and services is outside of their borders. Currently about eight percent of the world's *online* population resides outside of North America:

Figure 18. Stakeholder analysis (Brandon, 2006)

Stakeholder Analysis

	Internal		External		Stand			Power		Concern		Influence			
Stakeholder	Benefitor	Performer	Private	Public	For	Against	Neutral	High	Low	High	Low	Position	Resource	Political	Expert

Asia 36.5%
Europe 28.2%
North America 21.8%
Latin America 7.8%
Africa 2.3%
Middle East 1.7%
Australia/Oceania 1.7%

Today the majority of Fortune's 100's Web sites are available only in English (Betts, 2000). In our rush to get on the WWW, we sometimes forget that WW is for "World Wide" (Giebel, 1999). Wal-Mart (a $165 billion U.S. company) has a global work force of more than 1 million and runs more than 1000 of its 3406 retail outlets outside of the U.S.; yet its Web site (Wal-mart.com) is only for Americans (Sawhney, 2000). Today's average Website gets 30% of its traffic from foreign visitors, and today only 1% of small and midsize American businesses export overseas (Grossman, 2000).

Currently the breakdown of Internet user languages is roughly 50% English, 8% Japanese, 6% German, 6% Spanish, 6% Chinese, 4% French, and 20% other. That means if one does not localize their Web site, they will be ignoring more than half of the world. For the immediate future most of the Internet community will still understand English, but overall English is the native language to only 8% of the world. Most users in foreign countries prefer content in their own language; for example, 75% of users in China and Korea have such a preference (Ferranti, 1999). It was found that visitors spend twice as long, and are three times more likely to buy from a site presented in their native language (Schwartz, 2000). Multiple languages are used in many areas. Belgium has both French and Dutch. In Switzerland, German, French, and Italian are used. Also we have to take into account differing dialects that are used across various countries speaking a specific language. One cannot use "Classic German" in Germany, Austria, or Belgium, since they all speak a different German. The combination of language and dialect is called a "locale."

When one installs an operating system on their computer, they may specify a locale. Then to view content that has been localized for another language, one has to have their Internet browser properly equipped with the correct scripts (characters and glyphs/symbols). In some locals there may be one spoken language, but several writing systems for it such as in Japanese. The current versions of Netscape and Microsoft Internet Explorer support most languages directly or via a "download" of needed scripts. One still may have to adjust option settings in these products accordingly in order to associate the proper character set with the proper language (Brandon, 2002).

One can convert Web pages by hiring a translator or using a computer based translation product or service. Hiring a translator will provide the best localization, but is more costly than the automatic methods. Translators can easily be found in the Aquarius directory (http://aquarius.net) or Glen's Guide (www.gleensguide.com). It is best to use a translator that "lives" in the local region; if a translator has not lived in a region for a decade he has missed 10 years of the local culture. There are also many companies that provide translation services such as Aradco, VSI, eTranslate, Idiom, iLanguage, WorldPoint, and others. The cost of these services is about 25 cents per word per language (Brandon, 2002). One can extrapolate how high this translation cost can be for a company with 100 pages in its Web site, having to support 20 different locales worldwide, with many on going changes to content each month.

Automatic translation software is another option, but it is still in its infancy (Reed, 1999). Some popular software products for translation are: www.e-ling.com, www.lhs.com, and www.systransoft.com. The automatically translated text typically does not convey the full meaning of the original text. For example some English elevator signs translated to, then from, another language may read:

- **Bucharest:** "The lift is being fixed for the next day. During that time we regret that you will be unbearable."
- **Leipzig:** "Do not enter the lift backwards, and only when lit up."
- **Paris:** "Please leave your values at the front desk."

There are several Web sites which provide free translation services such as http://babelfish.altavista.com, http://translator.go.com, and www.freetranslation.com.

Another alternative, although certainly not optimal, is to provide a link on your English Web page to these free services so that visitors can translate your content themselves.

Shown in Figure 19 is the home page for FedEx (www.fedex.com). One can select from over 200 countries for specific language and content. Figure 20 shows the U.S. FedEx page, and Figure 21 shows the FedEx site for Mexico.

Another example is Nike's home page shown in Figure 22, and their Japan version shown in Figure 23.

Creating an effective foreign Website involves much more than just a good language translation. Not only do languages differ in other countries but semantics (the meaning of words and phrases) and cultural persuasions in a number of key areas are different. "Sensitivity to culture and national distinction will separate success from failure" (Sawhney, 2000). To be effective, a Website has not only to be understandable and efficient, but has to be culturally pleasing and inoffensive. To accomplish that, it may be necessary that not only language be localized, but that content, layout, navigation, color, graphics, text/symbol size, and style may be different. Many companies have put forth global Web sites simply by translation the English into the targeted language, but then had to pull back and re-plan and redesign the localized site due to cultural offenses. The Nike site shown in Figure 23 for example has some of these issues which are common to the localized version of Web sites for many companies; in that figure one can see that:

- Not all the English has been translated into Japanese (text inside figures for example);

Figure 19.

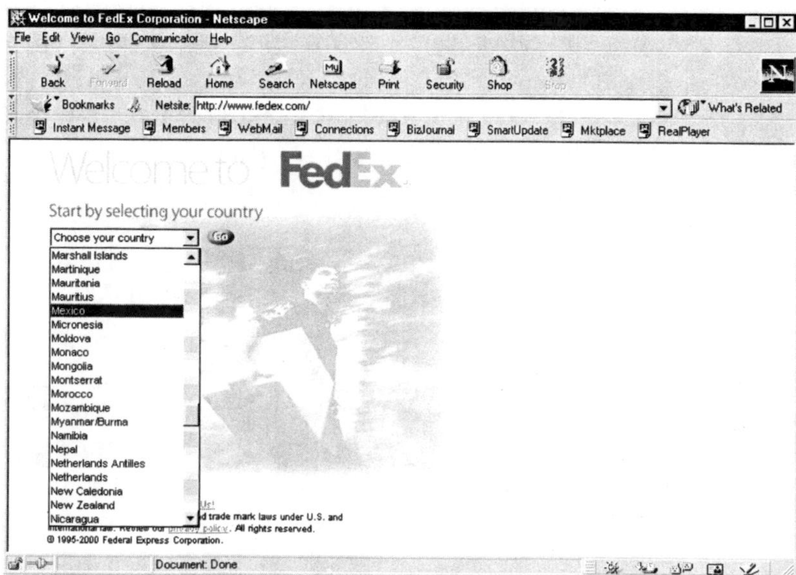

Figure 20.

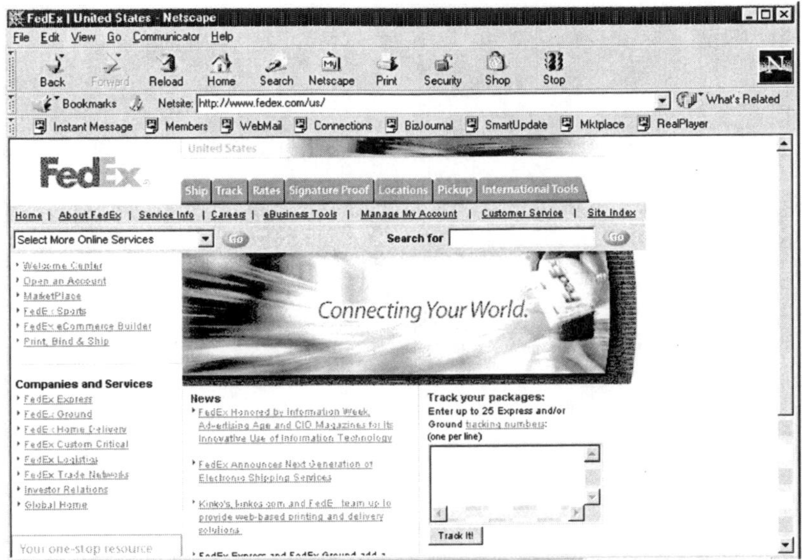

Figure 21.

- The actors are not Japanese;
- The clothes worn by the actors have American team names.

A country's humor, symbols, idioms, and marketing concepts may not send the same messages to other countries in the world. Oriental "manners" can be much different and more subtle than in other parts of the world (www.gwjapan.com); for example, avoid groups of four on Japanese sites. Sometimes even product names may be offensive or inappropriate. General Motors tried to market the Chevy Nova in Mexico (in Spanish "No Va" means "does not go") ! Some areas of global disagreement to avoid are equality of the sexes or races, body parts and sexuality, abortion, child

Figure 22.

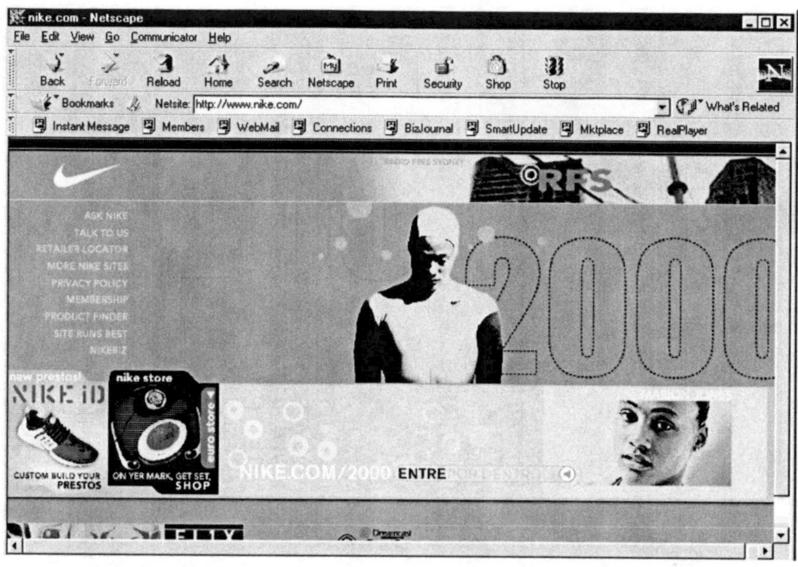

Figure 23.

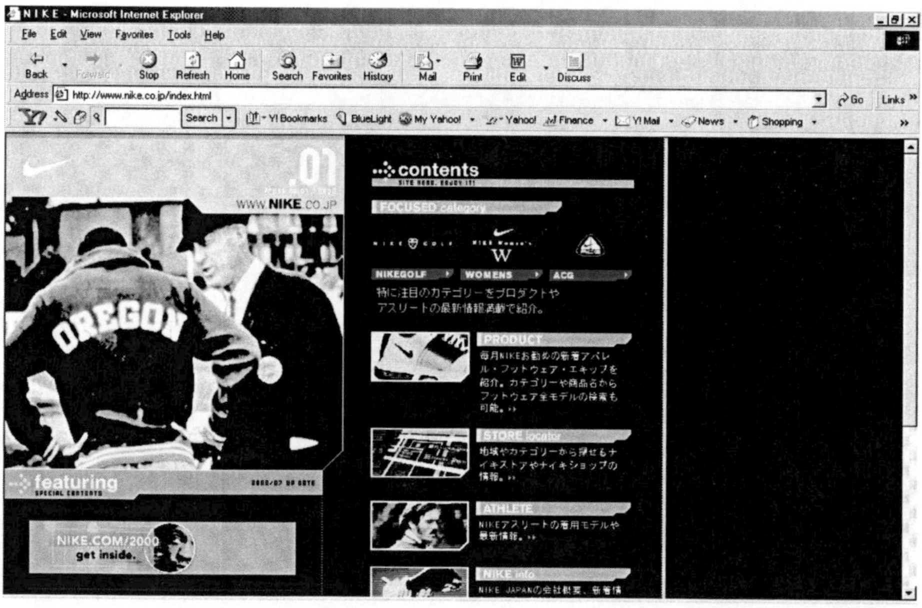

labor and majority age, animal rights, nudity, guns, work hours and ethic, capital punishment, scientific theories, and religious particulars (Brandon, 2002). Cultural persuasions work both ways. For example, many American Web sites offend other countries, but Americans are sometimes offended by foreign material. A European branch of a major U.S. software company ran an advertisement with a woman straddling a chair with her legs which said "Sometimes size is not important if you have the right tool." The advertisement did well in Europe but offended Americans.

Colors have symbolic and special meaning in most locals. In the U.S., the colors red, white and blue signify patriotism and red and green signify Christmas. In India, pink is considered too feminine. Purple is a problem in many places; it symbolizes death in catholic Europe and prostitution in the Middle East. Euro Disney had to rework its European sites after the first version used too much purple. Overall blue is the most culturally accepted color (Brandon, 2002). Much of the world is still using eight colors not 256 colors, thus it is best, for the immediate future, to use primary colors. An individual's perception of color depends not only on the ability to see it, but also on the ability to interpret it within the context of our emotional and cultural realities. "Ninety percent of (Web sites are colored poorly, they are simply overdone, and there is no sense of harmony" Holzschlag, 2000).

It is also very important to respect other cultures "symbols" (heroes, icons, etc.) both positive and negative (swastika). One guide site is Merriam Webster's Guide to International business (www.bspage.com/address.html). The classic books on these cultural subjects are excellent guides for Web pages also: *Kiss, Bow, or Shake Hands: How to do Business in 60 Countries* (Morrison, 1995), *Do's and Taboos Around the World* (Axtell, 1993), and Dun & Bradstreet's *Guide to Doing Business Around the World* (Morrison, 1997). This year a French court's ruling that Yahoo must make auctions of Nazi memorabilia unavailable in France indicates how uncertain and risky international e-business can be. "The troubling aspect of this case is that different countries can say that content not even targeted at their population breaks the law" (Perrotta, 2000). With the Internet it is not possible to know for sure where a user is logged in from due to "IP tunneling" possibilities. "Freedom" laws (such as the U.S. First Amendment) are not universal, and saying/printing some things can be illegal in some parts of the world. In the U.S. you can say what you like about "public figures," but not so in most of the rest of the world. There have even been several lawsuits in the U.S. concerning pornographic sites and the like due to different interpretations of laws in different states (different geographic/political parts of U.S.).Another legal issue concerns the privacy of personal data collected online. Many parts of the world have stricter laws than does the U.S., and U.S. companies have had judgments rendered against them in foreign courts. Recently an agreement has been reached between the U.S. and the European Union that would, among other things, mandate that all companies doing business in Europe notify users when personal data is being collected. Under that agreement, companies have four options in compliance to the new policy: register with the data-protection authority for the European Union, subscribe to a self-regulatory organization like Trust-e, prove they are subject to laws similar to the European Union, or agree to refer disputes to European regulators (Whiting, 2000).There are other areas that could cause legal problems, too. One is foreign advertising restrictions; for example, in Germany, one cannot directly compare your product with that of a competitor. In some other countries this comparison may not be illegal but may be in bad taste. Other areas consider safety, consumer protection laws, health, and other standards For example in the U.K., currently one cannot sell the drug Viagra, even though its sale is legal in the rest of the world; in Germany companies are not allowed to provide an unlimited return guarantee.

Nearly half of the U.S. Web sites refuse international orders because they are unable to process them (Grossman, 2000). One could always ask for advance payment in native currency (cash, cashiers check, International Money Order), but you had better have a very unique product for that approach to be successful. Foreign exchange rates vary daily so indicating that your prices are in your country's funds (exclusive of local taxes and custom duties) and using credit cards (so the credit card company does the conversion) is one way to deal with that issue. One can also link to a converter site (www.xe.net/ucc, www.oanda.com) or place a calculator on your page (www.xe.net/currency or www.bloomberg.com/markets/currency/currcalc.cgi). Also in many countries traditional pricing may be much lower or higher, so product pricing is important. For example, computer products are typically 50% more expensive in Europe than in the U.S.

However, credit cards are rare in Japan as is the use of checks. There postal workers collect cod's, and some companies send goods to brick and mortar places for consumers to pick up. In Germany only 5% of Web users (second to U.S. in overall net usage) use credit cards. Eighty-eight percent of European merchants use invoice billing (with a long net payment due time). So while credit cards are a convenient and popular mechanism in the U.S., it is not so in the rest of the world. To complicate matters even further, there are many (and always changing) international sales taxes, VAT (Value Added Taxes) in Europe, with different exempt items in each country. Selling in Europe may involve VAT registration in countries; one needs to get rulings and advise in writing from each country. Typically a company pays VAT based on the country it's based in, but that can depend on the country and item being sold. One approach to avoid all these problems is to use an escrow service such as Paymentech (www.paymentech.com) which now handles about 3 billion transactions/year; others are: www.tradesafe.com, www.internetclearing.com, www.iescrow.com, www.worldpay.com.

Dates are very important in e-commerce being used for events such as delivery dates, credit card expiration dates, product expire dates, and so forth. There is an international standard on dates (ISO 8601 Date Format), and even though you may not use it internally in your programs (for database operations and calculations), your Web display should be in the localized format. For example, the common U.S. format of 10/6/2000 is not uniformly understood; instead use Oct-6-2000. Major databases (i.e., Oracle) allow you to switch date formats per session or connection so the way a date is input (insert into table) or output (select from table) is automatically converted to the internal table representation of the date. Some popular Web databases (i.e., MySQL) do not provide this capability, so you will have to do the conversion in your own code (via client side JavaScript or Java Applets, or server side CGI programs or Java Servlets). Some popular Web programming languages have features to facilitate these conversions (i.e., Java's GregorianCalendar class or Perl's Date::Manip). Related to dates but not to the display problem, is the fact that each local has its own set of holidays; this will affect daily volumes and delivery schedules. Some locales may use a special calendar: Arabic Lunar Calendar, Jewish calendar, Iranian calendar, or the Japanese Imperial Calendar. In the U.S. a 12-hour clock is common, except in U.S. military establishments. The rest of the world uses mostly a 24-hour clock, so it is best to display time in the 24 hour format. Of course time zones will be different, so include your time zone along with the phone numbers for personal customer support. It is best to spell out the time zone in the native language. You could instead give your support time in GMT (Greenwich 2000 Standard) and use or link to www.timeanddate.com for a customizable world clock and calendar.

In addition to dates and times, other units of measure will be different also. Only the U.S. and Canada (partially) still use the "English System," the rest of the world is on the metric system now, even Britain. This may or may not affect the goods you are selling, but overall the lack of adoption of the metric system will now begin to put the U.S. and Canada at a disadvantage in world e-commerce. Even for an English only Web site, to do business internationally it will prove advantageous displaying product information in both English and metric measurements, or allow the user to dynamically change units (Hickman, 1998).

"Addressing" a customer may be more involved; some foreign addresses may have longer and more address fields. "For Europeans, trying to buy from American e-commerce companies is a lot like shopping in the Third World. While delivery address forms let you specify any country, the forms demand an American state, a five-digit zip code, a 3-3-4 formatted phone number, and they assume your street address only takes up one line"(Grossman, 2000). There is a universal standard, of sorts, here called the "UPU" (Universal Address Formats). Generally, it is of good advice including a country code (and base validation of remaining fields upon this country code), at least three address lines (40 characters each), city field (30 characters), a "state/province/region" field (20 characters), a postal code/zip field (10 characters), and a contact phone number (20 characters). Figure 24 shows an example global order form.

"Language is often the least challenging aspect of customizing, or localizing, a Web site for a foreign audience. The hard part is all the technical challenges;" including date/currency formats, bandwidth capabilities, tagging HTML properly, correct character sets to use, managing multilingual pages on the server, directing users to the language specific content, and so forth. (Yunker, 2000). Bandwidth and response time are vastly different around the world. In China, the 28.8 Kbp is standard, so one must minimize graphics and/or have a text only version for China and similar bandwidth limited areas. In Europe and Japan "wireless" or Mobile-commerce is more popular than in the U.S. currently, and this effects bandwidth and display sizing (Brandon, 2002).

Whether your HTML pages are manually created, statically created by an HTML editor (i.e., FrontPage, DreamWeaver, etc.) or dynamically created on the server, the HTML code will have to identify both the character set and encoding. Character sets are the common ASCII, an ISO standard [i.e., ISO 2022-JP for Japanese], or a special set. The encoding to use is identified via the HTML META tag, such as: <META

Figure 24.

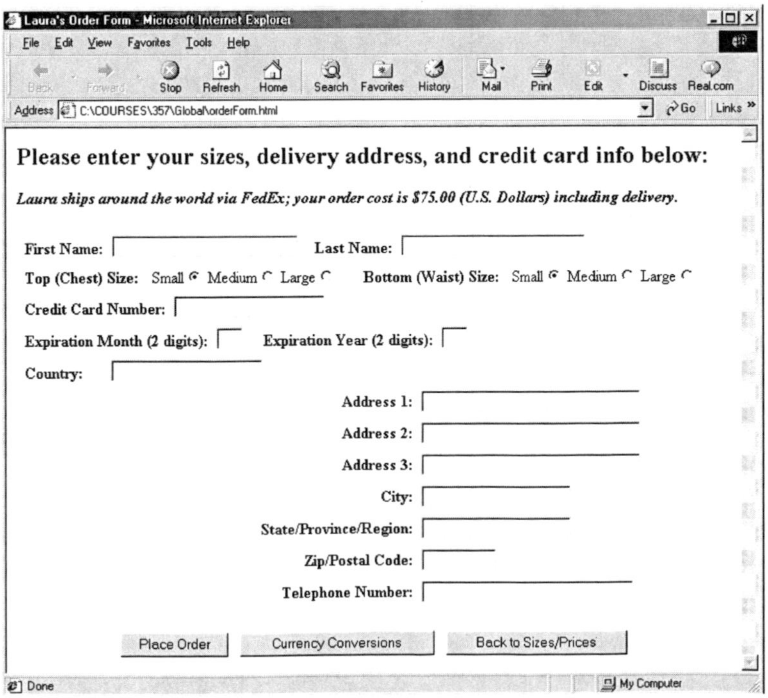
Laura's Order Form - Microsoft Internet Explorer
File Edit View Favorites Tools Help
Back Forward Stop Refresh Home Search Favorites History Mail Print Edit Discuss Real.com
Address C:\COURSES\357\Global\orderForm.html
Go Links
Please enter your sizes, delivery address, and credit card info below:
Laura ships around the world via FedEx; your order cost is $75.00 (U.S. Dollars) including delivery.
First Name:
Last Name:
Top (Chest) Size: Small Medium Large
Bottom (Waist) Size: Small Medium Large
Credit Card Number:
Expiration Month (2 digits):
Expiration Year (2 digits):
Country:
Address 1:
Address 2:
Address 3:
City:
State/Province/Region:
Zip/Postal Code:
Telephone Number:
Place Order
Currency Conversions
Back to Sizes/Prices
Done
My Computer

http-equiv="content-type" content="text/html; charset=Shift_JIS"><HTML Lang="ja"> for Japanese. You may also need to add ISO country codes to specify further dialect particulars (Brandon, 2002). The new standard is Unicode (ISO 10646, www.unicode.org) which uses 16 bits (double byte) to store up to 65,536 characters/symbols versus ASCII 8 bit codes (256 symbols). With Unicode you do not have both a character set and an encoding, it is one and the same ("charset=utf-8"). It probably is less of a problem with the Web browser's handling of international characters than with the database when order information and customer information is stored. Latest versions of database products also support Unicode, and those are the versions needed for full global support.

Navigation varies with some scripts from the more common left to right then top to bottom; Arabic & Hebrew are (usually) right to left, and Kana is vertical. The latest version of HTML contains tags to handle navigational direction. As well as navigational issues, other issues are hyphenation, stressing (underline, italics, bold in Roman, but different in other languages), bullet items, fonts, symbols above and below others, text justification, text sort orders, and GUI controls (text boxes and their labels, check boxes, radio buttons, drop downs, etc.) Field size is often a problem and the layout of graphical user interfaces may need to be redesigned; for example, German words are longer than other languages (Brandon, 2002).

When translating your content, you need to separate out the scripts (JavaScript, ASP, JSP, etc.) or just let the translators work from the displayed page, not the underlying HTML. Not all HTML editors support both displaying and saving "double-byte" characters/symbols, so be sure to choose one that does such as Frontpage 2000. Also with the symbolic Asian languages, you may need to add language support kits to the operating system (unless you have the latest version of Windows 2000, for example) for most graphics applications to work correctly. Also icons that have embedded text will be a problem, so it is best to separate the text from the icons. In a review of Howard Johnson's new Web site, Squier stated: "Hojo has made a big deal about this site being bilingual (English and Spanish), but I found little substance to back up the hype. The graphics, most of which contain text, are not translated into Spanish. This is sort of important, since we're talking about words like 'Reservations' and 'Free Vacation Giveaway'" (Squier, 2000). One can use both language specific text and visual international symbols to convey meaning and focus users. Common symbols in the world include light bulbs, telephones, books, envelopes, computers, flashlights, nature, tools, umbrellas, the globe, binoculars, eyeglasses, scissors, audio speakers, VCR/tape controls, microphones, arrows, magnifying glasses, cars/trains/boats/planes, a smile, and a frown (Fernandes, 1995).

For all of the above issues, it is evident that different Web content must be used in different locales. How to deploy and maintain these differences is a large and complex software architectural problem. The first consideration is directing users to the locale specific pages, and there are several methods that are typically used. One method is to put buttons, drop downs, or links on your native home page that the user can click to go to a locale specific page (see the earlier FedEx example). It is best to have the text on those buttons display the language name in the foreign language, although there are many sites that do not do it that way. For example on the button for Spanish say "Españoles" not "Spanish." The URL's of the locale version of your home page should be the same as your home page except end with the name of the country or locale, or end with the ISO standard country code abbreviation. That way it is easy for users to link directly to their native version also. For example with a home page URL of www.mycompany.com, have the Spanish version called www.mycompany.com/es or www.mycompany.com/espanole. Cookies can also be used to maintain a user's language choice so that when they return to the main URL they are switched to the locale specific version automatically (assuming most users of a specific PC will not be switching languages.) The FedEx site (Figure 6) works in this manner.

With the capabilities of modern operating systems and using the Java language, there is an automatic way of placing a user on the correct native page (Davis, 1999). When users install an operation system on their computer (such as Microsoft Windows 95/98/2000), they will specify a locale [via Control Panel/Regional Settings]; for most computers, the manufacturer sets this up upon assembly based upon the "ship to" address.

Your home page can simply be a container for a Java Applet which interrogates the operating system to find the regional setting. Then the Applet can load the correct locale/language version. The capability within the Java language for this is called "Resource Bundles" (Patten, 1999).

Instead of maintaining the URL's in the bundles, the actual phrases, codes, image filenames, video file names, and so forth, can be stored in the bundles. Then using Java server programming, dynamic HTML can be produced (under program control) "on the fly" to generate the native pages. "The biggest and most costly problem … is having to re-create Websites from scratch because the original was programmed with English text embedded in the code" (Disabatino, 2000). For dynamic HTML, this is typically done with a Java Servlet running on the server. Although technically more challenging, there are several advantages. First the HTML is generated dynamically and can be a function of time, date, or anything else as well as locale. Second when some information has to be changed, you do not have to open up and modify every language page; only the object that is being changed (phrase, image, etc.). Another key advantage is that the bundles can be classes, and as such an inheritance hierarchy can be set up. Dialects would be subclasses of the language and would inherit the properties of the language. In the subclasses, only the properties that were different in the dialect from the language would have to be maintained. There are products that facilitate this task of producing resource bundles or the like. Products such as Sun's Internationalization and Localization Toolkit (JILT), Multilizer Java Edition, or Catalyst Enterprise (Apicella, 2000) will capture all the textual references in a computer program (such as Java, C++, or PHP) and let you build a dictionary of translations in different languages. JILT uses resource bundles, and the other products take different approaches. This is a great aid in modern dynamic HTML, Java Applet, or Java Servlet based Web sites.

Then there is the enormous problem of version and configuration control with Web pages, just as there is in any software based system. Maintaining many language and or country/locale versions of a company's Web site will be a major task in the future. Over time, the English text changes as products, their features, and policies are changed. There must be a method to keep everything in synchronization. There are some "content management" products such as Idiom's WorldServer or BroadVision's Web-Publishing System that have some of those needed localization capabilities. For example, each text item, logo, graphic, and other items are tagged with a rule to indicate how it is to be handled in different languages and/or regions (Robb, 2000).

Some Web sites to aid in all these technical areas include Unicode (www.unicode.org), International Technical Issues (www.w3.org/International), Basis Technology (www.basistech.com), and the Microsoft Internationalization Whitepaper (http://msdn.microsoft.com/workshop/management/intl/locprocess.asp).

It can be very costly to build and maintain a foreign presence. A full business plan must be set up market analysis (product demand, pricing, and competition), total entry costs, then ROI must be considered (Tapper, 2000). Without doubt it is more expensive and time consuming to design and build an effective global Web presence than just a domestic site. Forbes has a list of 10 key general questions for companies considering going global (Klee, 2001):

- Do you have a good reason? Is exporting central to your company's strategy ?
- Do you have the right "stuff" to pull it off (talent, technology, leadership, …) ?
- Can you identify a market(s) ?
- Are you flexible ?
- Can you find a good distributor (partner) ?
- Can you cope with all the complexity ?
- Can you brave the "non-legal" barriers (ways of "doing business") ?
- Are you willing to extend credit and deal with currency turmoil ?
- Are you ready to run a much different kind of company ?
- Do the rewards outweigh the costs ?

"A company must have commitment from the top to make the endeavor of designing for international markets a success (Fernandes, 1995). Know your audience and see who your visitors are. Many companies are surprised when they analyze their log files and see who visits their site. There is software to facilitate this type of analysis and there is a new breed of application servers such as HitBox Enterprise from WebSideStory (www.websidestory.com) addressing visitor analysis. These application servers do not use log files (since they gather the information on-line from your static or dynamic Web pages) and thus do not require programming resources on your side.

Finally, to be most effective in the long run, an organization must get totally immersed in foreign and Web related matters. One can join global organizations like The Global Trading Web Organization (www.commerceone,com), subscribe to international trade newsletters (www.newsletteraccess.com/subject/intetrade.htm), and use other international services: www.worldbusiness.net/marketplace, www.digilead.com, ciber.bus.msu.edu/busres/tradlead.ht, Global Information Network (www.ginfo.net), Global Business Centre (www.glreach.com/gbc), GoingGlobal (www.going-global.com), WorldPoint (www.worldpoint.com), Internationalization of the Internet: (www.isoc.org:8080), InvestinEurope (www.investineurope.com).

As statistically shown earlier, U.S. Web users will play a smaller role each year in the "World Wide Web," China & Asian markets will grow dramatically. The "Euro" will become standard, and Europe may require U.S. based companies to charge VAT.

Communication infrastructures are building up in second and even third world countries (both government and private). Major communication build ups are currently occurring in the Pacific rim, Latin America, and South America (Ferranti, 1999). Companies such as FedEx will offer more sophisticated international shipping and logistic services to more parts of the world. More sophisticated software for translation, localization, and version control is being developed each month. In addition more companies will discover how to use the technology available within Java (JSP, Servlets, Applets, Beans). The Internet will become pervasive and become an integral part of our everyday lives via WevTV, Net "Applicances," Wireless devices, handheld devices, smart cards, and so forth.

In the not too distant future, the Web will be everywhere; and by "everywhere" we mean not only in all our electronic devices, but everywhere in the world. It has been said that the "Net brutally punishes latecomers." [Sawhney, 2000], so it is essential to start planning the internationalization and localization of e-commerce now. Also remember the Web is a two way street; foreign corporations will be coming after your customers soon!

SECURITY IN WEB APPLICATION DEVELOPMENT

With more powerful tools comes the potential for greater benefits including productivity increases, better cost/performance, and improved quality. However that power also brings a higher cost and damage potential when the tool is misused either accidentally or intentionally. Information technology is such a powerful tool, and that power in terms of computational speed is still doubling about every 18 months. Many other IT advances are also facilitating the possible misuse of IT including:

- Price for computational resources has dropped so low that even the smallest of organizations and countries can obtain massive power;
- Advances in data storage technology mean that huge amounts of data can be stored cheaply;
- Advances in data mining techniques mean that huge amounts of data can be analyzed in many ways;
- Advances in data networking mean that the cost and time of moving and accessing data has become very low, and that computers both inside and outside of an organization are increasingly connected.

As a result of these advances which facilitate IT misuse, computer security incidents are growing rapidly. The number of domestic US computer security incidents published by the CERT Coordination Center at Carnegie Mellon University has increased dramatically in recent years as shown in Figure 25. The number of these incidents has increased so much that CERT is no longer keeping detail information thereon.

These are just the security incidents that have been reported. Since IT is so prevalent in all the products and services that organizations offer

Figure 25. Reported security incidents (Brandon, 2006)

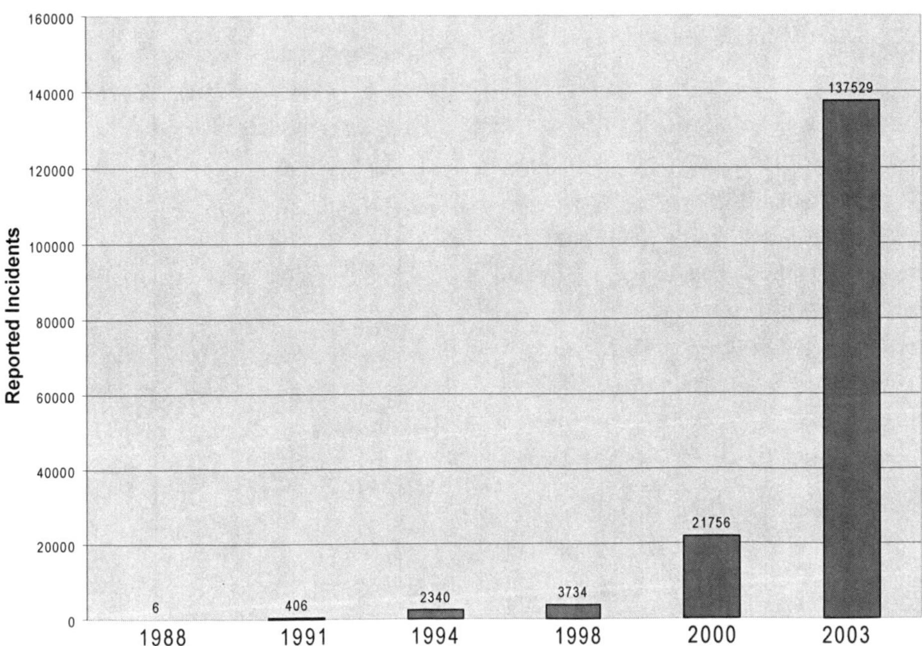

today and so prevalent in everything we do as individuals, the impact of these security problems is also quickly increasing. This problem could become enormous in the years ahead, and security breaches such as "Trojan horses" and "back doors" may already be in place within key software used by corporations and governments.

Security concerns are more prevalent in Web applications since the end users of the software are typically outside as well as inside the corporate virtual perimeter (LAN's and VPN's). Traditional applications have a number of security holes such as "buffer overflows," however Web applications have those holes and even more areas for security problems particularly with cookies and server side processing of client "get" or "put" data (including inadequate validation and parsing), SQL injections, "magic" quotes, variables inside of "includes" and "requires," etc.).

The security issue in IT project management is two fold:

- Being able to shield the project work, project team, and other resources from security threats;
- Being able to build adequate security protection into the product that is the subject of the project.

Security aspects should not be ignored in HR planning, and procedures such as background checks for all personnel (employees, contractors, consultants, etc.) involved with a project are becoming more appropriate. It is no longer sufficient simply to "secure the perimeter" physically and logically, but active security procedures need to be implemented for those objects already "inside of the perimeter." Figure 26 shows the types of issues that should be addressed in a simple project security plan.

CONCLUSION

In this chapter we have discussed the key issues involved with project management for developing Web applications. These included the fact that there are typically many more stakeholders involved with a Web application and these stakeholders may be globally dispersed; Web applications generally manage content and not just data; Web applications need to be deployed much quicker that traditional IT applications and they need to be engineered for more flexibility and adaptability in terms of changing requirements, changing content, changing presentation, mass user customization; and the potential for security problems are more prevalent in Web applications since the end users are outside as well as inside the corporate virtual perimeter.

FUTURE RESEARCH

There is both an opportunity and a need for more detailed and diverse research into many of these Web application project management issues, especially those involving security. *Being able to run projects and build Web-based systems that are impervious to both internal and external security threats will become vital to the success of organizations and survival of free world governments.*

REFERENCES

Apicello, Mario (2000). Multilizer for Java powers your apps to travel the Globe. *Infoworld*, January.

Axtell, Rodger (1993). *Do's and Taboos Around the World*, John Wiley & Sons.

Bauer, F. (1972). Software Engineering. *Information Processing*, *71.*

Figure 26. Project security plan (Brandon, 2006)

IT Project Security Plan

Personnel (employees, vendors, contractors, users ...)

- Work History Verification
- Background/Criminal Check
- Credit History Check
- Security Training
- Security Clearance

Access Control

- Log On (User Name & PW)
- Access Privileges
- Encryption
- Access Logs
- Review of Access Logs

Backup

- Incremental
- Full
- Recovery Testing

Physical

- Buildings
- Work Areas
- Central IT Facilities (servers, hubs, routers, firewalls ...)

Software

- Code Security Walkthrough
- Trojan/Backdoor Scans
- Intrusion Prevention Software
- Code Randomization
- Intrusion Prevention and Testing (Access, Session, Cookie, Buffer, XSS, Post/Get, SQL Injection, ...)
- Change Security Review

Beck, K. (1999). *Extreme programming explained: Embrace change.* Harlow: Addison-Wesley.

Bowen, B. (1997). *Software Reuse with Java Technology: Finding the Holy Grail.* Retrieved on from www.javasoft.com/features/1997/may/reuse.html

Boyatzis, R. (1982). *The competent manager: A model for effective performance*. Wiley.

Brandon, Daniel (2002). Issues in the globalization of electronic commerce. In V. K. Murthy and N. Shi (Eds.) *Architectural issues of Web-enabled electronic business.* Hershey, PA: Idea Group Publishing.

Brandon, Daniel (2006). *Project management for modern information systems*. Hershey PA: Idea Group Publishing .

Cleland, D. (1998). Stakeholder management. In J. Pinto. (Ed.) *Project management handbook*. Wiley.

Cockburn, A. (2001). *Agile software development*. Addison-Wesley.

Crawford, L. (2004). Global body of project management knowledge and standards. In Morris and Pinto (Eds.) *The Wiley guide to managing projects*. John Wiley & Sons.

Davis, M., & Smith , H., (1999). The Java International API: Beyond JDK 1.1. *Java Report*, February.

Disabatino, J. (2000). Web site globalization. *ComputerWorld, July*.

Eisenberg, R. (2004). Service-oriented architecture: The future is now. *Intelligent Enterprise, April 17*.

Fernandes, T. (1995). *Global interface design*. Academic Press.

Ferranti, M. (1999). From global to local. *Infoworld, October*.

Ferranti, M. (2000). Globalization tidal wave. *Infoworld, November*.

Giebel, T. (1999). Globalize your Web site. *PC Magazine, November*.

Grossman, W. (2000). Go Global. *Smart Business*, October

Grossman, W. (2000). The outsiders. *Smart Business, July*.

Hall, M. (2003). The Web services tsunami. *Computerworld, May 19*.

Hickman, N. (1998). Internationalizing your Web site. *WebTechniques, March*.

Highsmith, J. (2000). *Adaptive software development*. Dorset House.

Holzschlag, M. (2000). Color my world. *WebTechniques, September*.

Jacobson, I. et al. (1999). *The unified software development process*. Addison-Wesley Professional.

Klee, K. (2001). Going global: Out ten tests can help You Get Started. *Forbes Small Business, March*.

Korper, Steffano, & Ellis (2000). *The E-commerce book: Building the E-empire*. Academic Press.

Kruchten, P. (1998). *The rational unified process*. Addison-Wesley.

Morris, P. (2001). Updating the project management bodies of knowledge. *Project Management Journal, September*.

Morrison, T. (1997). *Dun & Bradstreet's Guide to doing business around the world*. Prentice Hall.

Morrison, T. (2000). Kiss, bow, or shake hands: How to do business in 60 countries. Adams Media.

Moschella, D. (2000). Ten key IT challenges for the next 20 years. *Computerworld, December*.

Neuman, C. (2000). Considering the color-blind. *Webtechniques, August*.

Palmer, S., & Felsing, J. (2002). *A practical guide to feature-driven development*. Prentice Hall.

Patten, B. & Grandlienard, G. (1999). Using resource bundles to international text. *Java Report, February*.

Perrotta, T/ (2000). Yahoo ruling exposes risks of being global. *InternetWorld, July*.

Peterson, C. (2000). Accessible Web sites matter. *Enterprise Development, June*.

PMI (2000), *The project management body of knowledge (PMBOK),* Project Management Institute.

Reed, S. (2000). Want to limit the audience for you Web site? Keep it English only. *Infoworld, August.*

Robb, D. (2000). Act globally, serve locally. *Information Week, July.*

Royce, W. (1970). Managing the development of large software systems. In *Proceedings of the IEEE WESTCON,* Los Angeles CA, IEEE Computer Society.

Sawhney, Mohanbir, & Sumant Mandai (2000). Go global. *Business, May.*

Schwaber, K., & Beedle, M. (2001), *Agile software development with schrum.* Prentice Hall.

Schwartz, H. (2000). Going global. *WebTechniques, September.*

Squier, J., & Nielsen, J. (2000). Deconstructing—Hojo.com. *Internet World, June.*

Standish Group (2004). Chaos chronicles. Retrieved on www.standisgroup.com

Stapleton, J. (1997). *DSDM dynamic systems development method.* Addison-Wesley.

Tapper, S. (2000). Is globalization right for you. *WebTechniques,* September

Uniscape Corporation (2000). Global content manager.

Whiting, R. (2000). U.S. companies to comply with European privacy rules. *Information Week, February.*

Williams, L., & Kessler, R. (2000). Strengthening the case for pair programming. *IEEE Software, 17*(4), 19-25

Winch, G. (2004). Managing project stakeholders. In Morris and Pinto (Eds.). *The Wiley guide to managing projects.* John Wiley & Sons.

Yunker, J. (2000). Speaking in charsets. *WebTechniques, September.*

Chapter XVI
Resources on Web-Centric Computing

Margaret Menzin
Simmons College, USA

ABSTRACT

This chapter consists of a comprehensive bibliography of web engineering resources. Since this list will constantly be changing, the author provides a web site for a current and more comprehensive listing: (URL="http://web.simmons.edu/~menzin/WebCentricResources.html") This print version of this resource shows the URL's in printed form, generally in the format as "(URL=...)". Be sure to check the General Tools section, in addition to the sections for specific topics.

GENERAL RESOURCES (FOR MANY WEB TOPICS)

Organizations

- World Wide Web Consortium (w3c) (URL='http://www.w3c.org'). The w3c develops most of the standards for the web. They are the source of definitive documentation on many subjects, and have a few primers. Generally it is better to learn a little about a topic before you dig into the documentation.
- OASIS (URL='http://www.oasis-open.org/committees/tc_cat.php?cat=ws') develops most of the other standards, along with ECMA (URL='http://www.ecma-in-

Certain sites recur frequently: they provide information on multiple topics, and often at multiple levels. This table summarizes those sites and is always a good place to start.. Please note that there are also many sites which are useful for only one or two categories (e.g. CSS or Web Services) and which are listed under those topics. Further, even for the general sites, the topical sections often have links to more specific useful resources.

Site	Comments	XHTML	CSS	Java-Script	CGI	Perl	PHP	XML	Web Services	Web2.0
w3c	Documentation for most technologies; hard to use unless you know something	Doc	N-I Doc		Doc			Doc	Doc	Doc
About	Hard to find with search	N	N	N				N		
w3schools	Superb tutorials; handy references well organized	N-I Ref	N-I Ref	N-I Ref			N-I		N-I	Ajax
Tizag	Similar to above tho' often less detailed; SQL tutorial too	N Ref	N Ref Scr	N-I		N-I	N-I		N	
U Minn. Duluth	Superbly organized links;Accessibility info too.	Link by search	Link	Link			Link	Link		
Earthweb *HTMLGoodies *JavaScripts *Developer *Gamelan *WebReference	Family of sites Java interfaces Advanced articles	 N; Ref	 I-A; Scr	 N Scr A: Link	 N I-A; Scr	 N I-A; Scr	 N I-A; Scr I-A; Link	 I-A I-A; Link	 I-A I-A I-A; Link	 I-A; Scr I-A I-A; Link
WebDeveloper	Links to N-I-E articles	Link	Link	Link		Link	Link	Link		
HotScripts -	Many advanced topics too			Scr	Scr	Scr	Scr			Scr
Web Developers Virtual Library	Clear, but some older	N-I	N-I	N-I	N	N; Link	N; Link	N-I-E; Link		
DevShed				I-E; Scr		A; Scr	A;Scr	A	A	A
IBM developerWorks	Wonderful articles if you have a specific topic	I-A	I-A	I-A	I-A	I-A	I-A	I-A	I-A	I-A
Menzin Links at Simmons College	Longer web page version of these references	Link	Link	Link	Link	Link	Link	Link	Link	Link

Legend:

N, **I** *and* **A:** *has Novice, Intermediate and Advanced articles and tutorials*

Doc: *has documentation on topic*

Ref: *has reference charts on syntax*

Link: *links to other articles and tutorials and books*

Scr: *has sample scripts and downloadable code*

ternational.org/'), ACM (URL='http://acm.org'), and ISO (URL='http://standards.iso.org/ittf/PubliclyAvailableStandards/').

- Articles and Tutorials on Many Subjects, starting with more elementary ones (URL="http://www.w3schools.com/browsers/browsers_stats.asp"). But beware that Netscape and Mozilla may be confounded; See also (URL="http://www.w3schools.com/browsers/browsers_mozilla.asp") Mozilla TIOBE Community Bulletin Board (URL="http://www.tiobe.com/tpci.htm") lists frequency of queries about various languages including web-centric ones.
- w3schools (URL="http://www.w3schools.com/default.asp") has on-line tutorials for many subjects - including CSS, XML, etc. Usually a very clear survey of the issues, features, and syntax. Read before you get into the grubby details for an excellent overview.
- Tizag (URL='http://www.tizag.com/') for elementary tutorials. Usually w3schools goes farther, but Tizag covers a few topics that w3schools miss.
- U. of Minnesota - Duluth organizes everything (URL='http://www.d.umn.edu/itss/support/Training/Online/webdesign'). This well-organized site has tutorials on CSS, JavaScript, XML, PHP, Usability and Accessibility. Tools for every topic, outstanding collection of resources; but no description of level of articles.
- DevPapers (URL='http://www.devpapers.com'), and About.com (URL='http://webdesign.about.com/') have many clear primers and introductory articles on XHTML, CSS, Perl, etc. Unfortunately hard to find information from previous newsletters.
- WebDeveloper.com (URL="http://www.webdeveloper.com/") for articles (some current, some dated; mostly intermediate) and forums to ask about Java, HTML, XHTML, JavaScript, Perl, CGI etc.
- EarthWeb (URL='http://www.htmlgoodies.com/introduction/about/article.php/3504566') is a useful family of sites covering all areas of web-centric computing; primers, tutorials, and newsletters ranging from elementary (e.g. HTML Goodies) to intermediate; all well-written, but some are dated.
- Includes WebReference: (URL="http://www.webreference.com/scripts/") for HTML, RSS, Perl etc. and also Developer.com (URL='http://www.developer.com/lang/') with very current technical articles on all the web-centric technologies.
- HotScripts (URL="http://www.hotscripts.com/") for articles on scripts for many languages and technologies - Perl, PHP, Python, JavaScript, XML. Well organized, for example, click on PHP, then books and you are led to short paragraphs describing over 60 PHP books.
- Web Developer's Virtual Library (URL="http://www.wdvl.com/") has tutorials on authoring, including intermediate level ones; JavaScript, CSS, XML, etc.
- DDJ (URL="http://www.ddj.com/currentIssuePage.html"). Dr. Dobbs Journal has links on many programming languages and platforms ; also news.
- ASPN (URL="http://aspn.activestate.com/ASPN/"). is a resource for programmers in Perl, PHP, Python, etc. in which Active State offers open source languages and tools. The Newsfeeds tab links to RSS feeds from many tech news sources.
- Major site. The Experts (URL="http://www.webreference.com/experts/") links to Dmitry's Design lab (a favorite), articles on XML, DHTML
- A List Apart (URL="http://www.alistapart.com/topics/code") — ALA. Mainly articles on CSS code, but also some on JavaScript, XML, layout and web site design; oriented to graphic designers.

More Advanced Resources

- Mozilla's Developer Center (URL="http://developer.mozilla.org/en/docs/DevEdge") links to articles and reference material on every topic, from the main page (URL="http://developer.mozilla.org/en/docs/Main_Page") in web programming and then some.
- Alphaworks from IBM (URL='http://www.alphaworks.ibm.com/') has articles on many cutting edge technologies; clear but not novice.
- DeveloperWorks (URL='http://www-128.ibm.com/developerworks') but also excellent articles on emerging standards and technologies.
- Amaya Home Page (URL="http://www.w3.org/Amaya/") editor which includes extensions to MathML; Amaya is a w3c.org tool for editing on the web.
- Apache (URL="http://www.apache.org/") is a very large, major open source organization with many projects. Apache is a web server; there are links to Xerces, Apache's XML editing tool, other XML projects, and many other projects and products.
- Source Forge (URL="http://sourceforge.net/docs/about"). Another large source for open source projects; you need to know what you are looking for.
- Sun Microsystems (URL="http://developers.sun.com/sitemap.jsp?requrl=/") Sun's developer site - information on Java, Web Services, etc.
- DevEdge (URL="http://devedge-temp.mozilla.org/toolbox/sidebars/").Sidebars for HTML, CSS, JavaScript etc. you may install in Firefox and Netscape 7+ for heavy duty developers. DevEdge was a wonderful site, now archived here. The sidebar tabs are professional tools for developers. For example, there are sidebars for JavaScript, CSS, DOM and XSLT.
- Zvon (URL='http://zvon.org/Output/bar_list.html'), sidebars for developers working with Mozilla and Opera browsers; tutorials on CSS, XML and extensions

FTP Utilities and PuTTY

- CoreFTPLite (URL="http://www.coreftp.com/") is my favorite free FTP client; supports both FTP and SFTP/SSH' easy to use and robust. Best of all, if you find a typo after you've uploaded a file you may use CoreFTP to edit it on the server.
- About choosing an FTP utility (URL="http://www.htmlgoodies.com/letters/292.html"),
- Setting up your FTP utility (URL="http://www.htmlgoodies.com/letters/292.html") or
- Filezilla (URL="http://sourceforge.net/projects/filezilla/")
- Free FTP and SFTP (Secure FTP) utility (from Source Forge)
- CyberDuck (URL="http://icu.unizh.ch/%7Edkocher/cyberduck/"), Free FTP and SFTP utility- this one for Macs
- Fugu (URL="http://rsug.itd.umich.edu/software/fugu/"), SFTP for Macs with a graphical front-end
- ZDNet (URL='http://downloads.zdnet.com/search.aspx?kw=FTP'), FTP clients, many with reviews. You may set the filter to 'free'
- PuTTY is a free implementation of Telnet for Windows and Unix systems. It may be downloaded at http://www.chiark.greenend.org.uk/~sgtatham/putty/ (Homepage for PuTTY) or http://www.softpedia.com/get/Network-Tools/Telnet-SSH-Clients/PuTTY.shtml. Easy installation instructions are at http://www.wnec.edu/~snarmont/putty/ and the user manual is at http://www.tartarus.org/~simon/puttydoc/index.html.

UNIX/LINUX

- A few reminders in case you need to brush up—while you are on a UNIX/LINUX machine: man commandName gives the manual's sections on that command.
- (URL="http://webreference.com/programming/unix/") the basic but goes to more topics; great resource
- (URL="http://wks.uts.ohio-state.edu/sysadm_course/html/sysadm-1.html") for systems admin a very detailed course.
- There are many LINUX (URL="http://groups.google.com/groups?hl=en&lr=&group=linux") (and UNIX) groups on Google.
- The following tutorials are all very, very basic:
- (URL="http://babbage.clarku.edu/%7Eachou/UnixPrimer/") University
- (URL="http://www.bridgewater.edu/cescc/acadcomp/UnixPrimer.htm") Unix Primer from Bridgewater College
- (URL="http://webreference.com/programming/unix/") the basic but goes to more topics great resource Linux Links (URL="http://www.linuxlinks.com/Beginners/")
- Includes a place for newbies, but most of the searches link to books at Amazon. BigNoseBird Tutorial on vi (URL="http://www.bignosebird.com/docs/vi.shtml")
- (URL="http://www.webmonkey.com/webmonkey/reference/unix_guide/") from Web Monkey has two dozen of the most frequently used commands.

Tech News Sources

- ACM (URL="http://technews.acm.org/current.cfm") for current news; you should also get their newsletter with weekly updates. If you are a student, you should join; student memberships are a bargain.
- IEEE (URL="http://www.ieee.org/web/aboutus/newsroom.html"). This site has more news about the IEEE than about technology. Local societies and local computer societies also frequently have newsletters.
- ZDNet (URL="http://reviews-zdnet.com.com/"). Reviews of software, hardware and tech toys; a good place to find free utilities (e.g. for FTP) as well as to comparison shop for new computers, digital cameras, etc.
- CNET Tech News (URL="http://news.com.com/2001-1_3-0.html"), TechWeb (URL="http://www.techweb.com/")
- Tech News World (URL="http://www.technewsworld.com/")
- SlashDot (URL="http://slashdot.org/")
- Wired.com (URL="http://www.wired.com/technology.html")
- The following sites offer newsletters geared to web-centric computing:
- Eweek (URL='http://www.eweek.com/article2/0,1895,1736489,00.asp')
- Informit (URL='http://www.informit.com')
- InfoWorld (URL='http://www.infoworld.com')
- ComputerWorld (URL='http://www.computerworld.com')
- Builder.com (URL="http://builder.com.com/5173-6389-0.html")

HTML and XHTML

XHTML/HTML

- HTML and XHTML: The Definitive Guide 6th Edition by Chuck Musciano & Bill Kennedy, published by O'Reilly 2006. Clear, with plenty of examples; read a short tutorial first, then use this for more detail and as a reference.
- HTML & XHTML: The Complete Reference 4th Edition by Thomas A. Powell, published by Osborne 2003. Excellent both to learn

from and as a reference; includes CSS1 and CSS2 and how to link with Java applets.

- HTML, XHTML, and CSS Bible (Bible) 3rd Edition by Bryan Pfaffenberger, Bill Karow, Chuck White, and Steven M. Schafer published by Wiley, 2004. The 'Bible' reference books are usually the most complete and usually invaluable tables of special characters (accented letters, apostrophes, etc.)
- About.com (URL="http://webdesign.about.com/library/bl_htmlcodes.htm")
- w3schools (URL='http://www.w3schools.com/tags/ref_entities.asp')
- w3c.org (URL="http://www.w3.org/TR/1999/REC-html401-19991224/sgml/entities.html") which includes mathematical characters, tables of XHTML (and HTML4) tags, elements and attributes.
- Complete documentation on XHTML (URL='http://www.w3.org/MarkUp/#recommendations')
- w3schools table of all XHTML and HTML tags, attributes, colors, etc.; (URL='http://www.w3schools.com/xhtml/default.asp')
- (URL='http://www.devguru.com/Technologies/xhtml/quickref/xhtml_index.html') reference on all XHTML tags from DevGuru.com. Very easy to use and helpful, but doesn't warn you about usage dropped in going from transitional to strict Doctype.
- (URL="http://webdesign.about.com/cs/htmltags/a/bl_index.htm") Library - HTML Tag Library from About.com. Links to summary of tags for HTML/XHTML by function (at bottom) and alphabetically (to right of sponsored links box) and links to summary of modules (major chunks by function) for XHTML
- The following articles refer to (the older) HTML, but still have some useful information: HTML 4.01 Specification (URL='http://www.w3.org/TR/html4/')
- You can also find the (URL='http://www.w3.org/TR/html4/interact/forms.html') forms specifications here (URL="http://www.w3.org/TR/1999/REC-html401-19991224/index/elements.html") elements from w3c.org
- (URL="http://www.w3.org/TR/1999/REC-html401-19991224/index/attributes.html") attributes from w3c.org Includes information on what is deprecated.
- IDocs References on HTML (URL='http://helpguide.inmotionhosting.com/html_faq/')
- Modularization of XHTML; further articles are in the More Advanced Articles section. (URL="http://www.w3.org/TR/xhtml11/") includes information on changes from XHTML1.0
- (URL="http://www.w3.org/TR/xhtml-basic/") documentation from w3c; XHTML Basic is a minimal subset for 'small user agents' (cell phones, smart watches, TV's).
- HTML Working Group Roadmap (URL="http://www.w3.org/MarkUp/xhtml-roadmap/"), what modules and versions will be coming your way and when

Other XHTML Documentation

- (URL="http://www.w3.org/TR/xhtml1/#guidelines") HyperText Markup Language (Second Edition).
- Making sure your XHTML is displayed properly in (old) HTML browsers; from w3.org. This article also includes information about various DOCTYPEs. Setting the character set encoding (URL='http://www.w3.org/International/O-charset.html') More information on character set encodings is in the Tutorials section.
- The DOM model (URL="http://www.wsabstract.com/domref/index.shtml"). The DOM has become very important (e.g. for CSS and

AJAX). Tutorials and Articles on the DOM are listed separately in the JavaScript section. This reference is from the JavaScript Kit.

Tutorials and Articles for HTML and XHTML

The links below are tutorials on HTML and articles on why you should move to XHTML. Although you should be writing XHTML, rather than HTML, a complete novice might find it easier to start with HTML. These tutorials are particularly clear.

- Introduction to HTML (URL="http://www.w3.org/TR/html401/intro/intro.html")
- Dave Ragett's Tutorials on HTML (URL='http://www.w3.org/MarkUp/Guide/'). Even though you should start with XHTML, this is such a clear, concise tutorial it's a classic. There are also links to his (URL='http://www.w3.org/MarkUp/Guide/Advanced.html') advanced tutorial, his tutorial on CSS,and his style guide (URL='http://www.w3.org/MarkUp/Guide/Style.html').
- HTMLTidy (URL='http://www.w3.org/People/Raggett/tidy/')
- HTML Tutorial from EchoEcho.com (URL='http://www.echoecho.com/html.htm'). Another clear HTML tutorial; has some more advanced features - e.g. meta tags refresh, etc. The navigation is so clear that you can quickly zoom to any details you need.
- Why switch to XHTML? (URL='http://www.webreference.com/authoring/xhtml/')
- Transition from HTML to XHTML From About.com, (URL="http://builder.com.com/5100-6371_14-5237333.html?tag=e606") switch to XHTML from Builder.com.
- XHTML Tutorial from w3schools.com (URL='http://www.w3schools.com/xhtml/default.asp'). The w3schools tutorials are always a good place to begin.
- (URL="http://webdesign.about.com/library/nosearch/bl_xclass1-1.htm") tutorials from About .com. You can also change the 1-1 in the URL to 2-1, 3-1, etc. to step thru the tutorials You can also have it emailed to you as a course: (URL="http://webdesign.about.com/c/ec/9.htm").
- XHTML 101 - Free XHTML Course These are a good but basic introduction to XHTML (URL="http://www.wdvl.com/Authoring/Languages/XML/XHTML/") XHTML, with eXamples.
- Tutorials from w3c.org. (URL='http://www.w3.org/MarkUp/#tutorials') includes links to some advanced tutorials. Articles on Fonts—please also refer to the sections on Site Design and Browsers
- (URL="http://builder.com.com/5100-6371_14-5210803.html?tag=sc") using ems and percents for font sizing, pitfalls of relative sizing for fonts
- (URL="http://builder.com.com/5100-6371_14-5215705.html?tag=sc") answer for font sizing? Keywords work better than absolute or relative sizes for fonts
- (URL="http://webdesign.about.com/od/fonts/a/aa080204.htm") Decide Which Font Family to Use — Serif, Sans-Serif, Monospace, Script, Fantasy Fonts (URL="http://www.w3.org/TR/REC-CSS2/fonts.html"). w3c's definitions of the font characteristics you may specify, and how to do it.
- Unordered lists (URL='http://www.developershome.com/wap/wcss/wcss_tutorial.asp?page=listProperties'). Beginning with XHTML1.0 strict this must be done through styling. This also works on mobile devices.
- Ordered lists (URL='http://www.developershome.com/wap/wcss/wcss_tutorial.asp?page=listProperties2'). Beginning with XHTML1.0 strict this must be done through

styling. This also works on mobile devices. The starting point of the numbering/lettering may also be specified (URL='http://www.w3schools.com/tags/tag_ol.asp')

- Guidelines for formatting lists (URL='http://webdesign.about.com/od/writing/ss/aa110705.htm')
- (URL="http://webdesign.about.com/library/weekly/aa061801a.htm") Attributes and XML Introduction.
- The span and div Tags (URL='http://webdesign.about.com/od/htmltags/a/aa011000a.htm'), Clear summary of the similarities and differences and how to use div appropriately.
- What's in a Title (URL="http://webdesign.about.com/library/weekly/aa060801a.htm"), what the title tag does for you and how to find it.
- (URL="http://developer.mozilla.org/en/docs/Properly_Using_CSS_and_JavaScript_in_XHTML_Documents#Using_CSS_rules_in_external_file") using CSS and Scripts with XHTML - please also see section below on CSS.
- From Mozilla; with examples (URL="http://developer.mozilla.org/en/docs/Properly_Using_CSS_and_JavaScript_in_XHTML_Documents:Examples")
- (URL="http://webdesign.about.com/library/weekly/aa041000a.htm") For valuator, please also see section below on HTMLTidy
- (URL="http://webdesign.about.com/cs/doctype/a/aaquirksmode.htm?nl=1"), Using the DOCTYPE Tag and why you need the DOCTYPE. Information on DOM and DHTML is at DOM - advanced articles on JavaScript. Attributes for the Image tag and checklist for your web page
- (URL='http://webdesign.about.com/od/authoring/a/aa111698.htm') Good checklist on everything from correct spelling to making sure your image files are small enough to download rapidly (12Kb)

Forms, Post and Get—See also Section on CGI

- A usability checklist for forms (URL='http://www.alistapart.com/articles/sensible-forms'), eExcellent article
- (URL="http://www.wdvl.com/Authoring/Scripting/WebWare/Server/") Server-side Scripting, an excellent introduction to what CGI is and why it's needed.WDVL: Web Programming: GET, POST, etc.
- (URL='http://www.wdvl.com/Authoring/Scripting/Tutorial/toc.html'), The 'Day One'pages provide a particularly clear description of GET, POST, (when to use which), HTTP headers and forms.
- (URL="http://www.wdvl.com/Authoring/JavaScript/Begin/begin1-4.html") on forms. The focus() and blur() methods
- Checkboxes vs. Radio Buttons (URL='http://www.useit.com/alertbox/20040927.html')
- From Jakob Nielsen, the great guru on usability, the Disabled attribute in forms (URL='http://webdesign.about.com/od/forms/a/aa071805.htm?nl=1'). This allows you to fill in a field and prevent the user from changing it
- HTML Forms from w3c.org (URL='http://www.w3.org/TR/html4/interact/forms.html'). Includes some advanced methods - e.g. fieldset for grouping related elements and control elements; additions to (URL='http://www.w3.org/TR/xhtml-modularization/abstract_modules.html#s_forms') forms in XHTML are also available in the forms module.
- (URL='http://www.w3.org/MarkUp/Forms/2003/xforms-for-html-authors.html'), the w3c.org explains how to convert Forms to Xforms
- (URL='http://www.htmlgoodies.com/introduction/newsletter_archive/goodiestogo/article.php/3573771'). Scripts to send form data by email, if you don't want to use the preferred CGI script.

More Advanced Articles

- (URL="http://webdesign.about.com/library/weekly/aa071502a.htm") with Excel — How to set up your web page so that it may be downloaded into Excel
- (URL="http://www-128.ibm.com/developerworks/xml/library/x-futhtml2.html?ca=dgr-lnxw03XHTML2") eventually: XHTML2.0, what to expect
- (URL="http://webdesign.about.com/library/weekly/aa041601a.htm") of XHTML. Good introduction to modularization of XHTML for XHTML, Also discusses the Transitional, Strict, and Frameset subsets of XHTML
- (URL='http://www.w3.org/TR/xhtml-modularization/Overview.html#toc') of all the modules in XHTML 1.1.
- Should you abandon table-based layouts in favor of CSS? (URL='http://www.elijournals.com/premier/showArticle.asp?aid=21661&utm_source=etip&utm_medium=email&utm_content=Should'). Yes- almost always! And this articles explains why.
- The document character set (Unicode) (URL='http://www.w3.org/International/O-charset.en.php').
- What are charsets all about? (URL='http://www.w3.org/International/questions/qa-doc-charset')
- (URL='http://www.w3.org/International/tutorials/tutorial-char-enc/') How do I use them - a tutorial.

Links and Resources for HTML and XHTML(Validators, HTML Tidy, and the connections between HTML and XHTML)

HTML Tidy is the standard for XHTML validation. Originally developed at the w3c, it is now under the aegis of Source Forge, the huge open source organization. It will also turn HTML into XHTML, but it is obviously better to start with XHTML.

- HTML tidy service (URL="http://cgi.w3.org/cgi-bin/tidy"). HTML Tidy on-line (interactive). Fast and easy to use. You don't need to learn all the details of HTML Tidy to use this page. Also both file upload or URL versions (URL='http://validator.w3.org/').
- General information about HTMLTidy and its capabilities (URL='http://www.w3.org/People/Raggett/tidy/'). Although this page describes itself as 'somewhat dated', it has a lot of excellent information as well as a Link to Source Forge and current versions of HTML Tidy.
- HTML Tidy Project Page (URL="http://tidy.sourceforge.net/"). The actual HTML Tidy programs, which you may download if you wish to 'Tidy' your pages off-line. Also (URL="http://tidy.sourceforge.net/docs/faq.html"), questions and answers
- (URL="http://tidy.sourceforge.net/docs/quickref.html") Options Quick Reference. Options you may set for HTML Tidy at SourceForge. This is a huge site, with a library of Tidy that you can call in various languages, and tools for version management. Explains how to download your own copy of HTML Tidy and set the options (URL="http://www.w3.org/TR/xhtml1/#guidelines") HyperText Markup Language (Second Edition) — How to write HTML which is XHTML compatible so there is less to 'Tidy' up. There are other valuators
- (URL="http://www.htmlvalidator.com/php/onlinecheck.php?google=online+html+checker") HTML Validator Online Check
- On-line HTML Validator, here is a free simple validator (URL='http://www.html-validator.com/lite/')

- Mozilla's DevEdge (URL='http://devedge-temp.mozilla.org/toolbox/tools-validation/')
- HTML Kit (URL="http://www.chami.com/html-kit/"), customizable HTML editor with HTML Tidy and beyond for validation; free for personal use.
- Validator advice and common 'gotchas' (URL='http://www.alistapart.com/articles/betterliving/'), from AListApart
- Advice on validation and links to validators (URL='http://webtips.dan.info/validators.html'), many uscful links; explanation of linters vs validators.

XHTML Editors

- Links to 14 free XHTML editors (URL='http://webdesign.about.com/od/htmleditors/tp/aatp_frehtedwin.htm'). Brief description of each editor.
- Large parts of this page were composed on Evrsoft's excellent (URL='http://webdesign.about.com/gi/dynamic/offsite.htm?zi=1/XJ&sdn=webdesign&zu=http%3A%2F%2Fwww.evrsoft.com%2F1stpage2.shtml')
- NoteTab Lite (URL='http://webdesign.about.com/gi/dynamic/offsite.htm?zi=1/XJ&sdn=webdesign&zu=http%3A%2F%2Fwww.notetab.com%2Fntl.php')
- Amaya is the w3c's browser/editor (URL='http://www.w3.org/Amaya/')
- (URL="http://webdesign.about.com/od/editors/index.htm") for Web Development editors, image map tools, etc.
- Reviews of some. WYSISYG Editors (URL='http://webdesign.about.com/od/htmleditors/tp/aatpwyswindows.htm'). The list includes the usual commercial products - Dreamweaver, Adobe, etc. and links to them
- More links on editors (URL='http://webdesign.about.com/od/editors/index.htm')
- (URL='http://downloads.zdnet.com/search.aspx?kw=XHTML+editor') ZDNet's Download site is another source of editors. Some reviews. Editors range from free to expensive. Many have free trial or less fancy free version. Check ZDNet's general download site (URL='http://downloads.zdnet.com/')
- SourceForge, the huge open source software foundation (URL='http://sourceforge.net/softwaremap/trove_list.php?form_cat=63'). also maintains a list of editor projects (including editors for other languages.) Of course, everything is free, but the site is overwhelming.
- HyperTextBuilder2006 (URL='http://www.paksoft-productions.com/hb/hb1.asp'). Earlier versions of this freeware editor were highly reviewed. Includes tools for developing server-side PHP scripts

Link Checkers

- Link Valet (URL='http://www.htmlhelp.com/tools/valet/'). Free and easy to use. Color coded report makes it easy to find broken links, although some fancy 'error pages' don't get picked up as broken.
- w3c link checker (URL='http://validator.w3.org/checklink').
- ZDNet (URL='http://downloads.zdnet.com/search.aspx?kw=link+checker'). List is updated and includes prices, release date, and some reviews.

Browsers Including Issues on Standards Compliance

Because browsers vary in their support of standards, it is important to test on all browsers in common use and know where you may get into trouble.

- Why standards are important (URL='http://karlcore.com/articles/article.php?id=9') and

(URL="http://www.internetnews.com/dev-news/article.php/3422651").

- Why you need to design for all browsers (URL="http://www.w3schools.com/browsers/browsers_stats.asp") usage
- But beware that Netscape and Mozilla may be confounded; See also (URL="http://www.w3schools.com/browsers/browsers_mozilla.asp") Mozilla.
- Most recent version of Firefox (URL='http://www.mozilla.com/firefox/'), with access to other Mozilla products (URL='http://www.mozilla.org/download.html')
- Archive of old browsers (URL='http://browsers.evolt.org/'). This is really a one-stop place and is the easiest way to find old versions of Firefox. (URL='http://browsers.evolt.org/?mozilla/firefox'). Also has Bobby (for low vision viewers)
- Safari (URL='http://browsers.evolt.org/?safari/')
- Of course, all versions of IE (URL='http://browsers.evolt.org/?ie/')
- SillyDog (URL='http://sillydog.org/narchive/')
- Archive of old (and current) Netscape Navigator browsers for PC's, Macs, Linux, etc. Current Opera browser (URL='http://www.opera.com/'). Mobile version of browser also available here. Netscape archive of its browsers. Doctype switch and the difference between Quirks and Standard mode (URL='http://gutfeldt.ch/matthias/articles/doctypeswitch.html'). Also has useful table (URL='http://gutfeldt.ch/matthias/articles/doctypeswitch/table.html')
- Quirks mode for Mozilla browsers (URL='http://developer.mozilla.org/en/docs/Mozilla%27s_Quirks_Mode'). Clear description of why Quirks mode is needed and how it works
- How your page looks in Lynx (URL='http://www.delorie.com/web/lynxview.html')
- Major browsers (URL='http://css.nu/pointers/tools.html'), and their bugs (URL='http://css.nu/pointers/bugs.html')
- PNG and cross-browser problems (URL='http://www.alistapart.com/articles/pngopacity/'). PNG is a graphics format which is not supported in all browsers. This article tells you how to work around that.
- AListApart articles on browsers (URL='http://www.alistapart.com/topics/code/browsers/'). Many of these articles are on CSS and site design issues.
- Design for Firefox first and then IE (URL='http://webdesign.about.com/od/internetexplorer/a/aa082906.htm'), to make pages look good in both browsers. Don't tell them what browser to use (URL='http://webdesign.about.com/od/browsers/a/aa041006.htm?nl=1') and use the standards!

Other XHTML Links and Resources

The more general resources come before the specific ones. Please also check the section on accessibility rules and tools.

- In general www.w3c.org (URL="http://www.w3c.org/") has all the detailed documentation, and the sites below have easier introductions. Still, it's hard to underestimate the importance of the w3c site for documentation, status reports, and even some tutorials.
- About.com (URL="http://webdesign.about.com/")
- www.w3schools.com (URL="http://www.w3schools.com/") novice to intermediate tutorials
- (URL="http://www.wdvl.com/Authoring/Scripting/") has excellent introductory to intermediate tutorials.
- WebReference (URL="http://www.webreference.com/tools/"). Their Experts (URL="http://www.webreference.com/

experts/") links, including Dmitry's labs (URL="http://www.webreference.com/dlab/") with great design advice.

- The Mozilla Development Center has a variety of HTML tools at http://developer.mozilla.org/en/docs/Category:HTML:Tools (URL="http://developer.mozilla.org/en/docs/Category:HTML:Tools")
- Sidebars for developing in Firefox. Web Developer Extension (URL='http://chris-pederick.com/work/webdeveloper/') has a toolbar for Firefox, Flock, etc. Zvon (URL='http://zvon.org/Output/bar_list.html')
- Also tutorials, including on XML and CSS (URL="http://www.w3.org/MarkUp/2004/xhtml-faq") Questions
- Open Source Web Development - DevShed (URL="http://www.devshed.com/"), articles on XHTML,CSS, XML, Perl, etc. Tend to be more advanced and often looking for less main-stream effects.
- ΔΗΤΜΛ Σχριπτσ ανδ ΔΗΤΜΛ Τυτοριαλσ (URL="http://devedge-temp.mozilla.org/toolbox/tools/2001/tune-up/") Tune-Up Wizard, tests pages in Netscape and Firefox.

Issues in Site and Page Design Color and Graphics

- Information on Color and Graphics (URL='http://www.bergen.org/AAST/ComputerAnimation/CompAn_Graphix.html'). Good explanation of everything you might want to know about graphics - including bit depth, dithering, anti-aliasing etc.
- The RGB Monitors (URL='http://www.wdvl.com/Authoring/Graphics/Colour/'), WDVL explains how colors are specified in XHTML and why.
- The 6x6x6 Color Cube (URL='http://www.wdvl.com/Graphics/Colour/666.html'). What colors you can specify using that cube are browser safe, or non-dithering, colors (URL='http://www.lynda.com/hex! .asp'). This is the most famous site for browser safe colors, which you should always use browser safe colors to avoid dithering. You may view the colors arranged by hue or by value (similar intensity/impact.)
- Other pages with browser-safe colors may be found at Primeline (URL='http://www.primeshop.com/html/216colrs.htm') and (URL='http://www.cookwood.com/html4_4e/examples/appendices/colorcharthex.html')
- Guide on XHTML and CSS. WDVL has a site on graphics (URL='http://www.wdvl.com/Authoring/Graphics/').Links to free graphics, tools, resources etc. near the bottom of the page. Also some more advanced (URL='http://www.wdvl.com/Graphics/') graphics
- About.com (URL='http://webdesign.about.com/od/colorcharts/l/bl_colors.htm')
- Cloford.com (URL='http://cloford.com/resources/index.htm') country codes.
- Dmitry's Color Lab (URL='http://www.webreference.com/dlab/9704/index.html'), excellent advice and information about color and design; several pages
- Optimizing Web Graphics (URL='http://www.webreference.com/dev/graphics/index.html'). A bit dated, but many wonderful tools and links, and no one minds a fast download.
- Understanding Color and Accessibility (URL='http://evolt.org/node/60472') from evolt.org .
- Color Wheel and color theory (URL='http://webdesign.about.com/cs/color/a/aacolortheory.htm'). Many useful links to tools and other resources.
- Tools for re-sizing images (URL='http://downloads.zdnet.com/search.aspx?kw=image+resizer')
- Sources of free images and icons

- HTML Writers' Guild (URL='http://www.hwg.org/resources/?cid=25') buttons, rules, etc.
- Realm Graphics (URL='http://www.ender-design.com/rg/icons.html')
- Clip-art.com
- (URL='http://www.free-graphics.com/'), Graphic Element samples
- (URL='http://www.lirmm.fr/bib-icons/Stanford/') basic but useful icons
- Ψαηοο эσ λιστ οφ ιχον χολλεχτιονσ- Always a good place to start
- LauraMcCanna'sFreeArtPage(URL='http://www.mccannas.com/free/freeart.htm') similar feel
- Icon Bazaar (URL='http://www.iconbazaar.com/')
- Index of bullets (URL='http://www.nbcs.rutgers.edu/Images/bullets/').
- Free Stuff Center (URL='http://www.freestuffcenter.com/sub/cliparttop.html') collections
- Webshots Photos (URL='http://www.webshots.com/homepage.html') very well categorized; has several of everything you can think of; may also use it them as screen savers.
- GraphixKingdom (URL='http://www.graphxkingdom.com/shtml/compl.shtml') art by category.
- Icon Browser (URL='http://www.ibiblio.org/gio/iconbrowser/') categorized, so time-consuming to use.
- Barry's Clip Art (URL='http://www.barrysclipart.com/')

Page Design–Including Usability Books

- Don't Make Me Think: A Common Sense Approach to Web Usability 2nd Edition by Steve Krug 2005.Good place to start if you've never thought about navigation etc.
- Designing Interfaces by Jenifer Tidwell published by O'Reilly 2005.From author's experience designing interfaces for the Math Works; very sophisticated; will be most appreciated by tose with a visual deign sense.
- The Non-Designer's Design Book by Robin Willams published by Pechpit Press 1994. I like this outstanding book even better than her Non-Designer's Web Book; basic principles on layout, mixing fonts etc.
- Short and wonderful Web Style Guide: Basic Design Principles for Creating Web Sites; Second Edition by Patrick J. Lynch and Sarah Horton; Yale University Press 2002. Classic guide for graphic designers; also available on- line (see below).

The Basics of Page Layout

- Basic (which is a good place to start) links at the bottom to other information (URL='http://webdesign.about.com/od/layout/a/aa062104.htm').
- Usability checklist for forms (URL='http://www.alistapart.com/articles/sensible-forms'), good advice!
- Useable forms for an international audience (URL='http://evolt.org/node/15118'). From evolt.org
- Checkboxes vs. Radio Button (URL='http://www.useit.com/alertbox/20040927.html')
- Article by Jakob Nielsen, the great guru of usability. His web site (URL='http://www.useit.com/alertbox/') including
- The Top 10 mistakes of web design (URL='http://www.useit.com/alertbox/9605.html') the all-important Usability 101. Also annual lists of best and worst web sites.
- Articles on writing for the web (URL='http://www.alistapart.com/topics/content/writing/') from AListApart
- On-line writing style (URL='http://www.webstyleguide.com/style/online-style.html') from Lynch and Horton's classic book

- (URL='http://developer.mozilla.org/en/docs/Tips_for_Authoring_Fast-loading_HTML_Pages'). Tips for fast-loading pages; useful for large pages/sites; from Mozilla Slash
- Getting URLs right (URL='http://www.mattkruse.com/info/slash/'), and (URL='http://www.useit.com/alertbox/reading_pattern.html')
- From Jakob Nielsen (URL='http://webdesign.about.com/od/authoring/a/aa111698.htm') checklist for creating web pages
- (URL='http://webdesign.about.com/od/usability/a/aa051506.htm') The all important back button — Links near the bottom lead to more articles about navigation
- Writing Effective Links (URL='http://evolt.org/node/60343')
- (URL='http://www.useit.com/alertbox/screen_resolution.html') layout
- Usability.gov (URL='http://www.usability.gov/'). Many articles, templates and lots of good advice about usability. Includes information about the process (URL='http://www.usability.gov/process.html') as well as on planning, writing, programming and usability testing. A treasure trove of information. Please also see tools for accessibility below.

Site Design Books and Articles

- Information Architecture for the World Wide Web 3rd Edition by Peter Morville, Louis Rosenfeld, 2006.For designing very large sites; very current - tagging etc.
- Art and the Zen of Web Sites (URL='http://www.tlc-systems.com/webtips.shtml'). Humorous but very on-target advice.
- Dmitry's Design Lab (URL='http://www.webreference.com/dlab/'). Great advice on site design, navigation, etc. Great place to start
- Web Style Guide (URL='http://www.webstyleguide.com/index.html?/pages/font_face.html').On-line version of Horton and Lynch's classic book. Good advice on everything from typography to navigation and on to site design. A complete course! Also further references (URL='http://www.webstyleguide.com/style/refs.html')
- Navigation advice (URL='http://evolt.org/navigation'), from evolt.org.
- Where Am I? (URL='http://www.alistapart.com/articles/whereami'), good advice on navigation.
- What makes for a good site? (URL='http://evolt.org/node/60371')
- (URL='http://webdesign.about.com/od/webdesignbasics/tp/aa122101a.htm?nl=1') How to drive your readers away
- (URL='http://www.alistapart.com/topics/design') design from AListApart (site oriented to graphic artists).
- (URL="http://webdesign.about.com/cs/contentmgmt/a/aa051704.htm") - What you Need to Know
- Choosing a CMS (Content Management System): needed for large or active sites by Drupal (URL='http://drupal.org/about').
- David Mercer's book (URL='http://www.amazon.com/gp/product/1904811809/ref=pd_cp_b_title/002-1921142-8541656')
- Typo3 (URL='http://typo3.com/')

Tools for Accessibility (e.g. for Color-Blind and Low-Vision Users)

- Web Accessibility Initiative (URL="http://www.w3c.org/WAI/"). W3c's links to everything you could need to know about accessibility problems on-line and how to address them. The w3c.org also has a (URL='http://www.w3.org/WAI/wcag-curric/overint.htm') Content Accessibility Curriculum.

Color Deficient Vision (URL="http://www.visibone.com/colorblind/"). See how your site looks to someone who is color-blind with links (at bottom) to transforming tools.

- (URL="http://www.btplc.com/age_disability/technology/RandD/colours/index.htm) BTexact Technologies - Safe Web Colors and browser safe colors for color-blind and links to other references and tools. This is a good sized site and it is worth poking around here
- (URL='http://www.vischeck.com/examples/') How pictures look to someone who is color blind
- (URL="http://webdesign.about.com/od/accessibility/a/aa062804.htm") site will be accessible for the color-blind
- (URL="http://users.rcn.com/hwbingham/accessbl/dcsgmlug/tsld005.htm") aural style sheets. For pages which are read aloud.
- (URL="http://www.w3.org/Style/CSS/Speech/NOTE-ACSS") aural style sheets.
- From w3.org, Web Standards (URL="http://www.webstandards.org/"). Webstandards.org is devoted to affordable accessibility for all. Through their site you may obtain the British guide to accessible sites
- (URL="http://www.w3.org/TR/1999/WAI-WEBCONTENT-19990505/#gl-table-markup") accessible tables from the w3c.org
- How to create tables (for tabular information, not for layout) which work for Braille readers, cell phones, etc. (URL="http://webdesign.about.com/od/accessibility/a/aa062804.htm").
- Designing Web Pages that are Color-Blind Friendly. Rules so your site will be accessible for the color-blind Web Exact (URL="http://webxact.watchfire.com/"). This is a free service which validates for conformance to accessibility rules.
- Lynx Viewer (URL="http://www.delorie.com/web/lynxview.html"). Lynx is a text-only browser.

Search Engines and Getting Found

- (URL="http://www.searchengines.com/"). Largest site on search engines; newsletters etc.
- How to get found and news (URL="http://www.internetnews.com/dev-news/article.php/3353971") What's the Difference? Google vs. Yahoo vs. Ask
- Goodies to Go! (URL="http://www.htmlgoodies.com/letters/286.html")
- (URL="http://searchenginewatch.com/showPage.html?page=2168031") How Engines Work. Several pages, including good information you'll want to know on how they rank pages
- (URL="http://searchenginewatch.com/links/article.php/2156221") Engines and Directories. Major search engines and how to get listed
- (URL="http://webdesign.about.com/cs/metatags/a/aa101501a.htm") Meta Tags. Uses for meta-tags, including general and internal search engines
- (URL="http://builder.com.com/5100-6371_14-5231995.html?tag=sc") links and link text on search engine placement.

CSS or Cascading Style Sheets Books and Documentation for CSS

Please note that many sites are a combination of tutorials, articles, documentation, examples and information about tools. All these are found in the Links and Resources for CSS section, so be sure to check out that section too. Also, some browsers (IE7) do not fully supportCSS2; please check CSS in Different Browsers section for details. Books, in addition to CSS coverage in XHTML books and DOM coverage in JavaScript books.

- Cascading Style Sheets: The Definitive Guide 3rd Edition by EricA. Meyer published by O'Reilly 2006. Excellent both to learn from and as a reference.

- The Zen of CSS Design: Visual Enlightenment for the Web The Zen of CSS Design: Visual Enlightenment for the Web 2005. Book version of the famous web site; lots of information
- Cascading Style Sheets 2.0 Programmer's Reference by Eric A. Meyer published by Osborne, 2001. Strictly a reference.
- W3C on CSS2 (URL="http://www.w3.org/TR/REC-CSS2/cover.html#minitoc"). See especially section 5 on selectors, 6 on assigning property values and the cascade and @import rule, section 4.3.6 on colors, section 7 on media types, and Appendices F and G for charts on descriptors and property values (including default values).
- Style Activity Statement (URL="http://www.w3.org/Style/Activity.html")
- Status of various w3c projects on style sheets - including CSS and XSL; see also Style Homepage (URL='http://www.w3.org/Style/')
- CSS Validator (URL='http://jigsaw.w3.org/css-validator/')
- (URL="http://www.w3.org/TR/xhtml1/#guidelines") HyperText Markup Language (Second Edition).Guidelines so your XHTML will work in existing HTML browsers.
- CSS Reference Table (URL="http://builder.com.com/5100-31-5071268.html").Very useful table of attributes and values; from Builder.com
- (URL="http://www.htmlhelp.com/reference/css/properties.html") Properties table, Only CSS1, but a good clear place for a beginner to start, as is the guide.
- (URL="http://www.htmlhelp.com/reference/css/structure.html") CSS rules, a short tutorial with links to more tutorials at the bottom. CSS1 only, but very clear.
- (URL="http://users.rcn.com/hwbingham/accessbl/dcsgmlug/tsld005.htm")Aural style sheets (pages which are read aloud) and (URL="http://www.w3.org/Style/CSS/Speech/NOTE-ACSS"), covers CSS1 and CSS2.
- CSS2 Reference from w3schools (URL='http://www.w3schools.com/css/css_reference.asp'), their tutorials are very useful, and so are their references.
- CSS: The Definitive Guide - 1st edition (URL='http://www.mykiss.de/index.htm'). This is indeed the definitive guide. The on-line version, which has less explanatory material than the print version, is also available through Safari Books at the ACM (URL='http://acm.org'). Discusses CSS1 thoroughly and looks ahead to CSS2.
- HTML Goodies' Reference (URL='http://www.htmlgoodies.com/beyond/css/css-ref/'). May cover only CSS1, but easy to use and fine until all common browsers implement CSS2 or you use CSS2's features.

Tutorials & Articles for CSS

- Learning CSS (URL="http://www.w3c.org/Style/CSS/learning"). There are books, tutorials, etc. on CSS from w3c.org
- (URL="http://www.w3c.org/Style/Examples/011/firstcss.en.html"). A first look at CSS from w3c.org; good place to start.
- CSS Tutorial (URL="http://www.w3schools.com/css/default.asp"), from w3schools.com; another excellent introduction.
- Dave Raggett's Introduction to CSS (URL="http://www.w3c.org/MarkUp/Guide/Style")
- (URL="http://www.richinstyle.com/guides/css2.html") tutorial
- (URL="http://www.htmlhelp.com/reference/css/style-html.html") sheets to your html page
- (URL="http://www.callihan.com/cssbook/csslinks.html"). Fine set of tutorials and links

- From AlsaCreations (URL='http://tutorials.alsacreations.com/'). Good place to go after one of the brief introductions above; sections on tables s layout and on menus.
- Tutorial from EchoEcho.com (URL="http://www.echoecho.com/css.htm"). Basic, clear. Many links on site: tools, forums, reference links.
- (URL="http://www.html.net/tutorials/css/introduction.asp") from html.net. Useful index on the left side; through intermediate level.
- MaxDesign's tutorials (URL="http://css.maxdesign.com.au/index.htm"). On lists, floats, etc. Start with the (URL="http://css.maxdesign.com.au/selectutorial/") Select-Tutorial. These tutorials are intermediate level.
- Quick Tutorial (URL="http://www.htmlhelp.com/reference/css/quick-tutorial.html") on CSS. From WDG and htmlhelp.com. May not include CSS2, but a fast start.
- Articles on how to link, classes vs id etc (URL='http://www.htmlhelp.com/reference/css/style-html.html')
- (URL="http://www.westciv.com/style_master/academy/css_tutorial/") tutorials. From beginning to advanced
- (URL="http://www.communitymx.com/content/article.cfm?cid=B6A50") this one is from CommunityMX . All sections are (URL="http://www.communitymx.com/content/article.cfm?cid=3B56F&print=true ") in part 6; site also has tutorials on Flash, Dreamweaver, etc.
- (URL="http://webdesign.about.com/library/style/bl_csstutorial.htm") Style Sheets (CSS) Tutorial About.com 's CSS tutorials, including table of contents to various lessons. Also links to libraries, (URL='http://webdesign.about.com/od/css/a/aastylelibrary.htm') templates, and (URL='http://webdesign.about.com/od/beginningcss/Beginning_Cascading_Style_Sheets_CSS.htm') CSS articles. They also offer more detailed (URL="http://webdesign.about.com/library/nosearch/bl_cssclass1-1.htm") Lessons (Change the 1-1 in the URL to 2-1, etc. to go to next lesson.) and many articles (You'll need to search for CSS, as the site keeps growing and has no hierarchical organization. Does have articles on specificity, CSS3, CSS browser support etc.)
- Tutorial from Zvon.com (URL='http://zvon.org/index.php?nav_id=tutorials&mime=html'), Also has links to tutorials on XML etc
- (URL="http://www.cameronolthuis.com/2006/04/top-10-css-tutorials/") CSS tutorials, with brief descriptions. None is elementary.
- (URL="http://www.w3.org/International/tutorials/css3-text/") internationalization and CSS3 from w3c.org
- (URL="http://www.htmlgoodies.com/beyond/css.html") Tutorials. Well written tutorials, very clear, but maybe dated (e.g. CSS1 instead of CSS2)
- (URL="http://www.wdvl.com/Authoring/Style/Sheets/Tutorial.html") Introduction to Style Sheets. Tutorial on CSS - good top-down approach, but only CSS1; good links at bottom of page, including "up one level" for many articles.
- (URL="http://brainjar.com/css/positioning/"). There are other tutorials here too - e.g. on DOM and on using CSS (URL="http://brainjar.com/css/using/")
- Lists (URL="http://www.alistapart.com/articles/taminglists/")
- CSS & Design Dev Tips (URL="http://builder.com.com/1200-6388-5220010.html").
- CSS and XSL: When to use which (URL="http://www.w3.org/Style/CSS-vs-XSL"). The w3c says 'Use CSS when you can, use XSL when you must.'
- (URL='http://www.elijournals.com/pre-

mier/showArticle.asp?aid=21661&utm_source=etip&utm_medium=email&utm_content=Should') Should you abandon tables for layout? Almost always and here's why

- (URL="http://friendlybit.com/css/inline-css-should-not-be-allowed-in-strict-doctypes/") Why you shouldn't use in-line CSS with a strict DOCTYPE. Friendlybit.com also has many articles on CSS, HTML and JavaScript, Including a (URL="http://friendlybit.com/css/beginners-guide-to-css-and-standards/") Beginner's Guide to CSS, and some simple templates (URL="http://friendlybit.com/css/simple-css-templates/"); (URL="http://friendlybit.com/css/beyond-the-web-with-css-princexml-and-s5/") CSS, PDFs, LaTeX and PrinceXML (Advanced).
- In-line styling in CSS2 (URL='http://meyerweb.com/eric/css/inline-format.html') By Eric Meyer (Advanced)
- (URL="http://devedge-temp.mozilla.org/viewsource/2003/devedge-redesign-css/index_en.xml") Case study of how Mozilla used CSS to redesign its DevEdge site
- (URL="http://brainjar.com/dhtml/menubar/") JavaScript to design a menu bar.
- The CSS Anarchist (URL='http://www.oreillynet.com/pub/a/network/2000/07/21/magazine/css_anarchist.html'). Older article on how to use CSS to wreck poorly coded sites and turn off blinking ads.

Links and Resources for CSS (Including Sample Layout Sources and Browser-Specific Hacks)

- W3C home page on CSS (URL="http://www.w3.org/Style/CSS/")
- Links to articles, tutorials (URL="http://www.w3.org/Style/CSS/learning") documentation, roadmaps of what's coming, news, etc.
- Learning CSS (URL="http://www.w3c.org/Style/CSS/learning")
- Books and articles on learning CSS — W3C Core Styles (URL="http://www.w3.org/StyleSheets/Core/"), 8 style sheets on w3c's server which you can link to from your pages.
- CSS Based Design (URL='http://adactio.com/articles/1109/'), By Jeremy Keith, author of the wonderful DOM Scripting book
- Many references on CSS: (URL="http://www.htmlhelp.com/reference/css/")
- From Web design Group at htmlhelp.com; also a link to (URL="http://www.htmlhelp.com/reference/css/references.html") references and a CSS Checker.
- U. of Minnesota - Duluth organizes everything (URL='http://www.d.umn.edu/itss/support/Training/Online/webdesign/css.html'). This well-organized site has tutorials on CSS, information on (literally) everything from headers to footers and everything in between. They also have links to their sites on JavaScript, XML etc.
- (URL="http://dmoz.org/Computers/Data_Formats/Style_Sheets/CSS/") - Computers: Data Formats: Style Sheets: CSS. Many links for HTML, XHTML and CSS.
- CSS Pointers (URL='http://css.nu/'), another large compendium of articles, FAQs, tools, etc. For the Pros.
- (URL="http://www.wdvl.com/Authoring/Style/Sheets/") Designing CSS Web Pages Tutorials and recommended ' Designing CSS Web Pages', 2-part article on knowing who your audience is and how to design for it. Also (URL="http://www.wdvl.com/Authoring/Style/Sheets/Resources.html") Resources at other sites
- All CSS articles (URL='http://www.wdvl.com/Authoring/Style/Sheets/').
- Sidebars on CSS etc. for developers (URL='http://devedge-temp.mozilla.org/

toolbox/sidebars/'). These tools disappeared from Netscape's DevEdge site, and Mozilla has brought them back into its archives (URL='http://devedge-temp.mozilla.org/central/css/index_en.html')

- CSS Validator (URL="http://jigsaw.w3.org/css-validator/")
- SelectOracle (URL='http://gallery.theopalgroup.com/selectoracle/'), tells you what a complex selector does.
- Site for Eric Meyer (URL="http://meyerweb.com/"). He wrote the books on CSS (or at least two of the best). If you're going to be a power user it's worth checking his blog. You can also subscribe to the CSS-discuss (URL="http://www.css-discuss.org/") mailing list and check out his cutting edge site. (URL='http://meyerweb.com/eric/css/edge/')
- CSS Zen Garden (URL="http://www.csszengarden.com/"). This is the most amazing site of examples of what you can do with CSS. One page is transformed by over 900 style sheets. (Check the (URL="http://www.mezzoblue.com/zengarden/alldesigns/") archives.). These are all done by professional graphic artists, but we can all admire them. They also have an excellent list of resources.
- (URL="http://placenamehere.com/neuralustmirror/200202/") many stylesheets for one page, but I prefer the ZenGarden.
- (URL="http://www.dezwozhere.com/links.html") resources is at Holy CSS Zeldman! In addition to many CSS links, there are also links on JavaScript and AJAX. It seems to me that you can find everything here: tutorials, books, sites, galleries at other sites (e.g. CSS Zen Garden), hints, etc. www.friendlybit.com has many articles on CSS, HTML and JavaScript, including a (URL="http://friendlybit.com/css/beginners-guide-to-css-and-standards/") Beginner's Guide to CSS, and some simple templates (URL="http://friendlybit.com/css/simple-css-templates/")
- Advanced CSS Resources (URL='http://www.blooberry.com/indexdot/css/index4.htm'). Lots of good information, including properties, a clear explanation of the cascade, and tables on browser support. You may need to click through lots of pages to get to what you want.
- CSS Vista (URL="http://www.sitevista.com/cssvista/")
- Tool to edit your CSS in IE and Firefox at the same time. Demo tool is free. CSS Tools (URL="http://www.soxiam.com/Notes/CSSTools")
- Links to a variety of CSS tools and galleries. Since 6/06 this has been put on a wiki (URL="http://www.soxiam.com/Code/HomePage") CSS (URL="http://www.soxiam.com/Tags/Css")
- (URL="http://tecfa.unige.ch/formcont/ecole-ete-tunis2001/links/css.html") CSS Documentation and articles. Many topics and useful resources. Covers CSS1 and CSS2.
- (URL="http://homepage.mac.com/chrispage/iblog/C42511381/E20060806095030/index.html") A trick/tool for debugging CSS.
- Examples of well-designed sites and CSS Resources (URL='http://www.meryl.net/css/cat_css_resources.php'); many classic resources are here - e.g. CSS Zen Garden, w3c, etc. and many template galleries.
- New York Public Library Style Guide (URL='http://www.nypl.org/styleguide/')
- CSS Pointers Authoring Tools (URL='http://css.nu/pointers/tools.html'); also has information about browsers and XSL tools.
- Codestyle.org (URL='http://www.codestyle.org/index.shtml'). A site for developers; be sure to check their XHTML page too, as its FAQs include CSS questions. Sample layouts:

- (URL="http://www.code-sucks.com/css%20layouts/") columns using CSS. Over 50 layouts. Choose the number of columns you want, then the particular layout, and then download the code.
- Templates for lists from Max Designs (URL="http://css.maxdesign.com.au/"). Builds the layouts step-by-step, tutorial style.
- Layouts from Eric Costello (URL="http://www.glish.com/css/") A half dozen classic layouts with code, and links to articles.
- Layout Gala (URL='http://blog.html.it/layoutgala/'). 40 different layouts you can download individually or all zipped together. Also links to articles on using negative margins and other techniques.
- Free templates (URL="http://www.six-shootermedia.com/free-templates/"). A dozen elegant designs you can download
- CSS Layouts (URL='http://css-discuss.incutio.com/?page=CssLayouts'). On-going blog about types of layouts, problems and hacks. Not elementary.
- A few basic layouts with explanations (URL='http://www.cssplay.co.uk/layouts/index.html'), from the always elegant CSS Play.
- Searchable repository of layouts (URL='http://tools.i-use.it/'). Fill in form on number of columns you want, if you want a header or not etc. and template is returned. Very useful
- pMachines templates (URL='http://pm-template.kartar.net/templates.html'), half a dozen basic templates. Display un-styled in older browsers
- CSS Showcase (URL='http://www.alvit.de/css-showcase/'), gallery of very professional looking menus, tabs and layouts; also articles
- CSS Drive (URL='http://www.cssdrive.com/'), reviewed and unreviewed designs. You need to work a bit to get to the code.

CSS in Different Browsers

Please also see Browsers section under XHTML Resources for information on usage and places to download.

- All browsers (URL="http://centricle.com/ref/css/filters/") support which CSS features; very useful
- (URL="http://webdesign.about.com/library/weekly/aa123101a.htm") support in CSS and how to Display Stylesheets Dynamically. What browsers support what features and links on work-arounds for older browsers
- (URL="http://css-discuss.incutio.com/?page=CssHack") Hacks for various browsers and CSS. This enormous list of hacks also has (near the top of the page) links to sites which summarize hacks. It also lists various specific (recent) hacks with their effects and the browsers they target.
- The Box Model Hack (URL="http://css-discuss.incutio.com/?page=BoxModel")
- One of the oldest and most famous hacks. Links to Brain Jar's (URL='http://www.brainjar.com/css/positioning/default.asp') clear explanation, and others. IE7 and CSS2 (URL="http://www.publish.com/article2/0,1895,1842891,00.asp"). Of course, Microsoft didn't go with the standard immediately.
- CSS in Different Browsers (URL='http://www.webreference.com/authoring/style/sheets/browser_support/index.html'). Long article (current as of 2006) discussing the issue.
- No tabular summary, but links to those for current browsers at Webdevout (URL='http://www.webdevout.net/') and for 2003 and older browsers at Blooberry (URL='http://www.blooberry.com/indexdot/css/supportkey/syntax.htm')

- Devout.net (URL='http://www.webdevout.net/'). Browser support for CSS and CSS hacks.
- Fonts commonly found in various browser (URL='http://www.upsdell.com/BrowserNews/res_fonts.htm'). Useful chart - so your choices are more likely to not default to the browser's choices.
- Survey of font usage by browser (URL='http://www.codestyle.org/css/font-family/index.shtml')
- (URL='http://www.codestyle.org/css/font-family/sampler-CombinedResults.shtml'). Equivalent fonts for PCs and Macs (URL='http://www.ampsoft.net/webdesign-l/WindowsMacFonts.html'). Nice chart makes it easy to choose
- CSS2 Generated Content (URL="http://webdesign.about.com/cs/css/a/aa042604.htm"): before and :after; does not work in IE6 or IE7.

JavaScript

Books and Documentation for JavaScript.

- Teach Yourself JavaScript in 24 Hours, 4th edition by Michael Moncur published by SAMS, 2006. While the first 3 editions were clear (but for novices), this edition is clear but goes further. Highly recommended if you've never used JavaScript.
- JavaScript: The Complete Reference, 2nd Edition by Thomas Powell and Fritz Schneider published by Osborne 2004. Not quite as complete as the Bible(directly below), but easier to read. I use both.
- The JavaScript Bible 6th Edition by Danny Goodman and Michael Morrison published by Wiley 2007. An invaluable reference; very clear on even abstruse items; earlier edition was very clear on the DOM, including coverage on non standards compliant browsers.
- JavaScript: The Definitive Guide 5th Edition by David Flanagan, published by O'Reilly 2006.Another excellent book; new edition includes Ajax.
- DOM Scripting: Web Design with JavaScript and the Document Object Model by Jeremy Keith published by APress 2005. A breathtakingly wonderful description of the DOM and how to make it work for you; also builds the basis for Ajax.
- (URL="http://developer.mozilla.org/en/docs/JavaScript_Language_Resources") resources for JavaScript. Not for beginners.
- (URL="http://developer.mozilla.org/en/docs/Core_JavaScript_1.5_Guide") Core JavaScript 1.5 Guide. Description of the language (if you already know it.), and (URL="http://developer.mozilla.org/en/docs/Core_JavaScript_1.5_Reference") Core JavaScript 1.5 Reference. Contains information about what is implemented in various versions. Also contains information about LiveConnect, the way to link Java and JavaScript code.
- (URL="http://developer.mozilla.org/en/docs/New_in_JavaScript_1.6") from JavaScript 1.5 to 1.6.
- (URL="http://developer.mozilla.org/en/docs/JavaScript_Language_Resources") page for all Mozilla's documentation, forums and articles on JavaScript.
- The w3c documentation for the DOM (URL='http://www.w3.org/DOM/')
- The DOM model (URL="http://www.ws-abstract.com/domref/index.shtml")
- (URL="http://www.javascriptkit.com/jsref/index.shtml") JavaScript From JavaScript Kit
- (URL="http://www.w3.org/TR/DOM-Level-2-Core/Overview.html#contents") DOM Documentation for ECMA Script (basis for JavaScript), **for older versions of JavaScript.**

- (URL="http://devedge-temp.mozilla.org/central/javascript/index_en.html") (Guides and References for 1.3, 1.4 and 1.5). Documentation for JavaScript 1.3 - 1.5 with information on changes from earlier versions, links to both the Guide and the Reference Manual, dDate object, documentation on the date object, including constructors and methods.

Tutorials and Articles for JavaScript

- (URL="http://www.htmlgoodies.com/primers/jsp/") tutorials on JavaScript. First the 'Primers' and then the 'Basics' (which go thru AJAX)
- (URL="http://developer.mozilla.org/en/docs/A_re-introduction_to_JavaScript") "Reintroduction to JavaScript", friendlier than Mozilla's documentation
- (URL="http://developer.mozilla.org/en/docs/Category:JavaScript") JavaScript, from Mozilla's forums.
- Javascript tutorial from w3schools.com (URL='http://www.w3schools.com/js/default.asp'), always a good place to start
- (URL="http://www.webreference.com/programming/javascript/index.html") Articles from WebReference. Many tutorials from 'JavaScript for non-programmers' to AJAX.
- (URL="http://www.webreference.com/programming/javascript/jf/column12/") 12-15 inclusive). Excellent resource; includes links to Mozilla pages just above and other collections. Be sure to check out the collections and the archives. Some tutorials are older.
- (URL="http://www.wdvl.com/Authoring/JavaScript/Tutorial/") JavaScript for Programmers from WDVL
- (URL="http://www.ems.psu.edu/~young/meteo473/lecture_javascript.htm") Description of JavaScript for students who have programmed in C++ or Java. Older, but very useful and to the point.
- (URL="http://www.webdevelopersjournal.com/JavaScriptWeenie.html") scripts from JavaScript Weenie; Intermediate level and useful.
- Tutorials and references from U. Minn- Duluth (URL='http://www.d.umn.edu/itss/support/Training/Online/webdesign/javascript.html'). Enormous list of tutorials and articles from the general to the specific; includes some links to book excerpts; regrettably, no description of the links.
- Tutorials from WebMonkey (URL='http://www.webmonkey.com/webmonkey/programming/javascript/tutorials/tutorial2.html'). Clear but somewhat dated (Browser detection is now usually not needed.).
- JavaScript Primers (URL='http://www.htmlgoodies.com/primers/jsp/') 30 steps to go from complete tyro to intermediate level.
- Tutorials from WebKnowHow (URL='http://www.webknowhow.net/dir/JavaScript/Tutorials/index.html'). About 20 tutorials, mostly for the beginner.
- Programmer How-To's (URL="http://html.tucows.com/programmer/programmertutmain.html"). Short introductory tutorials from TUCOWs. Good for those who can program in another language and need to switch to JavaScript.

Articles on Specific Topics (Not Advanced)

- Focus on forms (URL="http://www.htmlgoodies.com/letters/287.html"), the focus() and blur() methods.
- (URL="http://www.mredkj.com/tutorials/reference_js_intro.html") Variables, clearly described from mredkj.com, other interesting examples (URL="http://www.mredkj.com/tutorials/htmljs.html") - e.g. to change tables, validate forms, etc.
- (URL="http://devedge-temp.mozilla.org/

library/manuals/2000/javascript/1.5/guide/ident.html#1009571") characters, escaping them and Unicode. From Mozilla's Core JavaScript 1.5 Guide

- (URL="http://htmlgoodies.earthweb.com/beyond/jspass.html") Passing variables between pages in JavaScript.
- (URL="http://html.tucows.com/programmer/jstut/js_i_globalLocal.html") Local. Global and local variables - from Tucows; elementary but reliable
- Navigation and JavaScript (URL='http://evolt.org/javascript_navigation_cleaner_not_meaner') Fancy footwork for lists of links.
- (URL="http://javascript.internet.com/") Tutorials, Example Code, Reference, Resources, and Help. This site has tutorials, public scripts, book reviews, etc.
- (URL="http://www.javascript.com/") Resource: JavaScript Tutorials and Free Java Scripts. Tutorials and fancy scripts. Pretty good.
- (URL="http://html.tucows.com/programmer/jstut/js_i_images.html") JavaScript Intermediate Tutorial: Image Swapping. Tutorial on image swapping with links to related (and less related) topics.
- JavaScriptWeenie (URL="http://www.webdevelopersjournal.com/JavaScriptWeenie.html") An older site with many articles, tutorials, most of which are relevant for older browsers.
- Time in JavaScript (URL='http://www.htmlgoodies.com/introduction/newsletter_archive/goodiestogo/article.php/3600841'). Goodies to Go newsletter show you how to set timeouts, etc.

Articles on Specific Topics (Advanced, DOM, DHTML, the Event Model, Cookies, etc.)

- The w3c documentation for the DOM (URL='http://www.w3.org/DOM/')
- What is the DOM? (URL='http://developer.mozilla.org/en/docs/Gecko_DOM_Reference:Introduction#What_is_the_DOM.3F'). Clear explanation from Mozilla
- (URL="http://www.devarticles.com/c/a/JavaScript/JavaScript-Remote-Scripting-Fetching-Server-Data-with-the-DOM") Fetching Remote Data (in XML) from a Server with DOM – advanced.
- DOM Scripting (URL='http://adactio.com/articles/search?query=DOM+Scripting'). Articles by Jeremy Keith, author of the DOM Scripting book and extremely readable authority on the DOM
- (URL="http://www.devarticles.com/c/a/JavaScript/Preloading-Images-with-the-DOM-The-Introductory-Process/") Preloading Images with DOM – advanced.
- The DOM (URL="http://www.prairienet.org/~sjmccaug/dom.htm"). Older (1999) article explaining the JavaScript prototype to Java programmers.
- (URL='http://icant.co.uk/articles/from-dhtml-to-dom/FromDHTMLtoDOMscripting_March2006.pdf') From DHTML to DOM Scripting. Excellent, thorough article from DHTML to DOM.
- (URL='http://www.htmlgoodies.com/beyond/javascript/article.php/3627691').Why you should switch, part 1; (URL='http://www.htmlgoodies.com/beyond/javascript/article.php/3630271') part 2.
- Digital Web Magazine (URL='http://www.digital-web.com/topics/dom/'). Also book reviews, product reviews, interviews etc.
- (URL="http://www.dynamicdrive.com/dynamicindex4/preloadimage.htm") DHTML Scripts- Preload Image (with progress bar) Script. Pre-loading images for swapping; part of huge site with JS and DHTML scripts, many very in your face, and forums. Other topics.
- The JavaScript Event Model (URL="http://www.webmonkey.com/05/02/index4a.

html")–advanced. From WebMonkey. Tells you all you could ever want to know (and maybe more) about event handler models. Although the article is from Jan 2005, there is a lot of time devoted to Netscape Navigator 4 and IE 4 - hopefully for only a little while longer. Useful for old pages. New pages should stick to the w3c Event Model. Exception Handling in JavaScript. Advanced. Includes information about browser incompatibilities

- (URL="http://builder.com.com/1200-31-5084860.html") Builder.com - many advanced.
- Προπερλψ υσιvγ ΧΣΣ ανδ ϑαϖαΣχριπτ ωιτη ΞΗΤΜΛ with examples (URL="http://ciac.llnl.gov/ciac/bulletins/i-034.shtml") and how to set them, from the Department of Energy.
- (URL="http://computer.howstuffworks.com/cookie1.htm") from HowStuffWorks; Be sure to see next page too.
- DevGuru tutorials (URL='http://www.devguru.com/features/tutorials/tutorials.asp') Very advanced - e.g. interfacing JavaScript with ASP.
- Case Study of how Mozilla redesigned its site (URL='http://devedge-temp.mozilla.org/viewsource/2003/devedge-redesign-js/index_en.html'), Shows both pure CSS and hybrid CSS/JavaScript approaches. Older and sophisticated but still worthwhile

Links and Resources for JavaScript

The easiest debugger to use is the JavaScript Console (available in Firefox, Netscape Navigator and Opera). Most of these references concern more advanced tools.

- (URL="http://dojo.jot.com/DebuggingJavascript") debugging tools for all browsers, using the JavaScript Console.
- (URL="http://www.webmonkey.com/06/19/index3a.html") Good introduction from WebMonkey.
- Mozilla's tools (URL="http://www.mozilla.org/docs/web-developer/js/debugging/"). Including the Console and Venkman and the DOM inspector; add-on and news on these tools (URL="http://www.mozilla.org/js/jsd/")
- (URL="http://www.webreference.com/programming/javascript/venkman/") Venkman from WebReference, and (URL="http://www.svendtofte.com/code/learning_venkman/") list of resources on it. Looks very valuable.
- (URL="http://www.csie.ntu.edu.tw/~piaip/docs/CreateMozApp/mozilla-app-b-sect-5.html") tutorial on Venkman
- (URL="http://www.hotscripts.com/Detailed/61484.html") Hot Scripts
- (URL="http://freshmeat.net/projects/dbgcons/") debugger - Microsoft's Script Debugger
- (URL="http://www.digitalmediaminute.com/article/1622/javascript-trace-window") Trace Window (uses GreaseMonkey in Firefox).
- JSLint (URL="http://www.jslint.com/") Looks for problems in JavaScript pages; check out the documentation first; from 2002 so may not support DOM.

Other JavaScript (Especially Tools and Sources of Scripts)

Frameworks are in the Ajax section of the Web 2.0 resources

- (URL="http://www.siteexperts.com/") Site Experts. Another community of web developers
- (URL="http://www.experts-exchange.com/Web/Web_Languages/JavaScript/") Exchange - another community

- (URL="http://www.webmonkey.com/webmonkey/reference/javascript_code_library/") Library of JavaScript functions from WebMonkey
- Dev articles on JavaScript (URL="http://www.devarticles.com/c/b/JavaScript/"). Enormous compendium of articles, many advanced
- Scripts may be found here (URL="http://www.scripts.com/javascript-scripts/") by category.
- JavaScript.com (URL="http://www.javascript.com/"). Free scripts - some a bit cutesy; free newsletter.
- Mozilla's tools (URL="http://www.mozilla.org/js/"). Mozilla has a large amount of information and tools, including the debuggers, etc. Their DOM Central (URL='http://devedge-temp.mozilla.org/central/dom/index_en.html') tools (URL='http://devedge-temp.mozilla.org/toolbox/tools/') Also Sidebars for CSS, JavaScript and XSLT (URL='http://devedge-temp.mozilla.org/toolbox/sidebars/') Mozilla's DOM Inspector (URL='http://www.mozilla.org/projects/inspector/')
- SourceForge Projects on Text Editors (URL='http://sourceforge.net/softwaremap/trove_list.php?form_cat=63'). SourceForge material is all open source. FCKEditor is here, for example.
- BrainJar (URL="http://www.brainjar.com/"). This is the site that brought you their (URL='http://brainjar.com/dhtml/domviewer/') DOM viewer; tutorials and tools, small but worthwhile.
- DevGuru Reference (URL='http://www.devguru.com/technologies/JavaScript/home.asp'). Bore down through the menus to get a clear explanation of various elements of JavaScript.
- comp.lang.javascript (URL='http://groups.google.com/group/comp.lang.javascript'). Google group still going strong.
- DynamicDrive (URL="http://www.dynamicdrive.com/"). Tools and scripts and links; focus on DHTML; threads and posts on their forums (URL='http://www.dynamicdrive.com/forums/') good place to look for answers when nothing else helps.
- JavaScriptKit (URL="http://www.javascriptkit.com/"). Tools, tutorials, scripts; quite up-to-date, including material on AJAX etc.
- JavaScript.internet.com (URL='http://javascript.internet.com/'). Large repository of scripts, tutorials, etc. Check out the FAQs first. Links to Web Reference, XML.files, and WDVL (Web Developer's Virtual Library).
- Scripts.com's (URL='http://www.scripts.com/javascript-scripts/'). Enormous repository of JavaScript scripts organized by type, also searchable. Scripts are rated by users. Most are free.
- DHTML Shock (URL="http://www.dhtmlshock.com/"). Scripts with a focus on DHTML.
- Cross-browser (URL="http://www.cross-browser.com/"). Scripts and articles on cross-browser JS and DHTML using the DOM.
- Builder.com articles on JavaScript (URL='http://builder.com.com/1200-31-5084860.html'). Not organized, but is searchable. Many quite advanced.
- Web Reference on JavaScript (URL='http://www.webreference.com/programming/javascript/'). Searchable archive of articles and tutorials from elementary to advanced intermediate.
- AListApartarticlesonscripting(URL='http://www.alistapart.com/topics/code/scripting/'). Focus is on implementing a smooth professional appearance
- WebStandards.org (URL="http://www.webstandards.org/"). A self-described grass-

roots organization trying to advocate for Web standards which ensure accessibility.

- WebDeveloper (URL="http://www.webdeveloper.com/"). Hosts many forums, including one on JavaScript. Also has a searchable archive of articles on JavaScript (URL='http://webdeveloper.earthweb.com/webjs/') which returns too many irrelevant articles.
- JavaScriptCity (URL='http://www.javascriptcity.com/other/webres.htm'). Tutorials, references etc. Appears to be geared toward the novice.
- (URL="http://www.sitepoint.com/forums/showthread.php?s=462aa45700774e13a666b693037bd9b6&t=90580") Sitepoint, has a forum with many useful links, as do the publishers.
- WROX. (URL="http://p2p.wrox.com/forum.asp?FORUM_ID=21") 2programmer)
- APress (URL="http://www.beginningjavascript.com/")
- CodingForum, Tek-Tips; (URL="http://www.tek-tips.com/threadminder.cfm?pid=216")
- Evolt (URL="http://lists.evolt.org/").
- Yahoo maintains links to similar sites. (URL="http://developer.yahoo.com/javascript/") My experience is that similar Yahoo lists have a goodly part of the top resources.
- The FAQts on JavaScript. (URL="http://www.faqts.com/knowledge_base/index.phtml/fid/53/") . Huge site with answers to many JS questions, by subject. Site does not work in Netscape, but does in Firefox and IE; ironic?
- DevShed (URL='http://www.devshed.com/c/b/JavaScript/'). Large collection of articles, scripts for web developers. Selections for Python, PHP etc. are more recent than the few on JavaScript.

CGI and Server-Side Scripting

General (Including Information on HTTP and TCP/IP and resources for multiple languages)

- HTTP (URL='http://ftp.ics.uci.edu/pub/ietf/http/rfc1945.html') Older (1996) but very useful explanation; well organized and probably has all you need to know and then some.
- TCP/IP Resources (URL='http://www.private.org.il/tcpip_rl.html'). Great list of FAQs, tutorials, and books. Kept up to date.
- Overview of TCP/IP and the Internet (URL='http://www.garykessler.net/library/tcpip.html'). Current description. Starts with the history (OK to skip) but gives all the details you need if you are not a network administrator.
- Primer on TCP/IP (URL='http://www.garykessler.net/library/rfc2151.html'). Older (1997), but a good place to start.
- Another older (1998) Primer on Internet technology (URL='http://www.anu.edu.au/people/Roger.Clarke/II/IPrimer.html')
- Daryl's TCP/IP Primer (URL='http://www.ipprimer.com/section.cfm'). Goes deeper than a primer; kept up to date.
- Older (1998) course on managing servers (URL='http://infomotions.com/musings/waves/handout/waves.txt'). Classic references; good description of client-server model, http, mime types etc.
- IANA (URL='http://www.iana.org/'). Who owns what domain name, which port is what, and what the language abbreviations are
- Which Scripting and Programming Languages are People Searching about? (URL='http://www.tiobe.com/tpci.htm'). Monthly survey on queries about Perl, PHP, JavaScript, Java etc.
- Developer.com (URL='http://www.developer.com/lang/'). Excellent very current

intermediate level articles, but you need to search. Better yet, subscribe to some of their free newsletters and be notified of new articles.

- ASPN (URL="http://aspn.activestate.com/ASPN/"). Resource for programmers in Perl, PHP, Python, etc. Major site.
- HotScripts (URL="http://www.hotscripts.com/"). Articles and scripts for many languages - Perl, PHP, Python, JavaScript, XML
- OnLamp.com (URL='http://www.onlamp.com/'). O'Reilly maintained site with articles and book excerpts on all the open source technologies - Linux Apache, Perl/PHP/Python, MySQL. Links to Safari books (also available free through the ACM is you are a member).
- CGI Resources (URL="http://www.cgi-resources.com"). Links to very many CGI scripts, mostly in Perl, and tutorials (rated by users).
- DevShed Perl Tutorials, Scripts, etc. (URL='http://www.devshed.com/c/b/Perl/'). Links to user submitted and rated scripts and articles in many languages; most free.
- WebReference (URL='http://www.webreference.com/programming/php/'). Intermediate to advanced articles. Includes the Mother of Perl site. Also PHP (with links at the bottom) and now starting on python (URL='http://www.webreference.com/programming/python/')
- Matt's HTTP Cookie Library (URL='http://www.scriptarchive.com/cookielib.html') and his CGI resource (URL='http://cgi.resourceindex.com/')

CGI (Including Regular Expressions - see also Perl section)

- CGI Programming with Perl by Gunther Birznieks, Scott Guelich, and Shishir Gundavaram 2nd edition publishd by O'Reilly 2000. Very useful book.
- Learning Perl 3rd edition by Randal L. Schwartz and Tom Phoenix, published by O'Reilly 2007. Excellent way to learn enough Perl to write CGI scripts easily.
- The Web Wizard's Guide to Perl and CGI, The Web Wizard's Guide to PHP, both by David A. Lash, 2002. These books won't make you a great guru, but they will get you going very quickly; well-written.
- Client and server (URL="http://www.html-goodies.com/letters/288.html"). Introduction to terms.
- (URL="http://www.wdvl.com/Authoring/Scripting/Tutorial/toc.html") Contents on Web Programming. Tutorial on server-side programming and sending info from HTML forms. Excellent introduction . Definitely the place to start if you don't know the difference between GET and POST or need a short introduction to Perl.
- These tutorials, from Selena Sol, are also available at Extropia (URL='http://www.extropia.com/tutorials/perl_cgi/pre_requisite_intro.html') which also has links to Perl tutorials for web developers.
- (URL="http://www.wdvl.com/Authoring/Scripting/Tutorial/request_headers.html") Request Headers. What's in an http header.
- What is CGI? (URL="http://www.devpapers.com/article/137"). Very elementary introduction.
- CGI Scripts - Writing and Using CGI (URL="http://webdesign.about.com/od/cgi/"). Another very elementary introduction to CGI; several short articles to read.
- Web Development Primer (URL="http://www.devpapers.com/article/47/"). Elementary introduction to the technologies you find server-side (ASP, Perl, PHP etc.).
- Gentle Introduction to CGI (URL='http://infomotions.com/musings/waves/gentlein-

troductiontocgi.html'). Also has many links to books and on-line resources. Updated in 2004

- The Web Development Environment (URL='http://www.extropia.com/tutorials/devenv/toc.html'). An excellent introduction; less cursory than the earlier ones; you will enjoy it more if you have already heard of some of the technologies mentioned (Java, HTTP, etc.) A shorter version (URL='http://www.extropia.com/development/web_ware_white_paper.html') is available for newbies. w3c.org
- Introduction to CGI (URL='http://www.w3.org/CGI/Overview.html')
- Other information from w3c about standards (URL='http://www.w3.org/CGI/')
- Why learn CGI? (URL="http://web.oreilly.com/news/cgi_0700.html"). From the people (O'Reilly) who publish a book on the subject - includes kind words about their own (excellent) book; this article is mainly motivational.

The next few articles will actually show you something you can modify and use; for more detailed information you need to go to the section on your language of choice (Perl, PHP or Python)

- CGI Developer's Guide (URL='http://www.webbasedprogramming.com/CGI-Developers-Guide/'). Older, but will get you going.
- CGI Tutorial (URL='http://www.comp.leeds.ac.uk/Perl/Cgi/start.html'). Good introduction to CGI, environment variables, etc.
- Easy introduction to CGI (URL='http://perl.about.com/gi/dynamic/offsite.htm?zi=1/XJ&sdn=perl&zu=http://www.jmarshall.com/easy/cgi/'). Another good short introduction
- CGI articles (URL='http://perl.about.com/od/cgiweb/') and more general CGI articles (URL='http://perl.about.com/od/cgiweb/'). There are many articles in the Perl and Web Design section, but you need to know what you want - e.g. guestbook - so you can search for it here.
- CGI Tutorials (URL="http://www.html-goodies.com/beyond/cgi.html"). Very well written tutorials, but maybe dated (1/2005). For example, you are more likely to FTP your server scripts to the server than to use telnet.
- (URL="http://www.acmqueue.com/modules.php?name=News&file=print&sid=361") Problem of Statelessness.
- Validating user sessions in PHP (URL='http://www.phpbuilder.com/columns/validating_php_user_sessions.php3?page=1'). Good introduction to the problem of statelessness; php implementation of solutions.
- Form validation (URL="http://www.html-goodies.com/letters/290.html"). Validating a form before sending it to a CGI script - mainly to be sure all fields are filled in.

The following links are to collections of articles

- WebKnowHow's Tutorials, articles, scripts and books on CGI (URL='http://www.webknowhow.net/dir/Other_Topics/HTTP/index.html'). Many (unsorted) links, many to old sites.
- Webliography on CGI, Perl, etc. (URL='http://infomotions.com/musings/waves/webliography.html'). Great list of links, even though many are to older articles. CGI articles from HTML Goodies. Includes codes for a guest book and for counters. Tutorials on regular expressions. Although some lean towards one language or another, regular expression syntax is pretty much the same in all the scripting languages.
- (URL="http://www.capescience.com/education/tutorials/cc5/cc5_regex_intro.pdf") on regular expressions

- LibraryofRegularExpressions(URL="http://www.regxlib.com/default.aspx")
- Regular expressions in JavaScript, (URL='http://ist.marshall.edu/ist263/js21.html#l2101') which has a regular expression finder (URL='http://ist.marshall.edu/ist263/examples/js_reg.html')
- (URL='http://gnosis.cx/publish/programming/regular_expressions.html') Regular expressions tutorial from a Python book; rather telegraphic
- Howtouseregularexpressions(URL='http://www.silverstones.com/thebat/Regex.html')
- The Perl 30-minute Regular Expressions Tutorial (URL='http://www.english.uga.edu/humcomp/perl/regex2a.html') >
- Regex tutorial (URL='http://www-static.cc.gatech.edu/classes/RWL/Projects/citation/Docs/Design/regex.intro.1.doc.html') and its continutation (URL='http://www-static.cc.gatech.edu/classes/RWL/Projects/citation/Docs/Design/regex.intro.2.doc.html').
- Advanced (Perl) Regex tutorials - Part I, (URL='http://www.perl.com/pub/a/2003/06/06/regexps.html') Part 2, (URL='http://www.perl.com/pub/a/2003/07/01/regexps.html') and documentation (URL='http://www.perl.com/doc/manual/html/pod/perltoc.html#perlre_Perl_regular_expression').
- PHP and regex tutorial (URL='http://weblogtoolscollection.com/regex/regex.php')
- How to do Regex in Python (URL='http://www.amk.ca/python/howto/regex/')
- Regular-expressions (URL='http://www.regular-expressions.info/') Language neutral, but not elementary. Links to books. Regular expressions and .Net (URL='http://www.codeproject.com/dotnet/RegexTutorial.asp')
- (URL='http://www.greenend.org.uk/rjk/2002/06/regexp.html') syntax summary details subtle differences among languages and Unix variants.
- (URL="http://www.perldoc.com/perl5.6.1/pod/perlretut.html") Expressions in Perl Very complete ; from PerlDoc
- (URL="http://www.perldoc.com/perl5.6.1/pod/perlre.html") Regular Expressions in Perl.
- Regular Expressions (URL="http://www.visibone.com/regular-expressions/")
- Server-Side Includes. Tutorial on Server-side includes (URL="http://www.mattkruse.com/info/ssi/"). Older (1995 article, last updated in 2002) but author says basic principles still apply (URL="http://cgi.resourceindex.com/Documentation/Server_Side_Includes/") to many resources of Server-side includes.
- More links (URL="http://bignosebird.com/ssi.shtml") on server-side includes

Perl (see also CGI)

- Downloading PERL - ActivePerl (free) and mod_perl. You want to write your scripts in Active Perl; mod_perl is the Perl interpreter which comes free with your Apache web server. (URL="http://www.activestate.com/Products/?_x=1") products open source programming languages tools. Download site for ActivePerl
- ActivePerl User Guide (URL="http://aspn.activestate.com/ASPN/docs/ActivePerl/5.8/index.html"). In table of contents, look for "Getting Started" to see how run example.pl. Also open it in Notepad.
- (URL="http://aspn.activestate.com/ASPN/docs/ASPNTOC-ACTIVEPERL-001/") Help - Online Docs : Getting Started
- (URL="http://aspn.activestate.com/ASPN/docs/ASPNTOC-ACTIVEPERL-003/") Help - Online Docs : ActivePerl FAQ
- ActiveState also has PHP, Python and XSLT products (URL='http://aspn.activestate.com/ASPN/')

- mod_perl: (URL='http://perl.apache.org/')
- Sourceforge.net (URL='http://sourceforge.net/softwaremap/trove_list.php')

PERL Documentation, Tutorials and Resources

For tutorials on regular expressions please see the CGI section.

- (URL="http://learn.perl.org/") The site for people learning Perl. Good listing of current books at all levels.
- Picking up Perl (URL='http://helpguide.inmotionhosting.com/perl_faq/'), Easy to use reference. More complete than elementary tutorials. 'The weekend crash course' is available through Books 24x7 e.g. at many university libraries or if you are a member of ACM.
- CGI and Perl Tutroial (URL='http://www.expertwebinstalls.com/cgi_tutorial/'). Another easy place to start, especially if you want all the facts quickly.
- Beginning Perl Tutorial (URL='http://www.pageresource.com/cgirec/index2.htm'). Great place to start. From pageresource.
- Perl 5 Tutorial (URL='http://www.extropia.com/tutorials/perl5/')
- Perl tutorial (URL='http://www.comp.leeds.ac.uk/Perl/start.html'),older, but still a good introduction.
- Tizag Tutorial on Perl (URL='http://www.tizag.com/perlT/'). Another fine place to begin learning Perl; has information on interfacing with databases (URL='http://www.tizag.com/perlT/perlmysqlquery.php')
- Webmonkey tutorials on Perl, CGI, etc. (URL='http://www.webmonkey.com/webmonkey/programming/perl_cgi/index.html').
- (URL="http://htmlgoodies.earthweb.com/primers/perl/"). Perl primers from HTML Goodies.
- Perl Monks (URL='http://www.perlmonks.org/index.pl?node=Tutorials'), These range from the elementary to the advanced. The useful site is very large, and, newbies will find it be easier to start with one of the places above (e.g. from pagereource.com).
- (URL="http://stein.cshl.org/WWW/software/CGI/cgi_docs.html") wanted to know about perl's module cgi.pm. From Lincoln Stein, who wrote the cgi module.
- Tutorial on Using Modules (URL='http://stein.cshl.org/genome_informatics/using_perl_modules/')
- CGI Resources (URL='http://cgi.resourceindex.com/Documentation/Programming_Languages/Programming_in_Perl/'). Links to many tutorials, references, sources of scripts.
- Perl Tutorial (URL='http://www.perl.com/pub/a/2000/10/begperl1.html'). For those who are sophisticated programmers and want to start in Perl
- Perldoc has intermediate level tutorials (URL='http://perldoc.perl.org/index-tutorials.html').

Tutorials related to Perl and Databases.

- Tutorial on databases and their perl interface (URL='http://www.extropia.com/tutorials/sql/toc.html').
- A gentle introduction. Database programming with Perl (URL='http://www.perl.com/pub/a/2003/10/23/databases.html')
- Perl and MySQL (URL="http://www.devpapers.com/article/42")
- Interfacing MySQL with Perl (URL='http://www.oreilly.com/catalog/msql/chapter/ch10.html'). O'Reilly book chapter. Down the Perl/DBI module to interface with MySQL (URL='http://dev.mysql.com/downloads/dbi.html')
- FreeBSD (URL='http://www.freebsddiary.org/mysql-perl.php').

- Tutorial on Perl MySQL functions (URL='http://www.thescripts.com/serversidescripting/perl/tutorials/perlandmysql-together/page3.html'). This is from thescripts.com, where you may also post questions to the developer community for various scripting languages and databases.
- Tutorial on Using Perl etc. to write software for genome research (URL='http://stein.cshl.org/genome_informatics/'). Includes such topics as subroutines.
- (URL='http://stein.cshl.org/genome_informatics/subroutines/') interfacing with databases.
- (URL='http://stein.cshl.org/genome_informatics/mysql-dbi/').

Documentation and Resources.

- Perl Documentation at Perl.org (URL='http://www.perl.org/docs.html') perldoc.perl.org (URL='http://perldoc.perl.org/'). Current release is Perl5;
- Documentation for Perl4 is at CMU (URL='http://www.cs.cmu.edu/cgi-bin/perl-man')
- Perldoc has (URL="http://perldoc.perl.org/CGI.html") including for older versions (URL="http://search.cpan.org/src/LDS/CGI.pm-3.23/cgi_docs.html"). Also a cheat sheet (URL='http://perldoc.perl.org/perlcheat.html').
- CPAN (URL="http://www.cpan.org/"). Where to find MODULES to add many useful functions to your Perl - e.g. the standard module, oraperl (for Oracle databases) etc. Also how Install CPAN Modules (URL="http://www.cpan.org/modules/INSTALL.html")
- (URL="http://www.perl.com/") The Source for Perl - perl development, perl conferences. Articles etc. for serious PERL programmers from the O'Reilly book folk
- (URL="http://perl.about.com/library/weekly/aa030500a.htm") building and installing Perl modules. They also link to more of their own tutorials (URL='http://perl.about.com/od/?once=true&') and (URL='http://perl.about.com/od/softwarescripts/tp/topcgisites.htm') other Perl resources.
- (URL="http://perl.about.com/gi/dynamic/offsite.htm?site=http://cgi-lib.berkeley.edu/") homepage. Library of CGI scripts in Perl; Also directly accesible at Berkely (URL="http://cgi-lib.berkeley.edu/"). Includes documentation, source files, examples, etc.
- Documentation for Perl CGI Module (URL="http://perldoc.perl.org/CGI.html"). Documentation is readable and it is easy to find what you are looking for.
- Downloading and installing additional Perl modules (URL='http://www.cpan.org/modules/INSTALL.html')
- Google groups on Perl (URL="http://groups.google.com/groups?q=Perl+Help"). This site includes several very useful groups (see green links): one for Perl beginners, (URL="http://groups.google.com/group/perl.beginners/topics") on Perl modules, (URL='http://groups.google.com/group/comp.lang.perl.modules/topics?lnk=lr') and another for (URL='http://groups.google.com/group/comp.lang.perl.misc/topics?lnk=sg') miscellaneous Perl questions. Can not be recommended enough. Search for key words in the Perl beginners group, for example, and then follow the whole thread (link in the upper right). Excellent resource.
- Perl community: (URL='http://www.perl.org/community.html') mailing lists and sites.
- CGI and Perl FAQs (URL='http://www.perl.com/doc/FAQs/cgi/perl-cgi-faq.html'). In addition to the FAQs, has a great list of books and other resources.

- Scripts for Perl (URL="http://www.webreference.com/perl/"). This is the Mother of Perl site ; intermediate to advanced articles.
- (URL="http://www.activestate.com/Products/Download/Download.plex?id=ActivePerl"). Build a custom installation of Perl and check system requirements - see warning on RedHat Linux v8
- (URL="http://perl.about.com/gi/dynamic/offsite.htm?site=http%3A%2F%2Fwww.ics.uci.edu%2Fpub%2Fwebsoft%2Flibwww-perl%2F") Perl libraries for www interactions - e.g. HTTP etc.
- (URL="http://builder.com.com/5100-6371_14-5363190-1-1.html") documents with Perl's XML::Simple. PERL module which has an XML parser (written in PERL)

Perl Examples, Scripts, Hints

- (URL="http://perl.about.com/library/weekly/aa070901c.htm") data in Perl scripts
- (URL="http://perl.about.com/library/weekly/aa070901a.htm") Perl Scripts
- (URL="http://perl.about.com/library/weekly/aa070901d.htm") reads and parses CGI data
- (URL="http://www.scriptarchive.com/") CGI Scripts. Links for free scripts.
- Scripts.com (URL='http://www.scripts.com/')
- (URL="http://www.getcruising.com/crypt/") Scripts From The Crypt / www.getcruising.com. More free scripts.
- Perl Meets Bio-informatics (URL="http://bio.oreilly.com/"). From O'Reilly - perhaps because they've published a book on this topic.
- Template Toolkit (URL='http://www.template-toolkit.org/info.html#Overview')
- Perl and XML (URL="http://www.webreference.com/perl/tutorial/1/")
- How to embed XML in an HTML Page (URL="http://tips.linux.com/tips/04/03/11/1620226.shtml?tid=85&tid=70&tid=43") edit all scripts with one line of Perl

PHP

Please also check the Multiple Languages listings in the General CGI section also.

- The main PHP site (URL="http://www.php.net/"). Home site for PHP, including documentation, download of php processor, and tutorial.
- PHP Tutorial (URL="http://us4.php.net/manual/en/tutorial.firstpage.php") from its makers.
- PHP: Introduction and Manual (URL="http://www.php.net/manual/en/").
- Zend makes PHP interpreters, etc (URL="http://www.zend.com/"). Their DevZone (URL='http://devzone.zend.com/public/view') also links to articles, tutorials and the manual. (URL='http://devzone.zend.com/manual/view/page/')
- (URL="http://www.developer.com/lang/php/article.php/3345121") Patterns within PHP.
- What can PHP do? (URL="http://php.net/manual/en/intro-whatcando.php"). Commentary from the PHP folks.
- PHP tutorial from w3schools (URL='http://www.w3schools.com/php/default.asp'). Always an excellent start; includes lessons on interfacing with XML, MySQL and AJAX.
- Working with PHP datatypes (URL="http://www.devpapers.com/article/184"). Several short tutorials. Gets you going quickly.
- PHP-Entry Point-Lesson One (URL="http://www.devpapers.com/article/162/") includes some information on pro's and con's; links to to various PHP sites
- PHP4 tutorial (URL='http://www.thesitewizard.com/php/index.shtml')

- (URL="http://php.net/manual/en/tutorial.forms.php") in PHP
- (URL="http://builder.com.com/5100-6374_14-5216364.html?tag=e606") document statistics with PHP.
- Using PHP to extract stats about XML documents. (URL="http://www.google.com/search?q=site%3Asearch.cpan.org+Help+php+perl") help group on PHP installation etc.
- Apache guide to installing and configuring PHP with Apache (URL='http://mpcon.org/apacheguide/index.php')
- Firepages (URL='http://www.firepages.com.au/'). Large site with downloads for PHP development, forums, etc.
- (URL='http://www.d.umn.edu/itss/support/Training/Online/webdesign/php.html') U. of Minnesota Duluth. Large excellent set of PHP resources including tutorials, articles on regex, using PHP with databases, and PHP and Ajax.
- (URL='http://www.phpbuilder.com/columns/validating_php_user_sessions.php3?page=1'). Session validation.
- Tips on Debugging in PHP (URL='http://centricle.com/archive/2005/08/php-debugging-tips')
- Unit Testing in PHP (URL='http://phpunit.de/'). This is also available at older sites.
- (URL='http://pear.php.net/package/PHPUnit2') The most current site includes a tutorial on writing (regression) tests.
- Simpler testing utilities are available for JUnit both from (URL='http://www.junit.org/') JUnit and from SourceForge (URL='http://sourceforge.net/projects/jwebunit/')
- PHP port testing for mock objects. (URL='http://www.mockobjects.com')
- LastCraft (URL='http://www.lastcraft.com/simple_test.php')
- Implementing Design Patterns in PHP (URL='http://php.net/manual/en/introduction.php')
- PHP Projects at SourceForege (URL='http://www.firepages.com.au/')
- DevX Resources and Script in PHP (URL='http://www.devx.biz/showcats.php?sbcat_id=2&PHPSESSID=1b55ae8f691d0adee2a77c94e86bf715'). Well categorized, but many have very few scripts.
- DevPlug is a PHP Developer's Forum (URL='http://www.devplug.com/index.php')
- PHP Resource Index (URL='http://php.resourceindex.com/'). Scripts, snippets of code, and tutorials; large site, hierarchically organized

Python

- Whetting Your Appetite (URL="http://www.python.org/doc/current/tut/node3.html")
- Python tutorial
- (URL="http://www.computerworld.com/newsletter/0,4902,101390,00.html?nlid=APP"). Also follow the links on the right hand side
- (URL="http://www.computerworld.com/softwaretopics/software/story/0,10801,104484,00.html?source=NLT_APP2&nid=104484") on why Python, from the horse's mouth.
- Tutorial on Python from Developer.com (URL='http://www.developer.com/open/article.php/625901')
- (URL="http://www.amk.ca/python/howto/regex/regex.html") regular expressions in Python Be sure to also check the general references on regular expressions in the CGI section.

And More Server Side Processing

- ASP Tutorial (URL="http://www.w3schools.com/asp/default.asp") from w3schools - usually excellent introduction.

- (URL="http://builder.com.com/5100-6389-5035160.html?tag=e606") flexible shopping cart with XML and ASP.

XML, WEB SERVICES, AND RELATED TECHNOLOGIES

XML

- Inside XML by Steve Holzner published by New Riders 2000. Excellent for both learning and as a reference; includes XSL and useful material on interfaces with Java and JDOM and SAX.
- XML and Web Services Unleashed by Ron Schmelzer, Travis Vandersypen, Jason Bloomberg, and Madhu Siddalingaiah published by SAMS 2002. Another book useful for learning and as a reference; many examples; note that this book includes web services.
- Charles F. Goldfarb's XML Handbook, 5 Edition by Charles F. Goldfarb and Paul Prescod published by Prentice Hall 2003. Includes some material on web services; for a long time this was the standard; well-written, it is still very useful
- Learning XML, Second Edition by Erik T Ray published by O'Reilly 2003. More of a book to learn from than the ones above; very clear
- XML 1.1 Bible by Elliotte Rusty Harold published by IDG 2004. Includes coverage of CSS, but very limited coverage of important topic of schemas, so I would use the other books first.
- W3C HTML Home Page (URL="http://www.w3c.org/MarkUp/")
- XML homepage from the w3c (URL="http://www.w3c.org/XML/"). Includes description of the many working groups and links to them, to various languages specifications and versions, and to many other resources.
- XML Schemas (URL="http://www.w3.org/2001/XMLSchema"). What are the components of an XML Schema and here their definition in a DTD
- (URL="http://www.w3.org/2001/XMLSchema.dtd"). Chart of Built-In DataTypes (URL='http://www.w3.org/TR/2004/REC-xmlschema-2-20041028/datatypes.html#built-in-datatypes')
- XML Recommendations (URL="http://www.w3.org/TR/2000/REC-xml-20001006").
- XML Specs (URL="http://www.w3c.org/XML/Core/#Publications")
- XML Namespaces (URL="http://www.w3.org/TR/1999/REC-xml-names-19990114/").
- The Cover Pages (URL="http://xml.coverpages.org/"). Important collection of on-line reference material on SGML/XML languages and various standards, now hosted by OASIS. You may want to start at (URL='http://xml.coverpages.org/AboutXMLCoverPages.html') About XML Cover Pages. Site is for the knowledgeable (except possibly the news stories on front page).
- DTD's attributes (URL='http://www.w3schools.com/dtd/dtd_attributes.asp')

Tutorials and Articles (Note: Tutorials on SOAP and WSDL are under the 'Web Services' heading, but information on SAX, DOM, JDOM etc. is here)

- XML Tutorial from w3schools (URL="http://www.w3schools.com/xml/default.asp"). Great introductory tutorial. See also their description of XML syntax (URL="http://www.w3schools.com/xml/xml_syntax.asp")
- TopXML: XML Tools, XML Articles and XML Learning Tutorials (URL="http://

www.topxml.com/"). Many tutorials - XML, SOAP,XSLT, etc and also links to XML parsers. Great site.

- Another very basic tutorial (URL="http://www.spiderpro.com/bu/buxmlm001.html"). So you don't get scared away.
- (URL="http://www.wdvl.com/Authoring/Languages/XML/Tutorials/Intro/toc.html") basic tutorial from WDVL. More in-depth than some of the other elementary tutorials, but doesn't get to schemas
- (URL="http://www.programmingtutorials.com/xml.aspx") tutorials from IBM, Microsoft, etc. at various levels and on sub-topics (e.g. security). A great resource
- (URL="http://tecfa.unige.ch/guides/xml/pointers.html#section5") links to tools, tutorials, and many resources for XML. Also has info on links to databases and server-side uses of XML
- (URL="http://www.zvon.org/index.php?nav_id=tutorials&mime=html") on everything from HTML to XML
- (URL="http://www.xml.com/") XML From the Inside Out — development, resources, specifications. O'Reilly site with many articles
- What is XML? (URL="http://www.htmlgoodies.com/tutors/xml.html")
- XML Terms (URL="http://webdesign.about.com/library/weekly/aa070102a.htm"). What the basic terms mean
- (URL="http://webdesign.about.com/od/xml/a/aa071601a.htm") About.com Very basic
- XML and DOM (URL="http://www.w3schools.com/dom/default.asp"). Usual excellent introduction from w3schools.com.
- XML Namespaces (URL="http://webdesign.about.com/library/weekly/aa070802a.htm"). Introductory tutorial from the About.com site.
- XML Schema Primer from w3.org — (URL='http://www.w3.org/TR/xmlschema-0/'). Very good, especially if you know a little bit. Three pieces:
- Primer (URL="http://www.w3c.org/TR/2004/REC-xmlschema-0-20041028/")
- Structures (URL="http://www.w3c.org/TR/2004/REC-xmlschema-0-20041028/")
- DataTypes (URL="http://www.w3c.org/TR/2004/REC-xmlschema-2-20041028/")
- XFront has links to many great tutorials (URL="http://www.xfront.com/"). including ones on XML Schema (URL="http://www.xfront.com/xml-schema.html"), (URL=http://www.xfront.com/canonical/sld001.htm). Canonical XML, (URL="http://www.xfront.com/BestPracticesHomepage.html") Practices etc. and also XSDs for all countries, all currencies, etc. Great source!
- This is from Roger Costello (one of the authors of the w3c.org Primer) and is a great resource. (URL="http://www.wdvl.com/Authoring/Languages/XML/Conferences/XML2000/schemaBest.html") on Best Practices for XML Schema.
- (URL="http://www.xfront.com/BestPracticesHomepage.html") the same best practices tutorial at another link.
- Zvon tutorials (URL='http://nwalsh.com/docs/tutorials/'). Tutorials on XML, Schemas, Namespaces, XPath and XSLT, RDF, XInclude, XUL etc.
- DevGuru tutorials (URL='http://www.devguru.com/features/tutorials/tutorials.asp'). Tutorials tend to be oriented towards using Microsoft technology rather than standards based.
- (URL="http://pd.acm.org/course_slc.cfm?pre=XMLPWS&trk=1153&crs=WP-1503-90") from the ACM There are several.
- The next group of tutorials is from IBM:

- (URL="http://www-106.ibm.com/developerworks/edu/x-dw-xusax-i.html") understanding SAX
- (URL="http://www-106.ibm.com/developerworks/edu/x-dw-xusax-i.html") DOM
- (URL="http://www-106.ibm.com/developerworks/edu/x-dw-xusax-i.html") XML in Java (see also a search for JDOM) Tutorial looks at SAX, DOM, JDOM and JAXP
- (URL="http://www-128.ibm.com/developerworks/views/xml/libraryview.jsp") from IBM Many are quite advanced
- (URL="http://www.topxml.com/tutorials/main.asp?id=jdom") more XML tutorials.
- (URL='http://nl.internet.com/ct.html?rtr=on&s=1,1xua,1,8cvb,gojs,hgol,6qvo') Introspection for XML data to map it to Java Beans; advanced.
- Reading and Writing XML in .Net (URL='http://www.15seconds.com/issue/050601.htm')
- (URL="http://www.developer.com/xml/article.php/3388311") Schemas

Other XML Resources

- The w3c has links to all its working groups, for example, the Schema Working Group (URL='http://www.w3.org/XML/Schema')
- You can also find new Technical Reports (URL='http://www.w3.org/TR/')
- xml.apache.org (URL="http://xml.apache.org/"). Link to Xerces parser, implementations of XSL, etc. A very important site.
- (URL="http://tecfa.unige.ch/guides/xml/pointers.html") Enormous and very useful set of links on validators and parsers, XML, XSL etc, and, tutorials and news about XML languages.
- XML Developer Center Home (URL="http://msdn.microsoft.com/xml/"). Microsoft's XML developer center, including articles, novice to advanced, and links.
- Open Source Web Development - Dev Shed (URL="http://www.devshed.com/"). Articles on XHTML,CSS, XML, Perl, etc.
- Developer.com articles on XML and related technologies (URL='http://www.developer.com/xml/'). Current and clear articles.
- Links from Moller and Schwartzbach book (URL='http://www.brics.dk/ixwt/links.html'). Useful links for everything to do with XML and Web Technologies, tho' it looks like most are not for newbies
- (URL="http://www.jars.com/classes/jresout.cgi?resource=12045") Resource Listing. Java class to automate the encoding of properties from Java into XML.
- (URL="http://builder.com.com/5100-6373_14-5258218.html?tag=e606") what you always wanted to know about SAX (and XmlReader).
- When to use SAX and when to use DOM (URL="http://builder.com.com/5100-6389_14-5165682.html?tag=sc") Design: Use DOM to create data-driven HTML documents.
- Using DOM to get conditional XML (URL="http://www.developer.com/db/article.php/3413151") the XML Alternative.
- Configuring / designing your database with XML. XML-Dev (URL="http://xml.org/xml/xmldev.shtml").An open unmoderated discussion list on development of various XML languages; Now managed by OASIS. For the knowledgeable.
- Perl and XML (URL="http://www.webreference.com/perl/tutorial/1/")
- How to embed XML in an HTML Page XML Tools (URL="http://www.w3.org/XML/Schema"). Long and wonderful list from w3.org.
- Free XML Tools (URL="http://www.garshol.priv.no/download/xmltools/plat_ix.html"). By platform, or vendor or name etc.

- Alphaworks is IBM's site (URL="http://www.alphaworks.ibm.com/xml"). This is the link for xml; many links to subtopics and tutorial on newer technologies (e.g. AJAX)
- (URL="http://www.w3.org/Consortium/Offices/Presentations/Schemas/all.html") of schema topics: Slides on why schemas are better than DTDs
- (URL="http://www.it-analysis.com/article.php?articleid=11384") The user interface. With, we hope better solutions to accessibility issues
- (URL="http://www.hyfinity.co.uk/Hyfinity%20-%20Bloor%20XML%20Machine.pdf"). Interesting view of the ubiquity and power of XML.
- (URL="http://www-106.ibm.com/developerworks/xml/library/x-injava/") The Document Object Model. This IBM site has many links to tutorials, information and documentation on using Java to manipulate XML files, etc. Very valuable.
- XML meets JavaScript in Firefox (URL='http://developer.mozilla.org/webwatch/?p=204'). Series of articles on using JavaScript to manipulate XML; many Firefox1.5 issues
- (URL="http://www-106.ibm.com/developerworks/edu/x-dw-xusax-i.html") XML in Java (see also a search for JDOM).
- Tutorial looks at SAX, DOM, JDOM and JAXP (URL="http://www.internetnews.com/dev-news/article.php/3443541") to get very large. Which can cause problems ?Which some companies have worked on.
- Business Communications Review (URL='http://www.bcr.com/bcrmag/2004/12/p35.php')
- ZDNet (URL='http://news.zdnet.com/2102-9588_22-5534249.html?tag=printthis').
- (URL="http://www.w3schools.com/dom/dom_validate.asp") from XML courtesy of w3schools.
- (URL="http://www.developer.com/xml/article.php/3081641") well-formed XML.
- (URL="http://www-128.ibm.com/developerworks/xml/library/x-c14n/?ca=drs") Canonical Form. Canonical form standardizes possible equivalent renditions of XML to one (canonical) form; this is important for checking digital signatures, etc. In addition to the subject in the title, there is a link to Ogbuji's Thinking XML columns at IBM?s developerWorks (URL="http://www-128.ibm.com/developerworks/views/xml/library.jsp?sort_order=desc&expand=&sort_by=Date&show_abstract=true&view_by=Search&searc_by=thinking%20xml:")
- (URL='http://www.xfront.com/canonical/CanonicalXML.html'). Very clear tutorial from Roger Costello at xfront.com
- Xforms Working Draft (URL='http://www.w3.org/TR/xforms11/'). More powerful handling of data and events than with HTML forms. See also their (URL='http://www.w3.org/MarkUp/Forms/2003/xforms-for-html-authors.html') XForms for HTML Authors
- XForms tutorial from w3schools (URL='http://www.w3schools.com/xforms/default.asp')
- Will XForms Matter? (URL='http://www.transformmag.com/showArticle.jhtml?articleID=17100027').
- Fight over XForms clouds future of Net (URL='http://news.zdnet.com/2100-9588_22-5581106.html?tag=nl.e539'). See also links on XUL at Mozilla.org and on Flash MX in the miscellaneous section below.
- XML and IDs: what is new (7.06) (URL='http://news.com.com/2100-1032_3-5785604.html)
- w3c addresses addressing in XML (URL='http://www.eweek.com/article2/0,1759,2004454,00.asp')

XML Discusses XHTML

- (URL="http://www.w3.org/TR/2002/REC-xhtml1-20020801/") HyperText Markup Language (Second Edition).Discussion of XHTML1.0 as a re-formulation of HTML
- (URL="http://www.w3.org/MarkUp/2004/xhtml-faq") Questions and FAQs about why bother going to XHTML.
- XHTML 1.1 - Module-based XHTML (URL="http://www.w3.org/TR/xhtml11/"). XHTML1.1 which is now a strict reformulation of HTML in XML
- XHTML Modularization Overview (URL="http://www.w3.org/MarkUp/modularization")
- DTDs for XHTML (URL="http://www.w3.org/TR/xhtml1/dtds.html#dtds"). See how HTML (XHTML) is formulated in an XML DTD. There are several DTDs (depending on the version of XHTML you want)
- XHTML tag index (URL='http://www.devguru.com/Technologies/xhtml/quickref/xhtml_index.html'). See also the section on XHTML

XMLSpy (and Other Parsers)

- (URL="http://www.altova.com/download.html") XMLSpy Tools. XMLSpy is a wonderful tool, and you may try their tools for 30 days free. Altova also makes tools for XSL (Stylevision) and for RDF editing (Semantic Works). There are free academic licenses available, and there is excellent documentation and a brief tutorial.
- xml.apache.org (URL="http://xerces.apache.org/"). Link to Xerces XML parser, (API in C++, Java or Perl) at Apache. Constantly evolving. Links to other XML-related tools on Apache Homepage (URL='http://xml.apache.org/')
- (URL="http://www.topxml.com/parsers/default.asp") Parsers
- XML parsers including Microsoft's (URL="http://topxml.com/parsers/default.asp#Microsoft%20Parsers")
- Other Parsers (URL="http://topxml.com/parsers/other_parsers.asp") links and brief descriptions for many XML parsers.
- StylusStudio (URL='http://www.stylusstudio.com/xml_download.html'). Like Altova (XMLSpy) they also have XL and XQuery engines, etc.
- (URL="http://downloads-zdnet.com.com/3000-2414-10295171.html?tag=sptlt") - ZDNet Downloads. Module (in Java) to edit XML and XHTML
- (URL="http://www.jars.com/classes/jresout.cgi?resource=12040") Resource Listing XML/XSLT editor (java based)
- (URL="http://builder.com.com/5100-6371_14-5363190-1-1.html") documents with Perl's XML::Simple. PERL module which has an XML parser (written in PERL).
- Scholarly Validator (URL="http://www.stg.brown.edu/service/xmlvalid/"). Free on-line validator from Brown University for small files
- Simple parser at w3schools (URL='http://www.w3schools.com/xml/xml_parser.asp')
- Parser Validator in Perl (URL='http://sourceforge.net/project/showfiles.php?group_id=89764')
- Parser in PHP (URL='http://us2.php.net/manual/en/ref.xml.php'). This parser does not validate; php.net also has tools to create your own parser.
- Many XML Tools (URL='http://downloads.zdnet.com/Windows/Developer+Tools+-+Windows/XML+-+Tools+-+Windows/'). ZDNet has a listing of many XML tools for Windows, often with free trials.
- css.nu (URL='http://sourceforge.net/project/showfiles.php?group_id=89764')

- XSmiles is an XML borwser for exotic devices (URL='http://www.x-smiles.org/xsmiles_objectives.html').
- StAX, (URL='http://javaboutique.internet.com/tutorials/stax/')

XSL / XSLT (including XPath)

XSL/XSLT and XPath - Articles, documentation and tutorials.

- Why both CSS and XSL - which should I use? (URL="http://www.w3c.org/Style/"). From w3c - who brought you both standards (URL="http://www.w3c.org/Style/CSS-vs-XSL") can) and when to us XSL (when you must). XSL (URL="http://www.w3c.org/Style/XSL/")
- What are the components of XSL (XSLT, XPath, etc.)? Very useful page, many good links.
- (URL='http://www.w3.org/TR/xsl/slice1.html#section-N629-Introduction-and-Overview') Introduction and Overview of XSL is also helpful.
- XSL Specs (URL='http://www.w3.org/TR/xsl/')
- XSL Transformations (XSLT) (URL="http://www.w3.org/TR/1999/PR-xslt-19991008").
- w3.org documentation on XSLT Web Style Sheets (URL="http://www.w3.org/Style/") w3c on style sheets - CSS and XSL
- (URL="http://www.w3.org/Style/XSL/") (XSL). Links to specs, tutorials, and articles on XSL, XSLT etc.
- XSLT client-side (URL="http://www.digital-web.com/articles/client_side_xslt"). Simple introduction, but you'll need the www.w3schools.com (URL="http://www.w3schools.com/") or other tutorial to go further like Tizag tutorial on XSLT (and earlier one on XPath). Like the w3schools tutorials, this is a gentle introduction.
- Client-side XSLT: Not just for server geeks any more (URL='http://www.digital-web.com/articles/client_side_xslt') Good introduction
- (URL="http://www.w3schools.com/xsl/xsl_transformation.asp") XML into XHTML.
- XSL Homepage of w3c (URL="http://www.w3c.org/Style/XSL/"). Including links to XPath, XQuery, XSLT, XSL-FO etc. There is a large list of links to specs on all these languages, mailing lists, and software. A great starting place for these technologies. You can also find links to examples of XSL style sheets (e.g. at TopXML)
- Style Activity Statement (URL="http://www.w3.org/Style/Activity.html") style sheets - including CSS and XSL.
- (URL="http://www.computerworld.com/developmenttopics/development/webdev/story/0,10801,92787,00.html?nas=APP-92787") – Computerworld. What XSL is and why you should use it. Good start. See also (URL="http://www.computerworld.com/developmenttopics/development/webdev/story/0,10801,92797,00.html?nas=APP-92797").
- (URL="http://www.developer.com/xml/article.php/3348311") XSLT and JSF in Cocoon. Using XSLT with Java Beans etc. Advanced. Multi-part tutorial: Discover the Wonders of XSLT
- Part 1 -XSLT (URL="http://www.developer.com/xml/article.php/3314291")
- Part 2: XPath (URL="http://www.developer.com/xml/article.php/3325751")
- Part 3 of 3: Advanced Techniques: (URL="http://www.developer.com/xml/article.php/3339891") Tests and advanced techniques
- Part 4: XSLT Quirks (URL='http://www.developer.com/xml/article.php/3357231')
- Part 5: Workflows (URL='http://www.developer.com/xml/article.php/3368891')
- (URL="http://www.developer.com/xml/article.php/3354151") Data Model
- XPath (URL="http://www.w3schools.com/xpath/default.asp") and XSL (URL=http://

www.w3schools.com/xsl/xsl_languages.asp) tutorials. From w3schools.org. I think this is one of the clearest of the introductory tutorials.

- Tizag also has clear introductory tutorials on XPath (URL='http://www.tizag.com/xmlTutorial/xpathtutorial.php') and XSLT (URL='http://www.tizag.com/xmlTutorial/xslttutorial.php')
- Learning XSLT has implications for CSS (URL='http://copia.ogbuji.net/blog/2005-06-21/XSLT___CSS'). Tutorial; main interest is using CSS to style XML in browsers
- Tutorial on XSLT 2.0 (URL='http://nwalsh.com/docs/tutorials/extreme2006/')
- Norman Walsh has also posted (URL='http://nwalsh.com/docs/tutorials/')
- (URL="http://www.zvon.org/xxl/XPathTutorial/General/examples.html") tutorial with links to examples on the left side. From Zvon
- XPath Tutorial (URL="http://www.vbxml.com/xsl/tutorials/intro/default.asp")
- Several XSLT Tutorials (URL='http://css.nu/pointers/other.html'). On their home page (URL='http://css.nu/pointers/').
- Mozilla maintains a lot of information on XPath and XSL (URL='http://developer.mozilla.org/en/docs/XPath')
- (URL="http://www.developer.com/xml/article.php/3344421") 2.0? The primary purpose of XPath is to address parts of an XML document. XPath uses a compact, non-XML syntax to facilitate use of XPath within URIs and XML attribute values. XPath gets its name from its use of a path notation as in URLs for navigating through the hierarchical structure of an XML document.
- (URL="http://www.developer.com/java/other/article.php/3361261") Writing Java Code to Emulate an XSLT Transformation, Emulating XSLT transformations in Java XSLT and Java Server
- Faces (URL="http://webdesign.about.com/library/weekly/aa110501a.htm") XPath
- (URL="http://www.developer.com/xml/article.php/3383961") Queries Tutorial.
- About.com tutorial on XSLT (URL='http://webdesign.about.com/b/a/166143.htm')
- (URL="http://www.developer.com/xml/article.php/3420551") XML and XSL
- (URL="http://webdesign.about.com/b/a/116546.htm?nl=1") Stylesheet Language Formatting Objects NOTE: XSL-FO is the old name for XSLT.
- XSL-FO for formatting objects. Using FO with Java (URL='http://javaboutique.internet.com/tutorials/FOP/')
- (URL="http://webdesign.about.com/cs/xslinformation/a/aaintrotoxsl.htm") to XSL - XSL, XSLT, XPath, and XSL Formatting Objects.
- Introduction to XSL (URL='http://newmedia.purchase.edu/%7EJeanine/interfaces/xmlstuff.html').
- Java and XSLT (URL='http://www.developer.com/java/other/article.php/3313341')

Tools and Implementations of XSLT

Contemporary browsers all implement XSL for rending documents in a browser. These links are for more general implementations of XSL.

- XT (URL="http://www.blnz.com/xt/index.html"). Free implementation of XSLT written in java
- (URL="http://www.jars.com/classes/jresout.cgi?resource=12040") Resource Listing XML/XSLT editor (java based)
- (URL="http://www.w3schools.com/xsl/xsl_transformation.asp") into XHTML Stylus Studio sells a full line of tools for XML, XSLT, etc. (URL='http://www.stylusstudio.com/xslt.html')
- Many XML Tools (URL='http://downloads.zdnet.com/Windows/Developer+Tools+-

+Windows/XML+-+Tools+-+Windows/') ZDNet has a listing of many XML tools for Windows, often with free trials. XMLSpy includes an XSLT engine.

- Microsoft tool to create XSLTs (URL='http://doc.advisor.com/Articles.nsf/nl/15021')

Specific extended MLs–MathML etc. and also RDF and RSS

Note: This section contains XML and security first, then RDF and RSS, XML and Databases, then MathML and Amaya, XML in the financial services industry and then miscellaneous. The Semantic Web, Web 2.0, AJAX, mashups etc. are in their own section.

- (URL="http://www.looselycoupled.com/opinion/2004/oneill-trust-infr1213.html") XML and security, which comes first.
- (URL="http://fonpc18.hum.uva.nl/Taal-Generatie/WWW/xml.coverpages.org/ws-security.html") is the specification for web services security. This is from OASIS.
- (URL="http://sunxacml.sourceforge.net/guide.html") eXtensible Access Mark-up Control Language. This is Sun's implementation and programmer's guide.
- On-going news (URL='http://sunxacml.sourceforge.net/')
- (URL="http://www.w3.org/2001/03/WSWS-popa/paper23/") secure web services. From w3.org, of course. SAML stands for Security Assertion Mark-up Language. (URL="http://www.w3.org/2004/08/ws-cc/aaccws-20040827") XACML . This area is still changing? Suggest you search the w3.org site for these two extensions: URL="http://www.w3.org/Encryption/2001/") and working group at w3.org
- (URL="http://www.internetnews.com/dev-news/article.php/3344041"), added to Apache Project
- (URL='http://buyersguide.eweek.com/bguide/whitepaper/WpDetails.asp?wpId=MTk0NQ&category=&fromlogin=yes&EWS=NO') Security issues and XML. Requires free login at ZDNet

RDF (Resource Description Framework–is a Resource for Describing Catalogs)

- RDF Primer (URL="http://www.w3.org/TR/rdf-primer/")
- RDF Spec (URL="http://www.w3.org/RDF/")
- RDF is of interest (URL="http://search.dmoz.org/cgi-bin/search?search=RDF") to librarians, etc.
- RDF Validator (URL='http://www.w3.org/RDF/Validator/')
- Mozilla site on RDF (URL='http://developer.mozilla.org/en/docs/RDF')
- And some examples (URL='http://www.mozilla.org/rdf/doc/examples.html')
- What is RDF? (URL='http://www.xml.com/pub/a/2001/01/24/rdf.html/')
- From O'Reilly site XML.com. Zvon tutorials (URL='http://nwalsh.com/docs/tutorials/')
- Tutorials on XML, Schemas, Namespaces, XPath and XSLT, RDF, XInclude, XUL etc. Relational databases on the semantic web and RDF (URL='http://dig.csail.mit.edu/2006/dbview/dbview.py')
- See also Tim Berners-Lee paper which started this subject (URL='http://www.w3.org/DesignIssues/RDB-RDF.html')

RSS (Really Simple Syndication–is how an XML Feed is used to Update News and Blogs Continuously)

- What is RSS? (URL="http://www.xml.com/pub/a/2002/12/18/dive-into-xml.html"). From O'Reilly - good casual introduction; includes a sample to get you started.

- (URL="http://www.webreference.com/authoring/languages/xml/rss/intro/2.html") What is RSS? From webreference.com - lots of information.
- (URL="http://www.webreference.com/authoring/languages/xml/rss/intro/3.html") Introduction to RSS.
- (URL="http://www.webreference.com/authoring/languages/xml/rss/intro/2.html") Syndication and Aggregation.
- (URL="http://www.webreference.com/authoring/languages/xml/rss/") on RSS.
- (URL="http://www.learningcircuits.org/2004/may2004/0405_Trends.htm") XML to feed info on sites to subscribers. RSS monitors (new) content on sites and feeds it to subscribers. May also be used for on-going education.
- (URL="http://www.newarchitectmag.com/archives/2000/02/eisenzopf/") headlines with RSS.
- Many RSS scripts (URL="http://www.webreference.com/scripts/")
- Scripts for Perl and RSS (URL="http://www.webreference.com/perl/"). Mother of Perl site.
- Exploring RSS and XML in Flash (URL="http://www.webreference.com/xml/column82/")
- (URL="http://www.garshol.priv.no/download/xmltools/tools-rss.rdf") site listing free xml tools.
- (URL="http://www.internetnews.com/dev-news/article.php/3431901") Exposure or pod-casting. An audio RSS feed
- (URL="http://www.w3.org/2001/10/glance/doc/howto") news feeds from w3.org, but with a quick 'cheat sheet' on how to do it.
- IBM maintains a site called alphaworks (URL="http://www.alphaworks.ibm.com/rss").
- This site provides current information about RSS and other new technologies. (URL='http://developer.mozilla.org/en/docs/RSS')
- All the following information on RSS is from one of my students: (URL="http://essaysfromexodus.scripting.com/xml/scriptingnews2.xml")
- XML News by Dave Winer; Winer designed this format at Userland and (URL="http://davenet.scripting.com/1997/12/15/scriptingNewsInXML") and included it in his blog.
- The history of RSS (URL="http://blogs.law.harvard.edu/tech/rssVersionHistory")
- RSS 2.0 Spec (URL="http://blogs.law.harvard.edu/tech/rssVersionHistory")
- (URL="http://www.purplepages.ie/RSS/netscape/rss0.90.html") Using RSS 0.9, clear explanation, though RSS 2.0 is current spec
- RSS Aggregators / readers. Full list at (URL="http://blogspace.com/rss/readers").
- Mac OS X: NetNewsWire (URL=http://ranchero.com/netnewswire/).
- Windows: SharpReader (URL="http://www.sharpreader.net/")
- Linux: Straw (URL="http://www.nongnu.org/straw/").
- Web: Bloglines (URL="http://www.bloglines.com/")
- RSS Validators: (URL="http://feedvalidator.org/")
- (URL="http://rss.scripting.com/"). Selected Sites with RSS Feeds
- British Broadcasting Corporation (http://news.bbc.co.uk/shared/bsp/hi/services/htmlsyndication/html/default.stm)
- DevX (http://www.devx.com/DevX/Article/16190)
- LiveJournal Syndicated Feeds (URL="http://www.livejournal.com/syn/list.bml")
- United States Department of Defense (http://www.dod.gov/news/rss/ (URL="http://www.dod.gov/news/rss/")
- WebReference.com (http://www.webreference.com/services/news/), For a more exhaustive list, go to the RSS Compendium (http://allrss.com/rssfeeds.html)

- (URL="http://www-106.ibm.com/developerworks/webservices/demos/weather/index.html") IBM's RSS feed on weather.
- Quick Start on RSS (URL='http://software.techrepublic.com.com/abstract.aspx?docid=89736')Requires free registration, but worth it.
- RSS Tutorial (URL='http://www.mnot.net/rss/tutorial/'). Gentle introduction with links to more advanced information.
- RSS Workshop (URL='http://rssgov.com/rssworkshop.html'). Links to places to learn how t create your own feed in the language of your choice; lots of examples.
- WebKnowHow articles on RSS (URL='http://www.webknowhow.net/dir/Other_Topics/XML/index.html'). Introductory.
- O'Reilly has articles on RSS (URL='http://www.oreillynet.com/rss/'). Also check out their tutorials (URL='http://www.oreillynet.com/topics/rss/rss').
- (URL="http://itmanagement.earthweb.com/columns/executive_tech/article.php/3617901") for CSS poor in most RSS readers (7/06)
- Building a Generic RSS Class in PHP (URL='http://www.devarticles.com/c/a/MySQL/Building-a-Generic-RSS-Class-With-PHP/'). Requires some knowledge of MySQL and PHP.
- Working with RSS and Oracle JDeveloper (URL='http://www.developer.com/java/web/article.php/3524171').If you prefer Oracle and Java (URL='http://www-128.ibm.com/developerworks/library/x-wxxm36.html?ca=dgr-lnxw03RSSExtensions')

XML and Databases, including XQuery; Please also see RDF

- XQuery for database querying (URL='http://www.w3.org/TR/xquery/'). This is now the standard.
- XQuery Tutorial (URL='http://www.w3schools.com/xquery/default.asp')
- What is XQuery? (URL='http://www.xml.com/pub/a/2002/10/16/xquery.html')
- Introduction at O'Reilly site xml.com. XQuery tutorial (URL='http://www.brics.dk/~amoeller/XML/querying/').
- An Introduction to XQuery (URL='http://www-128.ibm.com/developerworks/xml/library/x-xquery.html'). A look at the w3c schema with many links to other resources. Using XML with Databases Tutorial
- SQLX.org is devoted to SQL and XML working together (URL='http://www.sqlx.org/'). Because SQL is an ISO standard its specs are copyrighted; a copy may be bought through SQLX/XML
- Programming with SQL and XML (URL='http://www.research.ibm.com/journal/sj/414/reinwald.pdf'). Excellent and comprehensive article.
- XViews for implementing Views on XML databases (URL='http://dl.alphaworks.ibm.com/technologies/xviews/ViewsForXML-Whitepaper.pdf') . This is an IBM research paper, not a w3c standard. The web standard has moved to RDF (see above) and the Data Access Language (next item).
- The Data Access Working Group (URL='http://www.w3.org/2003/12/swa/dawg-charter'). It is also working on issues of querying on the semantic web (URL='http://www.w3.org/2001/sw/DataAccess/')
- Relational Databases on the Semantic Web (URL='http://www.w3.org/DesignIssues/RDB-RDF.html')
- Does RDF implement E-R? by Tim Berners-Lee. Integrating web services with SQL Server (URL='http://www.developer.com/db/article.php/3547866')
- Also using the SQLXML classes (URL='http://www.developer.com/db/article.php/3341881')
- XML and DB2 (URL='http://itmanagement.earthweb.com/datbus/article.php/3455461')

- (URL='http://www.eweek.com/article2/0,1759,1747224,00.asp') IBM's SOA plans
- (URL='http://www.eweek.com/article2/0,1759,1747224,00.asp'). XMLSpy includes an XQuery engine
- Oracle's site (URL='http://www.oracle.com/technology/tech/xml/index.html'). Example: (URL='http://www.oracle.com/technology/pub/articles/vasiliev_xquery.html') Oracle version of XQuery and Using XQurey with XSL (URL='http://www.oracle.com/technology/oramag/oracle/05-sep/o55xquery.html')

MathML and Amaya

- MathML (URL="http://www.w3.org/Math/"). Home page at w3.org. Includes a complete description of MathML (URL='http://www.w3.org/TR/MathML2/chapter4.html')
- Amaya Home Page (URL="http://www.w3.org/Amaya/"). Web authoring tool which includes support for MathML
- MathML in Mozilla (URL='http://www.mozilla.org/projects/mathml/'). A rich resource, including transling from TEX, SOAP interfaces, etc.

Financial Service Industry

- (URL="http://www-128.ibm.com/developerworks/xml/library/x-think22.html?ca=drs"). A good introduction to eBXML, XBRL, FIXML, MDDL, FpML, and assorted standards groups for this industry
- (URL="http://www.developer.com/java/other/article.php/2204681") Introducing the Vision. What is eBXML? Follow links at bottom for series of articles - for whole series of articles
- (URL="http://www.eweek.com/print_article/0,1761,a=129009,00.asp") Sharpen Banks' Biz Reports. XBRL (eXtended Business Reporting Language) must be used by banks and other public companies (per Sarbanes-Oxley Law) to consolidate units when they report their financial results; will also migrate to internal reports.
- (URL="http://www.eweek.com/print_article/0,1761,a=111879,00.asp") Ease Financial Reporting. How XMRL will help
- (URL="http://www.eweek.com/print_article/0,1761,a=119178,00.asp") Road to Compliance. What's involved in compliance for Sarbanes-Oxley (SOX) from an IT point of view. SOX compliance is a big topic.
- (URL="http://zdnet.com.com/2100-1107-5267008.html") into objects - News – ZDNet. BPEL (Business Process Execution Language) for the non-techie
- (URL="http://www.infoworld.com/article/04/07/02/27TCwsibpel_1.html") Is BPEL the real deal?: July 02, 2004: By Phillip J. Windley : APPLICATION_DEVELOPMENT : APPLICATIONS : WEB_SERVICES
- (URL="http://www.developer.com/xml/article.php/2247851") ebXML
- The interaction of eBXML (URL="http://www.webservicesarchitect.com/content/articles/irani02.asp") and (URL="http://www.webservicesarchitect.com/content/articles/irani03.asp") Services
- (URL="http://xml.coverpages.org/ni2004-11-08-a.html") Uses XML to define a vocabulary for common business forms, such as purchase orders etc.
- (URL="http://news.zdnet.com/2100-3513_22-5444901.html") to work with eBXML

Miscellaneous: VoiceML, XUL, WML (wireless), etc.

- (URL='http://www.w3schools.com/xml/xml_technologies.asp'). Short description of the main XML Technologies.

- MetaMap of All the XML Technologies (URL='http://www.mapageweb.umontreal.ca/turner/meta/english/index.html'). A wonderful map of how these are all related and what the acronyms mean.
- The Mark-up Languages (URL='http://www.computerworld.com/news/2005/story/0,11280,103711,00.html?source=NLT_APP&nid=103711'). Description of what is around (as of August 2005) including less widely used ones (URL='http://www.computerworld.com/action/article.do?command=viewArticleBasic&articleId=103712&intsrc=article_pots_side')
- (URL="http://builder.com.com/5100-6371_14-5034846.html?tag=e606") adding functionality to your Flash movies (FlashXML)
- (URL="http://www.alphaworks.ibm.com/tech/xviews") designing GUIs
- XUL (URL='http://developer.mozilla.org/en/docs/Introduction_to_XUL')
- Tutorial on creating application with Mozilla (URL='http://www.csie.ntu.edu.tw/%7Epiaip/docs/CreateMozApp/index.html'). Includes chapter on XUL
- (URL="http://www.wdvl.com/Authoring/Tutorials/xml.html") VoiceXML etc. from WDVL.com
- Introduction to WML (URL='http://webdesign.about.com/b/a/166148.htm?nl=1'). WML is Wireless Mark-up Language.
- Wireless Web Initiative (URL='http://www.w3.org/Mobile/').
- Mastering Wireless Programming: First Steps (URL='http://www.developer.com/net/csharp/article.php/3429261')
- (URL="http://www.microsoft.com/office/xml/default.mspx") Schemas
- (URL="http://www.cs.brandeis.edu/%7Ejamesp/arda/time/") language for temporal events.
- XInclude (URL="http://www.w3.org/TR/2004/REC-xinclude-20041220/"). An evolving specification for including &merging XML documents (URL="http://news.zdnet.com/2100-3513_22-5498458.html?tag=nl.e539") surprisingly it works only with schemas (no DTDs). Here is a brief introduction (URL='http://www.devxnews.com/article.php/3449981')
- Zvon tutorials (URL='http://nwalsh.com/docs/tutorials/')
- Tutorials on XML, Schemas, Namespaces, XPath and XSLT, RDF, XInclude, XUL etc. Primer on OpenXML (Microsoft) and Open Document Format (everone else) (URL='http://www.baselinemag.com/article2/0,1540,1933690,00.asp')
- (URL="http://www.eweek.com/article2/0,1895,1934910,00.asp") documents format - vs. Microsoft (3/06).
- OpenXML gaining (URL="http://www.eweek.com/article2/0,1895,1950375,00.asp") (4/06... especially after acceptance by ECMA (URL='http://www.eweek.com/article2/0,1895,1900768,00.asp')
- ODF (Open Document Format) is (URL='http://www.eweek.com/article2/0,1895,1900768,00.asp')
- (URL="http://www.devsource.com/article2/0,1895,1988024,00.asp") between Microsof's OpenXML and Open Document Format (ODF) (7/06). Microsoft?s plug-in for conversions from Office To OpenXML (7/06)
- (URL="http://linux.webbuyersguide.com/news/4182-wbglinux_news.html") behind ODF; You should also check out Massachusetts's push for ODF.
- (URL='http://downloads.zdnet.com/Windows/Developer+Tools+-+Windows/XML+-+Tools+-+Windows/'). ZDNet has a listing of many XML tools for Windows, often with free trials.

WEB SERVICES (PLEASE NOTE THAT MANY LINKS RELATED TO WEB SERVICES - E.G. SECURITY, MOBILE WEB, ETC. ARE IN SECTION DIRECTLY ABOVE, CHOREOGRAPHY AND BPEL ARE IN THIS SECTION)

Web Services - General information

- Overview of Web Services (URL='http://www.webservicescenter.com/frame_Overview_WS.htm'). A brief overview; not much detail; links to implementing web services with Java on Linux.
- What are web services? (URL="http://looselycoupled.com/glossary/web%20services"). The 'home' links to many articles about web services.
- Client and server (URL="http://www.htmlgoodies.com/letters/288.html")
- Web Service Demos (URL='http://www.mindreef.net/tide/scopeit/start.do')
- Fun to try. w3c.org definition of a web service (URL='http://www.w3.org/TR/ws-arch/#whatis'). This is in their document on web service architecture.
- (URL="http://mapageweb.umontreal.ca/turner/meta/english/index.html") How the different technologies are related. May be more useful after you know a little about the area, and then invaluable. There is also an information link (URL='http://mapageweb.umontreal.ca/turner/meta/english/index.html')
- Relationship of Web Service's Major Components (URL='http://www.ws-standards.com/'). Simpler than the MetaMap; good place to start, then go to MetaMap. This site also has an excellent glossary (URL='http://www.ws-standards.com/glossary.asp')
- (URL="http://www.developer.com/java/web/article.php/2207371") Service Oriented Architecture (SOA)? What are Web Services? Very good introduction to SOA, for those who already know the lingo. 8 pages!
- SOA (without web services) in plain language (URL='http://itmanagement.earthweb.com/article.php/3629561')
- The difference between SOA and Web Services' (URL='http://ct.enews.baselinemag.com/rd/cts?d=189-336-2-16-170148-36925-0-0-0-1')
- (URL="http://www.infoworld.com/article/04/02/20/08OPstrategic_1.html") Web services alphabet soup: Application Development. Web services alphabet soup - and modularity.
- The Web Services Family Tree (URL='http://www.modernlifeisrubbish.co.uk/article/web-tech-family-tree').
- From XHTML to AJAX, Understanding Service-Oriented Architecture (URL='http://www.developer.com/java/web/article.php/2207371'). Great introduction (actually a book chapter). 'Composition' has the same meaning as composition of functions in math - i.e. one followed by another
- (URL='http://www.computerworld.com/newsletter/0,4902,94720,00.html?nlid=AM'). Another book excerpt, slightly less detailed than the previous one.
- Web Services Architecture (URL='http://www-128.ibm.com/developerworks/webservices/library/w-ovr/?dwzone=webservices'). Although this claims to describe only IBM's architecture, it is a good description of all the web service architectures.
- Web Services Architecture (URL='http://www.webservicesarchitect.com/content/articles/webservicesarchitectures.pdf'). Description of basic and more complex web services stacks; covers major vendors: Sun, IBM, Microsoft, Oracle, HP, BEA and Borland.

- Web Services Essentials (URL='http://developers.sun.com/techtopics/webservices/essentials.html'). From Sun, this emphasizes a role for Java. Of course, designed to be platform neutral, web service may be implemented in other languages, too.
- Introduction to Web Services (URL='http://www.altova.com/documents/aot/webserviceswebbasedworkbook/ModelPage_MOD01_01.html'). Very clear, but basic. From Altova, creators of XMLSpy. Starts general and then moves to their tool. Service Oriented Architecture: How and Why. Arguments of for IT managers and why SOA is the way to go. Good survey.
- New Rules Govern the SOA Lifecycle (URL='http://www.looselycoupled.com/opinion/2005/rodri-rules-gov0701.html').
- As Ye SOA So Shall Ye Reap (URL='http://www.computerworld.com/action/article.do?command=viewArticleBasic&articleId=269329&source=NLT_APP&nlid=48') , Humorous story.
- (URL='http://www.acmqueue.com/modules.php?name=Content&pa=showpage&pid=246') Fuzzy boundaries: Objects, Components and Web Services. Which to use when; further discussion (URL='http://www.acmqueue.com/modules.php?name=Content&pa=showpage&pid=327').
- Web Services Hurdles (URL='http://www.computerworld.com/newsletter/0,4902,95201,00.html?nlid=AM'). Including security and reliability. Also cultural changes (URL='http://www.computerworld.com/developmenttopics/development/story/0,10801,110771,00.html?source=NLT_APP&nid=110771') and peoples' customs. Quality control issues (URL='http://www.computerworld.com/managementtopics/management/story/0,10801,110436,00.html?source=NLT_APP&nid=110436')
- No warranties, or QoS issues (URL='http://www.looselycoupled.com/stories/2004/warrant-dev1004.html').
- Securty issues (URL='http://blogs.zdnet.com/service-oriented/index.php?p=62&tag=nl.e539').
- Performance issues (URL='http://blogs.zdnet.com/service-oriented/index.php?p=59')
- The Rise and Fall of CORBA (URL="http://www.acmqueue.com/modules.php?name=Content&pa=showpage&pid=396"). What we can learn from CORBA, one of the ancestors of web services.
- The Top 5 Myths about SOA (URL='http://www.intelligententerprise.com/channels/appmanagement/showArticle.jhtml?articleID=186700581').
- Capitalizing on SOA (URL='http://www.developer.com/design/article.php/3588361'). Good discussion of SOA patterns, philosophy, pros and cons
- TestingQuality in an SOA (URL='http://it-management.earthweb.com/columns/print.php/3416971')
- The Value of SOA (URL='http://www.developer.com/java/ent/print.php/3424671')
- Close the loop: Using SOA to automate human interaction (URL='http://www.developer.com/java/ent/print.php/3606031'). With specific detailed example.
- White Papers from Ziff Davis (URL='http://www.webbuyersguide.com/bguide/Whitepaper/Wplibrary.asp_Q_sitename_E_webbuyersguide'). Including on ROI, strategies for success, costs and benefits, how to adopt etc.
- Four Steps to SOA (URL='http://www.computerworld.com/newsletter/0,4902,109387,00.html?nlid=AM')
- (URL='http://www.baselinemag.com/article2/0,1540,1938218,00.asp) Lessons from Early SOA Adopters.
- SOA FAQs (URL='http://www.mcpressonline.com/mc/.6b3cde85'). Free registration at MC Press On-line required.
- Building SOA Your Way (URL='http://www.infoworld.com/archives/emailPrint.

jsp?R=printThis&A=/article/05/09/12/37FEsoaevolve_1.html'). Making it work for you; case studies.
- Finding Your Way to SOA (URL='http://www.sun.com/software/media/flash/tour_soa/index.html'). Sun introduction in either a business or technical version.
- (URL='http://www.baselinemag.com/print_article2/0,1217,a=182675,00.asp') SOA: Integrating Applications. Case studies of who did what and how much they saved

Web Services: Books and Documentation

Please note that standards in this area are issued by both W3C (for XML and many of its extensions, SOAP, WSDL) and OASIS (URL='http://www.oasis-open.org/committees/tc_cat.php?cat=ws'), also WS-I (URL='http://www.ws-i.org/').

- The Executive's Guide to Web Services by Eric A. Marks and Mark J. Werrell published by Wiley, 2003. A lot of managerial discussion of the advantages of SOA, but a very clear description of the protocol stacks and what each part of web services does.
- Understanding Web Services by Eric Newcomer, Addison Wesley, 2002. Excellent introductions; second book is an update of the first.
- Web Services Essentials by Ethan Cerami published by O'Reilly 2002. , Excellent look at enough of the technology so you understand what is really happening; also detailed information about installing and interfacing with the server if you will be responsible for that too.
- XML Protocols Matrix (URL='http://www.w3.org/2000/03/29-XML-protocol-matrix'). Who is in charge of what protocol.
- Web Services Architecture (URL='http://www.w3.org/TR/ws-arch/').
- Web Services Addressing Standards: Core (URL='http://www.w3.org/TR/2006/REC-ws-addr-core-20060509/').
- SOAP Binding (URL='http://www.w3.org/TR/2006/REC-ws-addr-soap-20060509/'). Explanation of what the addressing standard do (URL='http://www.eweek.com/article2/0,1895,1960351,00.asp')
- Standards to Improve Performance (URL='http://www.devxnews.com/article.php/3463881') and (URL='http://www.eweek.com/article2/0,1759,1763645,00.asp')
- Web Services Polices (URL='http://www.w3.org/2004/08/ws-cc/gdfpwsp-20040904'). w3c compares their version with that from IBM and Microsoft. Includes discussion of how to specify Quality of Service (QoS)
- Table of all policy groups and standards on web services at w3c.org (URL='http://www.w3.org/2002/ws/')
- WS-I (URL='http://www.ws-i.org/'). WS-I, a consortium of major software provides such as IBM, SAP, and Microsoft.
- (URL='http://www.infoworld.com/article/06/04/21/77668_HNwsimessage_1.html?source=NLC-WS2006-04-26') asynchronous messaging standard.
- OASIS (URL='http://www.oasis-open.org/specs/index.php'). They have issued standards on UDDI, reliability, security, transactions, and eBXML. They develop the UDDI Specs (URL='http://lists.oasis-open.org/archives/tc-announce/200501/msg00000.html'). They also have three security related standards: for exchanging security related information (URL='http://www.oasis-open.org/committees/tc_home.php?wg_abbrev=security') for secure web services (URL='http://www.oasis-open.org/committees/tc_home.php?wg_abbrev=wss') for secure exchange of multiple messages

(URL='http://www.oasis-open.org/committees/tc_home.php?wg_abbrev=ws-sx')

- Web Service Management (WSDM) (URL='http://www.internetnews.com/dev-news/article.php/3488811') which includes both Management of Web Services (MOWS) and Management Using Web Services (MUWS).
- WSDM homepage (URL='http://www.oasis-open.org/committees/tc_home.php?wg_abbrev=wsdm') links (near the bottom) to primers on both MOWS and MUWS.
- Web Service Modeling Ontlology (URL='http://www.w3.org/Submission/WSMO/') Semantic
- Annotations for WSDL Working Group (URL='http://www.w3.org/2002/ws/sawsdl/') are developing standards for semantic web services.
- Links to SOAP, Web Services and related specifications (URL='http://www.soaprpc.com/specifications.html'), unfortunately last updated in 9/05.

Web Services: Articles and Tutorials

- Web Services Primer (URL='http://www.xml.com/lpt/a/760'). Excellent place to start; from O'Reilly and XML.com; many links at bottom of article.
- Web Services Architecture (URL='http://www-128.ibm.com/developerworks/webservices/library/w-ovr/?dwzone=webservices'). Although this claims to describe only IBM's architecture, it is a good description of all the web service architectures. This is a very general article; if you want more information at this level, please look in the section on Web Services - general information.
- Web Services tutorial (URL='http://www.w3schools.com/webservices/default.asp')
- After this tutorial, go to their tutorials on SOAP (URL='http://www.w3schools.com/soap/default.asp') and WSDL (URL='http://www.w3schools.com/wsdl/default.asp')
- (URL='http://www.webreference.com/js/column96/') columns 96 to 106 (each of which is multi-page) includes coverage of calling web services from IE and of writing DTDs for XML.
- (URL="http://mapageweb.umontreal.ca/turner/meta/english/index.html") of how the different technologies are related. May be more useful after you know a little about the area, and then invaluable. There is also an information (URL='http://mapageweb.umontreal.ca/turner/meta/english/index.html')
- xmethod.com tutorials (URL='http://www.xmethods.net/ve2/ViewTutorials.po')
- Web Services (URL='http://msdn.microsoft.com/webservices/webservices/')
- Large collection of articles on Web Services at MSDN. Everything from Web Services Basics (URL='http://msdn.microsoft.com/webservices/webservices/understanding/webservicebasics/default.aspx') to very advanced. For example, start with the (URL='http://msdn.microsoft.com/webservices/webservices/understanding/webservicebasics/default.aspx?pull=/library/en-us/dnwebsrv/html/webservbasics.asp') XML Web Services Basics or the non- technical (URL='http://www.microsoft.com/net/basics/webservicesoverview.asp') See What Web Services Can Do for You and move on to articles on security, problems with SOAP encoding, etc.
- ACM Portal (URL='http://portal.acm.org/toc.cfm?id=1139922&coll=portal&dl=ACM&type=issue&idx=J79&part=periodical&WantType=periodical&title=Communications%20of%20the%20ACM&CFID=485357545&CFTOKEN=485357545') Special Issue of CACM on 'Services Science', which includes Web Services: vol. 49, July 2006. Issue has many articles, including

(URL="http://delivery.acm.org/10.1145/1140000/1139944/p30-spohrer.pdf?key1=1139944&key2=7572863511&coll=portal&dl=ACM&CFID=485357545&CFTOKEN=485357545") Introduction, (URL="http://delivery.acm.org/10.1145/1140000/1139945/p35-chesbrough.pdf?key1=1139945&key2=0203863511&coll=portal&dl=ACM&CFID=485357545&CFTOKEN=485357545") Research Manifesto, (URL="http://delivery.acm.org/10.1145/1140000/1139946/p42-sheehan.pdf?key1=1139946&key2=7113863511&coll=portal&dl=ACM&CFID=485357545&CFTOKEN=485357545") Understanding Service Sector Innovation, (URL="http://delivery.acm.org/10.1145/1140000/1139947/p48-zysman.pdf?key1=1139947&key2=6523863511&coll=portal&dl=ACM&CFID=485357545&CFTOKEN=485357545") Algorithmic Decomposition of Services, and (URL="http://www.acmqueue.com/modules.php?name=Content&pa=showpage&pid=396")

- What we can learn from CORBA, one of the ancestors of web services. Design of Web Services (URL='http://xfront.com'). Thought provoking tutorial from Roger Costello. Costello posts many excellent tutorials here, too.
- Web Services: a B2B demonstration (URL='http://builder.com.com/5100-6389_14-5141373.html?tag=sc')
- Complete sophisticated example. XML-RPC tutorials (URL='http://www.xmlrpc.com/directory/1568/tutorialspress'). See also their homepage (URL='http://www.xmlrpc.com/') is ancestor of SOAP and many SOAP messages imitate this process.
- Combine Polymorphism with Web Services (URL='http://www.developer.com/design/article.php/3459001')
- (URL='http://www-128.ibm.com/developerworks/webservices/edu/ws-dw-ws-wsrm-i.html?S_TACT=105AGX19&S_CMP=ZHP') Understand Web Services Reliable Messaging.
- Google Directory on Web Services and SOAP (URL='http://www.google.com/Top/Computers/Programming/Internet/Service-Oriented_Architecture/Web_Services/SOAP/FAQs,_Help,_and_Tutorials/')
- (URL="http://www.webservicesarchitect.com/content/articles/samtani04.asp") Web services, SOA and Application Frameworks. Making them work together to address issues such as scalability, security, and transaction and state management. Links to a reading list on the left. Site, focuses on web services architecture. Securing Web Services the Low-Tech Way (URL='http://www.developer.com/services/article.php/3649576')

SOAP

Be sure to look in the sections on general information and tutorials (directly above) first; this section includes examples and language-dependent/platform-dependent information.

- SOAP Primer (URL='http://www.w3.org/TR/2003/REC-soap12-part0-20030624/#L1149').
- SOAP tutorial (URL='http://www.w3schools.com/SOAP/default.asp')
- Understanding Web Services: Part I SOAP (URL='http://www-128.ibm.com/developerworks/edu/ws-dw-ws-understand-web-services1.html').You need an IBM password for this, so it is good for academic use and those who use IBM products.
- Archives of SOAP (URL='http://discuss.develop.com/archives/wa.exe?A0=soap&D=0') at Discuss.Develop.com. Discussion list for problems implementing SOAP applications. Searchable.
- SOAP tutorial (URL='http://www.altova.com/documents/aot/webserviceswebbased-

workbook/modelpage_mod04_01.html). From Altova, developer of XMLSpy. First few slides are general and then moves to how to use XMLSpy to implement a web service.

- (URL='http://msdn.microsoft.com/webservices/webservices/understanding/webservicebasics/default.aspx?pull=/library/en-us//dnsoap/html/understandsoap.asp') Understanding SOAP. From MSDN. Excellent article for those with some familiarity from more basic entries.
- Brief SOAP Primer (URL='http://discuss.develop.com/archives/wa.exe?A2=ind0007&L=soap&F=&S=&P=9777')
- (URL='http://www.wdvl.com/Authoring/Scripting/Tutorial/request_headers.html'). Useful reminder as SOAP messages are transported with HTTP
- Web Service Demos (URL='http://www.mindreef.net/tide/scopeit/start.do'). You can see the SOAP messages by clicking the Invoke tab and then the Edit/Preview button.
- AMQP:Advanced Message Queuing Protocol (URL='http://www.computerworld.com/action/article.do?command=viewArticleBasic&articleId=9001328&source=NLT_APP&nlid=48'). From Cisco, RedHat, et al; to interoperate with SOAP and other messaging protocols.
- SOAP tutorials (URL='http://www.soaprpc.com/tutorials/'). About 20 different tutorials with brief comments. Older with some broken links.
- Try A Busy Developer's Guide to SOAP 1.1 (URL='http://www.soapware.org/bdg')
- PerlandSOAP.SOAPexamples(URL='http://www.soaprpc.com/examples/')
- (URL='http://guide.soaplite.com/#quick start guide with soap and soap::lite') Includes an excellent Quick Start Guide, which may also be found at O-Reilly's Perl.com site. (URL='http://www.perl.com/pub/a/2001/01/soap.html') The O-Reilly version links also to more complicated examples (URL='http://www.perl.com/pub/a/2001/04/24/soap.html')
- SOAPLite (URL='http://www.soaplite.com/'). Much of this duplicates the site above, but this one is easier to navigate and includes news.
- Client-side SOAP (URL='http://www.soapuser.com/client3.html'). Also links to table of SOAP packages (URL='http://www.soapuser.com/client1.html') for UNIX and Windows and information on (URL='http://www.soapuser.com/server1.html') server-side SOAP. Good tutorials.
- Archives of SOAP (URL='http://discuss.develop.com/archives/wa.exe?A0=soap&D=0') at Discuss.Develop.com. Discussion list for problems implementing SOAP applications. Searchable.
- PHP and SOAP. Zend offers a SOAP implementation with PHP Extension. Devzone at Zend (URL='http://devzone.zend.com/public/view/tag/tutorials'). New articles and tutorials are in the news (URL='http://devzone.zend.com/public/view/tag/SOAP'). Also more advanced articles, such as on PHP SOAP extension (URL='http://devzone.zend.com/node/view/id/689')
- Tutorial on Web Services and PHP (URL='http://www.topwebnews.com/2006/08/16/web-services-and-php/')
- NuSOAP, older and now less common, implements SOAP in PHP without PHP extensions (URL='http://sourceforge.net/project/showfiles.php?group_id=57663') Download NuSOAP at SourceForge.org (URL='http://64.233.161.104/custom?q=cache:KN6cZ-tqnXhUJ:www.zend.com/zend/tut/tutorial-campbell.php%3Farticle%3Dtutorial-campbell%26kind%3Dt%26id%3D5444%26open%3D1+tutorials+NuSOAP&hl=en&gl=us

&ct=clnk&cd=3') Beginner tutorials from NuSOAP originally from Zend.

- Links to more information on NuSOAP (URL='http://dietrich.ganx4.com/nusoap/')
- eZ (URL='http://ez.no/').
- PEAR (URL='http://pear.php.net/').
- Java and SOAP. Java and XML: SOAP (URL='http://www.onjava.com/pub/a/onjava/excerpt/java_xml_2_ch2/index.html?page=1'). Excellent discussion of SOAP and RPC; first half of multi-page book excerpt is general and second half is Java-specific.
- (URL='http://java.sun.com/webservices/docs/1.5/api/javax/xml/soap/package-summary.html') Java classes to create the SOAP API. Package from Sun. The Sun ONE Studio 4 tutorial.(URL='http://docs.sun.com/source/816-7860/index.html')
- Writing a Handler Class to Process (SOAP)Messages (URL='http://www.developer.com/services/article.php/3503766')
- Web Services Using JavaScript and .NET (URL='http://www.codeguru.com/vb/vb_internet/webservices/article.php/c7781/').
- (URL='http://www-128.ibm.com/developerworks/edu/i-dw-r-jsfwebclient.html?S_TACT=105AGX19&S_CMP=ZHP") Develop Web Services Using JSF Web service tools in Rational Application Developer
- Water (URL='http://www.waterlanguage.org/'). Some people are using it for teaching as it is designed to have a short learning curve.
- Apache web server. Writing a SOAP client for Apache (URL='http://www.xmethods.com/gettingstarted/apache.html#writingapacheclients'). The xmethod site also includes a guide to installing Apache SOAP (URL='http://www.xmethods.com/gettingstarted/apache.html') examples of SOAP services with their WSDL descriptions.
- Introducing Axis2 (URL='http://www.developer.com/services/article.php/3525481'). Axis2 is the web service framework from Apache. A search at this site will lead you to many articles about Axis2, including Building a non-Java web service on Axis2. (URL='http://www.developer.com/services/article.php/3570031')
- Apache (URL='http://ws.apache.org/'). Apache site for SOAP (URL='http://ws.apache.org/soap/').
- Microsoft platform (as your server platform) (URL='http://www.4guysfromrolla.com/webtech/070300-2.shtml') from 4GuysFromRolla.
- (URL='http://www.webreference.com/js/tips/011030.html').A Young Person's Guide to SOAP
- (URL='http://msdn.microsoft.com/msdnmag/issues/0300/soap/'). What SOAP is and how it is better than CORBAS and DCOM
- Microsoft's .NET SDK (URL='http://www.microsoft.com/downloads/details.aspx?FamilyId=9B3A2CA6-3647-4070-9F41-A333C6B9181D&displaylang=en'). This replaces Microsoft's SOAP Toolkit 3.0, officially retired in3/05 and on extended support until 3/08; Also see How to Migrate to the .Net Framework (URL='http://msdn.microsoft.com/webservices/webservices/building/soaptk/default.aspx?pull=/library/en-us/dnsoap/html/stkmigration.asp')

WSDL

- Introducing WSDL (URL='http://www.altova.com/documents/aot/webserviceswebbasedworkbook/ModelPage_MOD02_01.html'). Tutorial from Altova; first few slides are general and then moves to using XMLSpy to create WSDL documents.
- WSDL Tutorial (URL='http://www.w3schools.com/wsdl/default.asp'). This is

the most detailed of the 3 basic introductions listed here.

- Writing WSDL (URL='http://www.webreference.com/js/column96/10.html')
- Understanding WSDL (URL='http://msdn.microsoft.com/webservices/webservices/understanding/webservicebasics/default.aspx?pull=/library/en-us/dnwebsrv/html/understandwsdl.asp'). Excellent detailed explanation from MSDN. Easier if you read one of the previous entries first. From Aaron Skonnard, one of the authors of the book
- Essential XML. (URL='http://www.xmethods.com/ve2/ViewListing.po;jsessionid=S2WT_OQ7zQgg3lZrKESAR15w(QHyMHiRM)?key=uuid:7EAC299F-4D1C-D852-9A9B-B98F25F26092') XMethods lists many web services available publicly for demos and examination. It is helpful to work through several. Highlighted links allow you to examine the WSDL entry and try it.(See also the full list (URL='http://www.xmethods.net/ve2/Directory.po')
- (URL='http://www.xmethods.net/ve2/index.po;jsessionid=XgTwyl97YJjmSCRuGhxH0rSW(QHyMHiRM)') Weather - Temperature is an easy place to start. .(Cape Clear offers a more complex global version of this.)
- (URL='http://www.xmethods.com/ve2/ViewListing.po;jsessionid=S2WT_OQ7zQgg3lZrKESAR15w(QHyMHiRM)?key=uuid:7EAC299F-4D1C-D852-9A9B-B98F25F26092') US Yellow Pages, with its WSDL entry, (URL='http://ws.strikeiron.com/ypcom/ypl?WSDL') is more complicated as it provides several web services in one WSDL.
- Web Service Demos (URL='http://www.mindreef.net/tide/scopeit/start.do'). You can see the WSDL by clicking on the Analyze tab.
- WSDL Editors (URL='http://msdn.microsoft.com/msdnmag/issues/02/12/xmlfiles/default.aspx'). Article is from 12/02, but many of these (e.g. XMLSpy from Altova and the one from Cape Clear (URL='http://www.capescience.com/soa/')

UDDI

- UDDI4J (URL='http://www-128.ibm.com/developerworks/library/ws-uddi4j.html?dwzone=xml').
- IBM, Microfot SAP had problems implementing UDDI (URL='http://weblog.infoworld.com/article/05/12/16/HNuddishut_1.html')
- UDDI-based models of eCommerce (URL='http://www.webservicesarchitect.com/content/articles/siddiqui02.asp')
- (URL='http://www.webservicesarchitect.com/content/articles/irani02.asp') An Introduction to eBusiness
- (URL='http://www.webservicesarchitect.com/content/articles/irani03.asp') eBXML and web services
- eBXML:thevision(URL='http://www.developer.com/java/other/article.php/2204681')

Choreography, BPEL and REST

- w3c Group on Choreography (URL='http://www.w3.org/2002/ws/chor/'). Choreography refers to sequences of web services interactions.
- There is a Overview of the Choreography Model (URL='http://www.w3.org/TR/ws-chor-model/')
- Requirements (URL='http://www.w3.org/TR/ws-chor-reqs/')
- Choreography Description Language (URL='http://www.w3.org/TR/2005/CR-ws-cdl-10-20051109/')
- Primer(URL='http://www.w3.org/TR/2006/WD-ws-cdl-10-primer-20060619/')
- Understanding ebXML, UDDI, XML/EDI (URL='http://www.xml.org/xml/fea-

ture_articles/2000_1107_miller.shtml') eBXML and BPEL and modeling business processes

- SOAP vs eBXML (URL='http://lists.w3.org/Archives/Public/xml-dist-app/2000Dec/0090.html').
- (URL="http://www.internetnews.com/dev-news/article.php/3346001") Web Services Language. To coordinate a series of services - WS-CDL and also see about BPEL. Language may be used to describe peer-to-peer interactions
- (URL="http://www.w3.org/TR/2004/WD-ws-cdl-10-20040427/") Choreography Description Language (WS_CDL).
- Working draft of language spec from w3.org. Orchestration of Rule-Based Web Services (URL='http://www.w3.org/2005/rules/wg/wiki/UCR/BPEL_Orchestration_of_Rule-Based_Web_Services').
- Web Services Integration Patterns (URL='http://webservices.xml.com/pub/a/ws/2004/06/16/patterns.html')
- More patterns (URL='http://webservices.xml.com/pub/a/ws/2004/06/30/patterns.html')
- Reconciling Web Services with REST (URL='http://www.w3.org/2005/Talks/1115-hh-k-ecows/#(1)')
- REST and the Rest of the World (URL='http://www.xml.com/lpt/a/923')
- REST Tutorial (URL='http://www.xfront.com/REST.html')

Web Services: Other Resources

- OASIS and its subsidiary XML.org (URL='http://xml.org')
- The Cover Pages (URL='http://xml.coverpages.org/news2006Q1.html')
- ACM Portal (URL='http://portal.acm.org/portal.cfm')
- WS-I (URL='http://www.ws-i.org/'). Standards, articles and news,
- WebServices.org (URL='http://www.webservices.org/'). Group of about a dozen providers of enterprise-level software for web services, of which the most notable are CA and Systinet. Lots of news about who signed what deal, but most of the useful information is restricted to members of those companies.
- Apache Software Foundation (URL='http://apache.org/'). Source for Apache server, Tomcat java servlet server, Xerces XML parser and many other packages.
- Apache web services projects (URL='http://ws.apache.org/') for SOAP and many others
- Sourceforge.net (URL='http://apache.org/'). Major source for over 100,000 open source projects in enormous range of categories. Search for web services and specific language or environment.
- Eclipse (URL='http://www.eclipse.org/resources/'). Has an entire section on web tools; known primarily for Java IDE, now also offering a PHP IDE and web services tools (URL='http://dev2dev.bea.com/pub/a/2005/09/eclipse_web_tools_platform.html')
- IBM (URL='http://www-128.ibm.com/developerworks/opensource')
- w3c open source tools (URL='http://www.w3.org/Status'). Open Source Web Services Tools in Java.
- DevShed (URL='http://www.devshed.com/'). Sources for good articles. Please also check the tools for many topics for a list of newsletters.
- Management of performance of web applications (URL='http://www.webappmanagement.techweb.com/thankyou.jhtml?_requestid=304797').
- IBM's alphaworks (URL='http://www.alphaworks.ibm.com/rss'). Many excellent articles, some very sophisticated, and also RSS feeds on XML, webservices, etc.

- IBM's Developer Works (URL='http://www-130.ibm.com/developerworks/webservices/'). The Web Services and SOA site has many excellent tutorials, articles and technical forecasts, all of high quality. A few of the longer tutorials require an IBM password.
- XML.com from 'Reilly (URL='http://www.xml.com/'). Organized site with entries ranging from blogs to book excerpts. High quality. O'Reilly also owns Safari on-line books (also available through the ACM if you are a member).
- Web Services Architect (URL='http://www.webservicesarchitect.com/resources.asp'). Magazine with some news, high quality downloadable articles, links to books and other resources
- Microsoft Web Services (URL='http://msdn.microsoft.com/webservices/') General articles and links to their magazine, as well as tutorials on their own products. High quality, but not as many current articles as the IBM sites.
- Sun Developer Network (URL='http://developers.sun.com/techtopics/webservices/index.html'). Mostly focused on Java and on Sun developer kits, though some more general articles.
- Google Directory on Web Services (URL='http://www.google.com/alpha/Top/Computers/Programming/Internet/Service-Oriented_Architecture/Web_Services/'). More focused than a straight search; more specifically: FAQs, tutorials, etc.
- SOAP, (URL='http://www.google.com/Top/Computers/Programming/Internet/Service-Oriented_Architecture/Web_Services/SOAP/FAQs,_Help,_and_Tutorials/')
- Web services (URL='http://www.google.com/Top/Computers/Programming/Internet/Service-Oriented_Architecture/Web_Services/FAQs,_Help,_and_Tutorials/')
- SOA architecture (URL='http://www.google.com/alpha/Top/Computers/Programming/Internet/Service-Oriented_Architecture/Web_Services/').
- XMethods.net (URL='http://www.xmethods.net/ve2/ViewTutorials.posmethods'). Many tutorials (for various platforms) and articles
- XFront (URL='http://xfront.com'). Tutorials and articles from Roger Costello
- (URL='http://devedge-temp.mozilla.org/viewsource/2003/wsdl/01/index_en.html') Mozilla and Web Services. Many articles, including on (URL='http://devedge-temp.mozilla.org/viewsource/2003/wsdl/01/index_en.html') how to access web services in Mozilla agents. Also check their project (URL='http://www.mozilla.org/projects/') related to XML technologies
- SOA Systems (URL='http://www.soasystems.com/ws-standards/') focus on web service standards and on SOA. Links to SOA Magazine (URL='http://www.soamag.com/'). Also useful glossary (URL='http://www.soasystems.com/ws-standards/glossary.asp')
- SOAPRPC (URL='http://www.soaprpc.com/'). News, many links to articles and other resources.
- Developer.com (URL='http://www.developer.com/services/'). Many useful articles, intermediate to advanced; this links to their web services section.
- Gamelan (URL='http://www.developer.com/java/') articles for java developers. Primarily news.
- SearchWebServices (URL='http://searchwebservices.techtarget.com/home/0,289692,sid26,00.html').
- eweek (URL='http://eweek.com'). Regular sources of news for techies; announcements of new standards and products, summaries of conference speeches, etc.

- Loosely Coupled (URL='http://www.looselycoupled.com/'). Similar to above, but more technical and with an SOA focus.
- JackBe (URL='http://www.jackbe.com/NewsEvents/in_the_news.php'). News and short articles on SOA and Ajax.
- Application Development Trends (URL='http://www.adtmag.com/index.aspx'). Primarily news; some whitepapers.
- Ziff Davis Buyer's Guides and White Papers (URL='http://www.webbuyersguide.com/bguide/Whitepaper/Wplibrary.asp_Q_sitename_E_webbuyersguide'). Downloadable after free registration; oriented towards IT management. Similar to above, but more technical and with an SOA focus.
- Fusion Zone: SOA (URL='http://www.fusion-zone.com/soa.html'). Languages and Tools.
- IBM WebSphere (URL='http://www-130.ibm.com/developerworks/websphere/')
- Sun Web Services Developer Kit (URL='http://developers.sun.com/techtopics/webservices/downloads/index.html'). XMLSuite, XMLSpy, etc.Tools for XML, web services, etc. Beautiful interfaces. Academic license available or 30 day trial.
- Stylus Studio (URL='http://www.stylusstudio.com/web_services.html#'). Tools for XML, web services, etc. 7-day trial
- SOAPLite toolkit (URL='http://www.soaplite.com/').
- SOAP client for PHP (URL='http://pear.php.net/package/SOAP')
- Water (URL='http://www.waterlanguage.org/').
- Mindreef SOAPScope Toolkit (URL='http://www.mindreef.net/tide/scopeit/start.do')
- Web services toolkit; free trial (URL="http://www.looselycoupled.com/stories/2003/skills-dev0210.html#content") for web services. The bottom has links to vendor training, such as that at CapeClear (URL="http://www.capescience.com/")
- Collaxa (URL="http://www.collaxa.com/developer.welcome.html")
- (YRL=http://www.grandcentral.com/view/dev.page.home/home/) Grand Central. These sites have lots of information: e.g. tutorials on regular expressions.
- (URL='http://developer.capeclear.com/files/Regular_Expressions.pdf') and Java Web Services at CapeClear (you need to search for tutorials; there are many more than are suggested by the tab on their home page), but some require logins.

Examples of Web Services

- Amazon, eBay and Google (URL="http://internetweek.com/e-business/showArticle.jhtml?articleID=20900637") Amazon Reaches 50,000-Developer Mark With Its Web Services Program: May 24, 2004
- How Amazon has attracted developers to write for the stores which sell on Amazon, or how Amazon Web Services counts itself as having 50,000 developers (URL="http://www.amazon.com/gp/browse.html/ref=smm_sn_aws/002-7644346-5341611?node=3435361") : Help / AWS home page
- Home page for Amazon Web Services - all you need to know is XML and SOAP. (URL='http://www.amazon.com/AWS-home-page-Money/b/ref=gw_br_websvcs/104-7340309-9227160?%5Fencoding=UTF8&node=3435361'). Help / AWS home page / FAQ. Amazon Web Services FAQs - including what services are available etc. AWS includes a web services platform, (URL='http://www.awszone.com/?'), a 'scratch pad' (Turk), a search engine and many e-commerce web services (URL="http://kosmoi.com/Computer/Internet/Web/XML/Amazon/").
- (URL="http://www.websitepublisher.net/article/aws-php/2.") Amazon Web Services:

A Brief Introduction Using PHP - Website Publisher.

- (URL="http://www.xml.com/pub/a/2004/08/04/tr-xml.html") Services and XSLT. Article from O'Reilly describing both the web services and the REST interface to Amazon

Web 2.0 and Related Technologies - The Semantic Web

- Semantic Web initiative at w3c (URL='http://www.w3.org/2001/sw/ ')
- Microformats (URL='http://www.xfront.com/microformats/Purpose-of-Microformats.html'). Costello's slides on the subject; as always a clear discussion.
- Microformats.org (URL='http://microformats.org/about/')
- State of the Art on (Semantic) Modeling in XML (URL='http://www-128.ibm.com/developerworks/xml/library/x-think30.html')
- A Simple Linkage from Web Services to the Semantic Web (URL='http://www.w3.org/2006/Talks/0306-ep-spdl/#(1)')

AJAX (Asynchronous JavaScript and XML)

- AjaxPatterns.org (URL='http://ajaxpatterns.org/Books')
- Ajaxian.com (URL='http://ajaxian.com/by/topic/books/') reviews (URL='http://ajaxian.com/by/topic/book-reviews/')
- The XMLHttpRequest Object (URL='http://www.w3.org/TR/XMLHttpRequest/')
- Fields in the HTTPRequest Object (URL='http://www.w3.org/Protocols/HTTP/HTRQ_Headers.html')
- The Beginning of Ajax Standardization (URL='http://www.devsource.com/print_article2/0,1217,a=176625,00.asp')
- What is Ajax? (URL='http://www.ajaxinfo.com/default~viewart~5.htm'). Very basic introduction.
- What is Ajax? (URL='http://webdesign.about.com/od/ajax/a/aa101705.htm')
- AJAX introduced (URL='http://adaptivepath.com/publications/essays/archives/000385.php'). The article which named it - with a great introduction to what it is
- About.com's introduction to Ajax (URL='http://webdesign.about.com/b/a/210999.htm?nl=1')
- Keeping Up with the Ajax Trend (URL='http://www.developer.com/java/ent/article.php/3562876'). Slightly older but very clear introduction. Links to various Ajax debugging tools. Ajax - the issues in using. Good introduction
- Ajax - Cut thru the hype (URL='http://www-128.ibm.com/developerworks/web/library/wa-ajaxtop1/index.html'). Excellent discussion of pros and cons of using Ajax; from IBM DeveloperWorks.
- Demystifying the Buzz about Ajax (URL='http://www.webforefront.com/archives/2005/05/ajax_demystifyi.html'). Assumes you are familiar with browser objects and intricacies
- Ajax and REST (URL='http://www-128.ibm.com/developerworks/web/library/wa-ajaxarch2.html'). A further discussion of what it takes to develop Ajax sites.
- Ajax and alternatives (URL='http://www.ajaxinfo.com/default~viewart~8.htm'). Pros and Cons of Ajax vs. applets, XUL, etc. Useful discussion, but you won't be surprised to learn these folks love Ajax.
- Measuring the Benefits of Ajax (URL='http://www.developer.com/java/other/article.php/10936_3554271_1').
- From an Ajax developer, who also wonders (URL='http://www.developer.com/java/web/article.php/3574116')

- Will Ajax Replace the Desktop? (URL='http://www.publish.com/article2/0,1895,1900778,00.asp')
- Top 10 Reasons Ajax is Here to Stay (URL='http://www.developer.com/java/other/article.php/356'7706')
- Ajax Is No Overnight Success (URL='http://www.eweek.com/article2/0,1895,1911715,00.asp')
- Including Security and Performance Risks (URL='http://www.eweek.com/article2/0,1895,1916673,00.asp')
- EWeek (URL='http://www.eweek.com/article2/0,1895,1998795,00.asp?kc=EWWSUEMNL080906EOAD')
- Ajax has some problems (URL='http://www.eweek.com/article2/0,1895,1949606,00.asp')
- Wikipedia article (URL='http://en.wikipedia.org/wiki/Ajax_%28programming%29')
- Ajax has accessibility issues (URL='http://www.eweek.com/article2/0,1895,1987300,00.asp'), but developers are working to overcome them (URL='http://www.eweek.com/article2/0,1895,1987300,00.asp')
- Putting Ajax to Work (URL='http://www.infoworld.com/article/05/10/17/42FEajaxcase_1.html'), examples and a summary of some of the toolkits.
- A detailed Explanation of the XMLHttpRequestObject (URL='http://www.devx.com/webdev/Article/33024/1954?pf=true'). This very clear article tells you all you need to know. (You can even omit the last part about interfacing with Java.).
- The XML HttpRequest Object (URL='http://developer.apple.com/internet/webcontent/xmlhttpreq.html'). Clear with nice summary tables; read the previous article first.
- Ajax Tutorial (URL='http://www.xul.fr/en-xml-ajax.html'). Very basic. Uses innerHTML and not yet aware of IE 7, which supports XMLHttpRequest objects.
- How to Use Ajax (URL='http://www.peachpit.com/articles/article.asp?p=425820&rl=1'), from Peachpit, which caters to graphic designers.
- Ajax Tutorial (URL='http://www.tizag.com/ajaxTutorial/'), from Tizag: level similar to w3schools tutorials.
- Getting Started with Ajax (URL='http://www.alistapart.com/articles/gettingstartedwithajax'). From A List Apart. One example uses innerHTML (not supported by assistive browsers), but a richer example than most tutorials.
- XUL.fr/en/ (URL='http://www.xul.fr/en/'). How to Develop Web Applications with Ajax (URL='http://www.webreference.com/programming/javascript/jf/column12/'). Series starts in column12, continues on to column13, 14 and 15. Good explanation.
- Understanding Ajax (URL='http://www.computerworld.com/action/article.do?command=viewArticleBasic&articleId=9002507&source=NLT_APP&nlid=48'). Longer tutorials: first is basic and second uses frames and non-cross browser examples, but also gets to more advanced tricks
- Alternate Ajax Technologies (URL='http://www.webreference.com/programming/ajax_tech/')
- Ajax and the Java Platform (URL='http://java.sun.com/developer/technicalArticles/J2EE/AJAX/index.html?cid=59754'). Good article to get started, which uses almost no Java (use the server response of your choice). Note that Sun's Java Studio Creator 2 provides Ajax-like functionality for those who don't want to use Ajax directly.
- Αφαξ ανδ Συν϶σ ϑαϖα ΔΒ (URL='http://www.johnwiseman.ca/blogging/?p=61')
- Usng Ajax from Perl (URL='http://www.perl.com/pub/a/2006/03/02/ajax_and_perl.html')
- Ajax Active Tables (URL='http://www.soxiam.com/Code/AJAXActiveTableLibrary')

- Symfony is a PHP toolkit (URL='http://www.symfony-project.com/')
- AHAH is a subset of Ajax (URL='http://microformats.org/wiki/rest/ahah')
- HowtoUseAjax(URL='http://www.informit.com/articles/article.asp?p=425820&rl=1'). It leads to Creating Smart Forms with Ajax which includes form validation and downloadable code.
- Nitty Gritty Ajax (URL='http://www.webmonkey.com/06/15/index3a.html'). Shows how one drop-down menu can control the choices in the next drop-down menu.
- U. Minnesota Duluth References on JavaScript (URL='http://www.d.umn.edu/itss/support/Training/Online/webdesign/javascript.html#top'). This site is a great reference. For example, there are links to a number of articles here about making Ajax accessible as well as to tutorials.
- Tutorials from Ajax Matters (URL='http://www.ajaxmatters.com/blog/ajax-tutorials/').Many tutorials, with descriptions, from the basic to advanced. Ajax Links found useful by developer/blogger John Wiseman

More Advanced Articles and Other Resources, Including Examples and Frameworks

- IBM Resource Center on Ajax (URL='http://www-128.ibm.com/developerworks/ajax'). Many articles and more advanced tutorials. A very rich site, worth checking regularly. For example, see Part 1 (URL='http://www-128.ibm.com/developerworks/web/library/wa-ajaxarch/index.html') and Part 2 (URL='http://www-128.ibm.com/developerworks/java/library/wa-ajaxarch2.html'). See also (URL='http://www-128.ibm.com/developerworks/search/searchResults.jsp?searchType=1&searchSite=dW&searchScope=dW&query=Mastering+Ajax&Search.x=0&Search.y=0&Search=Search') Mastering Ajax - 8 part series. This excellent series starts at (URL = http://www-128.ibm.com/developerworks/web/library/wa-ajaxintro1.html). Part 2 is at (URL=http://www-128.ibm.com/developerworks/web/library/wa-ajaxintro2/) and changing the '2' at the end of the URL will get you to the other articles, except for Parts 7 and 8 start at (URL=http://www-128.ibm.com/developerworks/web/library/wa-ajaxintro7.html). (Parts 4-6 also explain how to manipulate the DOM).
- Ajax Transport (URL='http://www-128.ibm.com/developerworks/edu/x-dw-x-ajaxtrans-i.html?S_TACT=105AGX59&S_CMP=GR&ca=dgr-lnxw03AjaxTransport') may be done by methods other than XMLHttp, as explained in this tutorial.
- Mozilla's Ajax Center (URL='http://developer.mozilla.org/en/docs/AJAX').
- AjaxInfo.com (URL='http://www.ajaxinfo.com/'). Ajaxian is one of the oldest and largest Ajax resources. Forums on many topics, book lists and reviews, articles, reviews of frameworks for PHP, Perl, etc. Not to be missed
- Ajax Patterns (URL='http://ajaxpatterns.org/'). Wiki on all things related to Ajax - patterns, frameworks, libraries, books etc. Rich resource for programmers. Check out the popular pages. (URL='http://ajaxpatterns.org/Special:Popularpages')
- Open Ajax (URL='http://www.infoworld.com/article/06/02/01/74989_HNvendorspromoteajax_1.html?source=NLC-AD2006-02-02').
- Ajax Matters (URL='http://www.ajaxmatters.com').
- AjaxDNA (URL='http://www.ajaxdna.com/').
- Ajax World Magazine (URL='http://ajax.sys-con.com/')
- (URL='http://www.developerzone.biz/index.php?option=com_content&task=view

&id=82&Itemid=9'), Extending Ajax with Flash (URL='http://www.informit.com/articles/article.asp?p=418664&rl=1')

- Jackbe articles (URL='http://www.jackbe.com/Resources/resources.php')
- JSON may be used to speed up AJAX (URL='http://www.developer.com/lang/jscript/article.php/3596836'). JSON is JavaScript Object Notation (URL='http://www.json.org/'). See also JSON for the Masses (URL='http://www.dustindiaz.com/json-for-thc-masses/'). JSON vs XML/XMLHttpRequest (URL='http://web2journal.com/read/203935.htm')
- Using Ajax to generate random codes (URL='http://www.devarticles.com/c/a/JavaScript/JavaScript-Remote-Scripting-An-AJAXbased-Random-Code-Generator-in-Action/'). Uses this example to explore Ajax. The same author also show how to use Ajax to develop a chat application (URL='http://www.devarticles.com/c/a/XML/Building-an-AJAXBased-Chat-The-Barebones-Structure/').
- Using XML HTTP Request to include Javascript files and functions (URL='http://www.webreference.com/programming/javascript/mk/')
- Ajax Design Strategies from Sun (URL='http://whitepapers.zdnet.com/thankyou.aspx?authId=zasVTTxor1p0/GoW+gO8x7lFYCZVhWt1&&promo=590&tag=nl.e590&docid=264562&view=264562&load=1'). Callback and Ajax (URL='http://www.devsource.com/print_article2/0,1217,a=170236,00.asp'). Open Ajax Alliance (URL='http://www.openajax.org/')
- EWeek (URL='http://www.eweek.com/category2/0,1874,1949411,00.asp')
- Using Ajax and JSON with Oracle (URL='http://www.oracle.com/technology/pub/articles/cioroianu-ajax-data.html').
- Google Group on Ajax (URL='http://groups.google.com/group/ajaxpro?lnk=gschg&hl=en')
- Pre-loading Images with Ajax and the DOM (URL='http://www.devarticles.com/c/a/JavaScript/Preloading-Images-with-the-DOM-The-Introductory-Process/'). How to pre-load thumbnails and then fetch all the full-size images with the first click on a thumbnail.
- WebReference (URL='http://www.webreference.com/programming/javascript/')
- Creating a double drop-down list with Ajax (URL='http://www.developer.com/services/article.php/3575081')
- Including (URL='http://www.developer.com/services/article.php/3577826')
- Expanding a navigational menu with Ajax (URL='http://www.getelementbyid.com/scripts/index.aspx?CodeID=40')
- How to improve your search engine ratings with Ajax (URL='http://www.informit.com/articles/article.asp?p=517207&rl=1'). (or at least not have them get worse when using Ajax).
- Ajax and mutual exclusion (URL='http://www.developer.com/lang/jscript/article.php/3592016'). How to avoid concurrency problems.
- Examples of Applications Which Use Ajax, Top 10 Ajax applications (URL='http://www.aventureforth.com/2005/09/06/top-10-ajax-applications/'). Zimbra, (URL='http://www.zimbra.com/')
- Google Search with Ajax (URL='http://code.google.com/apis/ajaxsearch/'). API for putting a Google search on your web page using Ajax.
- Using the XML Http Request object (URL='http://jibbering.com/2002/4/httprequest.html')
- Content Cafe (URL='http://contentcafe.btol.com/ClientApplication/ContentCafe.

aspx?UserID=quantum&Password=books &ItemKey=1590595823')

- Simple Quiz with Ajax (URL='http://www.codeproject.com/useritems/SimpleAjaxQuizUsingAtlas.asp').
- ExpandingDOMtreeinspector(URL='http://www.kawa.net/works/ajax/tips/dump/domtree.html')
- Frameworks. Sites with links to many frameworks. AjaxInfo.com (URL='http://www.ajaxinfo.com/default~area~components.htm')
- Αφαξιαν has articles on frameworks for Ajax (URL='http://ajaxian.com/by/topic/framework/')
- Ajax used with PHP (URL='http://ajaxian.com/by/topic/php/') Ruby etc. Includes reviews by users. Great resource, but it may help to start with AjaxInfo before delving into this rich site.
- Xul.com (URL='http://www.xul.fr/ajax-frameworks.html').
- Seven Ajax Frameworks to watch out for (URL='http://www.indicthreads.com/articles/447/seven_ajax_frameworks_toolkits.html').
- Ajax Patterns (URL='http://ajaxpatterns.org/Frameworks')
- Open Source Technology (URL='http://ashko.blogspot.com/2006/11/best-top-ten-open-sourceajaxdhtml.html').
- XUL.fr/en/ (URL='http://www.xul.fr/en/').
- 12 Days of Ajax (URL='http://www.developer.com/xml/article.php/3645666'). Useful set of links on the last page.
- Prototype with (URL='http://www.symfony-project.com/') the symphony framework for PHP—Symphony tutorial (URL='http://www.symfony-project.com/tutorial/symfony_ajax.html')
- This (URL="http://blogs.ebusiness-apps.com/jordan/pages/Prototype%20Library%20Info.htm") overview is a good place to start, then this (URL="http://www.digital-mediaminute.com/article/1629/using-prototypejs-v131"). Prototype Documentation (URL="http://wiki.script.aculo.us/scriptaculous/show/Prototype") is also available at script.aculo.us (URL="http://script.aculo.us/").
- Joseph Scott blogs on Prototype (URL='http://joseph.randomnetworks.com/tag/ajax')
- Script.aculo.us (URL='http://script.aculo.us/'). A large open source library, almost as popular as Prototype, on which it is based; integrates with Perl, PHP, Java, Ruby and everything else you can think of; scripts for animation, drag and drop, etc;
- DOM utilities. Easy to install. (URL='http://wiki.script.aculo.us/scriptaculous/show/Usage') A Treasure trove.
- DoJo (URL='http://dojotoolkit.org/'). Has a new documentation kit (URL='http://dojotoolkit.org/api/')
- Ajaxian (URL='http://ajaxian.com/by/topic/dojo/') JSPAN (URL='http://www.ajaxinfo.com/default~area~COMPONENTS~compid~20.htm'). Integrates with JavaScript and PHP
- AjaxTags (URL='http://ajaxtags.sourceforge.net/').
- Clean Ajax (URL='http://sourceforge.net/project/showfiles.php?group_id=145307')
- Mozilla's AJAX Tolkit (URL='http://developer.mozilla.org/en/docs/AJAX:Getting_Started')
- FlapJax (URL='http://www.flapjax-lang.org/tutorial/')
- Yahoo User Interface Library (URL='http://developer.yahoo.com/yui/')
- Yahoo's connection manager (URL='http://www.developer.com/lang/article.php/3600611')
- Google web toolkit for Ajax (URL='http://code.google.com/webtoolkit/')
- How to make your own Google Maps, host your own Gmail (URL='http://www.

informationweek.com/news/showArticle.jhtml?articleID=189400799').

- JackBe (URL='http://www.jackbe.com/Company/about.php')
- MochiKit (URL='http://mochikit.com/') with the DOM and AJAX. There is a Google Group on Mochikit (URL='http://groups.google.com/groups/search?hl=en&q=Mochikit&qt_s=Search')
- (URL="http://moofx.mad4milk.net/#intro") Special effects in JavaScript, integrates with prototype and mootools frameworks
- Tibco (URL='http://www.computerworld.com/action/article.do?command=viewArticleBasic&articleId=9003795')

Compilation of References

Abiteboul, S., Buneman, P., & Suciu, D. (2000). *Data on the Web - From relations to semistructured data and XML*. San Francisco: Morgan Kaufmann.

Ahlgren, R., & Markkula, J. (2005, June 13-15). *Design patterns and organisational memory in mobile application development.* Paper presented at the Sixth International Conference on Product Focused Software Process Improvement (PROFES '05), Oulu, Finland.

Ahluwalia, K. S., & Jain, A. (2006, October 21-23). *High availability design patterns.* Paper presented at the 13th Conference on Pattern Languages of Programs (PLoP '06), Portland, USA.

Almeida, V.A.F., & Menasce, D.A. (2002). Capacity planning for Web services: An essential tool for managing Web services. IT Professional, (July-August), 33-38.

Alonso, G., Casati, F., Kuno, H. & Machiraju, V. (2004). *Web services—Concepts, architectures and applications.* Springer Verlag.

Alur, D. Crupi, J., & Malks, D. (2003). *Core J2EE patterns: Best practices and design strategies*, 2nd Edition. Upper Saddle River, NJ: Sun Microsystems Press / Prentice Hall.

Ambler, S. W. (2003). *Agile database techniques.* Indianopolis, IN: Wiley.

Ambler, S. W. (2004) *The Object primer: Agile model-driven development with UML 2.0.* Cambridge University Press.

Ambler, S.W. (2007). Test-driven development of relational database. *IEEE Software*, *24*(3), 37-43.

American Electronics Association. (2004). *Offshore outsourcing in an increasingly competitive and rapidly changing world: A high-tech perspective.* Retrieved on January 1, 2007, from http://www.aeanet.org/publications/IDMK_AeA_Offshore_Outsourcing.asp

Anderson, D. (2004). *Agile management for software engineering: Applying the theory of constraints for business results.* Upper Saddle River, NJ: Prentice Hall PTR.

Andrews, T., F. Curbera, et al. (2003). *Business process execution language for Web services Version 1.1.*

Aoyama, M. (1998).Web-based agile software development, *IEEE Software*, 57-65.

Apache Software Foundation. (2006). *Apache XMLBeans.* Retrieved on July, 2007, from http://xmlbeans.apache.org/index.html

Apicello, Mario (2000). Multilizer for Java powers your apps to travel the Globe. *Infoworld*, January.

Arnowitz, J., Arent, M., & Berger, N. (2007). *Effective Prototyping for Software Makers.* San Francisco: Morgan Kaufmann Publishers.

Aspray, W., Mayadas, F., & Vardi, M. Y. (Eds.) (2006). *Globalization and offshoring of software: A report of the ACM Job Migration Task Force.* New York: Association for Computing Machinery. Retrieved on January 1, 2007, from http://www.acm.org/globalizationreport

Axelrod, R. (1984). *The Evolution of cooperation.* New York: Basic Books.

Axtell, Rodger (1993). *Do's and Taboos Around the World*, John Wiley & Sons.

Bailey, B. P., & Konstan, J. A. (2003). Are informal tools better? Comparing DEMAIS, pencil and paper, and Authorware for early multimedia design. In *Proceedings of the SIGCHI Conference on Human Factors in Computing Systems* (pp 313-320). New York: ACM Press.

Ballard, B. (2007). *Designing the mobile user experience*. John Wiley and Sons.

Barry, D., & Stanienda, T. (1998). Solving the Java object storage problem. *IEEE Computer, 31*(11), 22-40.

Battin, R. D., Crocker, R., Kreidler, J., & Subramanian, K. (2001). Leveraging resources in global software development. *IEEE Computer, 18*(2), 70-77.

Bauer, F. (1972). Software Engineering. *Information Processing , 71.*

Baumer, D., Bischofberger, W., Lichter, H., & Zullighoven, H. (1996). User interface prototyping-concepts, tools, and experience. In *Proceedings of the 18th International Conference on Software Engineering (ICSE'96)* (pp 532-541). IEEE Computer Society.

Baure, C., and King, G. (2004). *Hibernate in Action.* Greenwich, CT: Manning Publications.

Beaulieu, M. (2002). *Wireless Internet applications and architecture*. Boston: Addison-Wesley.

Beck, K. (1999). *Extreme programming explained: Embrace change.* Harlow: Addison-Wesley.

Beck, K. (2001). Aim, fire. *IEEE Software, 18*(5), 87-89.

Beck, K. (2003). *Test-driven development: By example.* The Addison-Wesley signature series. Boston: Addison-Wesley.

Beck, K. (2005). *Extreme programming explained.* Upper Saddle River, NJ: Pearson Education.

Beck, K., & Andres, C. (2004). *Extreme programming explained: Embrace change.* Boston: Addison-Wesley.

Beck, K., & Fowler, M. (2001). *Planning Extreme Programming,* Addison-Wesley..

Beck, K., Beedle, M, van Bennekum, A, Cockburn, A. et al. (2001). *Manifesto for agile software development.* Retrieved on January 1, 2007, from http://www.agilemanifesto.org

Beck, K., Crocker, R., Meszaros, G., Vlissides, J., Coplien, J. O., Dominick, L., & Paulisch, F. (1996, March 25-29)). *Industrial experience with design patterns.* Paper presented at the 18th International Conference on Software Engineering (ICSE 1996), Berlin, Germany.

Becker S., & Berkemeyer, A. (2002). Rapid application design and testing for usability. IEEE Multimedia, (Oct-Dec), 38-46.

Becker, S. A., & Bostelman, M. L. (1999). Aligning strategic and project measurement systems. *IEEE Software, 16*(3), 46-51.

Becker, S., & Mottay, F. (2001). A global perspective of Web usability for online business applications. IEEE Software, 18(1), 54-61.

Behravesh, N. & Klein, L. (2004). *The comprehensive impact of offshore IT software and services outsourcing on the US economy and the IT industry.* Report prepared by Global Insight for the Information Technology Association of America, Arlington, VA.

Berners Lee, T. (2004). *New top level domains .mobi and .xxx considered harmful.* Retrieved on January, 2007, from http://www.w3.org/DesignIssues/TLD

Bhat, J., Gupta, M., & Murthy, S. (2006). Overcoming requirements engineering challenges: Lessons from offshore outsourcing. *IEEE Software, 23*(5), 38-44.

Boehm, A., & Lowe, D. (2006). *Murach's ASP.NET 2.0 Web programming with VB 2005.* Fresno, CA: Mike Murach & Associates.

Boehm, B. (1976). Software engineering. *IEEE Transactions on Computers*, 25, 1226-1241.

Boehm, B. (1981). *Software engineering economics.* Upper Saddle River, NJ: Prentice Hall PTR.

Boehm, B. W., & Turner, R. (2003). *Balancing agility and discipline: A guide for the perplexed.* Addison-Wesley Professional.

Boehm, B. W., Abts, C., Brown, A. W., Chulani, S., Clark, B. K., Horowitz, E., Madachy, R., Reifer, D., & Steece, B. (2001). *Software cost estimation with COCOMO II.* Prentice Hall.

Boehm, B., & Hansen, W. J. (2001). *The spiral model as a tool for evolutionary acquisition.* Retrieved on September 27, 2007, http://www.stsc.hill.af.mil/crosstalk/2001/05/boehm.html

Boehm, B., & Turner, R. (2003). *Balancing agility and discipline: A guide for the perplexed.* Boston: Addison Wesley Professional.

Boehm, B., & Turner, R. (2004). *Balancing agility and discipline.* Boston: Pearson Education.

Booch, G., Rumbaugh, J., & Jacobsen, I. (2000). *The Unified Modeling Language user guide.* Boston: Addison-Wesley.

Bosak, J. (1997). *XML, Java, and the future of the Web.* Retrieved on July, 2007, from http://www.ibiblio.org/pub/sun-info/standards/xml/why/xmlapps.htm

Bowen, B. (1997). *Software Reuse with Java Technology: Finding the Holy Grail.* Retrieved on from www.javasoft.com/features/1997/may/reuse.html

Boyatzis, R. (1982). *The competent manager: A model for effective performance.* Wiley.

BPEL Coalition (2006). *Business process execution language for Web services version 1.1.* Retrieved on April 14, 2008 from http://www.ibm.com/developerworks/library/ws-bpel/

Brajnik, G. (2001, June 4-6). *Towards valid quality models for Web sites.* Paper presented at the Seventh Conference on Human Factors and the Web (HFWeb '01), Madison, USA.

Brandon, Daniel (2002). Issues in the globalization of electronic commerce. In V. K. Murthy and N. Shi (Eds.) *Architectural issues of Web-enabled electronic business.* Hershey, PA: Idea Group Publishing.

Brandon, Daniel (2006). *Project management for modern information systems.* Hershey PA: Idea Group Publishing .

Brehm, N., & J. M. Gómez (2006). *Distribution of ERP system components and security considerations.* Paper presented in the 17th IRMA International Conference - Managing Modern Organizations with Information Technology, Washington, USA.

Britton, C., & Bye, P. (2004). *IT architecture and middleware.* Addison-Wesley.

Brooks, F. P. (1995). *The mythical man-month: Essays on software engineering.* Boston: Addison Wesley Professional.

Brown, W. J., Malveau, R. C., McCormick, H. W., & Mowbray, T. J. (1998). *AntiPatterns: Refactoring software, architectures, and projects in crisis.* John Wiley and Sons.

Bry, F. & Schaffert, S. (2002). Towards a declarative query and transformation language for XML and semistructured data: Simulation unification. In *Proceedings Intl. Conference on Logic Programming.* LNCS 2401, Springer-Verlag.

Burkhardt, J, Henn, H., Hepper, S., Rintdorff, K., & Schack, T. (2002). *Pervasive computing technology and architecture of mobile Internet applications.* London: Addison-Wesley.

Burton-Jones, A., Storey, V. C., Sugumaran, V., & Ahluwalia, P. (2005). A semiotic metrics suite for assessing the quality of ontologies. *Data and Knowledge Engineering, 55*(1), 84-102.

Buschmann, F., Meunier, R., Rohnert, H., Sommerlad, P., & Stal, M. (1996). *Pattern-oriented software architecture: A system of patterns.* Chichester: Wiley.

Cagen, J. & Vogel, C. M. (2002). *Creating breakthrough products: Innovation from product planning to program approval.* Upper Saddle River, NJ: Prentice Hall.

Carmel, E., & Agarwal, R. (2001). Tactical approaches for alleviating distance in global software development. *IEEE Software*, *18*(2), 22-29.

Carzaniga, A., Fuggetta, A., Hall, R., Heimbigner, D., van der Hoek, A., & Wolf, A. (1998). *A characterization framework for software deployment technologies* (Tech. Rep. CU-CS-857-98).Boulder, CO: University of Colorado, Department of Computer Science.

Cavangess, C. (2004). *Programming Jakarta struts.* Sebastopol, CA: O'Reilly.

Cheesman, J. & Daniels, J. (2002). *UML components: A simple process for specifying component-based software.* Reading, MA: Addison-Wesley.

Chen, J. Q., & Heath, R. D. (2005). Web application development methodologies. in W. Suh (Ed), *Web engineering: Principles and techniques.* Hershey, PA: Idea Group Publishing.

Chen, J. Q., & Heath, R. D. (2005). Web application development methodologies. In W. Suh (Ed.), *Web engineering: Principles and techniques.* Hershey, PA: Idea Group Publishing.

Chen, M. (2004). A methodology for building mobile computing applications. *International Journal of Electronic Business, 2*(3), 229-243.

Christensen, C. M. (1997). *The innovator's dilemma: When new technologies cause great firms to fail.* Boston: Harvard Business School Press.

Chung, L., Nixon, B. A., Yu, E., & Mylopoulos, J. (2000) *Non-functional requirements in software engineering.* Boston:.Kluwer Academic Publishers.

CIO Insight (2003). Outsourcing: How well are you managing your partners?. *CIO Insight*, *1*(33), 75-85.

Clark, B. (2004). *Enterprise application integration using .NET.* Addison-Wesley.

Cleland, D. (1998). Stakeholder management. In J. Pinto. (Ed.) *Project management handbook.* Wiley.

Cloyd, M.H. (2001). Designing user-centered Web applications in Web time. IEEE Software, 18(1), 62-69.

Cockburn, A. (2003). *Agile Software Development.* Boston, MA: Addison Wesley Professional.

Cockburn, A. (2005). *Crystal clear, A human-powered methodology for small teams.* Boston: Addison-Wesley.

Coda, F., Ghezzi, C., Vigna, G., & Garzotto, F. (1998, April 16-18). *Towards a software engineering approach to Web site development.* Paper presented at the Ninth International Workshop on Software Specification and Design (IWSSD-9), Ise-shima, Japan.

Collins, R.W. (2002). Software localization for Internet software: Issues and methods. IEEE Software, (March/April), 74-80.

Conallen, J. (2003). *Building Web applications with UML, 2/E.* Addison-Wesley.

Conway, M. E. (1968). How do committees invent? *Datamation*, *14*(4), 28-31.

Cooper, A. (1994). The perils of prototyping. *Visual Basic Programmer's Journal*, August September 1994, 1.

Crawford, L. (2004). Global body of project management knowledge and standards. In Morris and Pinto (Eds.) *The Wiley guide to managing projects.* John Wiley & Sons.

Crnkovic, I. & Larsson, M. (2000). A case study: Demands on component-based development. In *Proceedings of the 2nd International Conference on Software Engineering,* (pp. 23-31). ACM Press.

Crnkovic, I. (2003). *Component-based software engineering – New challenges in software development.* Paper presented at the 25th International Conference on Information Technology Interfaces. Cavtat, Croatia.

Cusick, J., & Prasad, A. (2006). A practical management and engineering approach to offshore collaboration. *IEEE Software*, *23*(5), 20-29.

da Silva, P. P. (2001). *User interface declarative models and development environments: A survey.* (LNCS 1946, 207-226).

Dart, S. (2001). Configuration management: A missing link in Web engineering. Norwood, MA: Arttech House.

Davis, M. (2001, February) *Struts, an open-source MVC implementation.* Retrieved on January 3, 2007, from http://www-128.ibm.com/developerworks/java/library/j-struts/

Davis, M., & Smith , H., (1999). The Java International API: Beyond JDK 1.1. *Java Report,* February.

Davison, D. (2003). *Top 10 Risks of offshore outsourcing.* META Group Research.

DeMarco, T., & Lister, T. (1999). *Peopleware: Productive projects and teams.* New York: Dorset House.

Deshpande Y., Olsina, L., & Murugesan, S. (2002). Web engineering. Report on the Third ICSE Workshop on Web Engineering, ICSE2002, Orlando, FL, USA.

Deshpande, Y., & Hansen, S. (2001) Web engineering: Creating a discipline among disciplines. *IEEE Multimedia, April-June,* 82-87.

Deshpande, Y., Ginige, A., Murugesan, S., & Hansen, S., (2002). Consolidating Web engineering as a discipline. SEA Software, (April), 32-34.

Deshpande, Y., Murugesan, S., et al. (2002). WEB ENGINEERING. *Journal of Web Engineering, 1*(1), 003-017.

Di Lucca, G. A., A. R. Fasolino, et al. (2002). Testing Web applications. Software Maintenance, 2002. In *Proceedings.* International Conference on.

Dignan, L. (2003). Leaping, then looking. *Baseline, 1*(22), 17-29.

Disabatino, J. (2000). Web site globalization. *ComputerWorld, July.*

DoD C4ISR Architecture Working Group (1997). *C4ISR architecture framework.* Version 2. Retrieved May 18, 2007, from http://www.fas.org/irp/program/core/fw.pdf

Duarte, D. L., & Snyder, N. T. (2006). *Mastering virtual teams.* Hoboken, NJ: Jossey-Bass.

Dustin, E., Rashka, J., & McDiarmid, D. (2001). *Quality Web systems: Performance, security, and usability.* Addison-Wesley.

Eclipse (2007). *Eclipse process framework.* Retrieved May 18, 2007, from http://www.eclipse.org/epf

Eisenberg, R. (2004). Service-oriented architecture: The future is now. *Intelligent Enterprise, April 17.*

Engardio, P., Bernstein, A., Kripalani, M., Balfour, F., Grow, B., & Greene, J. (2003). The new global job shift. *Business Week*, February, 36-42,44,46.

Escalona, M. J., & Koch, N. Requirements engineering for Web applications—A comparative study. *Journal of Web Engineering, 2*(3), 193-212.

Evans, E. (2003). *Domain-driven design: Tackling complexity in the heart of software.* Addison-Wesley.

Farrell, D. (2003). *Offshoring: Is it a win-win game?* McKinsey Global Institute, Aug 2003.

Federal Office of Management and Budget (2007). *Federal enterprise architecture framework.* Retrieved May 18, 2007, from http://www.whitehouse.gov/omb/egov/a-2-EAModelsNEW2.html

Fenton, N. E., & Pfleeger, S. L. (1997). *Software metrics: A rigorous & practical approach.* International Thomson Computer Press.

Fernandes, T. (1995). *Global interface design.*Academic Press.

Ferranti, M. (1999). From global to local. *Infoworld, October.*

Ferranti, M. (2000). Globalization tidal wave. *Infoworld, November.*

Firesmith, D., & Henderson-Sellers, B. (2002). *The OPEN process framework: An Introduction.* Addison-Wesley.

Floyd, C. (1984). A systematic look at prototyping. In Budde, R., Kuhlenkamp, K., Mathiassen, L. & Zullighoven, H., *Approaches to prototyping*, (pp 1-18). Berlin: Springer Verlag.

Fowler, M. (2001). Separating user interface code, *IEEE Software*, *18*(2), 96-97.

Fowler, M. (2003). *Patterns of enterprise application architecture*. Boston: Addison-Wesley.

Fowler, M. (2006). *Using an agile software development process with offshore development.* Retrieved on January 1, 2007, from http://www.martinfowler.com/articles/agileOffshore.html

Frank, C. E., Naugler, D., & Traina, M. (2005). Teaching user interface prototyping. *Journal of Computing Sciences in Colleges*, *20*(6), 66-73. Consortium for Computing Sciences in Colleges.

Gamma, E., Helm, R., Johnson, R., & Vlissides, J. (1995). *Design patterns: Elements of reusable object-oriented software*. Addison-Wesley.

Garcia-Molina, H., Papakonstantinou, Y., Quass, D., Rajaraman, A., Sagiv, Y., Ullman, Y. D., Vassalos, V. & Widom, J. (1997). The TSIMMIS approach to mediation: Data models and languages. *Journal of Intelligent Information Systems*, *8*(2), 117-132.

Garrett, J. (2005). *Ajax: A new approach to Web applications.* Retrieved on July, 2007, from http://www.adaptivepath.com/publications/essays/archives/000385.php

Garzotto, F., Paolini, P., Bolchini, D., & Valenti, S. (1999, November 15-18). *Modeling-by-patterns of Web applications.* Paper presented at the International Workshop on the World Wide Web and Conceptual Modeling (WWWCM '99), Paris, France.

Geary, D. (2001). *Advanced JavaServer pages*. Upper Saddle River, NJ: Sun Microsystems Press / Prentice Hall.

German, D. M., & Cowan, D. D. (2000, January 4-7). *Towards a unified catalog of hypermedia design patterns*. Paper presented at the 33rd Hawaii International Conference on System Sciences (HICSS '00), Maui, USA.

Giebel, T. (1999). Globalize your Web site. *PC Magazine, November.*

Gilb, T. (1988). *Principles of software engineering management*. Addison-Wesley.

Gillenson, M., Sherrell, D. L., & Chen, L. (2000). A Taxonomy of Web site traversal patterns and structures. *Communications of the AIS*, *3*(4), 2000.

Ginige, A. & Murugesan, S. (2001). Web engineering: An introduction. *IEEE Multimedia*, *8*(1), 14-18.

Ginige, A. & Murugesan, S. (2001). The essence of Web engineering: Managing the diversity and complexity of Web application development. IEEE Multimedia, 8(2), 22-25.

Ginige, A. (2002). Web engineering: Managing the complexity of Web systems development. In *Proceedings of the 14th International Conference on Software Engineering and Knowledge Engineering*, (pp 721-729). New York: ACM Press.

Ginige, A., & Murugesan, S. (2001). Web engineering: A methodology for developing scalable, maintainable Web applications. *Cutter IT Journal, 14*(7), 24-35.

Glass, R. (2001). Who's right in the Web development debate? Cutter IT Journal, 14(7), 6-10.

Grady, R. & Caswell, D.L. (1987). *Software metrics: Establishing a company-wide program*. Upper Saddle River, NJ: Prentice-Hall.

Graham, I. (2003). *A pattern language for Web usability.* Addison-Wesley.

Grossman, W. (2000). Go Global. *Smart Business*, October

Grossman, W. (2000). The outsiders. *Smart Business, July.*

Gruber, T. R. (1993). Toward principles for the design of ontologies used for knowledge sharing. In N. Guarino & R. Poli (Eds.) *Formal ontology in conceptual analysis and knowledge representation*. Kluwer Academic Publishers.

Grünbacher, P. (2003). Requirements engineering for Web applications. In G. Kappel, B. Pröll, S. Reich, & W. Retschitzegger (Eds.), *Web engineering: The discipline of systematic development of Web applications*. Hoboken, NJ: John Wiley & Sons.

Gudgin, M., M. Hadley, et al., (2003). *SOAP version 1.2 part 1: Messaging framework*. World Wide Web Consortium.

Hafiz, M. (2006, October 21-23). *A Collection of privacy design patterns*. Paper presented at the 13th Conference on Pattern Languages of Programs (PLoP '06), Portland, USA.

Hall, M. (2003). The Web services tsunami. *Computerworld, May 19.*

Haller, A., Cimpian, E., Mocan, A., Oren, E. & Bussler, C. (2005). WSMX - A semantic service-oriented architecture. In *Proceedings of the Inernational Conference on Web Services ICWS'05*.

Hansen, S. (2002). Web information systems: The changing landscape of management models and Web applications. Proceedings of the 14th international conference on software engineering and knowledge engineering (pp. 747-753). ACM.

Hansen, S., Deshpande, Y. & Murugesan S. (2001). A skills hierarchy for Web-based systems development. In S. Murugesan & Y. Deshpande (Eds.), Web Engineering – Managing Diversity and Complexity of Web Application Development (LNCS Vol 2016, pp. 223-235). Berlin: Springer.

Harriman, A., Leo, M., & Hodegetts, P. (2004) Emergent database design: Liberating database development with agile practices. In *Proceedings of the Agile Development Conference*, (pp. 100-105).

Hasan, L. R., & Abuelrub, E. (2006, June 19-21). *Criteria for evaluating quality of Websites*. Paper presented at the Sixth IBIMA Conference on Managing Information in Digital Economy, Bonn, Germany.

Hawthorne Software Engineering Company, *Software Development*. Retrieved on September 27, 2007, http://www.hawthornesoftware.com/Software.htm

Henderson-Sellers, B., Lowe, D., & Haire, B. (2002). OPEN process support for Web development. *Annals of Software Engineering, 13*(1-4), 163-201.

Hendler, J. (2001). Agents and the Semantic Web. *IEEE Intelligent Systems, 16*(2), 30-37.

Hendler, J., Lassila, O., & Berners-Lee, T. (2001). The Semantic Web. *Scientific American, 284*(5), 34-43.

Hendrickson, E., & Fowler, M. (2002). The software engineering of internet software. *Software IEEE, 19*(2), 23-24.

Henkel, M., J. Zdravkovic, et al. (2004). *Service-based processes: Design for business and technology.* ICSOC '04: Proceedings of the 2nd international conference on Service oriented computing.

Herbsleb, J.D., & Mockus, A. (2003). An empirical study of speed and communication in globally distributed software development. *IEEE Transactions on Software Engineering, 29*(6), 1-14.

Hibernate (2006). *Hibernate reference documentation version 3.1.1.* Retrieved on January 2, 2007, from http://www.hibernate.org/5.html

Hickman, N. (1998). Internationalizing your Web site. *WebTechniques, March.*

Hieatt, E., & Mee, R. (2002). Going faster: Testing the Web application. *Software, IEEE, 19*(2), 60-65.

Highsmith, J. (2000). *Adaptive software development.* Dorset House.

Highsmith, J. (2002). *Agile software development ecosystems.* Addison-Wesley.

Highsmith, J. (2002). *Agile software development ecosystems.* Boston: Addison Wesley Professional.

Hoffman, A., & Neubauer, B. (2004). *Deployment and configuration of distributed systems.* Paper presented at the 4th International SDL and MSCWorkshop. Ottawa, Canada.

Hohmann, L. (2003). *Beyond software architecture: Creating and sustaining winning solutions.* Boston,: Addison-Wesley Professional.

Holter, E. (2006), *Client vs developer wars*. Chapel Hill, NC: Newfangled Web Factory.

Holzschlag, M. (2000). Color my world. *WebTechniques, September*.

http://www.ambysoft.com/unifiedprocess/agileUP.html

Httperf. Retrieved January 13, 2005, from http://www.hpl.hp.com/research/linux/httperf

HttpUnit. Retrieved January 13, 2005, from http://httpunit.sourceforge.net

Humphrey, W. S. (1994). *A discipline for software engineering*. Boston: Addison Wesley Professional.

Humphrey, W. S. (1996). *Introduction to the personal software process*. Boston: Addison Wesley Professional.

Humphrey, W. S. (1999). *Introduction to the team software process*. Boston: Addison Wesley Professional.

IEAD (Institute for Enterprise Architecture Developments) (2004). *Extended enterprise architecture framework*. Retrieved May 18, 2007, from http://www.enterprise-architecture.info

IEEE (1990). *IEEE Standard 610.12-1990, IEEE Standard Glossary of Software Engineering Terminology.*

Ivory, M.Y, & Hearst, M.A. (2002). Improving Web site design. IEEE Internet Computing, (March - April), 56-63.

Jacobs, I., & Walsh, N. (2004). *Architecture of the World Wide Web, volume one.* W3C Recommendation. World Wide Web Consortium (W3C).

Jacobsen, I., Booch, G., & Rumbaugh, J. (2000). *The unified software development process*. Boston: Addison-Wesley.

Jacobson, I., Booch, G. & Rumbaugh, J. (1999). *The unified software development process*. Reading MA: Addison-Wesley.

Janzen, D., & Saiedian, H. (2005). Test-driven development concepts, taxonomy, and future direction. *Computer, 38*(9), 43-50.

Jarvenpaa, S. L., Knoll, K., & Leidner, D. E. (1998). Is anybody out there? Antecedents of trust in global virtual teams. *Journal of Management Information Systems, 14*(4), 29-48.

Java Community Process (2002). *JSR 88: Java EE™ application deployment.* Retrieved on January 21, 2007, from http://jcp.org/en/jsr/detail?id=88

Java Community Process (2003). *JSR 175: A metadata facility for the Java™ programming language.* Retrieved on January 21, 2007, from http://jcp.org/en/jsr/detail?id=175

Jeenicke, M., Bleek, W.-G., & Klisechewski, R. (2003). Revealing Web user requirements through e-prototyping. In *Proceedings of the 15th International Conference on Software Engineering and Knowledge Engineering (SEKE 03).* Skokie, IL: Knowledge Systems Institute.

Jeffries, R. (2001, November). What is eXtreme programming? *XP Magazine.* [Electronic version] Retrieved on January, 2006, from http://www.xprogramming.com/xpmag/whatisxp.htm

Jeffries, R., Anderson, A., & Hendrickson, C. (2001). *Extreme programming installed.* The XP series. Boston: Addison-Wesley.

Jensen, B., & Zilmer, A. (2003). Cross-continent development using Scrum and XP. *Extreme programming and agile processes in software engineering.* Paper presented at the 4th International Conference. Berlin, Germany: Springer.

Jhingran, A.D., Mattos, D. & Pirahesh, N.H. (2002). Information integration: A research agenda. *IBM System Journal 41*(4), special issue on information integration. [Electronic version] Retrieved on April 14, 2008 from www.research.ibm.com/journal/sj/414/jhingran.pdf

Johnson, J., Boucher, K. D., Conners, K., & Robinson, J. (2001). Collaborating on project success. *Software Magazine*, February/March 2001.

Jones, T. C. (2000). *Software assessments, benchmarks, and best practices.* Boston: Addison Wesley Professional.

Kaiser, P. R., Tuller, W. L., & McKowen, D. (2000). Student team projects by internet. *Business Communication Quarterly, 63*(4), 75-82.

Kaminski, H., & Perry, M. (2007, May 27-30). *Open source software licensing patterns.* Paper presented at the Sixth Latin American Conference on Pattern Languages of Programming (SugarLoafPLoP '07), Porto de Galinhas, Brazil.

Kamthan, P. (2007). Towards a systematic approach for the credibility of human-centric Web applications. *Journal of Web Engineering, 6*(2), 99-120.

Kamthan, P. (2008). Patterns for improving the pragmatic quality of Web information systems. In C. Calero, M. Á. Moraga, & M. Piattini (Eds.), *Handbook of research on Web information systems quality.* Hershey, PA: Idea Group Publishing.

Kamthan, P., & Pai, H.-I. (2006, May 21-24). *Semantic Web-enabled Web engineering: The case of patterns.* Paper presented at the Seventeenth Annual Information Resources Management Association International Conference (IRMA ;06), Washington, D.C.

Kappel, G., Pröll, B., Reich, S., & Retschitzegger, W. (2003). An introduction to Web engineering. In G. Kappel, B. Pröll, S. Reich, & W. Retschitzegger (Eds), *Web engineering: The discipline of systematic development of Web applications.* Hoboken, NJ: John Wiley & Sons.

Kappel, G., Pröll, B., Reich, S., & Retschitzegger, W. (2006). *Web engineering.* John Wiley and Sons.

Karat, J., & Dayton, T. (1995). Practical education for improving software usability. *Conference on Human Factors in Computing Systems* (pp 162-169). New York: ACM Press.

Katzenbach, J. R., & Smith, D. K. (1993). *The wisdom of teams: Creating the high performance organization.* Boston: Harvard Business School Press.

Kendall, E. A. (1998). Utilizing patterns and pattern languages in education. *Annals of* Software *Engineering, 6*(1-4), 281-294.

Kishore, R., Rao, H. R., Nam, K., Rajagopalan, S., & Chaudhury, A. (2003). A relationship perspective on IT outsourcing. *Communications of the ACM, 46*(12), 87-92.

Klappholz, D. & Bernstein, L. (2001) Getting software engineering into our guts. *Crosstalk: The Journal of Defense Software Engineering, 14*(7). Retrieved on January 1, 2007, from http://www.stsc.hill.af.mil/crosstalk/2001/07/bernstein.html

Klee, K. (2001). Going global: Out ten tests can help You Get Started. *Forbes Small Business, March.*

Kloppmann, M., D. Koenig, et al., (2005). *WS-BPEL extension for people—BPEL4People* (A Joint White Paper by IBM and SAP), IBM/SAP.

Knight, A., & Dai, N (2002). Objects and the Web. *IEEE Software, 19*(2), 51-59.

Koch, N., & Kraus, A. (2002). *The expressive power of UML-based Web engineering.* Paper presented at the Second International Workshop on Web-Oriented Software Technology (IWWOST2), Malaga, Spain.

Koch, P. (2006). *ppk on JavaScript.* Berkeley, CA: New Riders.

Korper, Steffano, & Ellis (2000). *The E-commerce book: Building the E-empire.* Academic Press.

Kruchten, P. (1998). *The rational unified process.* Addison-Wesley.

Kruchten, P. (2003). *The rational unified process: An introduction.* 3rd Edition. Massachusetts: Addison Wesley.

Krutchen, P. (1995). The 4+1 View Model of software architecture. *IEEE Computer, 12*(6), 42-50.

Kumar, K., & Welke, R. J. (1992). Methodology engineering: A proposal for situation-specific methodology construction. In W. W. Cotterman & J. A. Senn (Eds.), *Challenges and strategies for research in systems development,* (pp, 257-269). John Wiley and Sons.

Kuniavsky, M. (2003). *Observing the user experience.* San Francisco: Morgan Kaufmann Publishers.

Kunze, M. (1998) Let there be light: LAMP : Freeware Web publishing system with database support. *c't*, Dec 1998, 230.

Kussmaul, C., & Jack, R. (2006). User interface prototyping: Tips & techniques. *Journal of Computing Sciences in Colleges*, *21*(6), 188-190. Consortium for Computing Sciences in Colleges.

Kussmaul, C., Jack, R., & Sponsler, B. (2004). Outsourcing and offshoring with agility: A case study. *Extreme programming and agile methods—XP / agile universe*, (pp 147-154). Berlin: Springer.

Lacity, M.C., & Willcocks, L.P. (2000). Relationships in IT outsourcing: A stakeholder perspective. In R. W. Zmud (Ed.), *Framing the domains of IT management: projecting the future through the past.* Cincinnati, OH: Pinnaflex Educational Resources, Inc.

Lam, W. (2001). Testing e-commerce systems: A practical guide. IT Professional, 3(2), 19-27.

Lander, M. C., Purvis, R. L., McCray, G. E., & Leigh, W. (2004). Trust-building mechanisms utilized in outsourced IS development projects: A case study. *Information and Management, 41*(4), 509-528.

Lassila, O., & Hendler, J. (2007). Embracing "Web 3.0". *IEEE Internet Computing, 11*(3), 90-93.

Lau, K. & Ukis, V. (2006). *Defining and checking deployment contract for software components.* Paper presented at the 9th International SIGSOFT Symposium on Component-based Software Engineering *(CBSE'06).* Stockholm, Sweden.

Layman, L., Williams, L., & Cunningham, L. (2004). Exploring eXtreme programming in context: An industrial case study. In *Proceedings of the Agile Development Conference* (pp. 32-41).

Le Hégaret, P., Whitmer, R., & Wood, L. (2006). *Document object model (DOM).* Retrieved on July, 2007, from http://www.w3.org/DOM/Overview

Lederer, A. L. & Prasad, J. (1992). Nine management guidelines for better estimating. *Communications of the ACM, 35*(2), 51-59.

Lenzerini, M. (2002). Data integration: A theoretical perspective. In *Proceedings of the Principles of Database Systems Conference PODS'02*, (pp. 233-246). ACM.

Levy, F., & Murnane, R. J. (2005). *The new division of labor: How computers are creating the next job market.* Princeton, NJ: Princeton University Press.

Lie, H. W., & Bos, B. (1999). *Cascading style sheets: Designing for the Web,* 2nd edition. Harlow, England: Addison Wesley Longman.

Lindland, O. I., Sindre, G., & Sølvberg, A. (1994). Understanding quality in conceptual modeling. *IEEE Software*, *11*(2), 42-49.

Lindvall, M., Basili, V., Boehm, B., Costa, P., Dangle, K., Shull, F., Tesoriero, R., Williams, L., & Zelkowitz, M. (2002). Empirical findings in agile methods. In *Proceedings of XP/Agile Universe 2002*, (pp. 197-207).

Liu, Y. & Smith, S. (2006). *A formal framework for component deployment.* Paper presented at the 21st ACM Conference on Object-Oriented Programming, Systems, Languages, and Applications (OOPSLA'06). Portland, OR.

Lowe, D. (2003). Web system requirements: An overview. Requirements Engineering, 8, 102-113.

Lübke, D. (2007). *User interface design styles in SOA applications.* Paper presented at the 8th Annual Global Information Technology Management Associaton World Conference, Napoli, Italy.

Lübke, D., T. Lüecke, et al., (2006). *Model-driven development of business applications using event-driven process chains.* GITMA 2006, Orlando Florida.

Lyardet, F., & Rossi, G. (1998, August 11-14). *Patterns for designing navigable information spaces.* Paper presented at the Fifth Conference on Pattern Languages of Programs (PLoP '98), Monticello, USA.

Lyardet, F., Rossi, G., & Schwabe, D. (1999, July 8-10)). *Patterns for adding search capabilities to Web information systems.* Paper presented in the Fourth European Conference on Pattern Languages of Programming and Computing (EuroPLoP 1999), Irsee, Germany.

Mahemoff, M. (2006). *Ajax design patterns.* O'Reilly Media.

Mahmoud, Q. (2002). MobiAgent: An agent-based approach to the wireless Internet. *Journal of Internet Computing, special issue on Wireless Internet, 3*(2), 157-162.

Manolescu, D., & Kunzle, A. (2001, September 11-15). *Several patterns for eBusiness applications.* Paper presented at the Eighth Conference on Pattern Languages of Programs (PLoP '01), Monticello, USA.

Martins, L. L., Gilson, L. L., & Maynard, M. T. (2004). Virtual teams: What do we know and where do we go from here? *Journal of Management, 30*(6), 805-835.

May, D., & Taylor, P. (2003). Knowledge Management with Patterns. *Communications of the ACM, 46*(7), 94-99.

McPhail, J. C., & Deugo, D. (2001, June 4-7). *Deciding on a pattern.* Paper presented at the 14th International Conference on Industrial and Engineering Applications of Artificial Intelligence and Expert Systems (IEA/AIE 2001). Budapest, Hungary.

Mellor, S., & Balcer, M. (2002). *Executable UML: A foundation for model driven architecture.* Addison-Wesley.

Menasce, D.A, & Almeida, V.A.F. (2002). Capacity planning for Web services: Metrics, models, and methods. Upper Saddle River, NJ: Prentice Hall.

Mendes, E., & Mosley, N. (2006). *Web engineering.* Springer-Verlag.

Mendes, E., Mosley, N. (2005). *Web Engineering.* Berlin, Germany: Springer.

Mendes, E., Mosley, N., & Counsell, S. (2006). The need for Web engineering: An introduction. In E. Mendes & N. Mosley (Eds), *Web engineering.* Berlin: Springer Verlag.

Mendling, J., & Nüttgens, M. (2005). EPC markup language (EPML) - An XML-based interchange format for event-driven process chains (EPC). *Information Systems and e-Business Management (ISeB) 4*(3), 245-263.

Merialdo, P. et al. (2003). Design and development of data-intensive Web sites: The Araneus Atzeni. ACM Transactions on Internet Technology, 3(1), 49-92.

Meszaros, G., & Doble, J. (1998). A pattern language for pattern writing. In R. C. Martin, D. Riehle, & F. Buschmann (Eds.). *Pattern languages of program design 3.* Addison-Wesley, (pp 529-574).

Mich, L., Franch, M., & Gaio, L. (2003). Evaluating and designing Web Site quality. *IEEE Multimedia. 10*(1), 34-43.

Mills, D. L., Sherrell, L. B., Boydstun, J., & Wei, G. (2006). Experiences using agile software development for a shopping simulation. In *Proccedings of IEEE Southeast Con 2006* (pp. 285-290).

Mockus, A., & Weiss, D.M. (2001). Globalization by chunking: A quantitative approach. *IEEE Software, 18*(2), 30-37.

Montero, F., López-Jaquero, V., & Molina, J. P. (2003, September 1-2). *Improving e-Shops environments by using usability patterns.* Paper presented at the Second Workshop on Software and Usability Cross-Pollination, Zürich, Switzerland.

Montero, F., Lozano, M., & González, P. (2002, August 5-7). *Designing Web sites by using design patterns.* Paper presented at the Second Latin American Conference on Pattern Languages of Programming (SugarLoafPLoP '02), Rio de Janeiro, Brazil.

Moore, S., & Barnett, L. (2004). *Offshore outsourcing and agile development.* Forrester Research, Inc.

Morein, R. (2005). Agile development of the database—a focal entity prototyping approach. In *Proceedings of the Agile Development Conference,* (pp. 103-110).

Mori, G., F. Paternò, et al., (2004). Design and development of multidevice user interfaces through multiple logical descriptions. *IEEE Trans. Software Eng. 30*(8), 507-520.

Morisio, M., & Oivo, M. (2003). Software engineering for the wireless Internet [Guest Editor's Introduction]. *IEEE Transactions on Software Engineering, 29*(12), 1057-1058.

Morris, P. (2001). Updating the project management bodies of knowledge. *Project Management Journal, September.*

Morrison, T. (1997). *Dun & Bradstreet's Guide to doing business around the world.* Prentice Hall.

Morrison, T. (2000). Kiss, bow, or shake hands: How to do business in 60 countries. Adams Media.

Moschella, D. (2000). Ten key IT challenges for the next 20 years. *Computerworld, December.*

Motorola Application Certification Program. (n.d.). Retrieved February 10, 2005, from http://qpqa.com/motorola/iden

Murugesan, S. & Ginige, A. (2005). Web engineering: Introduction and perspectives. In Suh, W. (Ed.), *Web engineering: Principles and techniques* (pp 1-30). Hershey, PA: Idea Group Publishing.

Murugesan, S. (1998). Web engineering. Presentation at the First Workshop on Web Engineering, World Wide Web Conference (WWW7), Brisbane, Australia.

Murugesan, S. et al. (1999). Web engineering: A New Discipline for Development of Web-based systems. In Proceedings of the First ICSE Workshop on Web Engineering, Los Angeles (pp. 1-9).

Murugesan, S., & Deshpande, Y. (Eds) (2001). Web engineering: Managing diversity and complexity of Web application development. Lecture Notes in Computer Science – Hot Topics, 2016. Berlin: Springer Verlag.

Myers, B. A. & Rosson, M. B. (1992). Survey on user interface programming. In *Proceedings of the ACM CHI'92 Conference* (pp 195-202). New York: ACM Press.

Myers, G. J., Badgett, T., Thomas, T. M., & Sandler, C. (2004). *The art of software testing.* Hoboken, N.J.: John Wiley & Sons.

Nam, T. J. & Gill, S. (2000). An effective prototyping method for delivering interaction design in industrial design education. In *Proceedings of the IDATER Conference*, 2000.

Neill, C., & Laplane, P. (2003). Requirements engineering: The state of the practice, *IEEE Software, 20*(6), 40-45.

Nelson, T. H. (1984). *Literary machines.* Mindful Press.

Neuman, C. (2000). Considering the color-blind. *Webtechniques, August.*

Newman, M. W., Lin, J., Hong, J. I., & Landay, J. A. (2003). DENIM: An informal Web site design tool inspired by observations of practice. *Human-Computer Interaction, 18*(3), 259-324.

Newward, T. (2006). *The Vietnam of computer science.* Retrieved on May 28, 2007, from http://blogs.tedneward.com/2006/06/26/The+Vietnam+Of+Computer+Science.aspx

Nguyen, H. Q., Johnson, R., & Hackett, M. (2003). *Testing applications on the Web: Test planning for mobile and Internet-based systems* (2nd Edition). John Wiley and Sons.

Nielsen, J. (1990). Paper versus computer implementations as mockup scenarios for heuristic evaluation. In *Proceedings of the IFIP Third International Conference on Human Computer Interaction* (pp 315-320). Amsterdam: North-Holland.

Nielsen, J. (1993). *Usability engineering.* San Francisco: Morgan Kaufmann Publishers.

Nikkanen, M. (2004). User-centered development of a browser-agnostic mobile e-mail application. In *Proceedings of the Third Nordic Conference on Human-Computer Interaction*, Tampere, Finland (pp. 53-56). New York: ACM Press.

Norman, D. (2002). *The design of everyday things.* New York: Basic Books.

O'Reilly, T. (2005). *What is Web 2.0: Design patterns and business models for the next generation of software.* Retrieved on July, 2007, from http://www.oreillynet.com/pub/a/oreilly/tim/news/2005/09/30/what-is-web-20.html

Oberle, D., Eberhart, A., Staab, S., & Volz, R. (2004). Developing and managing software components in an ontology-based application server. In *Proceedings of the 5th International Middleware Conference*, Toronto, Canada: Springer-Verlag.

Object Management Group (2002). *CORBA component model, version 3.0.*

Object Management Group (2006). *Deployment and configuration of component-based distributed applications specification, version 4.0.*

Ocampo, A., Boggio, D., Munch, J., & Palladino, G. (2003). Towards a reference process for developing wireless Internet services. *IEEE Transactions on Software Engineering, 29*(12), 1122-1134.

Offutt, J. (2002). Quality attributes of Web software applications. *IEEE Software, 19*(2), 25-32.

Olsina, L., & Rossi, G. (2002) Measuring Web application quality with WebQEM. *IEEE Multimedia, 9*(4), 20-29.

OMG (Object Management Group) (2007). *Model driven architecture*. Retrieved May 18, 2007, from http://www.omg.org/mda

Open Mobile Alliance. (2005). Retrieved from March 15, 2005, http://www.openmobilealliance.org

Oppenheimer, D., & Patterson, D.A. (2002). Architecture and dependability of large-scale Internet services. IEEE Internet Computing, September-October, 41-49.

Orriens, B., Yang, J. & Papazoglou, M. (2003). A framework for business rule driven Web service composition. In M. A. Jeusfeld & O. Pastor, (Eds). *Proceedings of the ER'2003 Workshops,* LNCS 2814, (pp. 52-64). Springer-Verlag.

Paasivaara, M., & Lassenius, C. (2004). Using interactive and incremental processes in global software development. In *Proceedings of the International Conference on Software Engineering (ICSE) Third International Workshop on Global Software Development*, (pp. 24-28).

Palmer, S. R., & Felsing, J. M. (2002). *A practical guide to feature-driven development.* Upper Saddle River, NJ: Prentice-Hall.

Passani, L., & Trasatti, A. (2002). *WURFL*. Retrieved on July, 2007, from http://wurfl.sourceforge.net/

Patten, B. & Grandlienard, G. (1999). Using resource bundles to international text. *Java Report, February.*

Patterson, J. (1994). *ISO 9000: Worldwide quality standard.* Stamford, CT: Crisp Learning.

Patton, S. (2002). Web metrics that matter. *Computer World.* [Electronic version]. Retrieved on September 27, 2007, from http://www.computerworld.com/databasetopics/data/story/0,10801,76002,00.html

Paulk, M. C. (2001). Extreme programming from a CMM Perspective. *IEEE Software, 18*(6), 19-26.

Paulk, M. C., Weber, C. V., Curtis, B., Chrissis, M. B., et al. (1994). *The capability maturity model: Guidelines for improving the doftware process*. Boston: Addison Wesley Professional.

Peak, P., & Heudecker, N. (2005). *Hibernate quickly.* Greenwich, CT: Manning Publications.

Peltier, M., Bezivin, J & Guillaume, G. (2001). MTRANS: A general framework, based on XSLT, for model transformations. In *Proceedings of the Workshop on Transformations in UML WTUML'01.*

Peltier, M., Ziserman, F. & Bezivin. (2002). On levels of model transformation. In *Proceedings of the XML Europe Conference* (pp. 1-17). Paris, France: Graphic Communications Association.

Perrotta, T/ (2000). Yahoo ruling exposes risks of being global. *InternetWorld, July.*

Perry, M., & Kaminski, H. (2005, July 6-19). *A pattern language of software licensing.* Paper presented at the Tenth European Conference on Pattern Languages of Programs (EuroPloP '05), Irsee, Germany.

Pertet, S. M., & Narasimhan, P. (2005). *Causes of failure in Web applications.* PDL Technical Report PDL-CMU-05-109. Carnegie Mellon University, Pittsburgh, USA.

Perzel, K., & Kane, D. (1999, August 15-18). *Usability patterns for applications on the World Wide Web.* Paper presented at the Sixth Conference on Pattern Languages of Programs (PLoP '99), Monticello, USA.

Peterson, C. (2000). Accessible Web sites matter. *Enterprise Development, June.*

Pinsonneault, A., & Caya, O. (2005). Virtual teams: What we know, what we don't know. *International Journal of e-Collaboration, 1*(3), 1-16.

PMI (2000), *The project management body of knowledge (PMBOK),* Project Management Institute.

Powell, A., Piccoli, G., & Ives, B. (2004). Virtual teams: A review of current literature and directions for future research. *ACM SIGMIS Database, 35*(1), 6-36.

Powell, T. A., Jones, D. L., & Cutts, D. C. (1998). *Web site engineering.* Prentice-Hall.

Preece, J., Rogers, Y., & Sharp, H. (2002). *Interaction Design: Beyond Human-Computer Interaction.* Hoboken, NJ: John Wiley & Sons.

Pressman, R. S. (2005). *Software engineering: A practitioner's approach* (6th ed.). New York: McGraw Hill.

Pressman, R.S. (2001). What a tangled Web we weave. *IEEE Software, 18*(1), 18-21.

Pressman, R.S. (2004). *Applying Web engineering, Part 3. software engineering: A practitioner's perspective* (6th ed.). New York: McGraw-Hill.

Purdue University (1989). *The Purdue enterprise reference architecture.* Retrieved May 18, 2007, from http://pera.net

Putman, J.R. (2001). *Architecting with RM-ODP.* New Jersey: Prentice Hall PTR.

Ramesh, B., Cao, L., Mohan, K., & Xu, P. (2006). Can distributed software development be agile? *Communications of the ACM, 49* (10), 41-46.

Reed, S. (2000). Want to limit the audience for you Web site? Keep it English only. *Infoworld, August.*

Reifer, D.J. (2000). Web development: Estimating quick-to-market software. IEEE Software, 17(6), 57-64.

Rettig, M. (1994). Prototyping for tiny fingers. *Communications of the ACM, 37*(4), 21-27.

Reynaud, C., Sirot, J. P. & Vodislav, D. (2001). Semantic integration of XML heterogeneous data sources. In *Proceedings of the IDEAS Conference (*pp. 199-208).

Ricca, F., & P. Tonella (2001). Analysis and testing of Web applications. In *Proceedings of the International Conference on Software Engineering,* (pp. 25-34).

Riggins, F. J., & Mitra, S. (2003). *A framework for developing net-enabled business metrics through functionality interaction.* Retrieved on September 27, 2007, from http://ids.csom.umn.edu/faculty/friggins/e-metrics.pdf

Rising, L & Janoff, N. S. (2000). The scrum software development process for small teams. *IEEE Software, July/August,* 2-8.

Robb, D. (2000). Act globally, serve locally. *Information Week, July.*

Romanosky, S., Acquisti, A., Hong, J., Cranor, L. F., & Friedman, B. (2006, October 21-23). *Privacy patterns for online interactions.* Paper presented at the 13th Conference on Pattern Languages of Programs (PLoP '06), Portland, USA.

Rosenberg, F. & Dustdar, S. (2005). Business rules integration in BPEL - A service-oriented approach. In *Proceedings of the 7th International IEEE Conference on E-Commerce Technology.*

Ross, M. (2002). Quality in Web design for visually impaired users. *Software Quality Journal, 10*(4), 285-298.

Rossi, G., & Koch, N. (2002, July 3-7). *Patterns for adaptive Web applications.* Paper presented at the Seventh European Conference on Pattern Languages of Programs (EuroPLoP '02). Irsee, Germany.

Rossi, G., Lyardet, F. D., & Schwabe, D. (1999). Developing hypermedia applications with methods and patterns. *ACM Computing Surveys. 31*(4es).

Rossi, G., Lyardet, F., & Schwabe, D. (2000, July 5-9). *Patterns for E-commerce applications.* Paper presented at the Fifth European Conference on Pattern Languages of Programs (EuroPLoP 2000), Irsee, Germany.

Rossi, G., Pastor, O., Schwabe, D., & Olsina, L. (2008). *Web engineering: Modelling and implementing Web applications.* Springer-Verlag.

Rossi, G., Schwabe, D., & Lyardet, F. (1999, May 11-14). *Improving Web information systems with navigational patterns.* Paper presented at the Eighth International World Wide Web Conference (WWW8), Toronto, Canada.

Roth, J. (2002). Patterns of mobile interaction. *Personal and Ubiquitous Computing. 6*(4), 282-289.

Royce, W. (1970). Managing the development of large software systems. In *Proceedings of IEEE WESCON* (pp 1-9). New York: The Institute of Electrical and Electronics Engineers.

Royce, W. (1970). Managing the development of large software systems. In *Proceedings of the IEEE WESTCON*, Los Angeles CA, IEEE Computer Society.

Rubin, J. (1994). *Handbook of usability testing.* Hoboken, NJ: John Wiley & Sons.

Rudd, J. and Isensee, S. (1994). Twenty-two tips for a happier, healthier, prototype. *interactions, 1*(1), 35-40.

Rudd, J., Stern, K., & Isensee, S. (1996). Low vs. high-fidelity prototyping debate. *interactions, 3*(1), 76-85.

Ruhe, M., Jeffrey, R., & Wieczorek, I. (2003, May). Cost Estimation for Web Applications. In *Proceedings of the 25th International Conference on Software Engineering, IEEE Computer Society,* (pp. 285-294).

Sanja, A. (2005). *Overview of agile management project perfect.* White Paper. Retrieved on January 2, 2007 from http://www.projectperfect.com.au/info_agile_programming.php

Satoh, I. (2003). A testing framework for mobile computing software. *IEEE Transactions on Software Engineering, 29*(12), 1112-1121.

Sawhney, Mohanbir, & Sumant Mandai (2000). Go global. *Business, May.*

Schach, S. R. (2003). *Object-oriented & classical software engineering.* New York: McGraw-Hill.

Schmidt, D. C., Stal, M., Rohnert, H., & Buschmann, F. (2000). *Pattern-oriented software architecture, Volume 2: Patterns for concurrent and networked objects.* John Wiley and Sons.

Schrage, M. (2000). *Serious play: How the world's best companies simulate to innovate.* Boston: Harvard Business School Press.

Schumacher, M., Fernandez-Buglioni, E., Hybertson, D., Buschmann, F., & Sommerlad, P. (2006). *Security patterns: Integrating security and systems engineering.* John Wiley and Sons.

Schwaber, K. (2003). *Agile project management with scrum.* Microsoft Press.

Schwaber, K., & Beedle, M. (2001). *Agile software development with SCRUM.* Upper Saddle River, NJ: Prentice Hall PTR.

Schwartz, H. (2000). Going global. *WebTechniques, September.*

Segerståhl, K., & Jokela, T. (2006, April 22-27). *Usability of interaction patterns.* Paper presented at the CHI 2006 Conference on Human Factors in Computing Systems, Montréal, Canada.

Selmi, S. S., Kraïem, N., & Ghézala, H. H. B. (2005, July 27-29). *Toward a comprehension view of Web engineering.* Paper presented at the Fifth International Conference on Web Engineering (ICWE '05), Sydney, Australia.

Seltsikas, P. & Currie, W.L. (2002). Evaluating the application service provider (ASP) business model: The challenge of integration. In *Proceedings of the 35th Annual Hawaii International Conference,* (pp. 2801-2809).

Seshadri, G. (1999). Understanding JavaServer pages model 2 architecture: Exploring the MVC design pattern. *JavaWorld, December.*

Shadbolt, N., Hall, W., & Berners-Lee, T. (2006). The Semantic Web revisited. *IEEE Intelligent Systems, 21*(3), 96-101.

Shan, T.C., & Hua, W.W. (2006). Solution architecture of N-Tier applications. In *Proceedings of 3rd IEEE International Conference on Services Computing* (pp. 349-356). California: IEEE Computer Society.

Sheth A. P. & Larson J. A. (1990). Federated database systems for managing distributed, heterogeneous, and autonomous databases. *ACM Computing Surveys, 22*(3), 183.

Siegel, D.A. (2003). The business case for user-centered design: Increasing your power of persuasion. Interactions, 10(3).

Siggelkow, B. (2005). *Jakarta struts cookbook.* Sebastopol, CA: O'Reilly.

Sillence, E., Briggs, P., Harris, P, & Fishwick, L. (2006). A framework for understanding trust factors in Web-based health advice. *International Journal of Human-Computer Studies, 64*, 697-713.

Simon, H. (1996). *The Sciences of the Artificial* (3rd Edition). The MIT Press.

Simons, M. (2002). Internationally agile. *InformIT*, March 15, 2002.

Sliski, T. J., M. P. Billmers, et al., (2001). *An architecture for flexible, evolvable process-driven user-guidance environments.* Paper presented at the8th European software engineering conference held jointly with 9th ACM SIGSOFT international symposium on Foundations of software engineering, Vienna, Austria.

Sneed, H. M., A. GmbH, et al. (2004). Testing a Web application. Web Site Evolution, 2004. WSE 2004. In *Proceedings of the Sixth IEEE International Workshop,* (pp. 3-10).

Snyder, C. (2003). *Paper prototyping: The fast and easy way to design and refine user interfaces.* San Francisco: Morgan Kaufmann Publishers.

Software Engineering Institute (2003). Process maturity profile: Software CMM® -CBA IPI and SPA Appraisal Results.

Software Engineering Institute (SEI) at CMU (2007). *Scenario-based architecture analysis method.* Retrieved on May 18, 2007, from http://www.sei.cmu.edu/architecture/scenario_paper

Spriestersbach, A., & Springer, T. (2004). Quality attributes in Mobile Web Application Development. In F. Bomarius & H. Iida (Eds.). *Product Focused Software Process Improvement.* (pp. 120-130).Springer-Verlag.

Squier, J., & Nielsen, J. (2000). Deconstructing—Hojo.com. *Internet World, June.*

Stamper, R. (1992, October 5-8). *Signs, Organizations, Norms and Information Systems.* Paper presented at the Third Australian Conference on Information Systems, Wollongong, Australia.

Standing, C. (2002). Methodologies for developing Web applications. *Information and Software Technology, 44*(3), 151-159.

Standish Group (1994). *The CHAOS report.* Retrieved January 1, 2007 from http://www.standishgroup.com/sample_research/chaos_1994_1.php

Standish Group (2004). Chaos chronicles. Retrieved on www.standisgroup.com

Standish Group (2007). *The Standish Group Chaos Report 2006.* Retrieved May 18, 2007, from http://www.standishgroup.com

Stapleton, J. (1997). *DSDM dynamic systems development method.* Addison-Wesley.

Steiner, D. H., & Palmer, D. W. (2004). *Extreme software engineering.* Upper Saddle River, NJ: Pearson Education.

Stern, A & Davis, J. (2003). A taxonomy of information technology services: Web services as IT services. In *Proceedings of the First International Conference on Service Oriented Computing.*

Stern, A. & Davis, J. (2004). Extending the Web services model to IT services. In *Proceedings of the IEEE International Conference on Web Services,* (pp. 824-825).

Stout, G. A. (2001). Testing a Website: Best practices. Retrieved from whitepaper on www. reveregroup. com

Struts (2005). *The Struts user guide.* Retrieved on January 2, 2007, from http://struts.apache.org/struts-doc-1.2.9/userGuide/index.html

Subramanian, N., & Chung, L. (2003, September). Process-oriented metrics for software architecture evolvability. In *Proceedings of the International Workshop on Principles of Software Evolution,* IEEE Computer Society, (pp. 65-70). Helsinki, Finland.

Subramanian, N., Chung, L., & Song, Y-t. (2006, June). An NFR-Based framework to establish traceability between application architectures and system architectures. In *Proceedings of the 7th ACIS International Conference on Software Engineering, Artificial Intelligence, Networking, and Parallel/Distributed Computing (SNPD 2006),* IEEE Computer Society, Las Vegas.

Subramanian, N., Puerzer, R., & Chung, L. (2005, September). A comparative evaluation of maintainability: A study of engineering department's Website maintainability. In *Proceedings of the International Conference on Software Maintenance,* (pp. 669-672). IEEE Computer Society. Budapest, Hungary.

Sun Microsystems J2ME Wireless Toolkit. (2005). Retrieved from http://java.sun.com/products/j2mewtoolkit

Sun Microsystems J2ME. (2005). Retrieved from http://java.sun.com/j2me

Sundstrom, E. & Associates (1998). *Supporting work team effectiveness: Best management practices for fostering high performance.* San Francisco, CA: Jossey-Bass.

Szyperski, C. (2002). *Component software: Beyond object-oriented programming,* 2nd Ed. Addison-Wesley.

Taft, D. K. (2005, November 11). Microsoft lauds 'Scrum' method for software projects. *eWeek.* [electronic version] Retrieved on January 3, 2007, from http://www.eweek.com/article2/0,1895,1885883,00.asp

Tapper, S. (2000). Is globalization right for you. *WebTechniques,* September

Tate, B. A., & Hibbs, C. (2006). *Ruby on Rails: Up and Running.* O'Reilly Media.

Taylor, M. J., McWilliam, J., Forsyth, H., & Wade, S. (2002). Methodologies and Website development: A survey of practice. *Information and Software Technology, 44,* 381-391.

The Open Group (2007). *The Open Group architecture framework.* Retrieved May 18, 2007, from http://www.opengroup.org/togaf

Tidwell, J. (2005). *Designing interfaces: Patterns for effective interaction design.* O'Reilly Media.

Trætteberg, H. (1999). *Modelling work: Workflow and task modelling.* CADUI.

Treasury Department CIO Council (2000). *Treasury enterprise architecture framework.* Version 1. Retrieved May 18, 2007, from http://www.eaframeworks.com/TEAF/teaf.doc

Uden, L. (2002). Design process for Web applications. IEEE Multimedia, (Oct-Dec), 47-55.

Uniscape Corporation (2000). Global content manager.

Van Duyne, D. K., Landay, J., & Hong, J. I. (2003). *The design of sites: Patterns, principles, and processes for crafting a customer-centered Web experience.* Addison-Wesley.

Van Eaton, J. (2005). *Outlook Web access - A catalyst for Web evolution.* Retrieved on July, 2007, from http://msexchangeteam.com/archive/2005/06/21/406646.aspx

Van Ommering, R. (2001). *Techniques for independent deployment to build product populations.* Paper presented at the 2001 Working IEEE / IFIP Conference on Software Architecture. Amsterdam, The Netherlands.

Verhoeven, J. (2003). Prototyping with PowerPoint. Retrieved on January 1, 2007, from http://www.jansfreeware.com/articles/misc-prototyping.html

Vijayaraghavan, G. V. (2003). *A taxonomy of E-commerce risks and failures*. Master's Thesis. Florida Institute of Technology. Melbourne, USA.

Virzi, R. A., Sokolov, J. L., & Karis, D. (1996). Usability problem identification using both low- and high-fidelity prototypes. In *Proceedings of the SIGCHI Conference on Human Factors in Computing Systems* (pp 236-243). New York: ACM Press.

Vogel, D. A., & Connelly, J. E. (2005). Best practices for dealing with offshore software development. *Handbook of Business Strategy*. Bradford, UK: Emerald Group Publishing Limited.

Walker, M., Takayama, L., & Landay, J. (2002). High-fidelity or low-fidelity, paper or computer medium? Choosing attributes when testing Web prototypes. In *Proceedings of the Human Factors and Ergonomics Society 46th Annual Meeting*, pp. 661-665. Santa Monica, CA: Human Factors and Ergonomics Society.

Wallace, D., Raggett, I., & Aufgang, J. (2002). *Extreme programming for Web projects*. Addison-Wesley.

Walsh, K. R. (2003). Analyzing the application ASP concept: Technologies, economies, and strategies. *Communications of the ACM, 46*(8), 103-107.

Weiss, M. (2003, September 8-12). *Patterns for Web applications*. Paper presented at the Tenth Conference on Pattern Languages of Programs (PLoP '03), Urbana, USA.

Wentzlaff, I., & Specker, M. (2006, July 10). *Pattern based development of user friendly Web applications*. Paper presented at workshop on Model-Driven Web Engineering (MDWE '06), Palo Alto, CA, USA..

Wesson, J., & Cowley, L. (2003, September 1-2). *Designing with patterns: Possibilities and pitfalls.* Paper presented at the Second Workshop on Software and Usability Cross-Pollination, Zürich, Switzerland.

Weyuker, E. (1998). Testing component-based software: A cautionary tale. *IEEE Software. September/October 1998.*

White, B. (1996, May 6-11). *Web document engineering*. Talk given at 5th International World Wide Web Conference, Paris, France. [Electronic version]. Retrieved on September 27, 2007, http://www.slac.stanford.edu/pubs/slacpubs/7000/slac-pub-7150.html

White, S. A. (2006). *Business process modeling notation specification, object management group standard.*

Whiting, R. (2000). U.S. companies to comply with European privacy rules. *Information Week, February.*

Whitson, G. (2006) WebHelix: Another Web engineering process. *The Journal of Computing Sciences in Colleges, 21*(5), 21-27.

Whitten, J. L., & Bentley, L. D. (2006). *Systems analysis and design methods.* McGraw Hill..

Widom, J. (1995). Research problems in data warehousing. In *Proceedings of the 4th International Conference on Information and Knowledge Management.*

Wiederhold, G. (1992). Mediators in the architecture of future information systems. *IEEE Computer, 25.* March, 38-49.

Willcocks, L. P. & Lacity, M. C. (1998). The sourcing and outsourcing of IS: Shock of the new? In L. P. Willcocks & M. C. Lacity (Eds.) *Strategic sourcing of information technology: Perspectives and practices*. Wiley.

Williams, C. (2006). Smart phones, stupid punters? *The Register*, July 13, 2006.

Williams, J. (2000). Correctly assessing the "ilities" requires more than marketing hype. IT Professional, 2(6), 65-67.

Williams, J. (2001). Avoiding CNN moment. IT Professional, 3(2), 68-70.

Williams, L., & Kessler, R. (2000). Strengthening the case for pair programming. *IEEE Software, 17*(4), 19-25

Winch, G. (2004). Managing project stakeholders. In Morris and Pinto (Eds.). *The Wiley guide to managing projects*. John Wiley & Sons.

Wu, Y., J. Offutt, et al. (2004). Modeling and testing of dynamic aspects of Web applications. Submitted for publication: 04-01.

Xu, L., B. Xu, et al. (2005). Testing Web applications focusing on their specialties. *ACM SIGSOFT Software Engineering Notes,* **30**(1).

Yoder, J., & Barcalow, J. (1997, September 3-5). *Architectural patterns for enabling application security.* Paper presented at the Fourth Conference on Pattern Languages of Programs (PLoP 1997), Monticello, USA.

Yunker, J. (2000). Speaking in charsets. *WebTechniques, September.*

Zachman, J.A. (1987). A framework for information systems architecture. *IBM Systems Journal, 26*(3), 276-295.

Zhang, Z. & Yang, H. (2004). Incubating services in legacy systems for architectural migration. In *Proceedings of the 11th Asia-Pacific Software Engineering Conference (APSEC'04)* (pp. 196-203).

Zhu, F., Turner, M., Kotsiopoulos, I., Bennett, K., Russell, M., Budgen, D., Brereton, P., Keane, J., Layzell, P., Rigby, M. & Xu, J. (2004). Dynamic Data Integration Using Web Services. In *Proceedings 2nd International Conference on Web Services (ICWS'04).*

Ziemer, S., & Stålhane, T. (2004, July 27). *The use of trade-offs in the development of Web applications.* Paper presented at the International Workshop on Web Quality (WQ 2004). Munich, Germany.

About the Contributors

Daniel M. Brandon. In a speech in Cape Town, South Africa in June of 1966 Robert F. Kennedy said "May you live in interesting times." He said the phrase was of Chinese origin, although even Chinese are not sure of this, but it certainly applies to us who have become deeply involved with information technology (IT). Dan Brandon took his first computer course in college in 1966 and has been fascinated by IT ever since.

As he started his career, IT was also really just starting. The disciplines of "Computer Science", "Software Engineering", and "IT Project Management" all were formulated and solidified during his working years. He has designed and developed software for every generation of software (assembly thru object oriented), and almost every generation of IT architecture and hardware (mainframe, mini, PC, client-server, web). He has also programmed in most major computer languages including ALGOL, FORTRAN, Pascal, Ada, COBOL, Basic, C, C++, Java, and modern web-based languages as PHP to develop applications in a wide variety of areas encompassing the business, engineering, and scientific fields.

As his career moved him from programmer to designer to manager, he was always involved with the application of new computer technology and concepts to the solution of application problems. The effective management of all the resources and stakeholders involved with building, integrating, and deploying IT applications has always represented a major interest and challenge to him. Along the way he managed to fit in some schooling, obtaining a BS (Case Western Reserve University), MS, and PhD in Engineering (University of Connecticut); his PhD specialization was in computer methods.

Before returning to the university environment, Dr. Brandon accumulated over thirty years of commercial experience in both the IT technical and management arenas. He was a Senior Engineer at Mobil Research, Manager of Application Development for Control Data Corporation, MIS Manager for several companies, and Director of Information Services at the NASA Stennis Space Center.

He is currently a professor of Information Technology Management (ITM) at Christian Brothers University (CBU) in Memphis, TN. CBU is a part of the De La Salle Christian Brothers global educational organization with 1000 plus schools in over 80 countries. Dr. Brandon's research interests include both the management and technical side of IT. At CBU he teaches undergraduate and graduate courses in IT, including: Information Systems Management, Project Management, Software Engineering, Database Design, Decision Support Systems, Internet Systems, and Programming.

Prior to this book, his most recent book (2006) is *Project Management for Modern Information Systems* by IRM Press (978-1591406945). He has been published in a number of other books, journals, and conference proceedings including *The Project Management Journal, The Wiley Guide to Managing Projects, Essentials of Project Control, Encyclopedia of Information Technology, Journal of Computing*

Sciences in Colleges, Successful Software Reengineering, Technologies & Methodologies for Evaluating Information Technology in Business, Issues and Trends of Information Technology Management in Contemporary Organizations, Architectural Issues of Web-Enabled Electronic Business, Managing Information Technology Resources in Organizations in the Next Millennium, and *Managing Information Technology in a Global Economy.*

He is a member of the Society of Information Management (SIM), the Information Resource Management Association (IRMA), the IEEE Computer Society, and the Project Management Institute (PMI). He also holds the PMP (Project Management Professional) Certification which is the highest certification granted from PMI. He continues to do consulting with a number of companies both locally and internationally. Currently he is involved in the design and development of comprehensive open source software systems for business applications.

* * *

Kevin A. Gary is an assistant professor in the Division of Computing Studies at Arizona State University's Polytechnic Campus. His research interests include automated workflow, software process, distributed software systems, and technology-supported learning. Dr. Gary joined ASU after spending four years in industry developing enterprise software solutions for e-learning. His observations mentoring junior software engineers led him to implement the Software Enterprise at ASU. He is a participant in the open source Image-guided Surgery Toolkit (IGSTK) and an industry consultant specializing in enterprise software systems. Dr. Gary earned his PhD from ASU in 1999.

Jorge Marx Gómez studied computer engineering and industrial engineering at the University of Applied Science of Berlin (Technische Fachhochschule). He was a lecturer and researcher at the Otto-von-Guericke-Universität Magdeburg where he also obtained a PhD degree in business information systems with the work Computer-based Approaches to Forecast Returns of Scrapped Products to Recycling. In 2004 he received his habilitation for the work Automated Environmental Reporting through Material Flow Networks at the Otto-von-Guericke-Universität Magdeburg. From 2002 till 2003 he was a visiting professor for business informatics at the Technical University of Clausthal. In 2005 he became a full professor of business information systems at the Carl von Ossietzky University Oldenburg. His research interests include business information systems, e-Commerce, material flow management systems, Federated ERP-Systems, Data Warehousing, recycling program planning, disassembly planning and control, simulation and neuro-fuzzy-systems.

Winnie Hua is a principal consultant in CTS Inc. She has more than 15-year project and consulting experience in a broad range of leading-edge technologies. She holds a graduate degree in Computer Science. As a solution architect/lead, she has led lifecycle design and development of large-scale eCommerce systems on diverse platforms using a variety of cutting-edge technologies and unified/agile methodologies. She has initiated/participated in advanced research on various emerging web technologies. She is a member of numerous professional associations, a frequent speaker in conferences/seminars, and also a co-founder of Charlotte Architecture and Technology Symposium (CATS).

Roger Jack is president of Elegance Technologies, Inc. Jack has experience in project management, and creating reliable and robust interfaces and architectures. He was vice president of U.S. software operations for NeST Technologies, where he managed many offshore projects. He has an MBA from Duke University's Fuqua School of Business, and an MS in computer science from Villanova University.

Pankaj Kamthan has been teaching in academia and industry for several years. He has also been a technical editor, participated in standards development, served on program committees of international conferences, and is on the editorial board of the *International Journal of Technology Enhanced Learning* and the *International Journal of Teaching and Case Studies*. His professional interests and experience include knowledge representation, requirements engineering, and software quality.

Harry Koehnemann is an professor of practice for the Division of Computing Studies at Arizona State University Polytechnic campus and a senior technical consultant for Rocket Gang. His interests include distributed software systems, software process, and modeling software-intensive systems. Harry has worked several years as a software architect and senior software developer on software systems ranging from large enterprise applications to embedded control systems. Harry is also a trainer and consultant in software tools and technologies, software modeling, and process. Harry received his PhD in computer science from Arizona State University in 1994.

Clif Kussmaul is chief technology officer for Elegance Technologies, which develops software products and provides product development services, and assistant professor of computer science at Muhlenberg College. Formerly, he was senior member of technical staff at NeST Technologies. He has a PhD from the University of California, Davis, an MS and MA from Dartmouth College, and a BS and BA from Swarthmore College. His interests include agile development, virtual teams, and entrepreneurship.

Daniel Lübke earned his diploma degree in business informatics at the TU Clausthal in 2004. He worked in software development and coaching. Currently, he is research assistant at the Software Engineering group at the Leibniz University Hannover. Areas of interests include modern approaches for distributed applications like Mobile Agents and Web service technologies and the software engineering paradigms behind them.

David Mills is an application developer and lecturer in mathematics and computer science at Stetson University. He is pursuing the MS degree in computer science at the University of Memphis, where he served for two years as an NSF Fellow on the Tri-P-LETS project. He holds both a BS degree in computer science and a BA in digital arts from Stetson University. His research interests are in software engineering, data mining, biotechnology, and computer science education.

Claus Pahl is a senior lecturer at Dublin City University's School of Computing, where he is the leader of the Web and Software Engineering group. Claus has graduated from the Technical University of Braunschweig and has obtained a PhD from the University of Dortmund. He has published more than 140 papers in various journals, books, conference, and workshop proceedings. He is on the editorial board of the International Journal on E-Learning and is a regular reviewer for journals and conferences in the area of Web and Software technologies and their applications. He is the principal investigator of several basic and applied research projects in Web software engineering. Claus' research interests cover a broad spectrum from service- and component technologies in software engineering to infrastructure and development technologies for Web applications such as e-learning.

David Parsons is a senior lecturer and coordinator for the information technology major within the Institute of Information and Mathematical Sciences at Massey University, Auckland, New Zealand. He also acts as a knowledge engineer for Software Education Associates, providing Java training and consultancy in New Zealand and Australia. His current research interests are in Web-based software architectures, mobile learning and agile software development. Beginning his academic career in the UK, he has worked as an educator/trainer, researcher and practitioner across Europe, North America and Australasia. Prior to his arrival in New Zealand in 2003, he was director of emerging technologies for international consultancy Valtech, based in the city of London, and before that that he was a principal technologist for BEA Systems' internal education. He has published widely on various aspects of software design and development and is the author of successful text books on C++, Java and Web application development.

Tony Shan is a renowned expert working in the computing field for 20+ years with extensive experience on architecture engineering, technology strategies, portfolio rationalization, and system designs in a number of multi-million-dollar IT projects in a broad range of industries. He has initiated advanced research on emerging computing technologies, resulting in an invention patent and several patent-pending initiatives as well as many unified methodologies and platform models for adaptive enterprise system development. He has played a principal strategist role in leading establishing IT strategies and architecture blueprints, coupled with pragmatic technology roadmaps and enterprise architecture standards/policies, for IT governance and portfolio/asset management in Fortune 100 international organizations. He serves as a mentor/advisor on leading-edge technologies, architecture, and engineering in various technical committees, and teaches a wide variety of courses as an adjunct professor and professional trainer. In addition to dozens of top-notch technical publications, he has authored several books on asynchronous Web services and heterogeneous business integration, and is working on multiple books on Internet technologies. He is a member of numerous professional associations and honorary society, a frequent speaker and chair/program committee member in key conferences/workshops, an editor/editorial advisory board member of IT research journals & books, as well as a founder of Greater Charlotte Rational User Group and Charlotte Architecture & Technology Symposium.

Nary Subramanian is currently an assistant professor of computer science in the computer science department at the University of Texas at Tyler, Tyler, Texas. Earlier he served as the assistant professor of computer engineering in the department of engineering at Hofstra University, New York. Dr. Subramanian received his PhD in computer science from the University of Texas at Dallas, an MSEE from Louisiana State University, Baton Rouge, and another MSEE from Delhi University, Delhi, India. Dr. Subramanian has about 15 years of experience in the industry in engineering, sales, and management. He has been a co-chair of the International Workshop on System/Software Architectures for 5 years, serves on the editorial board of the *International Journal of Software Architectures*, has been a guest-editor for conference proceedings and special journal issues, and has served on the Program Committees of several international conferences and workshops. His research interests include Web engineering, software architecture, software engineering, software metrics, software security, non-functional requirements, expert systems, computational biology, home appliance control systems, information systems, and legal systems. He has published more than 30 papers in journals, conferences, and workshops. Dr.

Subramanian has also served as the judge for several high-school science fairs. He has received awards from both the industry and the academia.

George M. Whitson III is currently a professor of computer science in the computer science department of The University of Texas at Tyler, Tyler, Texas. Whitson has been teaching mathematics and computer science for over 45 years at a wide variety of colleges and universities. Dr. Whitson received his bachelors' degree in applied mathematics at the University of Mississippi in 1961, his masters' degree in mathematics from the University of Massachusetts in 1963 and his PhD in mathematics, with an emphasis in Group Theory, at the University of Illinois in 1974. Initially doing research group theory Dr. Whitson migrated to computer science during the 70's working in numerical analysis, applications of the computer to abstract algebra, cryptography and denotational semantics. He has taught most undergraduate and masters' level courses in the Computer and Information Sciences. Dr. Whitson has been at the University of Texas at Tyler for the past 24 years serving in a number of roles including Coordinator of Computer Science, Lab Manager and professor. Dr. Whitson's current interests include Web applications program development, the history of computing and computer security.

Yaoling Zhu is a postgraduate research student at the School of Computing at Dublin City University. Yaoling is a graduate in Computer Science from the Zhengzhou Institute of Engineering, China. Yaoling has extensive experience in the software sector, working for several years as a senior software engineer for multinational companies such as Oracle, where he has been working on e-business outsourcing and Web service technologies in Oracle's European Development and Technology Centre. Yaoling's research focuses on data integration problems in Web-based software systems.

Index

Symbols

A

B

C

D

O

P

R

S

T

U

W

X

Y